Trends in Mathematics is a series devoted to the publication of volumes arising from conferences and lecture series focusing on a particular topic from any area of mathematics. Its aim is to make current developments available to the community as rapidly as possible without compromise to quality and to archive these for reference.

Proposals for volumes can be sent to the Mathematics Editor at either

Birkhäuser Verlag
P.O. Box 133
CH-4010 Basel
Switzerland

or

Birkhäuser Boston Inc.
675 Massachusetts Avenue
Cambridge, MA 02139
USA

Material submitted for publication must be screened and prepared as follows:

All contributions should undergo a reviewing process similar to that carried out by journals and be checked for correct use of language which, as a rule, is English. Articles without proofs, or which do not contain any significantly new results, should be rejected. High quality survey papers, however, are welcome.

We expect the organizers to deliver manuscripts in a form that is essentially ready for direct reproduction. Any version of $\TeX$ is acceptable, but the entire collection of files must be in one particular dialect of $\TeX$ and unified according to simple instructions available from Birkhäuser.

Furthermore, in order to guarantee the timely appearance of the proceedings it is essential that the final version of the entire material be submitted no later than one year after the conference. The total number of pages should not exceed 350. The first-mentioned author of each article will receive 25 free offprints. To the participants of the congress the book will be offered at a special rate.

Fundamental Problematic Issues in Turbulence

Albert Gyr
Wolfgang Kinzelbach
Arkady Tsinober
Editors

Springer Basel AG

Editors' addresses:

Albert Gyr and Wolfgang Kinzelbach
Institute of Hydromechanics
and Water Resources Management
ETH Hönggerberg
8093 Zürich
Switzerland

Arkady Tsinober
Faculty of Engineering
Tel Aviv University
Tel Aviv 69978
Israel

1991 Mathematical Subject Classification 76Fxx; 76D05; 35Q30

A CIP catalogue record for this book is available from
the Library of Congress, Washington D.C., USA

Deutsche Bibliothek Cataloging-in-Publication Data

Fundamentel problematic issues in turbulence / Albert Gyr ...
eds. - Basel ; Boston ; Berlin : Birkhäuser, 1999
(Trends in mathematics)
ISBN 978-3-0348-9730-3 ISBN 978-3-0348-8689-5 (eBook)
DOI 10.1007/978-3-0348-8689-5

Cover illustration taken from the article by Métais and Urbin in this volume.

Originally published by Birkhäuser Verlag in 1999
Softcover reprint of the hardcover 1st edition 1999
Printed on acid-free paper produced from chlorine-free pulp. TCF ∞

ISBN 978-3-0348-9730-3

9 8 7 6 5 4 3 2 1

Contents

3 Observational aspects

4 Two- and quasi-two-dimensional flows

5 Basic aspects of turbulence modelling

6 Statistical aspects

7 Passive objects

Preface

The Second Monte Verita Colloquium *Fundamental Problematic Issues in Turbulence* was held in Monte Verita, Switzerland, on March 23–27, 1998.

The main goal of the Colloquium was to bring together in the relaxed atmosphere of Monte Verita a group of leading scientists (consisting of representatives of different generations) and to discuss informally and free of the influence of funding agencies and/or other "politics" of nonscientific nature the basic issues of turbulence. The intention was to put major emphasis on the exposition of the problematic aspects and discussion(s) – not mere reporting of results, i.e. not having just one more meeting. For this purpose it was originally thought to leave all the afternoons free of formal presentations at all. However, this intention became unrealistic due to a number of reasons, and, in the first place, due to strong pressure from various parts of the scientific community and non-scientific constraints to broaden the scope and to increase the number of participants as compared to the First Colloquium held in 1991. This resulted in a considerable reduction of time for discussions. Nevertheless, the remaining time for discussions was much larger than usually allocated at scientific conferences.

On the scientific side the main idea was to bring together scientists working in turbulence from *different* fields, such as mathematics, physics, engineering and others. In this respect the Colloquium was definitely very successful and resulted in a number of interesting interactions and contacts. Clearly everybody was interested in all aspects of the problem which consisted of *mathematical issues* (morning session, March 23, 1998), *control and related* (afternoon session, March 23, 1998), *observational aspects* (morning session, March 24, 1998), *two- and quasi-two-dimensional flows* (afternoon session, March 24, 1998), *basic aspects of turbulence modelling* (morning session, March 25, 1998), *statistical issues* (morning and afternoon sessions, March 26, 1998) and *passive scalars* (morning session, March 27, 1998). A special and the last session of the Colloquium was devoted to the final discussion with the emphasis on the *physics of scaling in turbulence* and related issues.

Among the main questions addressed in the discussions were those related to the controllability of turbulent flows, possible qualitative difference between pure two-dimensional and real quasi-two-dimensional turbulent flows, common features of two-dimensional and three-dimensional turbulence, universality (or not) of small scale turbulence and their relation with the large scales, on how realistic are the prospects of reduced description of turbulent flows, necessity of dealing more with the physics of turbulence, in general, and in turbulence modelling and (beyond) scaling properties, in particular.

It seems that the last issue, i.e the *physics of turbulence*, is the key one for the progress in any aspect of the problem. This issue deserves a special topical meeting in the nearest future.

Zürich, October 1998 *A. Gyr, W. Kinzelbach, A. Tsinober*

1 Mathematical issues

Trends in Mathematics, © 1999 Birkhäuser Verlag Basel/Switzerland

Does turbulence saturate global transport estimates?

C. R. Doering

Abstract

The Navier-Stokes equations in three spatial dimensions are not generally known to possess unique, finite dissipation solutions for all times. Time averages of the viscous dissipation remain bounded, however, so limits on long time averages of global transport of mass, momentum and heat by turbulent incompressible flows may be rigorously estimated from the Navier-Stokes equations. For a number of fundamental flow geometries, the best available rigorous upper bounds are not all that far above the experimentally measured quantities in turbulent regimes. This observation naturally suggests the question in this paper's title, and it raises a number of other issues which are discussed here.

1 Introduction

The purpose of this paper is to review some open mathematical issues concerning the incompressible Navier-Stokes equations, as well as to point out some physically interesting things that we do know about solutions. While the question of uniqueness of solutions to the 3*d* Navier-Stokes equations remains one of the major unanswered questions in the field, weak solutions exist for many cases ([1], [2]). The uniqueness question is tied up with regularity issues, in particular whether or not the bulk energy dissipation rate remains bounded for all times ([3], [4], [5]). However, the time averaged energy dissipation rate is finite for weak solutions, and this allows for a mathematically rigorous exploration of the time averaged bulk transport of mass, momentum and heat by ostensibly turbulent flows. It is observed that in a number of cases, the best upper limits are not so terribly far from the experimentally measured quantites in turbulent regimes. While the rigorous bounds are not generally right on top of the experimental data, they often scale with Reynolds or Rayleigh numbers in accord with the data and with some accepted theories of turbulent transport (frequently modulo logarithmic factors).

The notion that turbulence may attempt to extremize global transport is an old thought going back at least to Malkus' seminal ideas in the fifties [6] and continuing through the fundamental work of Howard ([7], [8]) and Busse [9] and collaborators from the sixties to the present. To some extent that work may be considered a step between between the functional analysis of studying existence and regularity of solutions of the Navier-Stokes equations, and the statistical modelling of turbulence. In recent years, some functional techniques – in particular a mathematical device invented by E. Hopf in the forties [2] – have produced similar upper bounds straight from the equations of motion without any statistical hypotheses ([10]-[13]). To date these bounds have been observed to capture the essential scaling of bulk transport in turbulent flows.

In the rest of this paper we provide some details of these results. The following Section 2 reviews the situation regarding the existence, uniqueness, and regularity of solutions to the $2d$ and $3d$ Navier-Stokes equations, describing the distinction between strong and weak solutions and illustrating a connection between the mathematical challenges and the physics of turbulent dynamics. In Section 3 we recall some results for strong solutions: finite attractor dimension estimates and bounds on the energy spectrum. Section 4 describes results for limits on long time averaged energy dissipation rates and global transport for weak solutions, comparing them with experimental and theoretical expectations. Interestingly, the mathematical method for deriving these bounds shares much with the techniques of hydrodynamic stability theory although there is nothing at all "stable" about the solutions to which it applies. Section 5 reviews some results for convection problems, going beyond the Navier-Stokes equations to study related fluid dynamic systems. The concluding Section 6 is a summary and a list of some of the open questions in this mathematical analysis of turbulent dynamics.

2 Navier-Stokes equations and solutions

The incompressible Navier-Stokes equations are

$$\frac{\partial u}{\partial t} + u \cdot \nabla u + \nabla p = \nu \Delta u + f(x,t),$$

$$\nabla \cdot u = 0$$

where $x \in R^d$, $d = 2$ or 3, and $u(x,t) \in R^d$ is the velocity vector field. The pressure field is $p(x,t)$, the kinematic viscosity is ν, and $f(x,t) \in R^d$ is an applied body force. These equations are to be supplemented with initial and boundary conditions appropriate for the problem at hand. They are the continuum formulation of Newton's second law of motion and conservation of mass for a unit density incompressible Newtonian fluid.

Consider first the problem in the absense of rigid boundaries, i.e., on the torus T^d or on all of R^d. The global (in time) existence of weak solutions was first established by Leray in 1934 [1]. The appelation *weak* refers to the fact that these solutions only satisfy the Navier-Stokes equations in a smeared sense. For example, the incompressibility condition asserts in the weak sense only that for any smooth (and sufficiently rapidly decreasing on R^d) test function $\phi(x)$,

$$\int [(\nabla\phi) \cdot u] d^d x = 0.$$

Weak solutions are not generally known to possess enough regularity so that the customary integrations by parts may be performed; at each instant of time they are square integrable in space and hence possess finite kinetic energy, but their derivatives may not be square integrable. For weak solutions the usual conservation of energy relation is only an inequality:

$$\frac{1}{2}\int |u(x,t)|^2 d^d x \le \frac{1}{2}\int |u(x,0)|^2 d^d x + \int_0^t \left[\int u(x,t')\cdot f(x,t') d^d x\right] dt'$$
$$-\nu \int_0^t \left[\int |\nabla u(x,t')|^2 dx^d\right] dt'$$

That is, the kinetic energy at time t is not known to be equal to the sum of the initial kinetic energy plus the work done by the body force plus the total energy dissipated by viscous friction up to time t, it is only known to be *bounded* by that sum. In fact it turns out that in $2d$, the weak solutions are also strong solutions, i.e., unique smooth solutions that satisfy the equations in the usual sense. But as of this writing in 1998 we do not know whether these weak solutions are always unique in spatial dimensions $d = 3$, nor can we rule out divergences in the instantaneous bulk energy dissipation rate

$$\nu \int |\nabla u(x,t)|^2 d^d x$$

or singularities in the instantaneous *local* kinetic energy density,

$$\frac{1}{2}|u(x,t)|^2.$$

The problem with uniqueness is a problem of regularity for these weak solutions. If the weak solutions were smooth enough that the nonlinearity in the Navier-Stokes equations (essentially the $u \cdot \nabla u$) was a Lipschitz nonlinearity, then uniqueness would follow. But the ∇u factor is not known to be well behaved. Indeed, the rate of strain and vorticity ($\omega = \nabla \times u$) may be nonlinearly amplified by the evolution. The vortex stretching mechanism in the $3d$ Navier-Stokes equations, often thought of as the root of turbulent dynamics, is the mathematical stumbling block in the analysis of the solutions.

In certain restricted cases it can be shown that the weak solutions are in fact *strong* solutions of the $3d$ Navier-Stokes equations. Given an initial condition with finite energy dissipation rate (i.e., square integrable derivatives), the subsequent solution will be a strong solution for at least a finite time in finite volume. The solution will remain strong for all $t > 0$ if the initial kinetic energy and velocity gradients are small enough, the viscosity is large enough, the volume is small enough, and the body force is weak enough. In other words, global strong solutions may be shown to exist if some appropriate Reynolds and Grashof numbers are small enough. Small Reynolds numbers and Grashof numbers are precisely *not* parameter regimes for turbulence; the strongest regularity results are limited to parameter regimes corresponding to the physical conditions for the *absence* of turbulence.

The same general situation holds forth in the presence of rigid boundaries. Hopf extended Leray's construction of weak solutions to cases with rigid no-slip boundaries in 1941, producing similar less-than-complete results for uniqueness and conservation of energy *inequality* [2]. Hence at this point in time we may rigorously assert the existence of weak solutions for $3d$ Navier-Stokes equations, but not uniqueness and not the usual conservation of energy. We are justified in associating the source of the technical mathematical problems with the physics of turbulent fluid flows.

3 Results for strong solutions

The vortex stretching mechanism is absent in the $2d$ Navier-Stokes equations, as are the corresponding mathematical problems. In that case the weak solutions may often be elevated to strong solutions and correspondingly strong mathematical results may be obtained.

In many situations the $2d$ Navier-Stokes equations possess a finite dimensional attractor on which lie all time asymptotic solutions. For example, given a stationary body force $f(x)$ on a finite periodic domain of size $L \times L$, we may define the Grashof number

$$G = \frac{L^2 \sqrt{\int |f(x)|^2 d^2x}}{\nu^2}.$$

As first shown by Constantin, Foias and Temam in 1988 [14], the dimension D of the attractor is bounded from above according to

$$D \leq cG^{2/3}[1 + logG]^{1/3},$$

where c is an absolute constant (see also [15]). This result is remarkable because – within the logarithmic factor – the scaling is in accord with Kraichnan's "classical" $2d$ turbulence theory. Indeed, the time averaged enstrophy dissipation rate χ may be estimated in terms of the Grashof number:

$$\chi = \left\langle \frac{\nu}{L^2} \int |\nabla \omega(x, \cdot)|^2 d^2x \right\rangle \leq \frac{1}{\nu L^2} \int |f(x)|^2 d^2x = \frac{G^2 \nu^3}{L^6}$$

where $\langle \cdot \rangle$ means largest possible long time average. If the smallest effective length scale is determined by the enstrophy cascade cutoff by the viscosity, depending only on χ and ν, then it must be the Kraichnan length

$$\lambda_{Kraichnan} \sim (\frac{\nu^3}{\chi})^{1/6}.$$

If the attractor dimension corresponds to the effective number of degrees of dynamical degrees of freedom in the system, then one would expect it to be the number of "micro" areas that fit into the system volume, i.e.,

$$D \sim \frac{L^2}{\lambda^2_{Kraichnan}} \leq \tilde{c}G^{2/3},$$

(where $\tilde{c}$ is another absolute constant) agreeing within logarithms with Constantin, Foias and Temam's rigorous upper bound.

Another way to interpret the $G^{2/3}$ scaling is as follows. The number of degrees of freedom in the system at fixed driving "intensity" ought to be proportional to the system volume for large volumes. This means that there must be a (small) length scale $\sim \sqrt{\frac{L}{D}}$ that is independent of L as $L \to \infty$. If we consider the "intensity" to be proportional to the rms magnitude of the force,

$$\sqrt{\frac{1}{L^2}\int |f(x)|^2 d^2x},$$

then the only length scale that can be manufactured from this and ν is

$$\left(\frac{\nu^2}{\sqrt{\frac{1}{L^2}\int |f(x)|^2 d^2x}}\right)^{1/3},$$

i.e., a Kraichnan-like scale as above implying again the $G^{2/3}$ scaling. Hence we may interpret the $G^{2/3}$ scaling as corresponding to the existence of an "intensive" small length scale in the system which remains bounded and bounded away from zero in the infinite volume limit. (The logarithm in the rigorous result actually produces a small scale estimate that goes slowly to zero in the infinite volume limit, but we will ignore this for the sake of interpretation.) We will recall this interpretation in the next section in connection with some attractor dimension estimates for boundary-driven $2d$ flows.

Although we do not know *a priori* that the $3d$ Navier-Stokes equations have strong solutions for all times, a finite dimensional attractor, or a well defined small scale in the infinite volume limit, it is instructive to investigate what kinds of estimates arise for solutions which satisfy certain *apriori* bounds. That is, we assume some regularity for the solutions and ask for bounds on the attractor dimension or on some appropriate small length scale as a cutoff on the energy spectrum (dissipation scale).

Toward this end, define the following alternative "energy dissipation rates": the usual time averaged bulk energy dissipation rate

$$\epsilon = \left\langle \frac{\nu}{L^3}\int |\nabla u(x,\cdot)|^2 d^3x \right\rangle,$$

the larger maximum possible bulk energy dissipation rate,

$$\epsilon_1 = sup_t \, \frac{\nu}{L^3}\int |\nabla u(x,t)|^2 d^3x,$$

and an even more extreme value, the largest possible local energy dissipation rate

$$\epsilon_2 = sup_{x,t} \, \nu |\nabla u(x,t)|^2.$$

Generally $\epsilon \leq \epsilon_1 \leq \epsilon_2$ and we only know for a fact that $\epsilon < \infty$ for solutions of the $3d$ Navier-Stokes equations. Fluctuations and intermittency aggravate this heirarchy, and for turbulent solutions we expect $\epsilon << \epsilon_1 << \epsilon_2$.

The usual Kolmogorov scale is the unique length scale constructed from ϵ and ν:

$$\lambda_{Kolmogorov} \sim (\frac{\nu^3}{\epsilon})^{1/4}.$$

Clearly this definition of the Kolmogorov scale remains bounded away from zero in an infinite volume limit with finite ϵ, but the question is whether or not it corresponds to a small scale controlling the number of degrees of freedom via the attractor dimension, or to a dissipation scale controlling the cutoff of the energy spectrum of the flow, as is usually assumed to be the case in statistical turbulence theories. The mathematically rigorous state of affairs is only suggestive of these correspondences.

For example, a rigorous attractor dimension estimate for the $3d$ Navier-Stokes equations is

$$D \leq \left(\frac{L}{\lambda_2}\right)^3$$

with

$$\lambda_2 \sim (\frac{\nu^3}{\epsilon_2})^{1/4}.$$

We stress that it is not known *a priori* whether $\lambda_2 > 0$, but even if we assume that it is, we expect it to be very much smaller than $\lambda_{Kolmogorov}$.

In terms of the energy spectrum, the results to date [16] are of the form

$$\hat{u}(k,t) = O(P(k)e^{-\lambda_1 k}) \quad as \quad k \to \infty$$

where $\hat{u}(k,t)$ is the Fourier coefficient, $P(\cdot)$ is a polynomial and

$$\lambda_1 \sim (\frac{\nu^3}{\epsilon_1})^{1/4}.$$

Hence we again find a Kolmogorov-like length scale, but defined in terms of an overestimated energy dissipation rate. (Note that this result shows explicitely that a finite ϵ_1 is sufficent to ensure strong – in fact infinitely differentiable – solutions.)

In both of these cases we see that even if the regularity problem is overcome and the weak solutions of the $3d$ Navier-Stokes equations are really strong solutions, it is not at all clear that the usual association of small length scales would be rigorously valid. This illustrates that there are important physically significant questions that will still not necessarily be settled if regularity issues are solved.

4 Dissipation and global transport bounds

A major step in Hopf's construction of weak solutions to the Navier-Stokes equations is the presence of boundaries is the derivation of *a priori* estimates akin to the energy inequality [2]. Rigid surfaces may result in boundary terms from the necessary integrations by parts, and so Hopf introduced a mathematical device for eliminating such terms. This technique, when carefully developed, results in explicit bounds for the time averaged viscous energy dissipation rate which have direct implications for global transport in turbulent flows. The presence of rigid boundaries and driving from the boundaries usually defines an outer Reynolds number, and the goal would be to compare the upper estimates with experimentally measured transport and dissipation in the turbulent regime at high Reynolds number.

Consider, for example, a boundary-driven shear flow with $u = 0$ on the $x_3 = 0$ plane and $u = \hat{e}_1 U$ on the $x_3 = h$ plane, where $\hat{e}_1$ is the unit vector in the x_1 direction. For convenience let us impose periodic conditions on lenght scales L_1 and L_2 in the x_1 and x_2 directions respectively. The natural Reynolds number for this problem is

$$Re = \frac{Uh}{\nu}.$$

(We are interested in the flow in a layer of large horizontal extent, so we have in mind the limits $L_1 \to \infty$ and $L_2 \to \infty$; hence we do not consider Reynolds numbers based on L_1 and L_2.) A laminar stationary solution for the Navier-Stokes equations is planar Couette flow,

$$u = \hat{e}_1 \frac{Uz}{h},$$

which may be shown, using the methods of nonlinear energy stability, to be the global attractor for all initial conditions for small enough Reynolds numbers [17]. We are interested in turbulent flows at high Reynolds numbers.

The vertical flux of horizontal momentum across the layer is proportional to the wall shear stress, i.e., the frictional drag force/area felt by the rigid surfaces imposing the boundary conditions. A direct formal calculation from the Navier-Stokes equation (intergration by parts and a long time average) yields the relation

$$\epsilon = \frac{\tau U}{h}$$

where ϵ is the long time averaged energy dissipation rate per unit mass and τ is the long time averaged wall shear stress. Hence an estimate on ϵ produces an estimate on τ.

Hopf's method for deriving rigiorous bounds on weak solutions may be applied to this problem with the following result:

Theorem 1

$$\epsilon \leq c \frac{U^3}{h} \quad as \quad Re \to \infty,$$

where c is an absolute constant.

This result was first established ([10],[11]) with a constant $c \approx .09$, which has since been improved to $c \approx .01$ [18]. Note that this rigorous upper bound displays *turbulent Kolmogorov scaling*, i.e., it suggests ϵ independent of ν as $Re \to \infty$. This phenomena, wherein the global transport becomes independent of the microscopic material transport coefficient ultimately responsible for the transport, is a hallmark of turbulence [19]. The conventional interpretation of this effect is the cascade of energy to small scales where dissipation acts, persistent even in the vanishing viscosity limit $\nu \to 0$.

How does this result compare with experiment? Although there are not appropriate recent experiments for the simple plane geometry considered above, there are a number of recent experimental studies of shear driven turbulence in a cylindrical geometry, i.e., for turbulent Taylor-Couette flows ([20], [21]). A similar Kolmogorov scaling rigorous upper bound may also be derived for this geometry. In each experiment, for smooth surfaces driving the fluid the experiments are well fit (with one adjustable parameter) by the Prandtl-von Karman logarithmic law, a statistical closure theory, which predicts the large Re behavior

$$\epsilon \sim \frac{U^3}{h} \frac{1}{(log Re)^2}.$$

The rigorous upper estimate is about one order of magnitude above the experimental data at $Re = 10^6$. (For reference of scale, the experimental data is several orders of magnitude above the dissipation in the unstable laminar flow state at that Reynolds number so the rigorous upper bound is a far better indication of the energy dissipation and momentum transport than a laminar approximation.)

The upshot of these considerations, then, are that the best estimates appear to be within logarithmic factors of the experimental data. (In passing we note that the existing experimental data are also well fit by a $Re^{-1/4}$ correction to Kolmogorov scaling.) However, for less symmetric flow geometries the pure scaling seems to hold: when stirring by "perturbed" boundaries or paddles is applied, the $\epsilon \sim \frac{U^3}{h}$ scaling appears with a prefactor not far from the rigorous bound [21]. The pure Kolmogorov scaling is also observed in grid generated turbulence [22].

The upper bounds on energy dissipation and momentum transport in weak solutions descibed here have been generalized in a number of ways. Marchioro [23] showed that similar scaling results may apply for nonstationary boundary movement. Kerswell [24] applied Hopf's formalism to a completely unsymmetric geometry (flow in a rotating ellipsoid) and found similar scaling. Recently Wang [25] has shown that such Kolmogorov-like scaling is to be expected in all smooth domains with shearing boundaries. Wang's results apply to the less symmetric stirring experiments mentioned above.

It is interesting that Hopf's 1941 method yields rigorous upper bounds on dissipation for the weak solutions that display Kolmogorov scaling, proposed by Kolmogorov also in 1941, yet Hopf did not extract this from his analysis. Hopf's 1941 paper deals with very general cases and he produced explicit bounds that behave like e^{Re}, far beyond the physical values. (These kinds of exponential bounds are the best available for general boundaries with flux – i.e., sucking and blowing – even within Wang's modern analysis.) Had Hopf considered a few simple geomenties, it is likely that he would have discovered Kolmogorov's scaling from the Navier-Stokes equations directly. Hopf was in Leipzig in 1941, while Kolmogorov was in Moscow. Surely there was little scientific exchange at the time. One wonders what may have developed had there been better communication between those two men during that period.

Similar results have been obtained for pressure-driven shear turbulence in a pipe or channel geometry. In these problems it is the transport of mass that is to be estimated, and identities from the Navier-Stokes equations relate these quantites to the energy dissipation rate. Experiments and the conventional wisdom predict that the friction coefficient $C_f = \frac{\Delta p}{U^2}$, where Δp is the pressure drop and U is the velocity scale of mass flux in the pipe, should behave according to

$$C_f \sim \frac{1}{(logRe)^2}$$

for smooth pipes as $Re \to \infty$. Rigorous results [12] are

Theorem 2

$$C_f \leq c \quad as \quad Re \to \infty,$$

where c is an absolute constant.

This scaling ($C_f \sim Re^0$) is Kolmogorov scaling. The relation between experiment, theory and the rigorous results (i.e., the logarithmic discrepancy) is directly analogous to the situation for shear-driven turbulence. For rough pipes, pure Kolmogorov scaling is observed experimentally, but the prefactor depends on the degree of roughness. This author is not aware of what to expect experimentally for smooth pipes that are not straight.

These rigorous energy dissipation bounds on shear and pressure-driven flows also apply to the $2d$ Navier-Stokes equations. In that case, they may be used in the derivation of attractor dimension estimates for these flows. For $2d$ shear-driven flows in a channel of length L and width h, we have the bound [26]

$$D \leq c\, \frac{L}{h}\, Re^{3/2}$$

where c is another absolute constant and $Re = \frac{Uh}{\nu}$ is the Reynolds number based on the channel width. This bound is evidently sharp as regards its dependence on the channel length L; at fixed shear rate we expect the number of degrees of freedom to be proportional to the length of the channel. The Reynolds number dependence implies an intensive small length scale

$$\lambda = \left(\frac{Lh}{D}\right)^{1/2} \geq c^{-1/2}\, h\, Re^{-3/4}$$

which is the usual Kolmogorov scale of $3d$ turbulence theory. If the Kraichnan scale were to emerge this way, it would correspond to a $D \sim \frac{L}{h}\, Re$ scaling which we have not been able to prove.

Finally we remark that there is an interesting connection between Hopf's method, when it is optimized to give the best bound, and the mathematics of hydrodynamic stability theory. In effect, the implementation of Hopf's method involves producing a "background" flow field which is "stable" in a certain sense. The sense of stability is a slight generalization of nonlinear energy stability, and the best bounds result from the just marginally stable background field. To an extent, then, these calculations are a rigorous implementation of heuristic marginal stability arguments that have been applied to turbulence for a long time.

5 Convection problems

The classical Rayleigh-Bénard problem of a fluid layer heated from below provides another class of systems to which rigorous bounds on turbulent transport may be derived. In these problems it is the transport of heat that is connected to the energy dissipation rate. For a fluid layer of unit depth the usual nondimensional Boussinesq equations are

$$\frac{\partial u}{\partial t} + u \cdot \nabla u + \nabla p = \sigma \nabla^2 u + \sigma\, Ra\, \hat{e}\, T,$$

$$\nabla \cdot u = 0,$$

$$\frac{\partial T}{\partial t} + u \cdot \nabla T = \nabla^2 T.$$

where σ is the Prandtl number and Ra is the Rayleigh number. Here the Navier-Stokes equations are coupled to the advection-diffusion equation for the temperature field. These equations are to be supplemented with appropriate initial and boundary conditions, and weak solutions exist although we generally do not have uniqueness. The nondimensional (space and time averaged) heat flux is measured in terms of the Nusselt number Nu which is the ratio of the total heat flux to the purely conductive heat flux in the absence of fluid motion. The Nusselt number is a function of Ra and σ and the geometry of the container (sidewalls, aspect ratio).

For the Rayleigh-Bénard problem with rigid upper and lower boundaries, the large Ra bound on the energy dissipation for the weak solutions is

Theorem 3

$$Nu \le c\, Ra^{1/2} \quad as \quad Ra \to \infty,$$

where c is an absolute constant.

The theoretical and experimental situation is somewhat involved for intermediate values of Ra [27], but for the large Ra asymptotic behavior both theory [28] and recent experiments ([29], [30]) suggest

$$Nu \sim \sigma^{1/2} Ra^{1/2} \, \frac{1}{(log Ra)^{3/2}}.$$

This is encouraging as regards the Rayleigh number scaling, but the rigorous bound is uniform in the Prandtl number. The $(\sigma Ra)^{1/2}$ result is a generalized Kolmogorov scaling wherein the physical heat flux becomes independent of both the fluid viscosity *and* the fluid's heat conduction coefficient.

The Boussinesq equations have an infinite Prandtl number limit, namely

$$\nabla\tilde{p} = \nabla^2 u + Ra\ \hat{e}\ T,$$

$$\nabla \cdot u = 0,$$

$$\frac{\partial T}{\partial t} + u \cdot \nabla T = \nabla^2 T.$$

In this case a marginally stable boundary layer argument and experiments suggest $Nu \sim Ra^{1/3}$. Rigorous analysis of this problem yields [31]

Theorem 4

$$Nu \leq c\ Ra^{1/3}\ (log Ra)^{2/3} \quad as \quad Ra \to \infty,$$

where c is an absolute constant.

Again, the rigorous upper limit seems to be within a logarithmic factor of the expected turbulent behavior.

There is one convection problem where the rigorous upper bound agrees precisely with the heurstic marginally stable boundary layer argument. Consider the Rayleigh-Bénard problem for a fluid saturated porous layer where the fluid flow obeys Darcy'd Law:

$$\nabla\tilde{p} = -u + Ra\ \hat{e}\ T,$$

$$\nabla \cdot u = 0,$$

$$\frac{\partial T}{\partial t} + u \cdot \nabla T = \nabla^2 T$$

with appropriate boundary and initial conditions. In this case a marginally stable boundary layer argument, Kolomogorov scaling (heat flux independent of heat conduction coefficient) and some direct numerical simulations and experiments suggest $Nu \sim Ra^1$. Rigorous analysis of this problem [32] yields

Theorem 5

$$Nu \leq \frac{9}{256}\ Ra \quad as \quad Ra \to \infty.$$

6 Summary and discussion

The answer to the question posed in the title of this paper – does turbulence saturate global transport estimates? – appears to be "yes, sort of". In many of the examples described here the mathematically rigorous limits on bulk transport appears to be approached by turbulent flows, at least within logarithmic corrections. In most cases the leading scaling of the upper bounds is in accord with theoretical expectation, namely some sort of Kolmogorov-type scaling.

There remain a number of open questions and problems regarding these observations:

1. Are the logarithms really there? Are there corrections to scaling in the exponents? What is the true behavior of the energy dissipation rate as in the vanishing viscosity limit?

2. Can further information be extracted from in the Navier-Stokes equations to lower the rigorous results? Or are there solutions which exist mathematically and dominate the upper bounds but which never appear in real turbulent flows?

3. Often the bounds on weak solutions are independent of the spatial dimension. Are they more precise in higher dimensions?

4. Is it possible to bring the Prandtl number into play for the convection problems? Are bounds for other systems (e.g., Marangoni convection) similarly uniform in the Prandtl number?

Perhaps the most important question is whether such the rigorous limits on turbulent transport are precise enough to resolve outstanding scientific questions in applications. That remains to be seen.

7 Acknowledgements

I am grateful for many collaborators and friends with whom I have worked on these problems and learned much over the last decade. I would especially like to thank Professors Albert Gyr and Arkady Tsinober for their effort in organizing the excellent meeting in Monte Verita. This work was supported in part by United States National Science Foundation Awards PHY 96-96187 & DMS 97-09494.

References

[1] J. Leray, Acta Math. **63** (1934), 193.

[2] E. Hopf, Math. Annalen **117** (1941), 764.

[3] R. Temam, *Navier-Stokes Equations* (North Holland, 1984).

[4] P. Constantin and C. Foias, *The Navier-Stokes Equations* (University of Chicago Press, 1988).

[5] C.R. Doering and J.D. Gibbon, *Applied Analysis of the Navier-Stokes Equations* (Cambridge University Press, 1995).

[6] W. V. R. Malkus, Proc. R. Soc. London Ser. A **225** (1954), 185.

[7] L.N. Howard, J. Fluid. Mech. **17** (1963), 406.

[8] L.N. Howard, Annu. Rev. Fluid Dyn. **4** (1972), 473.

[9] F. H. Busse, Adv. Appl. Mech. **18** (1978), 77.

[10] C.R. Doering and P. Constantin, Phys. Rev. Lett. **69** (1992), 1648.

[11] C.R. Doering and P. Constantin, Phys. Rev. E **49** (1994), 4087.

[12] P. Constantin and C.R. Doering, Phys. Rev E **51** (1995), 3192.

[13] C.R. Doering and P. Constantin, Phys. Rev. E **53** (1996), 5957.

[14] P. Constantin, C. Foias and R. Temam, Physica D **30** (1988), 284.

[15] C.R. Doering and J.D. Gibbon, Physica D **48** (1991), 471.

[16] C.R. Doering and E.S. Titi, Phys. Fluids **7** (1995), 1384.

[17] D. Joseph, *Stability of Fluid Motions* (Springer, Berlin, 1976); B. Straughan, *The Energy Method, Stability and Nonlinear Convection* (Springer, Berlin, 1992).

[18] R. Nicodemus, S. Grossmann and M. Holthaus, Phys. Rev. Lett. **79** (1997), 4170.

[19] U. Frisch, *Turbulence: the Legacy of A.N. Kolmogorov* (Cambridge University Press, 1995).

[20] D.P. Lathrop, J. Feinberg and H.L. Swinney, Phys. Rev. Lett. **68** (1992), 1515; Phys. Rev. A **46** (1992), 6390.

[21] O. Cadot, Y. Couder, A. Daerr, S. Douady and A. Tsinober, Phys. Rev E **56** (1997), 427.

[22] K.R. Sreenivassan, Phys. Fluids **27** (1984), 1048.

[23] C. Marchioro, Physica D **74** (1994), 395.

[24] R.R. Kerswell, J. Fluid Mech. **321** (1996), 335.

[25] X. Wang, Physica D **99** (1997), 555.

[26] C.R. Doering and X. Wang, Physica D (in press, 1998).

[27] E. Siggia, Ann. Rev. Fluid Mech. **26** (1994), 137.

[28] R. Kraichnan, Phys. Fluids **5** (1962), 1374.

[29] X. Chavanne *et al*, Phys. Rev. Lett. **79** (1997), 3648

[30] S. Cioni, S. Ciliberto and J. Sommeria, J. Fluid Mech. **335** (1997), 111.

[31] P. Constantin and C.R. Doering, J. Stat. Phys. (submitted, 1998).

[32] C.R. Doering and P. Constantin, J. Fluid Mech. (submitted, 1998).

Trends in Mathematics, © 1999 Birkhäuser Verlag Basel/Switzerland

The closure problem for the Friedman-Keller infinite chain of moment equations, corresponding to the Navier-Stokes system

A. V. Fursikov

Abstract

A formulation of the closure problem for the Friedman-Keller infinite chain of moment equations, corresponding to the Navier-Stokes system is given. It consists of constructing a sequence of problems $\mathcal{A}^N(M^N)$ for N unknown functions M^N such that their solutions $M^N = (M_1^N, \dots, M_N^N, 0, \dots)$ approach the solution $M = (M_1, \dots, M_k, \dots)$ of the Friedmann-Keller chain containing infinite unknown functions M_k. A rigorous solution of the closure problem for small and for large Reynolds numbers is given.

More than 100 years ago O. Reynolds [1] proposed to study a turbulent fluid flow with the help of decomposing of its velocity $u(t,x)$ into the averaged part M_1 and the fluctuation $\tilde{u}$: $u = M_1 + \tilde{u}$.

After substitution of this decomposition into the Navier-Stokes equations one obtains an unclosed system of equations. In order to find and to investigate M_1 from these equalities it is necessary to close this unclosed system.

Many methods of closure were suggested since that time and most of them used certain additional physical conjectures. But there was one approach proposed in 1924 by Friedman and Keller which solves the closure problem in the frame of the theory of viscous incompressible fluids (see A.S. Monin, A.M. Yaglom [2] for details). They derived an infinite chain of equations for the infinite set of moments $(M_1, M_2, \dots, M_k, \dots)$ where M_1 is just the averaged part of the velocity vector field $u(t,x)$. After that the closure problem can be transformed to the following one:

Find a sequence of problems $\mathcal{A}^N(M^N)$ for N unknown functions $M^N = (M_1^N, \dots, M_N^N, 0, \dots, 0, \dots)$ such that its solution M^N tends to the solution M of the infinite chain as $N \to \infty$.

Below we present the solution of this problem.

§1. Definitions

Let us consider the Navier-Stokes equations

$$\partial_t u(t,x) - \Delta u + (u,\nabla)u + \nabla p(t,x) = 0, \quad \operatorname{div} u = 0, \quad u|_{t=0} = 0 \tag{1.1}$$

where $t \in (0,T)$, $x = (x^1, \dots, x^n) \in R^n$, $u(t,x) = (u^1(t,x), \dots, u^n(t,x))$, is the velocity of a fluid flow, ∇p is the gradient of pressure. We suppose that (1.1) is to be solved together with periodic boundary conditions with period l:

$$u(t,\dots,x^i+l,\dots)=u(t,\dots,x^i,\dots)\quad i=1,\dots,n. \tag{1.2}$$

Here n is the dimension of the Navier-Stokes equations: $n=2$ or 3.

We get rid of the term ∇p by acting to both parts of (1.1) the operator π of orthoprojection to solenoidal vector fields. (The explicit formula for π is written below in section 4). As a result we get the following equations which does not contain the gradient of pressure:

$$\partial_t u(t,x)-\pi\Delta u+B(u)=0,\quad \operatorname{div} u=0,\quad u|_{t=0}=0 \tag{1.3}$$

where

$$B(u)=\pi\left[(u,\nabla)u\right].$$

Let P be the space of initial conditions $u_0(x)$ of problem (1.1) supplied with the the norm $\|\cdot\|_P$. Evidently, P is also the phase space of the dynamical system generated by (1.1). For a rigorous definition of P with the help of Sobolev spaces see below in section 4. For each Borel subset $\omega\in P$ we define $\mu(\omega)$ as the probability of event: $u_0\in\omega$, where u_0 is an initial condition from (1.1). So $\mu(\omega)$ is a given probability measure. We will suppose that the measure μ is concentrated in a ball $B_R=\{u_0\in P:\|u_0\|_{H^3(\Omega)}<R\}$.

To define the moment functions of measure $\mu(du_0)$, remember that the tensor product of k copies of vector field $u_0(x)=(u_0^1(x),\dots,u_0^n(x))$ is defined by the formula:

$$u_0(x_1)\otimes\cdots\otimes u_0(x_k)=$$

$$\{u_0^{i_1}(x_1)\cdot u_0^{i_2}(x_2)\cdot\;\cdots\;\cdot u_0^{i_k}(x_k),\quad \text{where}\quad i_j=1,\dots,n,\quad j=1,\dots,k\}.$$

We fix $(x_1,\dots,x_k)$ and integrate the vector-valued function $f(u_0)\equiv u_0(x_1)\otimes\cdots\otimes u_0(x_k)$ over $u_0\in P$ with help of the measure $\mu(du_0)$. As a result we get the k-th moment of the initial measure μ:

$$m_k(x_1,\dots,x_k)\equiv\int_P(u_0(x_1)\otimes\ldots\otimes u_0(x_k))\,\mu(du_0)$$

To indicate that a solution of problem (1.3) depends on an initial condition u_0 we will write: $u(t,x)=u(t,x;u_0)$. The k-th moment at instant t is defined by the equality

$$M_k(t,x_1,\dots,x_k)=\int_P(u(t,x_1;u_0)\otimes\cdots\otimes u(t,x_k;u_0))\,\mu(du_0).$$

The Friedman-Keller infinite chain of the moment equations has the following form:

$$\partial_t M_k(t,x_1,\dots,x_k)+A_kM_k+B_kM_{k+1}=0,\qquad k=1,2,\dots \tag{1.4}$$

These equations are supplied with the initial conditions

$$M_k(t, x_1, \ldots, x_k) \mid_{t=0} = m_k(x_1, \ldots, x_k), \qquad k = 1, 2, \ldots . \tag{1.5}$$

Problem (1.4), (1.5) is derived from the Navier-Stokes equations (1.3). In (1.4) M_k are the moments at time t and m_k are the moments of an initial measure $\mu(du_0)$, $A_k = \sum_{j=1}^{k} (\pi\Delta)_{x_j}$ where the index x_j means that the operator $(\pi\Delta)_{x_j}$ acts along the vectorial variable x_j. The definition of the operator B_k from (1.4), generated by the nonlinear operator $B(u)$ from (1.3), is more complicated. We give it below in section 4.

Denote the vector of all initial moments by

$$m = (m_1(x_1), \ldots, m_k(x_1, \ldots, x_k), \ldots)$$

and denote the vector of all moments at instant t by $M(t, \cdot) \equiv M$:

$$M = (M_1(t, x_1), \ldots, M_k(t, x_1, \ldots, x_k), \ldots)$$

Designate the Cauchy problem (1.4), (1.5) by $\mathcal{A}(M)$. The closure problem for the infinite chain (1.4) supplied with initial conditions (1.5) is defined as follows: *Let M^N be an unknown vector-function whose first N components can differ from zero and the rest of the components is equal to zero:*

$$M^N = (M_1^N(t, x_1), \ldots, M_N^N(t, x_1, \ldots, x_N), 0, \ldots, 0, \ldots)$$

Find a sequence of problems $\mathcal{A}^N(M^N)$ such that the solutions M^N of $\mathcal{A}^N(M^N)$ approach the solution M of $\mathcal{A}(M)$:

$$\|M - M^N\| \longrightarrow 0 \quad \text{as} \quad N \longrightarrow \infty \tag{1.6}$$

where $\|\cdot\|$ is an appropriate norm.

Finally, we present in this section an important example of moment functions connected with averaging over time of a solution for (1.3). We recall that the time average is the function

$$M_1(t, \cdot) = \tau_0^{-1} \int_0^{\tau_0} u(t + \tau, x; u_0)\, d\tau, \qquad t \in [0, T],$$

where $u(t, x; u_0)$ is a solution of problem (1.3) and $\tau_0 > 0$ is a fixed number.

The initial moments in this case are defined by the formula

$$m_k(x_1, \ldots, x_k) = \tau_0^{-1} \int_0^{\tau_0} (u(\tau, x_1; u_0) \otimes \cdots \otimes u(\tau, x_n; u_0))\, d\tau\,.$$

One can determine easily the measure $\mu(du_0)$ corresponding to these moments (see Fursikov [8]).

The moments at time t can be written in this case as follows:

$$M_k(t, x_1, \ldots, x_k) = \tau_0^{-1} \int_0^{\tau_0} (u(t + \tau, x_1; u_0) \otimes \cdots \otimes u(t + \tau, x_n; u_0))\, d\tau.$$

§2. Solution of the closure problem for small Reynolds numbers

The simplest formal solution of the closure problem is as follows: we substitute the unknown vector M^N with N non-zero components into chain (1.4) and supply the obtained system with the first N initial conditions from (1.5). As a result we get the following Cauchy problem which we denote by $\mathcal{A}^N(M^N)$:

$$\partial_t M_k^N(t,x_1,\dots,x_k) + A_k M_k^N + B_k M_{k+1}^N = 0, \quad k=1,\dots,N-1 \tag{2.1}$$

$$\partial_t M_N^N(t,x_1,\dots,x_N) + A_N M_N^N = 0 \tag{2.2}$$

$$M_k^N|_{t=0} = m_k, \quad k=1,2,\dots,N \tag{2.3}$$

In order to formulate the result when relation (1.6) is true for solutions of problems $\mathcal{A}^N(M^N)$ defined in (2.1)–(2.3), we have to introduce appropriate spaces of functions. We denote by P_R the space of initial moments $m=(m_1,m_2,\dots)$. The space P_R is constructed by the phase space P of (1.3) and this is made below in Section 4. The following connection between spaces P, P_R is important.

Let $\mu(du_0)$ be an initial measure defined on P and $m=(m_1,\dots,m_k,\dots)$ be the set of its moments. Then 1.) *if $m\in P_R$ then* $\operatorname{supp}\mu_0 \subset \bar{B}_R = \{u\in P: \|u\|_P \le R\}$, 2.) *if* $\operatorname{supp}\mu_0\subset B_R$ *then $m\in P_R$.*

Recall that the Reynolds number Re is equivalent to the magnitude $\frac{lv}{\nu}$, where l is the characteristic length, v is the characteristic velocity and ν is the viscosity coefficient. In our case $\nu=1$, l is the period from (1.2), and we can treat the characteristic velocity as the radius of a ball B_R containing the support of measure μ_0. Therefore the magnitude of the parameter R can be interpreted as the Reynolds number corresponding to the set of moments $m\in P_R$.

The following results give a solution of the closure problem in the case of small Reynolds number:

1.) *Suppose that $n=2,3$, $M=(M_1,M_2,\dots)$ is a solution of* (1.3), (1.4) *which is defined for $t\in(0,T)$ and belongs to the space $L_2(0,T;P_R)$ with sufficiently small R. Then for solutions M^N of* (2.1)–(2.3) *we have:*

$$\int_0^T \|M^N(t,\cdot)-M(t,\cdot)\|_{P_R}^2\,dt \longrightarrow 0 \quad \text{as} \quad N\longrightarrow\infty$$

2) *If $R\gg 1$, then in the case of the Burgers equation*[1] *there exists an initial measure $\mu(du_0)$ concentrated in the ball B_R such that for the moments $M(t,\cdot)=(M_1,M_2,\dots)$ at instant t constructed by means of μ and for a solution M^N of* (2.1)–(2.3) *where m_k are the moments of μ the following relation is true:*

$$\int_0^T \|M^N(t,\cdot)-M(t,\cdot)\|_{P_R}^2\,dt \not\longrightarrow 0 \quad \text{as} \quad N\longrightarrow\infty$$

These results have been proved in M.I. Vishik, A.V. Fursikov [3].

[1] It is evident that all definitions and constructions given above can be easily taken over in the case of Burgers' equation $\partial_t u(t,x) - \partial_{xx}u + \partial_x u^2 = 0$.

§3. Solution of the closure problem for large Reynolds numbers

The part 2) of the result at the end of the previous section claims that the Cauchy problem (2.1)–(2.3) does not solve the closure problem in the case of large Reynolds numbers. Moreover, in the case of large Reynolds numbers the following property for solutions M^N of (2.1)–(2.3) can be true:

$$\int_0^T \|M^N(t,\cdot)\|^2_{P_R}\, dt \longrightarrow \infty \qquad \text{as} \qquad N \longrightarrow \infty \tag{3.1}$$

We suggest the following extremal problem $\mathcal{A}^N(M^N)$ which realises the solution of the closure problem for large Reynolds number. Denote

$$Q^N = \{M^N = (M_1^N, \ldots, M_N^N, 0, 0, \ldots), \quad M^N \quad \text{satisfies} \quad (2.1), (2.2)\}\,.$$

The problem $\mathcal{A}^N(M^N)$ is formulated as follows: To minimize the functional

$$\| M^N |_{t=0} - m \|_{P_R} \longrightarrow \inf \tag{3.2}$$

on the set of vector fields M^N satisfying the conditions

$$M^N \in Q^N\,, \tag{3.3}$$

$$\int_0^T \| M^N(t,\cdot) \|^2_{P_R}\, dt \le L\,, \tag{3.4}$$

where L is a constant bounding from above the norm of the approached solution M of (1.4),(1.5):

$$\int_0^T \| M(t,\cdot) \|^2_{P_R} < L$$

We preserve for a solution M^N of our extremal problem the property (3.3), i.e. we claim that M^N satisfies equations (2.1),(2.2). Besides, by means of inequality (3.4) we exclude for M^N a chance to satisfy property (3.1). But generally speaking, restriction (3.4) made impossible equality (2.3) and we are forced to replace this equality by relation (3.2).

One can prove easily the existence and uniqueness of a solution for problem (3.2)–(3.4).

The following theorem is true:

THEOREM 3.1 (A.V.FURSIKOV [4],[6]) *Let* $n = 2$, $R > 0$, $M = (M_1, M^2, \ldots)$ *be a solution of* (1.3), (1.4) *defined for* $t \in (0,T)$ *and belonging to* $L_2(0,T; P_R)$. *Let* $M^N \in Q^N$ *be the solution of* (3.2)–(3.4). *Then there exists a* $\varrho_0 = \varrho_0(R)$ *such that for* $\varrho \ge \varrho_0$

$$\int_0^T \| M(t,\cdot) - M^N(t,\cdot) \|^2_{P_\varrho}\, dt \longrightarrow 0 \qquad \text{as} \qquad N \longrightarrow \infty\,. \tag{3.5}$$

A similar assertion is true when the dimension of the Navier-Stokes equations equals 3 if we suppose that the approached moments are defined for $t \in (0, \infty)$:

THEOREM 3.2. (A.V.FURSIKOV [4], [5], [6]). *Let* $n=3, R>0, M=(M_1, M_2, \ldots) \in L_2(0,\infty; P_R)$ *be a solution of* (1.3), (1.4). *Let* $M^N \in Q^N$ *be the solution of* (3.2)–(3.4). *Then there exists a* $\varrho = \varrho_0(R)$ *such that for* $\varrho \geq \varrho_0$ *the relation* (3.5) *is true.*

In the case when the inclusion for M in Theorem 3.2 is changed to the weaker condition $M = (M_1, \ldots, M_k, \ldots) \in L_2(0, T; P_R)$ where $T < \infty$, it is also possible to solve the closure problem but with the help of an approximating problem $\mathcal{A}^N(M^N)$ differing from (3.2)–(3.4). To construct the approximating problem in this case we develop the theory of moments for the $3D$ Navier-Stokes equations with a random right-hand side (see A.V. Fursikov [8], A.V. Fursikov, O.Yu. Emanuilov [10]).

REMARK. Evidently, the closure problem can be solved only if a given solution M of Friedmann-Keller chain (1.3), (1.4) is defined unambiguously (otherwise, it is unknown what should be approached). In A.V. Fursikov [7] the uniqueness of solution M of problem (1.3), (1.4) is proved in the case when $M \in L_2(0, T; P_R)$ where $R > 0$. Hence, the assumption $M \in L_2(0, T; P_R)$ of Theorems 3.1, 3.2 formulated above is essential.

The assumption $M \in L_2(0, T; P_R)$ in Theorem 3.1 is not restrictive because in the $2D$-case one can construct the solution M of problem (1.3), (1.4) belonging to this space if initial moments are given.

On the other hand, in the $3D$-problem (1.3),(1.4) the theorem on existence of solution M belonging to $L_2(0, T; P_R)$ is not proved. Therefore from this point of view the Theorem 3.2 is conditional.

The following result on the rate of convergence $M^N \longrightarrow M$, where M^N is the solution of problem (3.2)–(3.4) is true:

Let the assertions of Theorem 3.1 *be fulfilled if dimension* $n = 2$, *and the assertions of Theorem* 3.2 *be true for* $n = 3$. *Then there exist a* $\varrho_0 = \varrho_0(R)$ *and* $0 < \lambda < 1$ *such that for* $\varrho \geq \varrho_0$

$$\int_0^T \| M^N(t, \cdot) - M(t, \cdot) \|_{P_\varrho}^2 \, dt \leq c\lambda^N$$

where $c > 0$ *depends on* λ *but does not depend on* N.

This result is proved in A.V. Fursikov, O.Yu. Imanuvilov [9].

§4. Addition: Certain formulas

In this section we present rigorous definitions of some objects used above.

i) The operator π of orthogonal projection on solenoidal vector fields is defined on periodic vector fields as follows (see M.I. Vishik, A.V. Fursikov [3, p. 216]): Suppose that $u(x) = (u^1(x), \ldots, u^n(x))$ is a vector field satisfying (1.2), $u(x) = \sum \hat{u}(\zeta) e^{ix\cdot\zeta}$

is its Fourier representation where

$$\hat{u}(\zeta) = (2l)^{-1} \int_0^l \cdots \int_0^l u(x) e^{-ix\cdot\zeta}\, dx$$

are Fourier coefficients of $u(x)$. Then the Fourier coefficients $\widehat{\pi u}(\zeta)$ of $\pi u(x)$ are defined as follows:

$$\widehat{\pi u}(\zeta) = \hat{u}(\zeta) - \frac{\zeta \cdot \hat{u}(\zeta)}{|\zeta|^2}$$

ii) In order to determine the phase space P of Navier-Stokes system (1.3) we recall that the Sobolev space $H^k(\Omega)$ is the space of functions $v(x), x \in \Omega$, for which the norm

$$\|u\|_{H^k} = \left(\sum_{0 \le i_1 + \cdots + i_n \le k} \int_\Omega \left| \frac{\partial^{i_1 + \cdots + i_n} u(x)}{\partial x_1^{i_1} \ldots \partial x_n^{i_n}} \right|^2 dx \right)^{1/2}$$

is finite. Then

$$P = \{v(x) \in H^3(\Omega) : \operatorname{div} v(x) = 0\}$$

where Ω is the torus, i.e. the direct product of circumferences having length l.
iii) In order to define the operator B_k from (1.4) we denote

$$(\hat{B}_{x_j} M_{k+1})(t, x_1, \ldots, x_k) = \pi_{x_j} \sum_{i=1}^n \frac{\partial}{\partial x_j^i} \left(M_{k+1}^{i_j}(t, x_1, \ldots, x_j, y, x_{j+1}, \ldots, x_k)|_{y=x_j} \right)$$

where

$$M^{i_j}(t, x_1, \ldots, x_{k+1}) =$$

$$\int u(t,x_1;u_0) \otimes \cdots \otimes u(t,x_{j-1};u_0) \otimes u^{i_j}(t,x_j,u_0) u(t,x_{j+1};u_0) \otimes \cdots \otimes u(t,x_k;u_0) \mu(du_0).$$

Then

$$(B_k M_{k+1})(t, x_1, \ldots, x_k) = \sum_{j=1}^k (\hat{B}_{x_j} M_{k+1})(t, x_1, \ldots, x_k)\,.$$

iv) Let $P(k) = P \otimes \cdots \otimes P$ (k times) be the tensor product of k copies of phase space P. Then

$$P_R = \prod_{k=1}^\infty P(k); \qquad \|m\|_{P_R}^2 = \sum_{k=1}^\infty R^{-2k} \|m_k\|_{P(k)}^2$$

is the phase space for chain of equations (1.4).

References

[1] Reynolds O. On the dynamical theory of incompressible viscous fluids and the determination of the criterion. Phil. Trans. Roy. Sec.; London, 186, (1894), pp. 123–161.

[2] Monin A.S., Yaglom A.M. Statistical Hydromechanics. vol. 1, 2, M.I.T. Press, Cambridge, MA, 1971, 1975.

[3] Vishik M.I., Fursikov A.V. Mathematical Problems of Statistical Hydromechanics. Kluwer Acad. Pub., Dordrecht, Boston, London, 1988.

[4] Fursikov A.V. On the closure problem for a chain of moment equations in the case of large Reynolds numbers. Nonclassical equations and equations of mixed type. Inst. of Math. SO AN SSSR, Novosibirsk, (1990), pp. 231–250 (in Russian).

[5] Fursikov A.V. The problem of closure of chains of moment equations corresponding to the three-dimensional Navier-Stokes system in the case of large Reynolds numbers. Soviet Math. Dokl. vol. 44, No. 1, (1992), pp. 80–85.

[6] Fursikov A.V. The closure problem for the chain of Friedman-Keller moment equations in the case of large Reynolds numbers. The Navier-Stokes equations II. Theory and Numerical Methods. (Oberwolfach, 1991), Lecture Notes in Math., vol. 1530, Springer-Verlag, Berlin, 1992, pp. 226–245.

[7] Fursikov A.V. On uniqueness of the solution of the chain of moment equations corresponding to the three-dimensional Navier-Stokes system. USSR Math. Sbornik, vol. 63, No. 2, (1989), pp. 465–490.

[8] Fursikov A.V. Moment Theory for Navier-Stokes equations with a random right side. Russian Acad. Sci. Izv. Math. Vol. 41, No. 3, (1993), pp. 515–555.

[9] Fursikov A.V., Emanuilov O.Yu. The rate of convergence of approximations for the closure of the Friedmann-Keller chain in the case of large Reynolds numbers. Russian Acad. Sci. Sb. Math. Vol. 81, No. 1, (1995), pp. 235–259.

[10] Fursikov A.V., Emanuilov O.Yu. Convergence Rate for the closure of the chain of moment equations corresponding to the Navier-Stokes system with stochastic right-hand side. Differential equations, Vol. 30, No. 4, (1994), pp. 646–658. Plenum Pub. Corp. 1995.

Trends in Mathematics, © 1999 Birkhäuser Verlag Basel/Switzerland

Vorticity alignment results for the 3D Euler and Navier-Stokes equations

J. D. Gibbon

Abstract

Preferential vorticity alignment is discussed theoretically for both the 3D Euler and Navier-Stokes equations. For the latter a rough theoretical picture is painted showing how approximate alignment can occur.

1 Introduction

In 3D isotropic incompressible Navier-Stokes turbulence the theorist is faced with explaining why the vorticity organizes itself into thin sets which fill a relatively small proportion of the flow volume and whose topology is dominated by quasi-one-dimensional tubes and quasi-two-dimensional sheets. This phenomena is intimately related to the preferential alignment of vorticity with eigenvectors of the strain matrix $S_{ij} = \frac{1}{2}\{u_{i,j} + u_{j,i}\}$, a process that has been discussed in detail in several papers (see [Ashurst, Kerstein, Kerr and Gibson 1987], [Vincent and Meneguzzi 1994], [Majda 1991], [Tsinober, Kit and Dracos 1992] and [Galanti, Gibbon and Heritage 1997] – see the latter for a list of references). It is clear that it is necessary to study the behaviour of vorticity pointwise in the flow instead of taking L^2-spatial averages which are insensitive to the geometric structure of the vorticity field. We therefore consider the behaviour of the two quantities

$$\alpha = \frac{\boldsymbol{\omega}\cdot S\boldsymbol{\omega}}{\boldsymbol{\omega}\cdot\boldsymbol{\omega}} \qquad\qquad \boldsymbol{\chi} = \frac{\boldsymbol{\omega}\times S\boldsymbol{\omega}}{\boldsymbol{\omega}\cdot\boldsymbol{\omega}}. \tag{1}$$

When $\boldsymbol{\omega}$ aligns with an eigenvector of S then α has the property of being an exact eigenvalue of S whereas $\boldsymbol{\chi} = 0$. To see why these two variables are interesting it is best to start with the 3D Euler equations for which we use a result of Ohkitani ([Ohkitani 1997])

$$\frac{D(S\boldsymbol{\omega})}{Dt} = -P\boldsymbol{\omega} \tag{2}$$

where $P = \{p_{,ij}\}$, the Hessian matrix of the pressure. Nonlinear terms $\sim |\nabla\mathbf{u}|^2\boldsymbol{\omega}$ have cancelled in the derivation of (2) leaving only the pressure Hessian term in which is buried nonlocal effects. Different concentrations of vorticity will affect one another through this term. This reduction in nonlinearity has interesting consequences.

In this paper, work in [Galanti, Gibbon and Heritage 1997] is summarized, together with recent work with A. Fokas ([Fokas and Gibbon 1998]). The Burgers vortex and shear layer solutions of both the Euler and Navier-Stokes equations play a pivotal role as they are solutions for which $\boldsymbol{\omega}$ aligns with an eigenvector of S and for which α is independent of the spatial co-ordinates and takes the value $\alpha = \gamma$, where γ is the strain parameter of a Burgers solution. Finally in §3 we summarize results for the Navier-Stokes equations and show that under certain approximations there is an attracting equilibrium state with a corresponding angle θ which is fairly close to zero.

2 The Euler equations

Let us begin with the vorticity form of the 3D Euler equations

$$\frac{D\boldsymbol{\omega}}{Dt} = S\boldsymbol{\omega} \qquad\qquad \operatorname{div}\mathbf{u} = 0 \tag{3}$$

where D/Dt is the usual total derivative and S the strain matrix defined in §1. Using (1), (2) and (3) we easily find that

$$\frac{D\alpha}{Dt} = \chi^2 - \alpha^2 - \alpha_p \qquad\qquad \frac{D\boldsymbol{\chi}}{Dt} = -2\alpha\boldsymbol{\chi} - \boldsymbol{\chi}_p \tag{4}$$

where α_p and $\boldsymbol{\chi}_p$ have the same definition as α and $\boldsymbol{\chi}$ in (1) but with P replacing S. At first glance α_p and $\boldsymbol{\chi}_p$ appear to behave as independent driving terms in (4). This is not the case, however, as P and S are related in other subtle ways. For instance, because $\operatorname{div}\mathbf{u} = 0$, the pressure p must satisfy $\Delta p = -u_{i,j}u_{j,i}$ which can be re-expressed as a trace condition

$$TrP = -Tr\left(S^2\right) + \frac{\omega^2}{2}. \tag{5}$$

Solutions of (4) must therefore also be compatible with (5). Clearly the tangent of the angle θ between $\boldsymbol{\omega}$ and $S\boldsymbol{\omega}$ can be defined as

$$\tan\theta = \frac{\chi}{\alpha}. \tag{6}$$

When $\boldsymbol{\omega}$ aligns (anti-aligns) with an eigenvector of S then $\theta = 0$ (π).

2.1 The orientation of χ with χ_p

More information about the relative orientation of $\boldsymbol{\chi}$ and $\boldsymbol{\chi}_p$ can be gleaned by deriving (4) an alternative way ([Fokas and Gibbon 1998]). Let us consider the unit vector $\hat{\boldsymbol{\xi}} = \boldsymbol{\omega}/\omega$, an arbitrary unit vector $\hat{\boldsymbol{\eta}}$ orthogonal to $\hat{\boldsymbol{\xi}}$ and a third orthogonal unit vector given by $\hat{\boldsymbol{\xi}} \times \hat{\boldsymbol{\eta}}$ in a spherical polar co-ordinate[1] system (r, θ, ϕ). The angles (θ, ϕ) allow us to express any other vector $S\boldsymbol{\omega}$ $(r \equiv |S\boldsymbol{\omega}|)$ lying in this 3-space as

$$\begin{aligned} S\boldsymbol{\omega} &= \hat{\boldsymbol{\eta}}\,|S\boldsymbol{\omega}|\sin\theta\cos\phi + (\hat{\boldsymbol{\xi}} \times \hat{\boldsymbol{\eta}})\,|S\boldsymbol{\omega}|\sin\theta\sin\phi + \hat{\boldsymbol{\xi}}\,|S\boldsymbol{\omega}|\cos\theta \\ &= \hat{\boldsymbol{\eta}}\,\omega\chi\cos\phi + (\hat{\boldsymbol{\xi}} \times \hat{\boldsymbol{\eta}})\,\omega\chi\sin\phi + \hat{\boldsymbol{\xi}}\,\omega\alpha \end{aligned} \tag{7}$$

[1]To keep to the convention of spherical co-ordinates, the notation here is reversed to that in [Galanti, Gibbon and Heritage 1997] where ϕ played the role of θ.

because $\omega\chi = |S\boldsymbol{\omega}|\sin\theta$ and $\omega\alpha = |S\boldsymbol{\omega}|\cos\theta$. Likewise we also have

$$P\boldsymbol{\omega} = \hat{\boldsymbol{\eta}}\,\omega\chi_p\cos\phi_p + (\hat{\boldsymbol{\xi}}\times\hat{\boldsymbol{\eta}})\,\omega\chi_p\sin\phi_p + \hat{\boldsymbol{\xi}}\,\omega\alpha_p. \tag{8}$$

In (7) and (8) the angles θ and θ_p are hidden in α, α_p, χ and χ_p. $\boldsymbol{\chi}$ lies in the horizontal plane at an angle of $\phi + \pi/2$ to the $\hat{\boldsymbol{\eta}}$-axis so

$$\boldsymbol{\chi}\cdot\boldsymbol{\chi}_p = \chi\chi_p\cos(\phi-\phi_p). \tag{9}$$

A simple computation shows that the orthogonal unit vectors $\hat{\boldsymbol{\eta}}$, $(\hat{\boldsymbol{\xi}}\times\hat{\boldsymbol{\eta}})$ and $\hat{\boldsymbol{\xi}}$ satisfy[2]

$$\frac{D}{Dt}\begin{pmatrix}\hat{\boldsymbol{\eta}}\\ \hat{\boldsymbol{\xi}}\times\hat{\boldsymbol{\eta}}\\ \hat{\boldsymbol{\xi}}\end{pmatrix} = \begin{pmatrix}0 & \beta & -\chi\cos\phi\\ -\beta & 0 & -\chi\sin\phi\\ \chi\cos\phi & \chi\sin\phi & 0\end{pmatrix}\begin{pmatrix}\hat{\boldsymbol{\eta}}\\ \hat{\boldsymbol{\xi}}\times\hat{\boldsymbol{\eta}}\\ \hat{\boldsymbol{\xi}}\end{pmatrix} \tag{10}$$

where β is an arbitrary function and the variables α, χ and ϕ must obey

$$\frac{D\alpha}{Dt} + \alpha^2 - \chi^2 + \alpha_p = 0, \qquad \frac{D\chi}{Dt} + 2\alpha\chi = -\chi_p\cos(\phi-\phi_p). \tag{11}$$

$$\left(\frac{D\phi}{Dt} - \beta\right)\chi = \chi_p\sin(\phi-\phi_p). \tag{12}$$

The equations in (11) are equivalent to (4) but with the additional information on the evolution of ϕ in (12). The eigenvalues of the matrix in (10) are given by $\lambda_\pm = \pm i\left(\chi^2+\beta^2\right)^{1/2}$ so constant values of χ and β represent rotations.

2.2 The Burgers vortex

The classic Burgers solution for the Euler equations takes the form of a strain field plus a superimposed 2D velocity field

$$\mathbf{u} = \gamma\left(-ax, -by, z\right)^T + \left(-\Psi_y, \Psi_x, 0\right)^T \tag{13}$$

where $a + b = 1$ and γ is the strain parameter, usually taken as constant ([Majda 1986]). Thus $\boldsymbol{\omega} = (0,0,\omega)^T$ and $\Delta\Psi = \omega$. In this case S and P take on a block diagonal form

$$S = \begin{pmatrix}S_{11} & S_{12} & 0\\ S_{21} & S_{22} & 0\\ 0 & 0 & \gamma\end{pmatrix} \tag{14}$$

and one eigenvector of S is $(0,0,1)^T$. Thus $\boldsymbol{\omega}$ is parallel to this and therefore $\theta = 0$ as $\boldsymbol{\chi} = 0$. The scalar vorticity ω satisfies

$$\omega_t - \gamma\left(ax\partial_x + by\partial_y\right)\omega + \left(\Psi_x\partial_y - \Psi_y\partial_x\right)\omega = \gamma\omega. \tag{15}$$

[2]Equation (10) can be reproduced using the Frenet-Serret equations ([Constantin 1994]) although its derivation is not dependent on them.

Axisymmetry occurs when we choose $a = b = \frac{1}{2}$ with the vorticity a function of r alone ($r^2 = x^2 + y^2$). This corresponds to the axisymmetric Burgers vortex. Solving $\Delta\Psi = \omega$ we have

$$\frac{d\Psi}{dr} = \frac{1}{r}\int_0^r s\,\omega(s,t)\,ds. \tag{16}$$

Now $x\omega_x + y\omega_y = r\omega_r$ and $x\Psi_y - y\Psi_x = 0$ so (15) becomes

$$\omega_t = \gamma\left(\omega + \frac{r}{2}\frac{d\omega}{dr}\right) \tag{17}$$

which has a solution of the form

$$\omega(r,t) = \exp(\gamma t)\,\omega_0\left(r\exp\left(\frac{\gamma t}{2}\right)\right) \tag{18}$$

where $\omega_0 = \omega(r,0)$. Hence the support collapses as the amplitude grows ([Majda 1991]).

The x-independent case (we choose $a = 0,\ b = 1$) corresponds to a Burgers shear layer for which $\boldsymbol{\omega}$ lies in the plane of the layer.

2.3 A finite time alignment criterion

Let us define

$$R^2 = \alpha^2 + \chi^2 \qquad\qquad A = \frac{R^2}{\chi}. \tag{19}$$

Then we obtain

$$\frac{DA}{Dt} = \frac{2R\dot{R}}{\chi} - \frac{R^2}{\chi}\frac{\dot{\chi}}{\chi}. \tag{20}$$

Using the $(\alpha, \boldsymbol{\chi})$ equations we obtain a cancellation of $2\alpha R^2/\chi$ leaving

$$\dot{A} = \left(\frac{\mu_p}{R^2}\right)A^2 - \left(\frac{2\alpha\alpha_p}{R^2}\right)A - 2\mu_p \tag{21}$$

where we have taken $\frac{DA}{Dt} = \dot{A}$ in the Lagrangian formulation of following a fluid particle or set of particles. In (21) μ_p is defined to be

$$\mu_p = \hat{\boldsymbol{\chi}}\cdot\boldsymbol{\chi}_p = \chi_p\cos(\phi - \phi_p)\,. \tag{22}$$

Provided we have initial conditions $A(t=0) = A_0$ such that $\dot{A}$ then $A \to \infty$ and therefore $\chi \to 0$ in a *finite* time t^* provided $\mu_p > 0$. This means that for alignment in this manner we need

$$\cos(\phi - \phi_p) > 0. \tag{23}$$

This combination of angles appears in (9). This result holds even though there is a time dependent coefficient $\mu_p R^{-2}$ on the A^2 term. In fact any positive function of t there will produce blow-up in A as $t \to t^*$ provided this function does not go to zero as $t \to t^*$. In fact in the limit $\chi \to 0$ we simply have $R \to \alpha^2$. If, however, $\mu_p < 0$, then (21) shows that for large times A reaches no more than a positive maximum value and the tendency is towards misalignment.

While alignment may occur for the Euler equations through this and perhaps other mechanisms, there is no reason why condition (23) should hold for all time; the system could easily fall out of alignment without any other process to hold it there. In the next section we consider the effect that viscosity has on the system and discuss the possibility that this can produce a mechanism which may hold the system roughly close to alignment in the local sense we have been discussing.

3 A mechanism for the formation of thin structures in Navier-Stokes flows

It is well known that the Burgers solutions of §2 have their Navier-Stokes equivalent

$$\mathbf{u} = \gamma\left(-ax, -by, z\right)^T + \left(-yf(r), xf(r), 0\right)^T \tag{24}$$

with $f(r) = \left(1 - \exp(-a_0 r^2\right) r^{-2}$ and $a_0 = \gamma/4\nu$. Alignment properties remain the same as in the Euler case; i.e. with $\alpha = \gamma$ and $\theta = 0$. Why should the Navier-Stokes equations favour such a topology? The extra terms that are produced in the $\alpha - \boldsymbol{\chi}$ equations by the viscosity are many in number and messy but the result can be summarized thus; if we define

$$\mu_i = \sum \varepsilon_{ijk} A_{kj}, \qquad \text{and} \qquad \lambda = \sum A_{jj}, \tag{25}$$

(see [Galanti, Gibbon and Heritage 1997] for the full formulae) where the tensor A_{ij} contains $\nabla\hat{\boldsymbol{\xi}}$ and ∇S, then equations for α and $\boldsymbol{\chi}$ become

$$\frac{D\alpha}{Dt} = \chi^2 - \alpha^2 + \nu\Delta\alpha + 2\nu|\nabla\hat{\boldsymbol{\xi}}|^2\alpha + \lambda, \tag{26}$$

and

$$\frac{D\boldsymbol{\chi}}{Dt} = -2\alpha\boldsymbol{\chi} + \nu\Delta\boldsymbol{\chi} + 2\nu|\nabla\hat{\boldsymbol{\xi}}|^2\boldsymbol{\chi} + \boldsymbol{\mu}. \tag{27}$$

Let us now discuss some approximations and put forward the following theoretical picture which is consistent with the formation of thin structures in isotropic Navier-Stokes turbulence. From now on we assume that the flow is regular and that all the necessary quantities are bounded. Now when the flow assumes a Burgers structure, λ and $\boldsymbol{\mu}$ take the simple values

$$\lambda = \gamma^2, \qquad \boldsymbol{\mu} = 0. \tag{28}$$

This highly simplified form assumed by λ and $\boldsymbol{\mu}$ in the Burgers case motivates us to *assume that the variables* α, $\boldsymbol{\chi}$, λ *and* $\boldsymbol{\mu}$ *have reached a simultaneous equilibrium in some connected region of vorticity whose growth has been controlled by dissipation.*

First we want to see if the 'fixed point' values taken by α and $\boldsymbol{\chi}$ at this equilibrium correspond to a small angle θ_0 and secondly whether this fixed point is stable. To investigate this stability question we begin by assuming that $\lambda = \lambda_0$ and $\boldsymbol{\mu} = \boldsymbol{\mu}_0$ are constant in the four equations in (26) and (27)

$$\frac{D\alpha}{Dt} = \chi^2 - \alpha^2 + \nu\Delta\alpha + \lambda_0, \tag{29}$$

$$\frac{D\boldsymbol{\chi}}{Dt} = -2\alpha\boldsymbol{\chi} + \nu\Delta\boldsymbol{\chi} + \boldsymbol{\mu}_0 \tag{30}$$

where, for reasons of simplicity, we have written these without the quantity $|\nabla\hat{\boldsymbol{\xi}}|^2$ which is the least important of the terms. Fixed points in $(\alpha_0, \boldsymbol{\chi}_0)$ occur at

$$2\alpha_0^2 = \lambda_0 + \sqrt{\mu_0^2 + \lambda_0^2}, \qquad \chi_{i,0} = \frac{\mu_{i,0}}{2\alpha_0}, \tag{31}$$

where $\mu_0 = |\boldsymbol{\mu}_0|$. Hence there are two fixed points in 4-space, corresponding to the two roots of α_0 in (31). Without the Laplacian term it is easy to show that the eigenvalue stability problem is

$$(\Lambda + 2\alpha_0)^2 \left[(\Lambda + 2\alpha_0)^2 + 4\chi_0^2\right] = 0, \tag{32}$$

thereby giving the four roots

$$\Lambda = -2\alpha_0 \ \ \text{(twice)}, \qquad \Lambda = -2\left(\alpha_0 \pm i\chi_0\right). \tag{33}$$

These roots correspond to an unstable fixed point for the negative root for α_0 and a stable one for the positive root. In the latter case, there is exponential contraction in two of the directions in the 4-space with a stable spiral in the other two directions. When the Laplacian is included we look at the stability of linearized solutions of the type $\exp(i\mathbf{k}\cdot\mathbf{x} + \lambda_0 t)$ around $(\alpha_0, \boldsymbol{\chi}_0)$. The only difference this makes to (33) is that $2\alpha_0 \to 2\alpha_0 + \nu k^2$. We conclude that provided $\alpha_0 > 0$ the equilibrium solution is stable to the disturbance of all wavenumbers. For the half-plane the two fixed points are

$$\sqrt{2}\alpha_0 = \pm\left[\lambda_0 + \sqrt{\mu_0^2 + \lambda_0^2}\right]^{1/2} \qquad \sqrt{2}\chi_0 = \left[-\lambda_0 + \sqrt{\mu_0^2 + \lambda_0^2}\right]^{1/2}. \tag{34}$$

Defining m to be

$$m = \frac{\mu_0}{|\lambda_0|}, \tag{35}$$

the angle θ_0 corresponding to the stable fixed point is (the $+$ sign is for $\lambda_0 > 0$ and the $-$ sign for $\lambda_0 < 0$)

$$\tan 2\theta_0^{\pm} = \pm m. \tag{36}$$

Now the exact Burgers solutions are represented by $m = 0$ and those with positive α_0 correspond to *attracting* fixed points of the system. We note however that when it takes nonzero values, m is dependent only on the ratio of μ_0 and $|\lambda_0|$. Although we have no hard information on the magnitude of m, in fact the angle θ_0^+ is clearly relatively insensitive to this magnitude. For $\lambda_0 > 0$, if $m \approx 0$ then $\theta_0^+ \approx 0$ but if $m \approx 1$ then $\theta_0^+ \approx \pi/8$. Even if $m = \infty$ then $\theta_0^+ = \pi/4$. Hence *at worst,* θ_0^+ *lies in a* 45° *cone.* Therefore, even when $m > 0$ we are still close to alignment. More generally, $\boldsymbol{\mu}_0$ and λ_0 derive respectively from forming the vector and scalar products of the same set of functions. In an isolated region of vorticity Ω, we might expect that on the spatial average over a random set of points in Ω, m would take the value $m \approx 1$. It is also possible that because $\boldsymbol{\mu}_0$ is formed from a cross product while λ_0 is formed from a dot product then at the natural angle of alignment the μ_0 term would be the weaker of the two making $m < 1$. We therefore conclude that over the spatial average within a small intense region for $\lambda_0 > 0$

$$\theta_0^+ \approx \pi/8 \tag{37}$$

but that θ_0^+ may indeed be somewhat smaller than this. Experiments and simulations ([Tsinober, Kit and Dracos 1992]) generally measure the cosine of the angle θ, often observing that a bunching around $\cos\theta \approx 1$ in their PDF's is a demonstration of alignment. For $\theta_0^+ \approx \pi/8$ we have

$$\cos\theta_0^+ \approx 0.92. \tag{38}$$

This value of m therefore produces fairly close alignment. This is consistent with the region finding an equilibrium shape near to a thin Burgers-like structure (for which the exact value of m is zero) which corresponds to $\lambda_0 > 0$. When $\lambda_0 < 0$, however, $\theta_0^- = \pi/2 - \theta_0^+$ and the equivalent value of θ_0^- is $3\pi/8$. Hence vortex lines are badly misaligned in this case.

Acknowledgements: I am thankful for conversations with Thanasis Fokas, John Gibbons, Trevor Stuart, Itamar Procaccia, Barak Galanti and Matthew Heritage.

References

[Ashurst, Kerstein, Kerr and Gibson 1987] Ashurst W., Kerstein A., Kerr R. and Gibson C., 1987, Alignment of vorticity and scalar gradient with strain rate in simulated Navier-Stokes turbulence, *Phys. Fluids A*, **30**, 2343.

[Vincent and Meneguzzi 1994] Vincent A. and Meneguzzi M., 1994, The dynamics of vorticity tubes of homogeneous turbulence, *J. Fluid Mech.*, **258**, 245–254.

[Majda 1991] Majda A., 1991, Vorticity, turbulence and acoustics in fluid flow, *SIAM Rev.*, **33**, 349–388.

[Tsinober, Kit and Dracos 1992] Tsinober A., Kit E. and Dracos T., 1992, Experimental investigation of the field of velocity gradients in turbulent flows, *J. Fluid Mech.*, **242**, 169.

[Galanti, Gibbon and Heritage 1997] Galanti B., Gibbon J.D. and Heritage M., 1997, Vorticity alignment results for the 3D Euler and Navier-Stokes equations, *Nonlinearity*, **10**, 1675–1694.

[Ohkitani 1997] Ohkitani K., 1997, Eigenvalue problems in three-dimensional Euler flows, *Phys. Fluids A*, **5**, 2570.

[Fokas and Gibbon 1998] Fokas A. and Gibbon J. D., 1998, in preparation.

[Constantin 1994] Constantin P., 1994, Geometric statistics in turbulence, SIAM Rev., **36**, 73.

[Majda 1986] Majda A., 1986, Vorticity and the mathematical theory of incompressible fluid flow, *Comm, Pure and Appl. Math.*, **39**, 187–220.

Trends in Mathematics, © 1999 Birkhäuser Verlag Basel/Switzerland

On the finite-time singularity problem in hydrodynamics

R. B. Pelz, Y. Gulak, J. M. Greene, O. N. Boratav

We propose a scenario for a spontaneous singularity in incompressible, frictionless flow and review the numerical evidence that supports it.

1 Introduction

Global regularity of solutions of the equations of incompressible inviscid fluid flow has not been established. It has been speculated that if solutions with spontaneous singularities, were to exist, they would have a profound impact on the dynamics and theories of turbulence [1]. For example, the purported nonzero rate of energy dissipation with increasing Reynolds number may be explained through the existence of singularities in enstrophy in the limit $\nu \to 0$. Another speculation is that intermittency is caused by the presence of nearly singular events, perhaps regularized by viscosity.

A hydrodynamic singularity will also have ramifications in general fluid mechanics. That a discontinuity forms from a smooth flow would require a modification in the mathematical formulation of the problem and perhaps the introduction of weak solutions. Physically, an incompressible flow which develops an infinite vorticity or velocity seems nonintuitive and hence quite fascinating. It would be a flow in which depletion of nonlinearity does not occur.

Recently, we have compiled numerical evidence and developed a plausible scenario for a finite time singularity. Key concepts are a local (not fully) self-similar collapse of a vortical structure, a crossover region which separates the inner collapse and the outer nonsingular regions, and coupling of strain rate and vorticity through nonlocal symmetries. In this paper, we review our work as well as supply an introduction to mathematical aspects of the problem. We further show that in a vortex filament solution which exhibits collapse, the directions of positive strain rate and the vorticity are the same and their magnitudes are related linearly for a Lagrangian point on the filament.

2 Mathematical aspects

The reason why global reglarity has not been shown for the Euler equations is that vorticity cannot be bounded for all time. The local existence (early time) theorem of the Euler equations states that if the initial flow $u(0)$ is in H^s, the Sobolev space in which all derivatives through s are in L^2, $s \geq 3$ and $|u(0)|_3 \leq N$, then up to time $t = t_0$ which depends on N, the solution $u(t)$ of the Euler equations is C^0 in H^s and C^1 in H^{s-1} [2].

Beale, Kato and Majda [3] then showed that

$$|u|_s \leq |u(0)|_s \exp C_s \int_0^t |\nabla u(\tau)|_{L^\infty} \mathrm{d}\tau. \tag{1}$$

In words, u(t) will continue to be in the local solution class if the integral of the max norm of the velocity gradient remains bounded. Conversely, if the integral first becomes unbounded at t_0, then the L^2 norm of the sth derivative of u becomes unbounded at t_0. They went on to relate the L^∞ norm of the velocity gradient to that of the vorticity, whereas Ponce [4] related it to the strain rate tensor. Since at any time before t_0 the integral is bounded, then the following is true

$$\lim_{t\to t_0} \sup |\nabla u(t)|_{L^\infty} = \infty. \tag{2}$$

Thus, there is early-time existence of smooth solutions, and if regularity of a high derivative of the velocity is lost subsequently, then this is indicated by the pointwise blowup of the first spatial derivative of the velocity.

If vorticity form is used, equation (2) means that the local rotation rate or frequency of a fluid element becomes singular. On dimensional grounds, the form of the blowup is $(t_0 - t)^{-1}$. Using conservation of kinetic energy and the Hölder norm, Constantin [5] showed that the length scale associated with the blowup region follows $(t_0 - t)^p$, where $p \geq 2/5$. Thus, a singularity of the Euler equations must manifest itself as a high-frequency, small-scale event.

3 The symmetric T^3 problem

In this section, evidence for the existence of a real-time singularity for a flow in a 3-torus (T^3) is presented from two different approaches: direct pseudo-spectral simulation of the initial value problem [6] and extended power series expansion about $t = 0$ [7].

Using the initial condition

$$u(x,y,z) = v(z,x,y) = w(y,z,x) = \sin x(\cos 3y \cos z - \cos 3y \cos z) \tag{3}$$

and 2π periodical symmetry, reflectional symmetry about planes $n\pi, n = 0, 1, \ldots$, and $\pi/2$ rotational symmetry about lines $x = y = \pi/2$, $y = z = \pi/2$ and $z = x = \pi/2$, the Fourier series kernel for u has the form $\sin kx \cos py \cos qz$ where k, p, q are either all odd or even. The symmetry $u(x,y,z) = v(z,x,y) = w(y,z,x)$, which is also preserved for all time, can be derived from the three symmetries above. We shall refer to the Euler equations with these symmetries and this initial condition as the symmetric T^3 problem.

Figure 1 shows the isosurface of vorticity from pseudo-spectral simulations of this problem at t=1.2, $|\omega| = 10$ ($|\omega|_{max} = 12.5$). The 12-vortex structure arranged as 6 dipoles moving towards the origin, shrinks in a visually self-similar manner, increasing in strength until it is on the order of the grid size at $t \approx 2$.

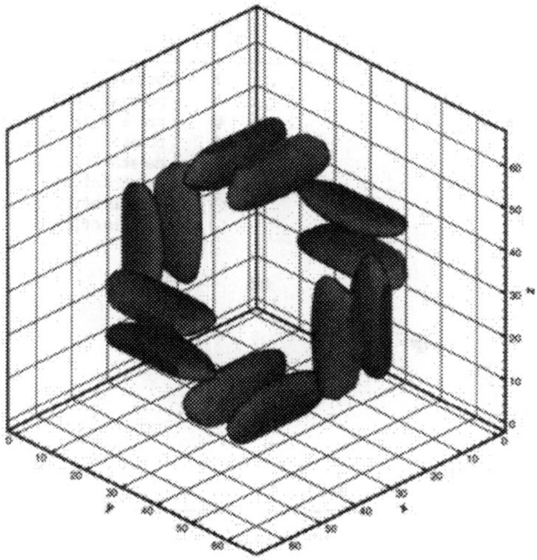

Figure 1: Vorticity isosurface $|\omega|/|\omega|_{max} = 0.8$ at time 1.2 from pseudospectral simulation of the Euler equations.

Since the collapse time is realtively early, a power series expansion about t=0 may reveal a singular behavior. A similar approach was used with the symmetric Taylor Green flow [8], but results were inconclusive.

The form for the velocity is $u = \sum_{p=0} u_p t^p$ where each coefficient is a Fourier series in space. After substituting into the Euler equations, the pth term can be written as the recursion $u_p = (\sum_{r=0}^{p-1} u_r \diamond u_{p-1-r})/p$ where the operation $\diamond$ is commutative but not associative. If the coefficients of the initial condition, u_0, are rational numbers, then the coefficients of subsequent terms are rational also, and the computation can, in principle, be done exactly. The number of digits in the rational expressions became prohibitively large, however, and extended precision (up to 154 digits) was used. The Euler equations become the quadratic equation $du/dt = u \diamond u$ on the $\diamond-$algebra. It is interesting to note that if there is a real-time singularity, then the $\diamond-$algebra should show it. The structure of this algebra is quite complicated, however, and needs to be better understood.

Currently, 36 terms of the series have been computed and the time series for enstrophy analyzed. Figure 2 shows the location of singularities of the enstrophy in the complex time plane as given by various Padé approximants. Despite a slight scatter in the location with order of approximant the results are consistent. The singularity nearest the origin is close to the imaginary axis, causing the Taylor series to diverge at a time $t \approx 0.7$. A number of defects, nearly coincident zeros and poles, exist, but do not effect convergence. In each approximant, an isolated

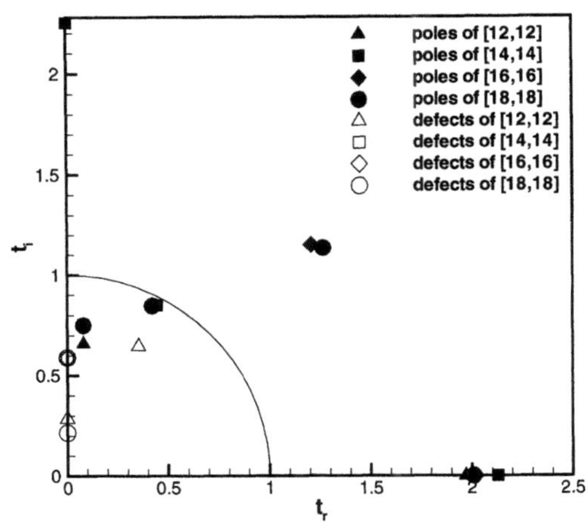

Figure 2: Location of the poles and defects (nearly coincident zeros and poles) of enstrophy in the complex time plane using Padé approximants. Note the isolated real-time pole around $t_r = 2$.

real-time singularity is predicted at a time about 2. Using an origin-preserving bilinear transformation and the ratio test, the biased order was found to be -1.

This work suggests that a spherical collapse of a symmetric vortex system occurs in a finite time in the T^3 problem.

4 The scaled Euler equations and asymptotic self-similarity

Almost all the scenarios that have been proposed for the singularity have involved the interaction and collapse of vortex tubes. In this section we examine the possibility of Euler solutions with self-similar collapse leading to singularity. Following Greene and Boratav [9], we assume that the collapse is centered at the origin and that the $|\omega|_{L^\infty} \sim (t_0 - t)^{-1}$ consistent with reference [3].

To the Euler equations $\partial v/\partial t = v \times \omega - \nabla \Pi$, $\quad \Delta \Pi = \nabla \cdot (v \times \omega)$, we introduce the following scaling transformation

$$\rho = \frac{r}{\lambda (t_0 - t)^{1-\alpha}}, \quad \tau = t - t_0, \quad v = \frac{\lambda V(\rho, \theta, \phi, \tau)}{(t_0 - t)^{\alpha}} \tag{4}$$

$$\omega = (t_0 - t)^{-1} \nabla_\rho \times V \qquad \text{and} \qquad \Pi = \frac{\lambda^2}{(t_0 - t)^{2\alpha}} P(\rho, \theta, \phi, \tau)$$

where $[\lambda] = LT^{\alpha-1}$. The Euler equations become

$$\tau\frac{\partial V}{\partial \tau} = (2\alpha-1)V - [V + (1-\alpha)\rho] \times (\nabla_\rho \times V) + \nabla_\rho[P + (1-\alpha)\rho\cdot V] \qquad (5)$$

Since τ appears explicitly only on the left, if the right-hand side is zero for some time $\tau < 0$, then there exists a singularity at $\tau = 0$.

As was pointed out by Constantin [5], conservation of energy in all R^3 restricts $\alpha = 3/5$. From this, he suggested that there are no nontrivial solutions that are fully self-similar. Conservation of circulation and helicity, however, give $\alpha = 1/2$.

To resolve these conflicts, we introduce the notion of asymptotic self-similarity – the existence of an asymptotic inner solution which is self-similar. Obviously, this requires a rather special matching region which can share and separate the inner fast scales and the outer slow ones.

If an inner solution exists then the ball over which the kinetic energy integration is done has a finite radius which may scale with the collapse. In this case, energy need not be conserved and the 3/5 scaling restriction is irrelevant.

We propose that the matching region is made up of purely radial vortex lines. The induced velocity has no radial component, thus separating the inner and outer regions. It is not hard to construct an outer solution in which vortex lines match to the radial. As we shall see in the next section, the wake of the collapse solution contains radial vortex lines

5 Filament model of the R^3 problem

It is obvious that the spherical collapse should be represented not in T^3, but in spherical coordinates in R^3. Motivated by numerical results of the T^3 problem, an associated symmetric R^3 problem can be defined. To keep the vorticity nulls and stable manifolds located on the axes, it is clear that the bilateral symmetry with respect to the zero planes must be present. For strain rate/vorticity feedback which will be explained further in the next section, we shall keep the velocity permutation symmetry $u(x,y,z) = v(z,x,y) = w(y,z,x)$ which is a $2\pi/3$ rotational symmetry about the diagonal $x = y = z$.

A filament model was used to study the R^3 problem [10]. (We are currently studying it using the Euler equations). In such a model vortex tubes are condensed to single filaments which are then discretized. Between nodes, a straight vortex segment approximates the filament. Movement of the nodes is then done by a summation over the segments. A core function (cubic Gaussian) is used as the local cutoff (regularization) of the Biot Savart integral. The core cross-sectional area, σ^2, evolves as follows: $D(\ln\sigma^2)/Dt = S_\omega$ where S_ω is the local strain rate in the direction of the filament.

The unique part of the model concerns the core model when part of a filament get close to a plane of symmetry, as it eventually will in anti-parallel collapse. The filament cannot cross the plane, but because there is limited core dynamics in a filament model, part of the core may cross in violation of the Helmholtz constraint.

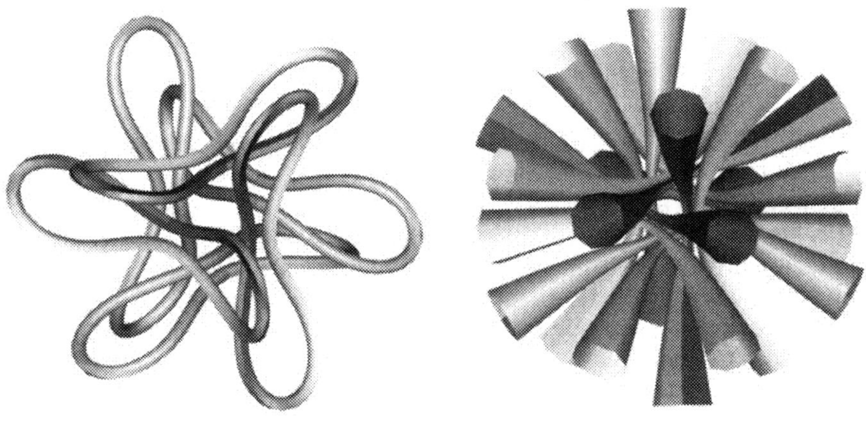

Figure 3: a.) Initial location and thickness of the vortex filaments. b.) Location and thickness of filaments at a time $t = 3 \times 10^{-8} t_0$ magnified 10640 times from a.

This was the non-Euler effect that nullified the filament collapse work of Siggia and Pumir [11]. In our model, the core is non-circular with one radius that is never greater than the distance to the symmetry plane, $\min(d, \sigma)$, where d is the distance to the symmetry plane. In the other direction orthogonal to this, the core has a value $\sigma^2 / \min(d, \sigma)$. This "flattening" process produces a two-length scale desingularization and reduces the velocity towards the symmetry plane so that d and σ vanish together in proportion at a finite time. Of course, this model is good only when the flattening is mild, which is the case in the following.

Figure 3a shows the location of the filaments initially as motivated by the structure of the solution in the T^3 problem. Half the core size is used to make the thickness. Figure 3b shows the position of the filaments at a time of 3.21. The solution has been magnified 10,640 times. The solution in the inner region is visually identical for magnifications greater than about 50. The inner solution has a scaling as predicted in Equation (5); the outer region (not shown) evolves very little on this time scale. The crossover region, the outer portion of Figure 3b, consists of nearly radial vortex lines. Since the circulation is preserved by definition in a filament model, the scaling exponent is $\alpha = 1/2$. Enstrophy blows up a $(t_0 - t)^{-1/2}$. Arc length is not a function of the inner scale.

In the filament model of the R^3 problem, a locally self-similar solution exists with the scaling and form given in the previous section.

6 Strain rate and vorticity

In this section, we examine the coupling of the strain rate and vorticity in the filament model. From the vorticity equation in the Lagrangian frame, $\dot{\omega_i} = \omega_j S_{i,j}$, where $S_{i,j} = (\partial u_i/\partial x_j + \partial u_j/\partial x_i)/2$, it is easy to see that if the strain rate component in the direction of the vorticity scales with the vorticity, then vorticity blows up in a finite time.

By differentiation of the Biot Savart law, Constantin [5] derived a formula which relates strain rate and vorticity. The component $S_\omega = S_{i,j}\hat{\omega}_i\hat{\omega}_j$ is the material logarithmic derivative of vorticity magnitude (α in Constantin's notation), and is written

$$S_\omega(x) = \frac{3}{4\pi} P.V. \int D[\hat{y}, \hat{\omega}(x+y), \hat{\omega}(x)] |\omega(x+y)| \frac{dy}{y^3} \tag{6}$$

where $D[e_1, e_2, e_3] = (e_1 \cdot e_3) \det(e_1, e_2, e_3)$ and the hat denotes unit vector. For discrete vortex filaments, the expression above takes the form

$$S_\omega(x) = \omega(x)\frac{3}{4}\sigma^2(x) \sum_i D[\hat{r}_i, \hat{\omega}_i, \hat{\omega}(x)] \frac{\Delta_i}{r_i^3} \tag{7}$$

where $r_i = x_i - x$, Δ_i is the length of the ith vortex segment and σ is the core thickness. We let the point x be on one of the filaments where it intersects the zero plane. Note that due to the symmetries, the direction of the vorticity in the zero plane is normal to the plane as is one of the eigenvectors of the strain rate tensor. We separate the contributions to $S_\omega(x)/\omega(x)$ into 3 parts, $S_\omega(x) = \omega(x)(S_{\omega,L} + S_{\omega,D} + S_{\omega,NL})$, where $S_{\omega,L}$ is the local contribution from the half of the loop containing x as the midpoint, $S_{\omega,D}$ is the contribution from the dipolar pair image (the other half of the loop), and $S_{\omega,NL}$ is the contribution from the remaining 5 loops.

Along this Lagrangian point x, each of these terms converge to constant nonzero values as $t \to t_0$ as follows: $(S_{\omega,L}, S_{\omega,D}, S_{\omega,NL}) = (-0.00075, 0.015, 0.11)$. The local contribution is negative indicating that this part of the filament is tending to decrease the vorticity magnitude. The contribution of the dipolar image is small but positive indicating growth. The main positive contribution to the integral, however, is from the nonlocal filaments due to the rotational symmetry.

Thus, the particular configuration of the filaments causes a strain-rate eigenvector with positive eigenvalue to be aligned with the vorticity, and causes this eigenvalue to be a linear function of the vorticity magnitude. Since behavior occurs along a Lagrangian trajectory, it leads to an inverse time blowup of vorticity.

7 Concluding remarks

Aided by numerical simulations, a possible spontaneous finite-time singularity in incompressible, frictionless flow has been constructed. Key features are an asymptotic self-similar inner solution and nonlocal symmetries which allow strain rate / vorticity coupling.

Questions concerning stability, viscous effects and realization of these symmetric initial conditions in a general turbulent flow are important and are currently being investigated. Our approach, however, has been to establish first a particular flow in which a spontaneous singularity develops, then to explore more generic situations.

References

[1] U. Frisch, "Turbulence," Cambridge, 1995

[2] R. Temam, "Navier-Stokes Equations and Nonlinear Functional Analysis," SIAM, Philadelphia 1995.

[3] J. T. Beale, T. Kato, and A. Majda, Commun. Math. Phys. **94** 61 (1985).

[4] G. Ponce, Commun. Math. Phys. **98** 349 (1985).

[5] P. Constantin, SIAM Review **36**, 73 (1994).

[6] O. N. Boratav and R. B. Pelz, Phys. Fluids **6** 2757 (1994).

[7] R. B. Pelz and Y. Gulak Phys. Rev. Letts. **79** 4998 (1997).

[8] R. H. Morf, S. A. Orszag and U. Frisch, Phys. Rev. Letts. **44** 572 (1980). and M. E. Brachet, D. I. Meiron, S. A. Orszag, B. G. Nickel, R. H. Morf and U. Frisch, J. Fluid. Mech. **130** 441 (1983).

[9] J.M. Greene and O. N. Boratav, Physica D **1663** 1 (1997).

[10] R. B. Pelz, Phys. Rev. E, **55** 1617 (1997).

[11] E. D. Siggia and A. Pumir, Phys. Rev. Lett. **55**, 1749 (1985).

Trends in Mathematics, © 1999 Birkhäuser Verlag Basel/Switzerland

The outer regions in singular Euler

R. M. Kerr

Abstract

New analysis of the scaling structure of a numerical solution consistent with a singularity of the three-dimensional incompressible Euler equations suggests two length scales dividing physical space into three scaling regimes surrounding the position of peak vorticity. Self-similar behavior is identified in the two interior regimes, but not exact self-similarity. The size of the inner regime decreases uniformly in all directions as $\rho = (T-t)$. The intermediate regime has an outer length scale that decreases as $R = (T-t)^{1/2}$ with strong deformation on the inner length scale $(T-t)$ and fully participates in the singular dynamics of the inner regime. The vortex structure of the intermediate regime is characterized as a vortex jump on the surface of a blunt cone, elongated in different directions, with the inner regime enclosed by the blunt tip. The outer regime constitutes the rubbish bin of the singular dynamics, and does not participate in the singular dynamics on the time scale remaining till the possible singularity.

1 Introduction

Properties associated with singularity behavior in a calculation of the three-dimensional incompressible Euler equations (Kerr, 1993a) included peak vorticity ω_p, strains $e_{ij,p}$, and the volume integral of the enstrophy production $\int dV \omega_i e_{ij} \omega_j$ all increasing as $1/(T-t)$, where T is the projected singular time, as well as enstrophy increasing as $\Omega \sim -\log(T-t)$. This contribution presents new analysis that has identified a collapsing structure defined by two length scales, an inner length scale associated with the positions of the peak vorticity and strain components that goes as $\rho \sim (T-t)$ and an outer length scale that goes as $R \sim (T-t)^{1/2}$, consistent recent mathematics (Constantin, 1994, Constantin et al., 1996).

The three dimensional structure of this flow has been discussed before (Kerr, 1993a, Kerr, 1996, Kerr, 1997). To illustrate the important points to be discussed here, a number of xz slices through the structure at $t = 17.5$ are shown in Fig. 1. Analysis at $t = 17.5$ is made possible by the application of a small, simple quartic hyperviscous term of the form $\exp(-\nu_4 k^4)$ in the x direction only (Kerr, 1997). This removes a build-up of energy in the highest modes, which can be seen in the spectra (Kerr, 1993a), that tended to obscure the results. Removal of these higher modes in analysis changed the kinetic energy, peak vorticity, and enstrophy by less than a per cent. Three times are used here, $t = 15, 17$ and 17.5, where the projected singular time is currently estimated to be 18.7. For a computational domain of $4\pi \times 2\pi \times 2\pi$, using symmetries in y and z, the mesh for

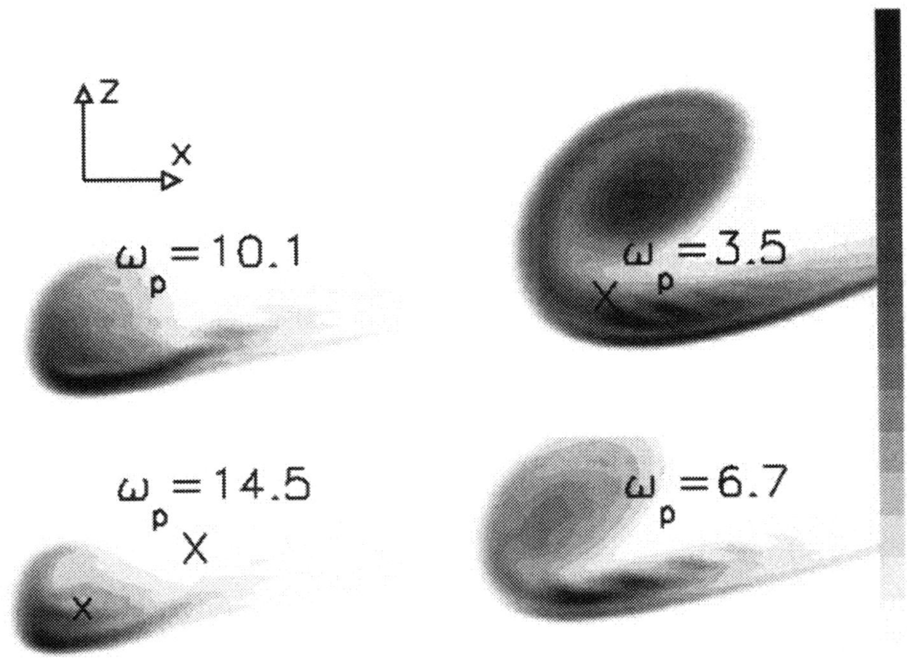

Figure 1: Contour plots of vorticity for 4 xz slices in y at $t = 17.5$. The xz domain is $0.7\pi \times 0.6\pi$ with y spacing of 0.06π. The vertical coordinate uses the Chebyshev mesh. That is, the non-uniform Chebyshev mesh is plotted with uniform spacing. This reduces noise in interpolating to a uniform mesh and resolves the sharp vertical structures better. The maximum vorticity in each slice is shown. Small x indicates position of $e_{yy,p}$ and large X represents the xz position of the extent of the intermediate scaling region and the peak vortical velocity v_p. The X in upper right represents the approximate y position of these.

$t = 15$ and 17 was $n_x \times n_y \times n_z = 512 \times 256 \times 128$ and for $t = 17.5$ the mesh was $n_x \times n_y \times n_z = 512 \times 256 \times 192$. More details are given in the original paper (Kerr, 1993a).

The lower left frame in Fig. 1 represents the xz slice in the symmetry plane studied earlier (Kerr, 1993a). The other frames in Fig. 1 represent different xz slices in order to show more of the 3D structure. The two perpendicular jumps in vorticity in the lower left frame are the thinnest dark regions. The original analysis (Kerr, 1993a) concentrated on the scaling in the vicinity of the juncture between the two sheets, which is the location of ω_p. It was shown that the maxima of all the components of the stress tensor in the symmetry plane are near this corner, with their positions converging to a single point as $\rho_i \sim (T - t) \to 0$ and their magnitudes growing as $1/(T-t)$, consistent with mathematical requirements (Ponce, 1985, Constantin et al., 1996). This paper will refer to the region where all positions scale as $\rho_i \sim (T - t)$ as the scaling of the inner region. The small x in the lower right frame of Fig. 1 locates one of these maxima, that of strain along the vorticity.

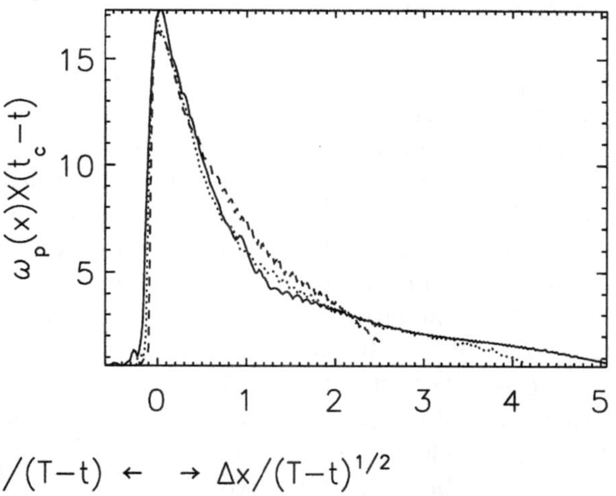

Figure 2: Profile of maximum vorticity for $y = 0$ as a function of $\Delta x = x - x|_{\omega_p}$ for three times with time scaling of vorticity indicated and two rescalings of position. For $\Delta x < 0$, $\Delta x/(T-t)$ and for $\Delta x > 0$, $\Delta x/(T-t)^{1/2}$. Times are: solid $t = 17.5$, dot $t = 17$, and dash $t = 15$.

2 New length scaling

To quantify the collapse better, Fig. 2 shows rescaled profiles of vorticity from the lower left frame of Fig. 1, the symmetry plane $y = 0$. Both vorticity and the position are rescaled. Each curve comes from the maxima of vorticity as a function of x for all z. Vorticity is rescaled by multiplying by $(T - t)$. The x-coordinate plotted is $\Delta x = x - x|_{\omega_p}$ rescaled in two ways, where $x|_{\omega_p}$ is the x position of

ω_p at any given time. To the left, $\Delta x < 0$, positions are scaled by $(T-t)^{-1}$, consistent with the proposed scaling of the inner region. However, to the right, $\Delta x > 0$, positions are scaled by $(T-t)^{-1/2}$. This new scaling applies out to at least $\Delta x/(T-t)^{1/2} = 2$. A similar rescaling of $\Delta z = x - z|_{\omega_p}$ can be done, with $\Delta z/(T-t)^{1/2}$ being applicable out to $\Delta z/(T-t)^{1/2} = 0.15$

In an incompressible flow there cannot be simultaneous compression in all three directions, there must be at least one direction of stretching. But there is no reason the scaling regime could not collapse in all directions if in the direction of stretching there were strong curvature of field lines. So far all serious the analysis of this flow, here and before (Kerr, 1993a), has been in the symmetry plane. Fig. 3 extends the scaling analysis just described in x and z to the vortical direction y, with position scaled only by the second length scale $R \sim (T-t)^{1/2}$. This applies out to $y/(T-t)^{1/2} \approx 0.5$, which in Fig. 1 goes nearly out to the upper right frame with the large X. Scaling y by $R \sim (T-t)^{1/2}$ also applies to y-velocity (Kerr, 1997), whose maximum is at the position of the large X in the upper right frame. Given that scaling by $R \sim (T-t)^{1/2}$ has been applied to all the analysis that has been done in y, it is likely that curvature should also follow this trend, that is $r_{curv} \sim (T-t)^{1/2}$.

The structural significance to $y/(T-t)^{1/2} \sim 0.5$ seems to be that from $y = 0$ out to this point, xz vorticity slices are best characterized as two vortex sheets meeting at a corner. In a three-dimensional visualization (Kerr, 1996) this region is approximately enclosed by an isosurface set at $.6\omega_p$ and within that isosurface vortex lines are relatively straight. For $y/(T-t)^{1/2} > 0.5$, helical vortex lines appear, which one can begin to see in the significant region of swirl in the upper right frame of Fig. 1. To show how far out in x and z scaling by $R \sim (T-t)^{1/2}$ works with respect to the position of ω_p, the large X in the lower right frame of Fig. 1 is at the position $(\Delta x, \Delta z)/(T-t)^{1/2} = (0.5, 0.15)$.

Where any of $(\Delta x, \Delta y, \Delta z)/(T-t)^{1/2} > (0.5, 0.5, 0.15)$ defines an outer region. It is in this outer region that most of the standard bulk properties such as kinetic energy, enstrophy, and helicity reside as the singular time is approached. Thus, while the outer region might not be intimately involved in the singular dynamics, the structure of this region should strongly affect the properties of reconnection that would occur in a viscous flow. These questions will be examined using new calculations and new analysis of existing calculations.

3 Analytic arguments

What can be said more definitively about the requirement for there to be two length scales for there to be a singularity of Euler? What literature exists (Constantin, 1994, Constantin et al., 1996) and a self-similar form for Euler proposed by J. Greene, says more about $R \sim (T-t)^{1/2}$ than $\rho \sim (T-t)$. Work on averages of angles between vorticity and strain must also be related (Gibbon and Heritage, 1997). The only definitive statement that there must be two length scales comes from a Hölder seminorm argument (Constantin, 1994).

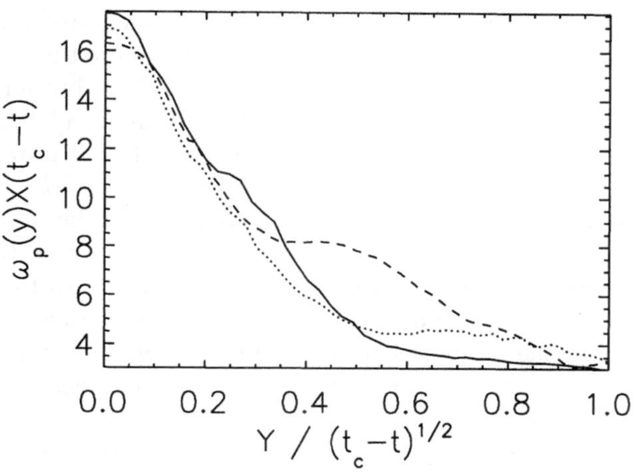

Figure 3: Profile of maximum vorticity over x and z as a function of y for three times with time scaling of vorticity and y indicated. Times are the same as in Fig. 2.

This is that if vorticity takes the form

$$\omega(x,t) = \frac{1}{T-t}\Omega\left(\frac{x}{L(t)}\right) \quad \text{where} \quad L(t) \sim (T-t)^p \tag{1}$$

then one can construct a length scale $\ell_\mu(t)$ using the Hölder seminorm that must collapse such that for there to be a singularity

$$\int_0^T (\ell_\mu(t))^{-5/2} dt < \infty \tag{2}$$

This implies first that if one or more length scales are found to describe the collapse, it is required that $p \geq \frac{2}{5}$ for there to be a singularity. This is satisfied by the ρ and R proposed here, where $p = 1$ and $1/2$, respectively. For there to be exact self-similarity, where there is only one length scale and all the energy is contained within that length scale, then a further requirement would be that $p \equiv \frac{2}{5}$. The suggestion of Greene for a self-similar form with $1/2$ therefore would not satisfy this requirement, although there are other reasons related to a singularity in trying to construct a smooth solution at any given time that would also preclude this form.

A more concrete reason for why two length scales are needed might come from another mathematical proof (Constantin et al., 1996). If the following integrals over time are controlled

$$\int_0^T N_1(s)\left(\frac{L}{R(s)}\right)^3 ds < \infty \qquad \int_0^T \frac{U(s)}{R(s)} ds < \infty \tag{3}$$

then there cannot be a singularity of Euler. In this case $R(s)$ is a radius controlling the size of the nonlinear stretching rate, that is essentially the radius of curvature of field lines. $N_1(s)$ and $U(s)$ have similarly obscure mathematical definitions, but for the anti-parallel case $N_1(s)$ is essentially the circulation σ in the symmetry plane, which are always finite, and $U(s)$ is a maximum velocity, which should not grow faster than the circulation divided by the largest length scale or $U(s) \sim \sigma/R(s)$. The second length scale comes into the proof because something must also be said about the "directedness" of field lines. That is how they spread as they curve. If they spread out too far, the effect of the Biot-Savart interaction that drives the vortex stretching over the scale $R(s)$ is too smoothed out. That is, some structure where the vorticity is confined, like the jump over ρ observed here, is needed. This is a very weak interpretation of the mathematical proof and some clarification from the mathematicians would be useful.

What can be done now for more physical understanding is circulation arguments that make the blowup of velocity (Kerr, 1997) at the position of the X in the upper right frame of Fig. 1 pausible and explain why $\int dV \omega_i e_{ij} \omega_j \rightarrow 1/(T-t)$. However, this cannot be explained properly in this short contribution.

4 New experimental evidence

Is there any evidence for two small length scales in turbulence? In refined self-similarity it is assumed that for a given r, all statistics of a given order depend on the statistics of dissipation over r. If for example, longitudinal and transverse structure functions have different scalings, it would represent a violation of refined self-similarity in the strictest sense and another length scale, besides the local r and the Kolmogorov scale η must have significance. Separate scaling laws for the longitundinal and transverse statistics were first suggested, then supported by numerical studies, for derivative statistics (Siggia, 1981a, Kerr, 1985). Fig. 4, taken from new experimental analysis by A. Praskovsky of his high Reynolds number Moscow wind tunnel data, shows scaling regions for the fourth- and sixth-order longitudinal and transverse structure functions that demonstrates this. Outer arrows define the traditional inertial subrange taken from the region going linearly as r for the third-order structure function. In the small r limit this is a very conservative definition of the lower bound of the inertial subrange, since for the second-order longitudinal and transverse structure functions obey the 4/3 rule for r must less than this. Taken over this entire subrange, the fourth and sixth order longitudinal and transverse structure functions exhibit different scaling in a way consistent with other recent studies (Boratav and Pelz, 1997, Herweiger and van de Water, 1995, Chen et al., 1997, Dhruva et al., 1997).

The important new observation is the inner arrows, which define a subregion within this traditional definition of the inertial subrange where refined similarity is still obeyed. That is, the fourth- and sixth-order longitudinal and transverse structure functions have the same scaling. The upper inner arrow is associated with where the 4/3 law for the second-order longitudinal and transverse structure

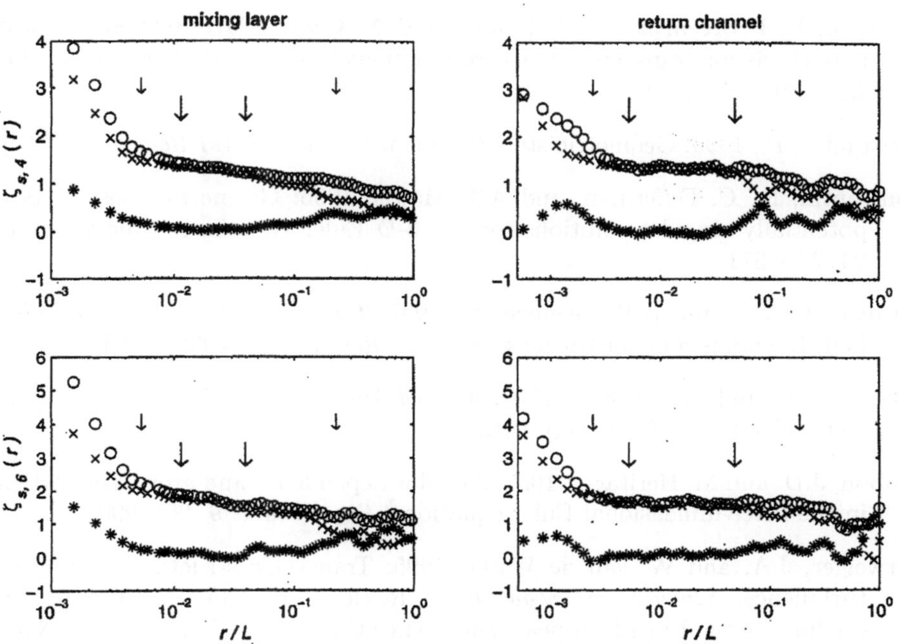

Figure 4: The local slopes $\zeta_{s,n}(r) = d\log[s_n(r)]/d\log(r)$, $n = 4, 6$. $\circ, s_n(4) =< |\Delta u_r|^n >$; $\times, s_n(r) =< |\Delta w_r|^n >$; $*, s_n(r) = R_n(r) =< |\Delta u_r|^n > / < |\Delta w_r|^n >$. The small and large vertical arrows indicate the classical inertial range bounds and the new bounds on the regime of complete self-similarity.

functions breaks down, and so this is not surprising. The surprising r is at the lower inner arrow. This new small length scale, if it proves robust, needs to be explained.

Acknowledgements

The National Center for Atmospheric Research is supported by the National Science Foundation.

References

Beale, J.T., T. Kato, A. Majda, 1984: Remarks on the breakdown of smooth solutions for the 3-D Euler equations. *Commun. Math. Phys.* **94**, 61.

Boratav, O.N. and R.B. Pelz, 1997: Structures and structure functions in the inertial range of turbulence. *Phys. Fluids* **9**, 1400–1415

Chen, S., K.R. Sreenivasan, M. Nelkin, and N. Cao, 1997: Refined similarity hypothesis for transverse structure functions in fluid turbulence. *Phys. Rev. Lett.* **79**, 2253–2256

Constantin, P., 1994: Geometric statistics in turbulence. *SIAM Review* **36**, 73–98.

Constantin, P., C. Fefferman, and A.J. Majda, 1996: Geometric constraints on potentially singular solutions for the 3-D Euler equations. *Comm in PDE* **21**, 559–571.

Dhruva, B., Y. Tsuji, K.R.Sreenivasan, 1997: Transverse structure functions in high-Reynolds-number turbulence. *Phys. Rev. E* **56**, R4928–R4930

Doering, C.R. and J.D. Gibbon, 1995: *Applied Analysis of the Navier-Stokes Equations.* Cambridge University Press.

Gibbon, J.D. and M. Heritage, 1997: Angular dependence and growth of vorticity in the three-dimensional Euler equations. *Phys. Fluids* **9**, 901–909

Herweiger, J.A. and W. van de Water, 1995: Transverse structure functions of turbulence. *Advances in Turbulence V*, R. Benzi ed., Kluwer, 210–216. (Proceedings of the Fifth European Turbulence Conference, Siena, Italy, 5–8 July 1995.)

Kerr, R. M., 1985: Higher order derivative correlations and the alignment of small-scale structures in isotropic numerical turbulence. *J. Fluid Mech.* **153**, 31–58

Kerr, R. M., 1993a: Evidence for a singularity of the three-dimensional, incompressible Euler equations. *Phys. Fluids A***5**, 1725–1746

Kerr, R.M. 1996: Cover illustration: vortex structure of Euler collapse. *Nonlinearity* **9** 271–272.

Kerr, R.M. 1997: Euler Singularities and Turbulence. In *19th ICTAM Kyoto '96* T. Tatsumi, E. Watanabe & T. Kambe eds., Elsevier Science, 57–70.

Ponce, G., 1985: Remark on a paper by J.T. Beale, T. Kato and A. Majda. *Commun. Math. Phys.* **98**, 349.

Siggia, E.D., 1981: Invariants for the one-point vorticity and strain rate correlation functions. *Phys. Fluids* **24**, 1934–1936. 1981: Numerical study of small scale intermittency in three-dimensional turbulence. *J. Fluid Mech.* **107**, 375–406.

2 Control and related

Trends in Mathematics, © 1999 Birkhäuser Verlag Basel/Switzerland

On controllability of the Navier-Stokes equations

A. V. Fursikov[1]

Abstract

Different statements of the controllability problem for the Navier-Stokes equations are given. Theorems on exact, on local exact and on approximate controllability of the 3D Navier-Stokes equations are obtained when control is concentrated on the boundary of a domain Ω filled by a fluid and in the case of periodic boundary conditions (i.e. these equations are defined on torus Π), the control is distributed and it is concentrated in a subdomain of Π.

In this paper we formulate different statements of the controllability problem for the 3D Navier-Stokes equations defined in a bounded domain Ω: the local exact controllability problem, the approximate controllability and the exact (i.e. global exact) controllability problems. We consider the case when the Navier-Stokes system is controlled from the boundary $\partial\Omega$ of Ω (Dirichlet boundary control) with periodic boundary conditions (i.e. it is defined on torus Π) and it is controlled by a local distributed control. This means that the control vector field is a right-hand-side of the Navier-Stokes equations concentrated in a given open subset of Π.

Zero controllability of the Navier-Stokes equations was established for the 2D case in A.V. Fursikov, O.Yu. Imanuvilov [1], and for the 3D case in A.V. Fursikov [2]. Local exact controllability of 2D Navier-Stokes equations was proved by A.V. Fursikov, O.Yu. Imanuvilov [3]. The same property for 3D Navier-Stokes and Boussinesq equations was established by A.V. Fursikov, O.Yu. Imanuvilov [4], [5], [6]. Approximate controllability for 2D Navier-Stokes equations was proved by J.-M. Coron [7]. Exact controllability for 2D Navier-Stokes equations was established by J.-M. Coron, A.V. Fursikov [8].

Here we establish local exact controllability of the 3D Navier-Stokes equations by a more constructive method than in [4], [5], [6]. Moreover, we set approximate controllability and exact controllability properties. The results of this paper were obtained by the author together with O.Yu. Imanuvilov. All complete proofs will be published in A.V. Fursikov, O.Yu. Imanuvilov [9].

[1]Department of Mathematics and Mechanics, Moscow State University, 119899 Moscow, Russia. Partially supported by RFBI under grant 96-01-00947.

§1. Exact controllability from the boundary of the Navier-Stokes system

Let $\Omega \subset R^3$ be a bounded domain with a boundary $\partial\Omega$ of the class C^∞, $T > 0$, $Q = (0,T) \times \Omega$ be the space-time cylinder, $\Sigma = (0,T) \times \partial\Omega$ be the lateral surface of Q. We consider the mixed boundary value problem for the Navier-Stokes system:

$$\partial_t v(t,x) - \Delta v + (v,\nabla)v + \nabla p = f(t,x), \quad \operatorname{div} v(t,x) = 0, \quad (t,x) \in Q \tag{1}$$

$$v|_\Sigma = \alpha(t,x'), \quad (t,x') \in \Sigma \tag{2}$$

$$v(t,x)\,|_{t=0} = v_0(x), \quad x \in \Omega \tag{3}$$

where $v(t,x) = (v_1(t,x), v_2(t,x), v_3(t,x))$ is a velocity vector field of a fluid flow, $\nabla p(t,x)$ is the pressure gradient, $f(t,x) = (f_1(t,x), f_2(t,x), f_3(t,x))$ is a given density of external forces, v_0 is a given initial vector field. The vector field α defined on the lateral surface Σ is not given, it is a control.

The exact controllability problem from the boundary Σ for the Navier-Stokes system is as follows: given a solution $(\hat{v}(t,x), \hat{p}(t,x))$ of (1):

$$\partial_t \hat{v}(t,x) - \Delta\hat{v} + (\hat{v},\nabla)\hat{v} + \nabla\hat{p} = f(t,x), \quad \operatorname{div}\hat{v} = 0, \tag{4}$$

to find a control $\alpha(t,x)$ defined on Σ such that the solution $v(t,x)$ of problem (1) - (3) coincides at instant $t = T$ with $\hat{v}(T,x)$:

$$v(t,x)|_{t=T} \equiv \hat{v}(T,x). \tag{5}$$

Here and anywhere below we suppose for definiteness that all given vector fields, i.e. $f, v_0, \hat{v}, \nabla\hat{p}$ are infinitely differentiable.

One of the main results of this paper is as follows:

THEOREM 1. *Suppose that for each connected component Γ_i of the boundary $\partial\Omega = \cup_{i=1}^M \Gamma_i$ the following equalities hold:*

$$\int_{\Gamma_i} (\hat{v}(t,x'), \nu(x'))dx' = 0, \quad a.a \quad t \in [0,T], \quad \int_{\Gamma_i} (v_0(x'), \nu(x'))dx' = 0$$

where $\nu(x')$ is the vector field of the outer normals to the boundary $\partial\Omega$.

Then there exists a control α such that the solution $(v, \nabla p)$ of the boundary value problem (1)–(3) with the boundary condition α in (2) satisfies equality (5).

REMARK. Note that we can make the assertion of the Theorem 1.1 more precise indicating the level of smoothness for the solution $(v, \nabla p)$ and for the control α. We can claim that $\alpha(t,x') \in (L_2(\Sigma))^3$, $v(t,x) \in V^{1,2}(Q)$, $\nabla p \in (L_2(Q))^3$ where, recall, $L_2(Q)$ is the space of scalar-valued functions $u(t,x)$ defined on Q with a finite norm:

$$\|u\|_{L_2(Q)} = \left(\int_Q |u(t,x)|^2\,dx\right)^{1/2}$$

and $(L_2(Q))^3$ is the space of vector fields $u = (u_1, u_2, u_3)$ whose components u_i belong to $L_2(Q)$. The space $V^{1,2}(Q)$ is the set of vector fields $v(t,x)$ defined on Q and satisfying equation $\operatorname{div} v = 0$ that possess a finite norm:

$$\|v\|_{V^{1,2}(Q)} = \left(\int_Q (|\partial_t v(t,x)|^2 + \sum_{i,j=1}^{n} |\partial^2 v(t,x)/\partial x_i \partial x_j|^2 + |v(t,x)^2|) \, dx \, dt \right)^{1/2}$$

§2. Exact controllability of the Navier-Stokes system with a local distributed control

We consider now the Navier-Stokes equations with a distributed control:

$$\partial_t v(t,x) - \Delta v + (v, \nabla) v + \nabla p = f(t,x) + u(t,x), \quad \operatorname{div} v = 0, \quad (t,x) \in Q \tag{6}$$

with periodic boundary conditions

$$v(t, \ldots x_i + L, \ldots) = v(t, \ldots x_i, \ldots), \qquad i = 1, 2, 3 \tag{7}$$

and with the initial condition

$$v(t,x)|_{t=0} = v_0(x), \quad x \in \Pi\,. \tag{8}$$

Here L is the period. The condition (7) is equivalent to the assumption that equation (6) is defined on the space-time cylinder $Q = (0,T) \times \Pi$ where Π is a torus, i.e. Π is the direct product of 3 circumferences with length L. Vector fields f, v_0 in (6), (8) are given, and $u(t,x)$ is a control. Let $\omega \subset \Pi$ be an open subset of the torus Π, $Q^\omega = (0,T) \times \omega$. We suppose that the control $u(t,x)$ is concentrated in the cylinder Q^ω ($\operatorname{supp} u \subset Q^\omega$), i.e. $u(t,x) = 0$ for $(t,x) \in Q \setminus Q^\omega$.

We define the space of controls

$$U_T(\omega) = \{u(t,x) \in (L_2(Q))^n : \operatorname{supp} u \in Q^\omega\}\,. \tag{9}$$

The following theorem on exact controllability of Navier-Stokes system (6) by a local distributed control is true:

THEOREM 2. *There exists a control $u \in U_T(\omega)$ such that a solution $(v, \nabla p)$ of problem* (6)–(8) *with this u exists in $V^{1,2}(Q) \times (L_2(Q))^3$ and satisfies equality* (5).

In order to obtain the Theorem 1 from Theorem 2 we choose a cube K containing the domain Ω. After identification of the cube's K opposite sides we obtain a torus Π which contains Ω. We denote $\omega = \Pi \setminus \Omega$ and extend the given vector fields $f, v_0, \hat{v}$ from Ω to Π. As a result, the conditions of Theorem 2 hold. Applying this Theorem we get the solution $v(t,x)$ of problem (6)–(8) which satisfies (5). Then the vector fields $v(t,x)|_{(0,T)\times\Omega}$ and $\alpha(t,x') \equiv v(t,x)|_{(0,T)\times\partial\Omega}$ satisfy all assertions of Theorem 1.

§3. Local exact controllability and approximate controllability of the Navier-Stokes system

We weaken the notion of exact controllability to two directions by introducing the notions of local exact controllability and approximate controllability. To do this we recall that $H^k(\Omega)$, where k is a natural number, is a Sobolev space of scalar functions defined and square integrable on Ω together with all the derivatives up to order k and $(H^k(\Omega))^3$ is the analogous Sobolev space of vector fields. We set

$$V^k(\Omega) = \left\{v(x) = (v_1, v_2, v_3) \in (H^k(\Omega))^3 : \quad \operatorname{div} v = 0\right\}.$$

Let the pair $(\hat{v}, \hat{p})$ satisfy Navier-Stokes equations (4) on Q. Navier-Stokes system (6) is called *local exact controllable* with respect to the controls space (9) if there exists $\varepsilon > 0$ such that for an arbitrary initial condition v_0 satisfying the inequality

$$\|\hat{v}(0,\cdot) - v_0\|_{V^1(\Pi)} \leq \varepsilon \tag{10}$$

there exists a control $u \in U_T(\omega)$ such that the solution $(v, \nabla p)$ of problem (6)–(8) exists in the space $V^{1,2}(Q) \times (L_2(Q))^3$ and satisfies the condition (5).

THEOREM 3. *The Navier-Stokes system is locally exact controllable with respect to controls space* (9).

Let an initial condition v_0 and a right side f be given and $\|v_0\|_{V^4(\Pi)} < R$. Consider the set W_R of vector fields $v_1 \in V^4(\Pi)$ satisfying

$$\sum_{i=0}^{1} \|v_i\|_{V^4(\Pi)} \leq R.$$

Navier-Stokes system (6) is called *approximately controllable* with respect to controls set (9) if for any $\varepsilon > 0, R > 0$ and for an arbitrary $v_1 \in W_R$ there exist an instant $T = T_{\varepsilon,R}$ and a control $u \in U_{T_{\varepsilon,R}}(\omega)$ such that the solution $(v, \nabla p)$ of problem (6)–(8) with the chosen control u satisfies the inequality

$$\|v(T_{\varepsilon,R}) - v_1\|_{V^1(\Pi)} \leq \varepsilon.$$

THEOREM 4. *Navier-Stokes system* (6)–(8) *is approximately controllable with respect to controls space* (9). *Moreover for each* $\varepsilon > 0$ *and* $v_1 \in W_R$ *one can choose the time* $T_{\varepsilon,R}$ *such that*

$$T_{\varepsilon,R} \to 0 \quad \text{for} \quad \varepsilon \to 0$$

for an arbitrary $R > 0$.

To prove Theorem 2 with the help of Theorems 3, 4 we choose an arbitrary point $T_1 \in (0,T)$ and sufficiently small $\varepsilon > 0$. We apply the Theorem 4 on interval $(0, T_1)$. After that we apply the Theorem 3 on interval (T_1, T) with the initial condition $v_0(x) = v(T_1, x)$, where $v(t,x)$ is the solution of the controlled Navier-Stokes equations constructed in Theorem 4. □

For the proof of Theorems 3, 4 see in A.V. Fursikov, O.Yu. Imanuvilov [9].

§4. Applications

Stabilization of an unstable steady-state solution. Let us consider the exact controllability problem for the Navier-Stokes equations with a control concentrated on the boundary $\partial\Omega$, i.e. problem (1)–(5). Suppose that the right-hand side $f(t,x)$ from (1) and (4) does not depend on t: $f(t,x) \equiv f(x)$. Assume also that a given solution $(\hat{v},\hat{p})$ of (4) is steady-state: $(\hat{v}(t,x), \nabla\hat{p}(t,x)) \equiv (\hat{v}(x), \nabla\hat{p}(x))$. Suppose for definiteness (although this is not necessary) that

$$\hat{v}(x)|_{\partial\Omega} = 0 .$$

We can treat this steady-state solution $(\hat{v}(x), \hat{p}(x))$ as a singular point of the dynamical system generated by system (1) with boundary condition

$$v(t,x)|_{\Sigma} = 0 . \tag{11}$$

Suppose that this singular point $(\hat{v}(x), \nabla\hat{p}(x))$ is unstable, i.e. for a dense set of initial conditions v_0 belonging to a neighbourhood of $\hat{v}$ the solution $v(t,x)$ of (1),(3),(11) does not tend to $\hat{v}(x)$. But if we switch on a boundary control (2) (instead of condition (11)) we can compel the solution $v(t,x)$ of (1) going out v_0 to reach $\hat{v}(x) : v(T,x) \equiv \hat{v}(x)$. This means that one can stabilize an unstable steady-state solution with the help of a boundary control.

Existence of chaotic flows. We consider the Navier-Stokes system (6) defined on $Q = (0,T) \times \Pi$ where Π is a torus and a control $u(t,x)$ is concentrated in $Q^\omega = (0,T) \times \omega$ with sufficiently "small" $\omega \subset \Pi$. For example, let ω be a ball with a radius small enough. Suppose that $f(t,x) \equiv 0$ in (6). We show that this problem can possess an arbitrarily complicated (chaotic) solution.

In fact, let us consider a subdivision $0 = t_0 < t_1 < \cdots < t_N = T$ of a time interval $(0,T)$ and associate an arbitrary solenoidal vector field $v_j(x)$ to each subdivision element t_j. Besides, we choose an arbitrary small $\varepsilon > 0$. Then the Theorem 4 on approximate controllability of the Navier-Stokes system yields that there exists a control $u \in U_T(\omega)$, such that the solution $(v(t,x), \nabla p(t,x))$ of problem (6) satisfies the condition

$$\|v(t_j,\cdot) - v_j\|^2_{V^1(\Pi)} < \varepsilon \quad \forall\, j = 1,\ldots,N. \tag{12}$$

Therefore, the right-hand side of the system (6) equals zero for $(t,x) \in Q \setminus Q^\omega$ and the solution of this system can be arbitrarily complicated in the sense of the estimate (12).

Reversibility property. The Theorem 2 implies immediately the following property: Let $(v_i(t,x), \nabla p_i(t,x))$, $i = 1,2$, be two solutions of the Navier-Stokes equations (1) defined on $Q = (0,T) \times \Pi$, where Π is a torus. Then for an arbitrary open set $\omega \subset \Pi$ and for an arbitrary time interval $(t_1,t_2) \subset (0,T)$ there exists a distributed control $u(t,x) \in (L_2(Q))^n$ concentrated on $(t_1,t_2) \times \omega$ which transforms the solution $(v_1(t,x), \nabla p_1(t,x))$ to the solution $(v_2(t,x), \nabla p_2(t,x))$. This property is called reversibility property. The reversibility property can be applied to a problem of the climate theory. (See for details J.-M. Coron, A.V. Fursikov [7].)

References

[1] Fursikov A.V. and Imanuvilov O. Yu. On exact boundary zero-controllability of two-dimensional Navier-Stokes equations. Acta Applicandae Mathematicae v. 37, 1994, pp. 67–76.

[2] Fursikov A.V. Exact boundary zero controllability of three dimensional Navier-Stokes equations. J. of Dynamical and Control Syst. v. 1, No 3, 1995, pp. 325–350.

[3] Fursikov A.V. and Imanuvilov O. Yu. Local exact controllability of two dimensional Navier-Stokes system with control on the part of the boundary. Sbornik. Math. v. 187, No 9, 1996, pp. 1355–1390.

[4] Fursikov A.V. and Imanuvilov O. Yu. Local exact controllability of the Navier-Stokes equations. C.R. Acad. Sci. Paris, v. 323, Ser. 1, 1996, pp. 275–280.

[5] Fursikov A.V. and Imanuvilov O.Yu. Local exact boundary controllability of the Navier-Stokes system. Contemporary Mathematics v. 209, 1997, pp. 115–129.

[6] Fursikov A.V. and Imanuvilov O.Yu. Local exact boundary controllability of the Boussinesq equation. SIAM Journal on Control and optimization, v. 36, No 2, (1998).

[7] Coron J.-M. On the controllability of the 2-D incompressible Navier-Stokes equations with the Navier slip boundary conditions. ESIAM Control, Optimization and Calculus of Variations, v. 1, 1996, pp. 35–75.

[8] Coron J.-M. and Fursikov A.V. Global exact controllability of the 2D Navier-Stokes Equations on a manifold without boundary. Russian J. of Math. Physics, v. 4, No 4, 1996, pp. 429–448.

[9] Fursikov A.V. and Imanuvilov O. Yu. Exact controllability of the Navier-Stokes and Boussinesq equations. Russian Math. Surveys (to appear).

[10] Lions J-L. and Magenes E. Problemes aux limites nonhomogenes et applications v. 1. Dunod, Paris, 1968.

[11] Alekseev V.M., Tikhomirov V.M. and Fomin S.V. Optimal control. Consultants Bureau, New York, 1987.

Trends in Mathematics, © 1999 Birkhäuser Verlag Basel/Switzerland

Low dimensional models, the minimal flow unit and control

J. L. Lumley, P. N. Blossey, B. Podvin-Delarue

Abstract

Near-wall turbulence is dominated energetically by a few large-scale, coherent flow structures which are responsible for the production of much of the turbulence in the boundary layer. We have studied the dynamics of these structures through direct numerical simulations of turbulent channel flow in the minimal flow unit(Jiménez & Moin 1991), the smallest computational box in which turbulence may be sustained. The minimal flow unit contains exactly one set of these coherent structures in each wall region and, as a result, provides an ideal setting for the study of their dynamics and control. We have constructed a low dimensional model for the dynamics of these coherent structures in the minimal flow unit, identifying the structures with the Proper Orthogonal (or Karhunen-Loève) Decomposition. Low dimensional models have been shown in the past to reproduce many of the qualitative features of boundary layer dynamics. Here we will explore the evidence for stronger, quantitative connections between the model and (a simulation of) the real flow, justifying the assumptions which have gone into the construction of the model and identifying intermittency in the model with the intermittent bursting events in the real flow. In addition, we will use the model to motivate control strategies for the turbulent boundary layer where the model may be applied both as an interpreter of the limited and noisy sensor information available at the wall and as a predictor of the dynamical behavior of the flow. Some simple control strategies based on preventing the instability of these coherent structures have been tested computationally, yielding reductions in skin friction (drag) as large as 20%.

1 Introduction

Over the past decade, our group has worked to develop low dimensional models as a tool for the study of turbulent flows. Our application of these tools has focused on the wall region of the turbulent boundary layer. There, most of the energy of the turbulence is contained, on average, in just a few coherent structures. Aubry et al. (1988) constructed a low dimensional model for the wall region of the turbulent boundary layer using the empirical eigenfunctions of the Proper Orthogonal Decomposition (POD) (Loève 1955, Lumley 1967). Their model severely truncated the eigenfunction expansion produced by the POD, including no streamwise variation and only five complex Fourier modes in the spanwise direction, yet the model still reproduces much of the dynamics of the wall region. In particular, intermittent jumps in the phase space of the model match qualitatively the breakdown of the coherent structures during the bursting process. In addition to displaying some of the features of the wall region, these models have been able to predict the response of the boundary layer to a variety of phenomena including adverse/favorable pressure gradients, polymer addition, and streamline curvature and have also suggested strategies for the control of near-wall flow.

In this work, we apply these tools in the setting of the minimal flow unit as a means to test both the models themselves and their ability to predict and interpret the dynamics of the turbulence. The use of the minimal flow unit in the study of near-wall turbulence was pioneered by Jiménez & Moin (1991). Their work focused upon finding the smallest computational domain which would support turbulence. The setting for this study was channel flow with solid, impermeable walls at the top and bottom of the domain and periodic boundaries in the streamwise and spanwise directions. The spanwise length of the domain (L_3^+) was treated as a bifurcation parameter, and the behavior of the flow was studied for various values of L_3^+. ($L_3^+ = L_3 u_\tau/\nu$, where u_τ is the friction velocity and ν the viscosity; the friction velocity is defined by $\mu \partial U/\partial y = \rho u_\tau^2$, with the gradient evaluated at the wall.) Jiménez & Moin (1991) found that below some threshold value, $L_3^+ \approx 100$, turbulence could not be sustained on both walls of the channel: the flow would relaminarize on one wall and for smaller values of L_3^+ would eventually become fully laminar. Despite the small size of the domain, the flow exhibits the proper turbulent statistics at both walls. The critical length scale ($L_3^+ \approx 100$) corresponds to the experimentally observed streak spacing. In fact, one observes exactly one high speed and one low speed streak in the minimal flow unit. These streaks break down in a process which very closely resembles the bursting seen in full simulations of turbulent channel flow. Webber et al. (1997) studied Karhunen-Loève Decomposition of the minimal flow unit, studying the dimension of the flow and tracking the energy in the (streamwise invariant) roll modes, (streamwise-dependent) propagating modes, and the representational entropy of the flow. They found that the dimension of the flow was approximately proportional to the size of the domain considered. In this light, the minimal flow unit provides a (relatively) low dimensional setting in which to study the dynamics of wall-bounded turbulence. The simplification of the dynamics through the consideration of only a single set of coherent structures may allow for insights which may be extended to the more general, higher-dimensional situation.

The low dimensional model of Aubry et al. (1988) and minimal flow unit both serve as simplifications of the full flow through the restriction of the dimension of the system. The minimal flow unit is more faithful to the true dynamics of the wall region, yet both are able to capture some of the features of the dynamics. In this work, we construct a low dimensional model for the minimal flow unit based on the eigenfunctions of the POD. In addition to constructing the model, we critically evaluate terms in the model equations through comparisons with the identical terms in direct simulations in the minimal flow unit. The model is further employed as an estimator of the flow in the minimal flow unit based only on wall shear stress measurements. The availability of information about the state of the flow based solely on measurements at the wall has direct bearing on the application of control of near wall flow. A simple algorithm for the control of near wall flow has been implemented, yielding nearly 20% drag reduction. We will also touch on current work employing the minimal flow unit in the study of polymer drag reduction.

2 Near-wall turbulence and the bursting process

Many turbulent flows display a combination of organized, or coherent, structures and apparently disorganized, or incoherent, structures. In the wall region of the turbulent boundary layer, coherent structures account for over 80% of the energy in the turbulent fluctuations. We want to develop a low dimensional model of the flow near the wall that resolves just these most energetic coherent structures. According to observations, these structures are vortices lying parallel to and close to the wall, oriented in the streamwise direction. Most often, these vortices appear as unequal pairs, but sometimes are found singly. They are responsible for the streak-like structures seen in flow visualizations and numerical simulations of the boundary layer: slow-moving fluid is lifted away from the wall on the updraft side of the vortex, giving rise to a low-speed streak, while a high-speed streak is produced by the downdraft on the opposite side.

The coherent structures are observed to burst, transferring energy from the large to the small scales and producing turbulent fluctuations. First, the updraft between the vortices strengthens, and the vortices move toward each other. The slow-moving fluid lifted away from the wall by the vortices induces an inflection in the mean velocity profile, which grows in intensity with the updraft and moves father out into the stream. While the updraft is strengthening, the wall shear stress, or drag, is approaching its peak owing to the high-speed fluid being drawn down toward the wall by the downdraft on the opposite side of the vortex. As the inflection in the mean velocity profile intensifies, it gives rise to a secondary instability and an accompanying burst of Reynolds stress. Through this instability, energy is transferred from the large scales of the turbulence, or the coherent structures, to the smaller scales.[1] As a result of the energy transfer, the vortices are weakened and move apart. Last, high speed fluid comes down from the outer part of the boundary layer, sweeping the wall clean.

3 A low dimensional approach for the minimal flow unit

With the desire to understand the dynamics of the large-scale coherent structures in the near-wall region, we constructed a low dimensional model in the setting of the minimal flow unit (Podvin & Lumley 1997). The Proper Orthogonal Decomposition was applied to several realizations of the flow to generate basis functions

[1]Webber et al. (1997) track the transfer of energy by means of the representational entropy, $S = -\sum_n a_n ln(a_n)$. The entropy is minimized when all energy resides in a single mode and is maximized when the energy is distributed evenly among all modes. During the bursting process, the entropy reaches a local minimum as the rolls reach their peak strength, then rises sharply as the energy from the rolls is distributed among the numerous smaller scale modes.

for the model. The velocity field in the channel is decomposed into a time-varying mean velocity, $U_1(y,t)$, and a fluctuating component, $u_i(x,y,z,t)$. The Proper Orthogonal Decomposition is applied to the fluctuating part of the velocity field, yielding an infinite set of eigenfunction-eigenvalue pairs (ϕ_n^i, λ_n). The fluctuating velocity field can then be represented as a combination of these eigenfunctions:

$$u_i(x,y,z,t) = \sum_n a_n(t)\phi_n^i(x,y,z) \tag{1}$$

where a_n are uncorrelated coefficient of variance λ_n. Since we are most interested in the dynamics of the near-wall region, we chose to apply the decomposition for $0 \leq y^+ \leq 40$. Note that the eigenfunctions of the POD reduce to Fourier modes in statistically homogeneous directions. Therefore, the POD is explicitly applied only in the wall normal direction: the streamwise and spanwise directions are decomposed using Fourier series. The first eigenfunction in the wall region of the minimal flow unit captures 60% of the kinetic energy of the turbulence. The first three structures capture 90% of the total kinetic energy.

We then construct the most basic low dimensional model possible for this flow, including only the first two Fourier modes of the most energetic eigenfunction. The velocity field is expanded in terms of these two modes, and the Navier-Stokes equations are projected onto these modes to yield a system of coupled, nonlinear ordinary differential equations. The construction of these models is described in detail in Aubry et al. (1988). The construction includes a gradient diffusion model with a fixed eddy viscosity to account for the effect of the small scales. We expect that, in reality, this eddy viscosity will be a function of time, peaking while energy is transferred from the large-scale coherent structures to the small scales during a burst.

The model equations display traveling wave behavior for values of the eddy viscosity equivalent to those employed in Aubry et al. (1988). In this regime, the model does not display intermittent bursting behavior: the traveling wave corresponds in physical space to the drift of the rolls in the spanwise direction. However, with an increase in the eddy viscosity, the traveling wave behavior will bifurcate in phase space to a fixed point, corresponding to stationary rolls. In fact, we observe this phenomenon in direct simulations of the minimal flow unit. Periodically, the regular drift of the rolls is interrupted as the rolls burst, with increased energy transfer to the small scales and a peak in the wall shear stress. The fixed eddy viscosity employed in our model prevents us from capturing this feature of the dynamics. However, evaluation of the energy transfer between specific modes in the DNS indicates that nearly half of the energy lost from the model goes to the first streamwise modes. Thus, the inclusion of only a few more modes in our model could give a more faithful picture of the bursting process, giving us hope that a relatively small model could capture the intermittent behavior of the bursting process.

4 State estimation

This low dimensional model for the minimal flow unit has been employed to estimate the state of the flow based on shear stress measurements at the wall (Podvin & Lumley 1997). The coefficients of the two modes contained in the low dimensional model are reconstructed based on shear stress measurements at the wall. The shear stress was measured on a grid of spacing 50 and 25 wall units in the streamwise and spanwise directions respectively. The two dimensional grid was found to be necessary to minimize aliasing errors. Neglecting the effect of higher modes on the shear stress, we compute the coefficients of the modes as a weighted sum of the instantaneous shear stress measurements:

$$a_k^{meas} = \beta \frac{D\hat{u}_k|_{wall}}{D\phi_k^1|_{wall}} + (1-\beta)\frac{D\hat{w}_k|_{wall}}{D\phi_k^3|_{wall}} \tag{2}$$

where $D \equiv d/dy$, ϕ^i refers here to the ith component of the first eigenfunction, $\hat{u}$ and $\hat{w}$ are the (Fourier transformed) streamwise and spanwise velocity fluctuations respectively, and k is the wavenumber in the spanwise direction. With these (still noisy) measurements of the model coefficients, the amplitude of the first mode was determined using a moving average of the measurements. The second mode was fixed through the enforcement of the traveling wave behavior predicted by the model equations. This technique performed well in predictions of the amplitude and phase of both modes. In particular, the measurements of the second mode were very noisy, so that the traveling wave ansatz allowed much better estimation than the measurements could provide. As in Lee et al. (1997), the spanwise shear stress proved to be more useful in the estimation of the state of the flow with weightings of β closer to zero improving our estimates.

5 Application to the control of turbulence

Through the bursting process, the coherent structures near the wall are responsible for much of the production of turbulence in the boundary layer; thus, interfering with the dynamics of the coherent structures provides a possible mechanism for the control of turbulence. Inhibiting the bursting process in the boundary layer also inhibits the production of turbulence, thereby decreasing momentum transport away from the wall (and, equivalently, turbulent skin friction). Our group has applied these ideas to the control of the low dimensional models themselves (Coller et al. 1994) and to the control of flow in the minimal flow unit (Carlson & Lumley 1996). The work of Carlson & Lumley (1996) relied on the implementation of a moving boundary to raise a Gaussian bump into the flow as an actuator. The bump was effective in moving the high speed streak away from the wall and diverting it into the low speed streak, but the velocity field induced by the bump was complicated, with a stack of three pairs of counter-rotating vortices. Our ideal actuator velocity field is simply a replica of the coherent structures themselves — the rolls — which we hope to use to inhibit the bursting of the naturally-occurring coherent structures. A sharp-edged flap can be expected to produce this

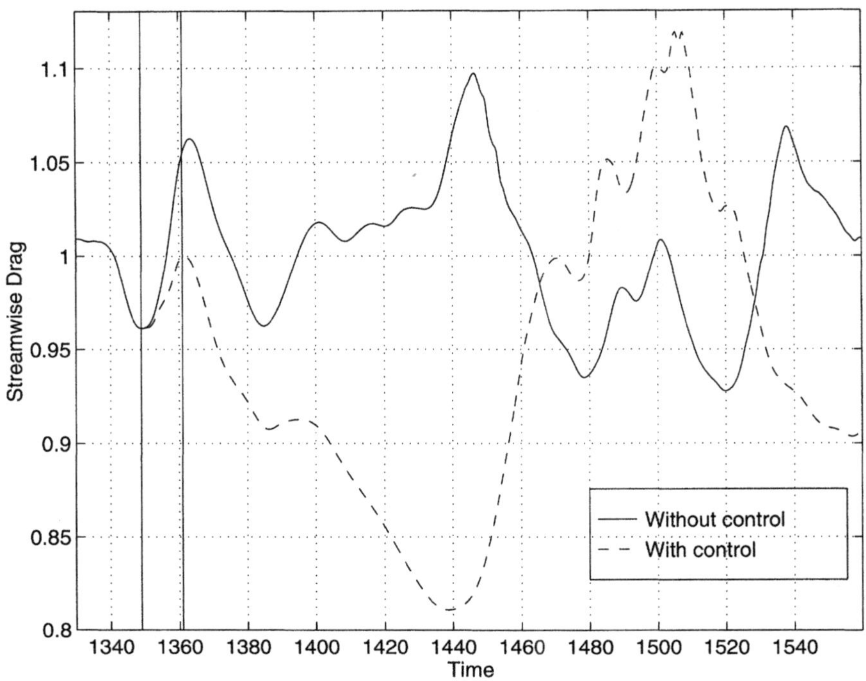

Figure 1: Mean streamwise wall shear stress with and without control. Control is turned on and then off at the times marked by the vertical lines.

sort of velocity field, with the flap inducing a pair of trailing vortices.[2] Since our pseudospectral codes are restricted to the computation of smooth boundaries and since the small scales induces by the sharp edges of the flap would require much higher resolution than the turbulence alone, we chose to simulate the effect of the flap on the flow through the use of a body force.

We added a body force to our simulations of the minimal flow unit and so induced a pair of counter-rotating vortices, thereby including the effect of the flap but not the flap itself. The body force, which acts solely on the wall normal component of velocity, added vorticity of opposite sign to the coherent structures present in the flow. The coherent structures and their associated streaks were weakened and the drag fell (see Figure 1). The weakening of the updraft delayed the ejection and the creation of the inflection point in the mean velocity profile. The streaks thickened in the wall-normal direction (much like in the polymer drag reduced flow) and seemed to stabilize for some time, before the coherent structures strengthened again and caused an ejection and burst. The action of the body force

[2]It should also be noted that groups at MIT (Rathnasingham & Breuer 1997) and Stanford (Jacobson & Reynolds 1998) have built actuators based on vibrating beams forcing air in and out of a cavity that produce just this sort of velocity field (in a time-averaged sense).

for a duration of 50 wall time units (u_τ/ν) had the effect of suppressing the burst for ten times its duration, resulting in drag reduction of 15% when averaged over 700 wall time units (Lumley & Blossey 1998).

6 Polymer drag reduction

Since polymers were first found to induce considerable reductions in turbulent skin friction, much study has been devoted to elucidating the mechanism by which the polymers effect this reduction. A theory has been developed to explain the mechanism through which the polymers affect the turbulence (Lumley 1969, Lumley 1973, Lumley & Kubo 1985). This theory relies on the extension of the polymers in the large strain rate fluctuations which occur outside of the viscous sublayer. The polymers, whose equilibrium state is something like a loosely coiled ball of string, act to increase the viscosity locally once they are extended. We have implemented this model for polymer drag-reduced flow in the minimal flow unit. The eigenvalues of the velocity gradient tensor are calculated. With the assumption that the polymers align themselves with the direction of the greatest extension, we choose the real part of the largest positive eigenvalue as an indicator of the local rate of extension. The stress due to the polymers is modeled using an isotropic viscosity which is a function of the local rate of extension.

$$\frac{du_i}{dx_j}w_j = \lambda w_i \tag{3}$$

$$\lambda_P \equiv max(Re(\lambda_i)), i = 1, 2, 3 \tag{4}$$

$$\nu_P = f(\lambda_P)\nu_M \tag{5}$$

$$(\tau_P)_{ij} = 2\nu_P s_{ij} \tag{6}$$

Here, λ and w are the (possibly complex) eigenvalues and eigenvectors of the velocity gradient tensor, and ν_P and ν_M are the polymer and molecular viscosity respectively. The one detail remaining is the shape of $f(\lambda_P)$. We choose a threshold value for λ_P such that the polymers have little effect below this value and the effect of the polymers will saturate for values of λ_P above this value. (See Figure 2.) This threshold value corresponds roughly to the inverse of the polymer relaxation time scale. When this model for the polymers was implemented in the minimal flow unit with a maximum value of ν_P/ν_M of 20, the additional viscosity did not result in drag reduction. Further increases in the ratio of ν_P/ν_M and reductions in the threshold for λ_P^+ down to 1/7 still did not result in drag reduction. The additional viscosity has some effect on the coherent structures, weakening them somewhat, but not sufficiently to significantly reduce momentum transport away from the wall.

Similar simulations in the setting of pipe flow have been performed by den Toonder et al. (1995). They were also unable to obtain drag reduction using an isotropic eddy viscosity, although they did use a somewhat different indicator of extension, $R = 3\mathbf{I}II_S/\mathbf{I}I_S$, where $\mathbf{I}I_S$ and $\mathbf{I}II_S$ are the second and third invariants

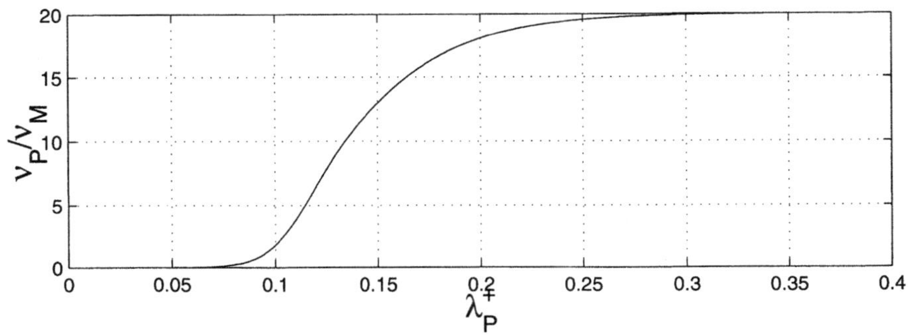

Figure 2: Polymer viscosity as a function of the extension rate of the flow, as characterized by λ_P^+ the largest positive eigenvalue of the velocity gradient tensor.

of the symmetric rate of strain tensor. Since they were unable to find drag reduction with an isotropic viscosity, they proceeded to implement a model which included an anisotropic polymer viscosity, assuming that the polymers were locally aligned with the flow velocity and that the local rate of extension could be characterized by: $R = \mathbf{n}\cdot\mathbf{S}\cdot\mathbf{n}$, where $\mathbf{n} = \mathbf{u}/|\mathbf{u}|$ and $\mathbf{S} = (\nabla\mathbf{u}+\nabla\mathbf{u}^T)/2$ (den Toonder et al. 1997). This anisotropic model did give rise to drag reduction on the order of 10% with a Trouton ratio of approximately 10.

When studying the local character of the strong extensional velocity fields ($\lambda_P \geq 0.15$), we find that in nearly 80% of the cases the velocity gradient tensor has purely real eigenvalues, with two extensional (positive) eigenvalues and one compressive (negative) eigenvalue.[3] Incompressibility implies that the sum of these three eigenvalues is zero; therefore, the compressive eigenvalue, which represents a compressive strain, is the strongest of the three. Since an isotropic viscosity is applied equally to all three of these strains, the strongest stress induced by the polymer viscosity in the direction of the compressive strain. This result contradicts our ideas about how the polymer acts in a turbulent flow. We expect the polymer to align itself with the direction of greatest extension and to damp the extensional component of the flow (through an extensional viscosity) rather than the compressive component of the flow. Although the model with an isotropic viscosity could conceivably lead to drag reduction if the threshold for the onset of the polymer effect was set low enough, the model does not properly capture the physics of the effect of the polymer on the flow, since the polymer is expected to act primarily on the extensional component of the flow.

[3]This situation has been predicted analytically by Cantwell (1992) who studied the evolution of the velocity gradient tensor under a restricted Euler equation which neglected viscous effects and the cross-derivatives of pressure.

7 Conclusions

The minimal flow unit and low dimensional models have been applied as a tool to study each other as well as state estimation based on wall shear stress measurement, control of the near-wall flow, and polymer drag reduction. Their low cost and simplified dynamics make the models and the minimal flow unit ideal testbeds for experimentation with models and algorithms for wall-bounded turbulence.

References

Aubry, N., Holmes, P., Lumley, J. L. & Stone, E. 1988. The dynamics of coherent structures in the wall region of a turbulent boundary layer. *J. Fluid Mech.* **192**: 115–173.

Cantwell, B. J. 1992. Exact solution of a restricted Euler equation for the velocity gradient tensor. *Phys. Fluids A* **4**(4): 782–793.

Carlson, H. A. & Lumley, J. L. 1996. Active control in the turbulent boundary layer of a minimal flow unit. *J. Fluid Mech.* **329**: 341–371.

Coller, B. D., Holmes, P. & Lumley, J. L. 1994. Control of bursting in boundary layer models. *Appl. Mech Rev.* **47 (6), part 2**: S139–S143. Mechanics USA 1994, ed. A. S. Kobayashi.

Jacobson, S. A. & Reynolds, W. C. 1998. Active control of streamwise vortices and streaks in boundary layers. *J. Fluid Mech.* **360**: 179–211.

Jiménez, J. & Moin, P. 1991. The minimal flow unit in near-wall turbulence. *J. Fluid Mech.* **225**: 213–240.

Lee, C., Kim, J., Babcock, D. & Goodman, R. 1997. Application of neural networks to turbulence control for drag reduction. *Phys. Fluids* **9**(6): 1740–1747.

Loève, M. 1955. *Probability theory.* Van Nostrand.

Lumley, J. L. 1967, The structure of inhomogeneous turbulence. In *Atmospheric Turbulence and Wave Propagation*, ed. A. M. Yaglom & V. I. Tatarski, Moscow: Nauka.

Lumley, J. L. 1969. Drag reduction by additives. *Ann. Rev. Fluid Mech.* **1**: 367–384.

Lumley, J. L. 1973. Drag reduction in turbulent flow by polymer additives. *J. Polym. Sci. Macromol. Rev.* **7**: 263–290.

Lumley, J. L. & Blossey, P. N. 1998. Control of Turbulence. *Ann. Rev. Fluid Mech.* **30**: 311–327.

Lumley, J. L. & Kubo, I. 1985, Turbulent drag reduction by polymer additives: a survey. In *The Influence of Polymer Additives on Velocity and Temperature Fields*, ed. B. Gampert, New York: Springer-Verlag, pp. 3–24.

Podvin, B. & Lumley, J. L. 1997. A low dimensional approach for the minimal flow unit. *J. Fluid Mech.* . Submitted.

Rathnasingham, R. & Breuer, K. S. 1997. Coupled fluid-structural characteristics of actuators for flow control. *AIAA J.* **35**(5): 832–37.

den Toonder, J. M. J., Hulsen, M. A., Kuiken, G. D. C. & Nieuwstadt, F. T. M. 1997. Drag Reduction by polymer additives in a turbulent pipe flow: numerical and laboratory experiments. *J. Fluid Mech.* **337**: 193–231.

den Toonder, J. M. J., Nieuwstadt, F. T. M. & Kuiken, G. D. C. 1995. The role of extensional viscosity in the mechanism of drag reduction by polymer additives. *Appl. Sci. Res.* **54**: 95–123.

Webber, G. A., Handler, R. A. & Sirovich, L. 1997. The Karhunen-Loève decomposition of minimal channel flow. *Phys. Fluids* **9**(4): 1054–1066.

Trends in Mathematics, © 1999 Birkhäuser Verlag Basel/Switzerland

Active control of turbulence in boundary layer flows

P. J. Ioannou, B. F. Farrell

1 Introduction

Preventing the transition to turbulence of a laminar flow and suppressing the variance of a turbulent flow, perhaps with the ultimate goal of inducing relaminarization, or alternatively controlling eddy fluxes produced by a given level of fluctuations are in themselves problems of great practical importance. In addition, understanding the physical mechanism of turbulence and turbulent transition should lead either to methods of control or to an explanation of why such control is not possible. From this perspective the control problem is seen as a test of physical theory. From the viewpoint of practical engineering, a comprehensive theory of the transition process and of the maintenance of fully developed turbulence that both implied new control mechanisms and provided a means of testing proposed mechanisms would be of great utility even if the result were only to discourage the search for e.g. a passive compliant membrane that relaminarized the turbulent boundary layer or an acoustic intervention to control turbulence in a free jet. Extensive attempts to reduce drag in turbulent boundary layer flow by imposing a variety of active and passive control measures have shown that in the absence of applicable theory it is very unlikely that an optimal method can be identified. Similar problems are faced if the aim is to control pressure fluctuations in order to reduce acoustic radiation or if we aim at increasing mixing in order to enhance chemical reactions in combustion problems.

At the outset we must address the question whether control can be achieved in principle. It would seem that the answer is affirmative: if it were possible to observe the entire disturbance field an appropriate force field could be chosen to produce any desired level of variance. Equally if we were able to continuously disrupt the disturbance field at constant energy then inspection of the expression for the total fluctuation energy tendency shows decrease of fluctuation energy at the rate of trace($\mathbf{A}$), where $\mathbf{A}$ is the linearized operator associated with the stationary flow, which for viscous flows is typically negative. And indeed such a limitation on the coherence of motions produced by self-interaction of the fluctuations restricts the temporal correlation of fluctuations and provides a limit on the variance that could be maintained by a turbulent flow.

However control is a practical problem. The question is whether feasible observations linked to feasible control actions would result in a desired level of fluctuation variance. For example it is not obvious that both observation and control of a single variable at the surface would be sufficient to produce suppression of turbulence throughout the boundary layer.

Further, it is not obvious, that success in controlling a flow implies the possibility of control of all flows. In fact it is quite likely that some flows may be more easily controlled than others. Specifically optimism concerning the control of boundary layer turbulence appears more justified in view of recent advances in theoretical understanding than does control of free jets and free shear layers in which robust inflectional instabilities exist and in which there is no natural platform for observations as in the case of boundary layer turbulence.

It is useful to draw a distinction between controlling transition to turbulence and intervening to control the fluctuations in a fully turbulent flow. In the first problem the control strategy derives from theoretical understanding of the transition process, while the second derives from understanding of how the energy injection occurs in turbulent flows.

While essentially nonlinear theories for transition have been advanced there is at least a large class of transition scenarios, commonly referred to as bypass transition, in which small but finite perturbations induce the transition and for which linear mechanisms are clearly implicated. An example is the rapid transition at high Reynolds numbers in flows with naturally free stream fluctuations. For these flows it has been demonstrated that the perturbation growth can be traced to non-modal transient mechanisms due to the non-normality of the linear operator associated with the highly sheared boundary layer (Farrell, 1988; Gustavsson, 1991; Butler & Farrell, 1992; Reddy & Henningson, 1993; Trefethen *et al.*, 1993; Farrell & Ioannou, 1993a,1993b). Consequently designing an optimal control strategy for such flows proceeds from identification and suppression of those perturbations in the free stream turbulence with the potential for growth.

Moreover, this theory of by-pass transition has recently been extended to a theory for the maintenance of the turbulent state in which the spectral transfer arising from the non-linear interactions of the fluctuations is parameterized as stochastic noise and the linear interaction of the fluctuations with the background flow predominantly determines the energy injection properties. Consequently, we believe that for fully developed stationary boundary layer turbulence the process of injection of energy is essentially linear and the optimal control strategy adopted for the transitional flow is in principle a guide for the control of the fully turbulent flow. In what follows we describe the method and provide an example of how the proposed control strategy is applied (*cf* Farrell and Ioannou, 1996).

2 Formulation of the active control problem

Consider the evolution of small perturbations imposed on a steady channel flow with streamwise (x) background velocity $U(y)$ varying only in the cross-stream direction (y). Harmonic perturbations with streamwise wavenumber k and spanwise (z) wavenumber l obey the linear equation:

$$\frac{d\phi}{dt} = \mathbf{B}\,\phi \;, \tag{1}$$

where the state variable is $\phi = [\hat{v}\ ,\hat{\eta}]^T$, in which $\hat{v}$ is the cross stream perturbation velocity, and $\hat{\eta} = il\,\hat{u} - ik\hat{w}$ is the cross stream perturbation vorticity ($\hat{u}, \hat{w}$ are the perturbation streamwise and spanwise velocities respectively). The operator is given by:

$$\mathbf{B} = \begin{bmatrix} \mathbf{L} & 0 \\ \mathbf{C} & \mathbf{S} \end{bmatrix}, \tag{2}$$

The components of the dynamical operator (2) are the Orr-Sommerfeld operator, **L**, the coupling operator between cross-stream velocity and vorticity, **C**, and the diffusion Squire operator, **S** (*cf* Butler & Farrell, 1992).

We choose to impose symmetric control at the channel walls $y = \pm 1$ in reaction to observations of a field variable at $Y_1^{ob} = -1+Y_o$ and at $Y_2^{ob} = 1-Y_o$. By cross-stream velocity control we mean that observations of the cross stream velocity at Y_1^{ob}, Y_2^{ob} are used to impose a cross stream velocity at $y = \pm 1$ according to:

$$\hat{v}(-1) = C\,\hat{v}(Y_1^{ob}) \quad , \quad \hat{v}(1) = C\,\hat{v}(Y_2^{ob}), \tag{4}$$

where C is a complex control constant. Clearly, alternative controls can be imposed in a similar manner.

The remaining boundary conditions for the case of active specification of the cross-stream ($\hat{v}$) velocity at the boundaries are the vanishing of the streamwise ($\hat{u}$) and spanwise velocity ($\hat{w}$) components at the walls. Consequently at the channel wall we have the following boundary conditions:

$$\left.\frac{d\hat{v}}{dy}\right|_{y=\pm 1} = 0 \quad , \quad \hat{\eta}(\pm 1) = 0\,. \tag{5}$$

The perturbation evolution equation (1) together with boundary conditions (4) and (5) form a linear system with homogeneous boundary conditions and the imposition of feedback control constitutes a change in the boundary conditions of the flow. Therefore the control action can not be understood using arguments about cancellation or reinforcement of perturbations that exist in the unmanipulated flow. Instead, suppression of turbulence occurs because control parameters alter the boundary conditions so as to constrain the perturbations to exhibit reduced growth compared to that found with the standard boundary conditions in the unmanipulated flow.

We will determine the magnitude and phase of the control C and the observation level Y_o that reduces the growth of perturbations. Plane Poiseuille flow with $U = 1 - y^2$ is used as an example.

In order to proceed it is necessary to have a measure of perturbation growth. We choose the perturbation energy and we denote with **M** the energy metric. We transform (1) into generalized velocity variables $\psi = \mathbf{M}^{1/2}\,\phi$ so that the usual L_2 norm corresponds to the square root of the mean energy. Under this transformation a perturbation ψ_0 at $t = 0$ evolves to time t according to:

$$\psi^t = e^{\mathbf{A}t}\,\psi_0 \tag{6}$$

in which the dynamical operator has been transformed to $\mathbf{A} = \mathbf{M}^{1/2}\,\mathbf{B}\,\mathbf{M}^{-1/2}$.

An appropriate measure of perturbation growth at time t is the square of the Frobenius norm of $e^{\mathbf{A}t}$. This quadratic measure is equal to the sum of the squares of the singular values of $e^{\mathbf{A}t}$. This measure is proportional to the growth over an interval t of the mean perturbation when all perturbations are forced equally initially. The time integral of this measure is proportional to the perturbation variance maintained in the channel flow under white noise forcing i.e. the accumulated variance over an interval t for unit forcing of each degree of freedom is given by:

$$< E^t > = \text{trace} \left(\int_0^t e^{\mathbf{A}t} \, e^{\mathbf{A}^\dagger t} \, dt \right) , \tag{7}$$

where the brackets denote ensemble averaging. The steady state maintained variance $< E^\infty >$ is given for asymptotically stable systems as the limit of (7) as $t- > \infty$.

The maintained variance for asymptotically stable flows is found by solving the Liapunov equation for the correlation matrix $\mathbf{V}^\infty$:

$$\mathbf{A} \, \mathbf{V}^\infty + \mathbf{V}^\infty \, \mathbf{A}^\dagger = - \mathbf{I} , \tag{8}$$

with $\mathbf{I}$ the identity matrix corresponding to unitary forcing. The asymptotic variance can be identified with the trace of the correlation matrix, $< E^\infty > =$ trace $(\mathbf{V}^\infty)$.

The variance maintained by unbiased forcing in an unmanipulated Poiseuille flow peaks at the roll axis ($k = 0$). For $R = 2000$ there is a broad maximum at $K = O(1)$. For large Reynolds number ($R > 1000$) the peak wavenumber increases linearly with Reynolds number. Oblique harmonic perturbations also build energetic streaks and maintain substantial variance. Consequently, in our investigation of optimal control parameters we include oblique perturbations.

Effective controls, C, are those that minimize

$$\frac{< E_C^\infty >}{< E_0^\infty >} , \tag{9}$$

in the complex C plane where $< E_0^\infty >$ is the variance maintained under stochastic forcing with no control applied ($C = 0$). We investigate the magnitude of the variance suppression as a function of the amplitude $|C|$ and phase Θ of the control for roll and oblique perturbations and for observation at various distances from the wall, Y_o. An effective control must lead to robust suppression of both roll and oblique perturbations.

3 In phase and out of phase control

We first constrain the control parameter C to be real. As expected, in phase control ($C > 0$) leads for small control amplitudes to increased variance. In the vicinity of $C = 1$ the flow becomes unstable (this instability is diffusive in nature and occurs also in the absence of flow for any finite Reynolds number). We find further that, remarkably, higher amplitudes of in phase control lead to robust variance suppression. This surprising suppression will be referred to as overdriving suppression. For example in phase overdriving at an amplitude $C \approx 2$ leads to variance suppression of the order of 60-70 % when observations are made at $Y_o = 0.2$ from the wall. This control robustly suppresses the variance of both roll and oblique perturbations.

Out of phase control ($C < 0$) of roll perturbations leads to robust reduction of variance with the suppression becoming more effective the farther the observation level is located from the controlled boundary (at least for $Y_o < 0.5$). Maximum suppression requires amplitudes $|C| > 4$ and is of the order of 90%. Unfortunately, this promising control strategy does not generalize to oblique perturbations. For distant observation levels i.e. $Yo > 0.2$ and with out of phase control of amplitude $|C| \approx 1$ oblique perturbations become unstable leading to variance increase. This instability appears at low Reynolds numbers (typically $R = O(500)$) and analysis of the energetics of the instability reveals that the control injects only a small amount of energy while the predominant energy source is the downgradient Reynolds stress term. The most unstable perturbations occurs for purely 2-D perturbations in agreement with predictions of Squire's theorem. The direct numerical simulation experiments of Choi *et al.* (1994) showed that out of phase control of unit amplitude ($C = -1$) leads to drag reduction for observations at locations less than 20 wall units and to drag increase for observations at greater distances from the wall. The cause of this drag increase is presumably inception of the instability described above. We performed the same stability analysis on the Reynolds-Tiederman profile. The results of the stability analysis of the Orr-Sommerfeld operator for this profile indicates that inception of the instability occurs for observations located at 30 wall units. The calculations reported here and the experiments of Choi *et al.* (1994) were carried out in a channel flow so that the possibility remains that the instability occurring in the vicinity of out of phase control ($\Theta = 180$) would not occur in boundary layer flows. To check this a stability analysis was performed on the one sided Reynolds-Tiederman profile. Although the instability occurs at a higher value of observation locations (for observations located at 40 wall units from the boundary) it is qualitatively similar to that found in channel flow.

4 Boundary control in quadrature with observations

We consider control actions in which the boundary response is in quadrature with the observation ($\Theta = \pm 90^o$). For roll perturbations ($k = 0$) a robust suppression of variance as a function of the control amplitude is found for various observation levels near the wall i.e. $Y_o < 0.4$. For example for $Y_o = 0.4$ the variance suppression reaches $70-80\%$ for control amplitudes $|C| \approx 4$. For observation levels $Y_o > 0.5$ roll perturbations show an increase in variance. Despite robust suppression of variance at $\Theta = 90^o$ the greatest suppression for roll perturbations occurs at $\Theta = 180^o$ corresponding to exactly out of phase control. Unfortunately, as we have seen, this out of phase control fails to similarly suppress oblique perturbations because of the existence of an unstable mode. Analysis of the resulting flow shows that the constraint imposed by the control leads to the development of a much weaker doublet of opposing streaks.

We turn now to variance suppression for oblique perturbations. We find that for observations at $0.3 < Y_o < 0.7$ robust suppression of the order of $60 - 70\%$ occurs at $\Theta = 90^o$.

We have already seen that out of phase control suppresses variance optimally for streamwise roll perturbations. However, even slightly oblique perturbations may become unstable for out of phase controls. Consequently, the most robust strategy is out of phase control of roll perturbations and $\Theta = 90^o$ control of oblique perturbations.

Physically the variance reducing control action inhibits the formation of streaks in the vicinity of the wall where high shear would lead to substantial build up of streak amplitude. The control boundary condition induces a time varying cross-stream velocity near the boundaries which inhibits the formation of the energetic near wall streaks. Analysis of energetics confirms that the energy growth arising from the Reynolds stress is reduced in the controlled flow.

An implication of these results is that the rms amplitude of the streamwise and cross-stream velocities in controlled flows peak at greater distance from the wall. This can also be seen to be the case in the numerical simulations presented by Choi *et al.* (1994).

Acknowledgement

This work was supported by NSF ATM-96-25453.

References

Butler, K. M., and B. F. Farrell, 1992: Three dimensional optimal perturbations in viscous shear flow. *Phys. of Fluids A*, **4**, 1637–1650.

Choi, H., P. Moin, and J. Kim, 1994: Active turbulence control for drag reduction in wall bounded flows. *J. Fluid Mech.* , **262**, 75–98.

Farrell, B. F.,1988: Optimal excitation of perturbations in viscous shear flow. *Phy. Fluids*, **31**, 2093–2102.

Farrell, B. F., and P. J. Ioannou, 1993a: Optimal excitation of three dimensional perturbations in viscous constant shear flow. *Phys. Fluids A* , **5**, 1390–1400.

Farrell, B. F., and P. J. Ioannou, 1993b: Perturbation growth in shear flow exhibits universality. *Phys. Fluids A* , **5**, 2298–2301.

Farrell, B. F., and P. J. Ioannou, 1996: Turbulence suppression by active control. *Phys. Fluids A* , **8**, 1257–1268.

Gustavsson, L. H., 1991: Energy growth of three-dimensional disturbances in plane Poiseuille flow. *J. Fluid Mech.* , **224**, 241–260.

Reddy, S. C., and D. S. Henningson, 1993: Energy growth in viscous channel flows. *J. Fluid Mech.* , **252**, 209–238.

Trefethen, L. N., A. E. Trefethen, S. C. Reddy, and T. A. Driscoll, 1993: Hydrodynamic stability without eigenvalues. *Science*, **261**, 578–584.

Trends in Mathematics, © 1999 Birkhäuser Verlag Basel/Switzerland

Origin and growth of structures in boundary layer flows

B. F. Farrell, P. J. Ioannou

1 Introduction

A central goal of high Reynolds number fluid dynamics is to gain a comprehensive understanding of the origin and growth of perturbations in shear flows in which the rate of strain of the background velocity provides the source of energy of the perturbations. In all cases the availability of energy for perturbation growth can be determined by linearizing the equations of motion about the appropriate background flow and searching for growing perturbations. If all possible perturbations are examined and only decaying ones are found then it is certain that the background flow will persist when subjected to a sufficiently small disturbance. However, determining the potential for growth of all possible perturbations has not been the historical course of inquiry in stability theory. Rather, traditional stability theory follows the program of Rayleigh (1880) according to which instability is traced to the existence of exponentially growing modes of the linearized dynamic equations. The classical application of the normal mode paradigm envisions unstable modes growing exponentially from infinitesimal beginnings over a large number of e-foldings so that the exponential mode of greatest growth eventually emerges as a finite amplitude wave. This assumption of undisturbed growth is necessary to ensure the asymptotic dominance of the most rapidly growing normal mode which in turn permits the theory to make predictions concerning the structures of finite amplitude. Acceptance of the theory of small oscillations was encouraged by its success in application to problems such as the Rayleigh-Benard and Rayleigh-Taylor problems.

Despite the wide acceptance accorded the normal mode theory there remained difficulties of correspondence in boundary layer shear flows between the observed temporal variation and spatial structure of growing perturbations and the time independent structure of the normal modes. While modal theory predicts that the most unstable perturbations should be 2-dimensional (Squire, 1933) transition proceeds in practice by amplification of fully three-dimensional structures.

Such discrepancies and lingering theoretical questions involving the need to complete the normal modes in the case of linear inviscid dynamics by inclusion of a continuous spectrum of singular neutral modes led to reexamination of the results of Kelvin (1887) and Orr (1907) on the stability of the continuous spectrum by Case (1960). These inquiries showed that the continuous spectrum is stable in the sense that it fails to produce unbounded growth in the limit $t->\infty$ and this negative result was generally interpreted as a proof assuring that the stability of a flow could be determined solely from inspection of its modal spectrum for exponential instabilities. However, it is now more widely appreciated that the modal spectrum only determines stability in the $t- > \infty$ limit and that a more general analysis is necessary to determine the stability properties at finite time. Given that all experiments are conducted in finite time and that the time scale for formation of energetic structures in the boundary layer is rapid finite time stability analysis would seem to be the more appropriate and indeed inquiry shows that in the boundary layer there are non-modal transient disturbances with large growth rates on rapid time scales. The most rapidly growing of these disturbances exhibit transient structural evolution during development that characterizes the observed development of the coherent structures in the boundary layer. Freed of concentrating on the $t - > \infty$ asymptotic, this generalized stability theory allows a much closer correspondence to be made with observed structures which are highly variable both temporally and structurally (Farrell, 1988; Gustavsson, 1991; Butler & Farrell, 1992; Reddy, Schmid & Henningson, 1993; Trefethen *et al.*, 1993; Farrell & Ioannou, 1993a; Reddy & Henningson, 1993; Farrell & Ioannou, 1996).

The existence of a subspace of growing disturbances suggests a mechanism by which the aggregated growth of individual structures supports the statistically steady variance of the fully turbulent flow. It can be shown that the net source of energy to the perturbation field attributable to nonlinear interactions among waves vanishes and it follows that extraction of energy from the forced background flow by the subspace of growing disturbances, which is fully described by linear dynamics, must be responsible for maintaining eddy energy in the fully developed turbulent state (Joseph, 1976 ; Henningson and Reddy, 1994). This observation suggests a mechanistic model for the turbulent state in which the mean flow is subjected to continuous perturbative forcing (Farrell and Ioannou, 1993b, 1994, 1995). The appropriate method of analysis for such a turbulence model is the stochastic dynamics of non-normal linear systems.

The elements of generalized stability theory are reviewed below.

2 Response of the non-normal operator associated with boundary layer flows to impulsive excitation

The equation governing first order perturbation dynamics in the boundary layer is a special case of the general linear dynamical system:

$$\frac{du}{dt} = \mathbf{A}\, u, \tag{1}$$

where $u = [\hat{v}\ , \hat{\eta}]^T$, is the state variable for each streamwise and spanwise Fourier component, $\hat{v}$ is the cross stream perturbation velocity and $\hat{\eta}$ is the cross-stream perturbation vorticity. The dynamical operator in (1) is obtained from the linearized Navier-Stokes equations by eliminating the pressure field using the continuity equation. The operator **A** is given by:

$$\mathbf{A} = \begin{bmatrix} \mathbf{L} & 0 \\ \mathbf{C} & \mathbf{S} \end{bmatrix}, \tag{2}$$

with:

$$\mathbf{L} = \Delta^{-1}\,(\,-i\,k\,U\,\Delta + i\,k\,U'' + \Delta\Delta/R\,)\,, \tag{3.a}$$
$$\mathbf{S} = -i\,k\,U + \Delta/R\,, \tag{3.b}$$
$$\mathbf{C} = -i\,l\,U'\,, \tag{3.c}$$

and in which the Laplacian operator is given by: $\Delta \equiv \frac{d^2}{dy^2} - K^2$, with K being the total horizontal wavenumber: $K^2 = k^2 + l^2$, and k, l the sreamwise and spanwise wavenumbers respectively. The mean velocity in the streamwise direction which varies only in the cross-stream direction, y, is U and cross-stream derivatives of the mean fields are denoted with a dash. The equations have been rendered non-dimensional with the maximum background velocity in the channel, U_o, and the channel half-width, L so that the Reynolds number is $R \equiv \frac{U_o L}{\nu}$; ν denoting the kinematic viscosity. A well posed inversion of the Laplacian in (3a) requires incorporating the boundary conditions at the channel walls $y = \pm 1$.

The components of the dynamical operator (2) are the Orr-Sommerfeld operator, **L**, the coupling operator between cross-stream velocity and vorticity, **C**, which corresponds physically to the generation of cross-stream vorticity by tilting of the mean spanwise vorticity; and the advection-diffusion Squire operator, **S**.

In the following it is assumed that (2) has been discretized so that **A** is the associated linearized dynamical matrix operator. If the background state is steady so that **A** is not a function of time then the solution to (1) is explicit:

$$u(t) = e^{\mathbf{A}t}\,u(0)\;. \tag{4}$$

The central distinguishing attribute of **A** that determines its transient dynamics is its normality, i.e. whether or not $\mathbf{A}\,\mathbf{A}^\dagger = \mathbf{A}^\dagger\,\mathbf{A}$. If **A** commutes with its Hermitian transpose, here indicated by the superscript dagger, then **A** is normal and has a complete set of orthogonal eigenvectors. Perturbation growth rate for normal **A**'s is bounded above by the member of the eigenspectrum of **A** with maximum real part.

Because the finite time perturbation dynamics of a non-normal operator can not be ascertained from the spectrum of the operator it is necessary to generalize ideas of perturbation growth by considering the growth, σ, of an arbitrary perturbation u(0) over time t:

$$\sigma^2 = \frac{(u(t)\,,\,u(t))}{(u(0)\,,\,u(0))} = \frac{(e^{\mathbf{A}t}u(0)\,,\,e^{\mathbf{A}t}u(0))}{(u(0),u(0))} = \frac{(e^{\mathbf{A}^\dagger t}\,e^{\mathbf{A}t}u(0)\,,\,u(0))}{(u(0)\,,\,u(0))}\,. \tag{5}$$

The inner product $(\cdot,\cdot)$ generates the Euclidean norm for the vector space: $||\cdot|| = (\cdot,\cdot)^{1/2}$. It follows from (5) that a complete set of orthogonal perturbations $u(0)$ can be ordered in growth over time t by eigenanalysis of the matrix: $e^{\mathbf{A}^\dagger t}\, e^{\mathbf{A}t}$. In particular, the greatest growth over time t as measured by the square of the Euclidean norm is given by the maximum eigenvalue of $e^{\mathbf{A}^\dagger t}\, e^{\mathbf{A}t}$ which is also equal to $||e^{\mathbf{A}t}||^2$ as can be also seen immediately from the singular value decomposition of $e^{\mathbf{A}t}$. The initial condition that gives the maximum growth at a given time is referred to as the optimal perturbation at that time.

There are two asymptotic limits of interest in connection with the excitation of the propagator. In the limit $t -> \infty$ maximum growth is obtained by the eigenfunction associated with the eigenvalue with maximum real part just as normal mode theory would suggest. To see this consider the matrix $\mathbf{E}$ constructed by arranging the eigenvectors of $\mathbf{A}$ as columns in order of growth rate together with the diagonal matrix, Δ, of the associated modal growth factors, from which the following similarity transformation of the propagator can be constructed:

$$e^{\mathbf{A}t} = \mathbf{E}\, e^{\Delta t}\, \mathbf{E}^{-1} . \tag{6}$$

In the limit $t -> \infty$ the first column of $\mathbf{E}$ and the first row of $\mathbf{E}^{-1}$ exponentially dominate with amplification factor $e^{\mathrm{Real}(\Delta_{1\ 1})t}$:

$$\lim_{t\ -> \infty} e^{\mathbf{A}t}{}_{\alpha\beta} = \mathbf{E}_{\alpha 1}\ e^{\Delta_{11}t}\ \mathbf{E^{-1}}_{1\beta} . \tag{7}$$

It can be seen by appeal to Schwartz's inequality that the initial condition of unit norm producing maximum growth over time t is the complex conjugate of $\mathbf{E^{-1}}_{1\beta}$ which is the conjugate of the biorthogonal of the leading eigenvector rather than the leading eigenvector itself:

$$\lim_{t\ -> \infty} e^{\mathbf{A}t}{}_{\alpha\beta}(\mathbf{E}^{-1}{}_{1\beta})^* = \mathbf{E}_{\alpha 1}\, e^{\Delta_{11}t}. \tag{8}$$

Modal theory correctly predicts that in the limit $t -> \infty$ that eigenvector which has associated eigenvalue with maximum real part dominates. Not so obvious is the fact that the optimal initial condition with which to excite that mode is the conjugate of the biorthogonal of the dominant mode rather than the mode itself.

Given the observed mean time scale for the formation of coherent structures the $t -> \infty$ asymptotic is not likely to provide a realistic precursor for the formation process. Of greater utility for this purpose is analysis of the $t -> 0$ limit of (5) which controls the initial growth of perturbations. Analysis of this limit provides the maximum possible instantaneous growth rate and the structure that produces this maximum growth rate. The growth rate and the perturbation of maximum instantaneous growth itself also provides a constructive nonlinearly valid bound on the potential for perturbation growth in the flow (Joseph, 1976).

The limit as $t -> 0$ is easily obtained by Taylor expansion of the matrix $e^{\mathbf{A}^\dagger t}\, e^{\mathbf{A}t}$ in (5):

$$\begin{aligned} e^{\mathbf{A}^\dagger t}\ e^{\mathbf{A}t} &\approx (\mathbf{I} + \mathbf{A}^\dagger t)\,(\mathbf{I} + \mathbf{A}t) \\ &= \mathbf{I} + (\mathbf{A} + \mathbf{A}^\dagger)t + \mathbf{0}(t^2), \end{aligned} \tag{9}$$

where $\mathbf{I}$ is the identity. It follows that a tight upper bound on instantaneous growth rate, and the structure producing this maximum instantaneous growth rate can be found by eigenanalysis of the matrix $\mathbf{A} + \mathbf{A}^{\dagger}$. The maximum eigenvalue of $(\mathbf{A} + \mathbf{A}^{\dagger})/2$ and its associated eigenvector provide the required growth rate and structure. Eigenanalysis of $\mathbf{A} + \mathbf{A}^{\dagger}$ typically reveals that high growth rates over short times can be realized in boundary layer flows even for low Reynolds numbers for which all normal modes of $\mathbf{A}$ are damped.

The most relevant time scales for the development of coherent structures in the boundary layer are between the asymptotic limits $t -> 0$ and $t -> \infty$ and for these intermediate time scales the initial and final structures are found most easily from the SVD analysis of the propagator $e^{\mathbf{A}t}$. Given that both asymptotic limits are subsumed it is appropriate to refer to this analysis as the generalized stability analysis of the system (1).

3 Response of non-normal dynamical systems to continuous excitation

Transient growth of disturbances in shear flow can be traced to a substantial subspace of perturbations that extract energy from the background flow. In section 2 analysis of these growing perturbations was framed as an initial value problem involving as a parameter the physically relevant interval in time over which growth occurs. The appropriate method of analysis for studying the maintenance of time mean variance by continuous incoherent forcing is the stochastic dynamics of the associated non-normal system.

The stochastically forced linear dynamical system can be written in the form:

$$\frac{du}{dt} = \mathbf{A}\, u + \mathbf{F}\, \eta(t)\,, \tag{10}$$

in which $\eta(t)$ is a temporally Gaussian white-noise forcing componentwise δ-correlated with zero ensemble mean and unit ensemble covariance:

$$< \eta_i(t_1)\, \eta_j^*(t_2) > \;=\; \delta_{\rm ij}\;\; \delta(t_1 - t_2)\,. \tag{11}$$

The spatial distribution of the forcings is provided by the matrix $\mathbf{F}$, and if it is chosen to be unitary the resulting statistics become independent of the particular choice of $\mathbf{F}$.

To obtain the stochastic growth of perturbations we first write the forced solution of (10) as:

$$u(t) = \int_0^t e^{\mathbf{A}(t-s)}\, \mathbf{F}\, \eta(s)\, ds\,. \tag{12}$$

The variance maintained by this stochastic forcing is given in the Euclidean norm by:

$$
\begin{aligned}
< ||u(t)||^2 > &= \left\langle \int_0^t ds \int_0^t ds' \eta^\dagger(s) \mathbf{F}^\dagger e^{\mathbf{A}^\dagger (t-s)} \, e^{\mathbf{A}(t-s')} \mathbf{F} \eta(s') \right\rangle \\
&= \mathrm{Trace}\left(\mathbf{F}^\dagger \ \int_0^t e^{\mathbf{A}^\dagger (t-s)} \ e^{\mathbf{A}(t-s)} \, ds \ \mathbf{F} \right) \\
&= \mathrm{Trace}\left(\mathbf{F}^\dagger \ \mathbf{B} \mathbf{F} \right) ,
\end{aligned}
\tag{13}
$$

revealing that the hermitian operator

$$
\mathbf{B}^t \ = \ \int_0^t e^{\mathbf{A}^\dagger s} \ e^{\mathbf{A} s} \, ds \ , \tag{14}
$$

accumulates the perturbation growth when all perturbations are stochastically excited. This operator should be compared with the operator $e^{\mathbf{A}^\dagger t} \ e^{\mathbf{A} t}$ eigenanalysis of which reveals the optimal perturbation growth as we have seen in the previous section. An alternative and computationally preferable method for calculating the stochastic dynamical operator $\mathbf{B}^t$ results from differentiating (14) with respect to time to obtain:

$$
\frac{d\mathbf{B}^t}{dt} \ = \mathbf{I} \ + \ \mathbf{A}^\dagger \, \mathbf{B}^t \ + \ \mathbf{B}^t \, \mathbf{A} \ , \tag{15}
$$

in which $\mathbf{I}$ is the identity matrix.

In direct analogy with the analysis of optimal growth in the previous section a complete set of orthogonal forcings forming the columns of a unitary $\mathbf{F}$ can be found for the stochastic variance at time t in (13) by eigenanalysis of the positive definite hermitian $\mathbf{B}^t$. If the operator $\mathbf{A}$ is asymptotically stable a stationary solution is obtained in which the eigenfunctions of $\mathbf{B}^\infty$ are ordered according to their contribution to the variance of the statistically steady state. The forcings ordered in this way will be referred to as stochastic optimals.

The stochastic optimals most effectively excite the stationary variance and should be contrasted with the orthogonal structures that most effectively span the maintained variance, which are commonly referred to as the EOF's of the dynamical system. The stochastic optimals bear a relationship to the EOF's in the stochastic analysis analogous to that between the optimal excitation and the optimal response in the SVD analysis of the propagator of the initial value problem. To obtain the EOF's we need first to form the correlation matrix:

$$
\begin{aligned}
\mathbf{C}_{ij}^t \ &= \ < u_i(t) \, u_j^*(t) > \\
&= \ \left(\int_0^t e^{\mathbf{A}(t-s)} \ \mathbf{F}\mathbf{F}^\dagger \ e^{\mathbf{A}^\dagger (t-s)} \, ds \right)_{\mathrm{ij}} \ ,
\end{aligned}
\tag{16}
$$

which satisfies:

$$
\frac{d\mathbf{C}^t}{dt} \ = \ \mathbf{F}\mathbf{F}^\dagger \ + \ \mathbf{A} \, \mathbf{C}^t \ + \ \mathbf{C}^t \, \mathbf{A}^\dagger \ . \tag{17}
$$

Each eigenvalue of the positive definite hermitian operator $\mathbf{C}^t$ equals the variance accounted for, under unbiased forcing and at time t, by the pattern of its corresponding eigenvector, and the pattern that corresponds to the largest eigenvalue contributes most to the perturbation variance at t.

If $\mathbf{A}$ is normal and the forcing unitary ($\mathbf{F}\mathbf{F}^\dagger = \mathbf{I}$) then $\mathbf{A}$, $\mathbf{B}^t$, $\mathbf{C}^t$ commute and the stochastic optimals, the EOF's, and the modes of the dynamical system coincide. For such a system eigenanalysis of $\mathbf{A}$ suffices for understanding the statistics of the perturbations in the linear limit. In contrast, for non-normal systems the stochastic optimals, the EOF's and the modes of the dynamical operator are all distinct (Farrell & Ioannou, 1993b; 1995).

If $\mathbf{A}$ is asymptotically stable the system approaches a statistically steady state as t increases in which $\mathbf{B}^\infty$ and $\mathbf{C}^\infty$ satisfy the Lyapunov equations :

$$\begin{aligned} \mathbf{A}\mathbf{C}^\infty + \mathbf{C}^\infty\mathbf{A}^\dagger &= -\mathbf{F}\mathbf{F}^\dagger \\ \mathbf{A}^\dagger\,\mathbf{B}^\infty + \mathbf{B}^\infty\,\mathbf{A} &= -\mathbf{I}\,. \end{aligned} \tag{18}$$

The Lyapunov equations (18) are readily solved for $\mathbf{B}^\infty$ and $\mathbf{C}^\infty$ given the asymptotically stable operator $\mathbf{A}$ and the forcing correlation matrix $\mathbf{F}$.

4 Conclusions

The turbulent state of wall bounded pipe and channel flows is characterized by energetic interactions between the highly sheared mean flow of the boundary layer and a subset of coherent disturbances having the form of streamwise streaks and associated streamwise vortices. Because of the very high shears found in the boundary layer, perturbation dynamics may plausibly be anticipated to be dominated by the mean straining field which forms the primary interaction between the mean flow and the perturbations and which is incorporated in the linear dynamical operator. This fundamental linearity of boundary layer turbulence is demonstrated by comparisons performed between simulations with and without inclusion of nonlinear wave interactions (Lee *et al.*, 1990). Despite this evident simplicity of the dynamics, straightforward calculation of the eigenspectrum of the linearized dynamical operator fails to produce structures with the observed form of streamwise vortices. This failure of correspondence can be understood from the perspective of analysis of the non-normal operator associated with the linear dynamics as due to the fact that while the perturbations of maximal growth in the linear problem take the form of streamwise rolls these are not the eigenfunctions of the linearized operator, rather they are the optimal structures identified with the first singular vectors of the propagator arising from the dynamical operator in an appropriate norm and with an appropriate time interval for development.

Ackowledgments

This work was supported by NSF ATM-96-25453.

References

Butler, K. M., and B. F. Farrell, 1992: Three dimensional optimal perturbations in viscous shear flow. *Phys. of Fluids A*, **4**, 1637–1650.

Case, K. M., 1960: Stability of inviscid plane Couette flow. *Phys. Fluids*, **3**, 143–149.

Farrell, B. F., 1988: Optimal excitation of perturbations in viscous shear flow. *Phy. Fluids*, **31**, 2093–2102.

Farrell, B. F., and P. J. Ioannou, 1993a: Optimal excitation of three dimensional perturbations in viscous constant shear flow. *Phys. Fluids A*, **5**, 1390–1400.

Farrell, B. F., and P. J. Ioannou, 1993b: Stochastic forcing of the linearized Navier-Stokes equations. *Phys. Fluids A*, **5**, 2600–2609.

Farrell, B. F., and P. J. Ioannou, 1994: Variance maintained by stochastic forcing of non-normal dynamical systems associated with linearly stable shear flows. *Phys. Rev. Lett.*, **72**, 1188–1191.

Farrell, B. F., and P. J. Ioannou, 1995: Stochastic dynamics of the midlatitude atmospheric jet. *J. Atmos. Sci.* , **52**, 1642–1656.

Farrell, B. F., and P. J. Ioannou, 1996: Generalized stability theory *J. Atmos. Sci.* , **53**, 2025–2040.

Gustavsson, L. H., 1991: Energy growth of three-dimensional disturbances in plane Poiseuille flow. *J. Fluid Mech.*, **224**, 241–260.

Henningson, D.S., and S. C. Reddy, 1994: On the role of linear mechanism in transition to turbulence. *Phys. Fluids A*, **6**, 1396–1398.

Joseph, D.D., 1976: *Stability of Fluid Motions I.* Springer Verlag, 282 pp.

Kelvin, Lord, 1887: Stability of fluid motions: Rectilinear motion of a viscous fluid between two parallel plates. *Phil. Mag.*, **24**, 188–196.

Lee, M. J., Kim J. & Moin P., 1990: Structure of turbulence at high shear rate. *J. Fluid Mech.*, **216**, 561–583.

Orr, W. Mc F., 1907: Stability or instability of the steady motions of a perfect liquid. *Proc. Roy. Irish Acad.*, **27**, 9–69.

Rayleigh, J. W. S., 1880: On the stability, or instability, of certain fluid motions. *Proc. London Math. Soc.*, **9**, 57–70.

Reddy, S. C., and D. S. Henningson, 1993: Energy growth in viscous channel flows. *J. Fluid Mech.*, **252**, 209–238.

Reddy, S. C., P. J. Schmid, and D.S. Henningson, 1993: Pseudospectra of the Orr-Sommerfeld operator. *SIAM J. Appl. Math.*, **53**, 15–47.

Squire, H. B., 1933: On the stability of three-dimensional disturbances of viscous flow between parallel walls. *Proc. Roy. Soc. A*, **142**, 621–628.

Trefethen, L. N., A. E. Trefethen, S. C. Reddy, and T. A. Driscoll, 1993: Hydrodynamic stability without eigenvalues. *Science*, **261**, 578–584.

Trends in Mathematics, © 1999 Birkhäuser Verlag Basel/Switzerland

Vortex control in a spatially growing round jet

O. Métais, G. Urbin

1 Introduction

Due to the diversity of their coherent structures, axisymmetric jets constitute a prototype of free shear flows of vital importance from both a fundamental as well as a more applied point of view. Indeed, a better understanding of the jet vortex structures should make possible the active control of the jet (spreading rate, mixing enhancement, ...) for engineering applications (see e.g. Zaman *et al.* [25]). In the last five years, the progress in the experimental methods for detection and identification has made possible a detailed investigation of the complex three-dimensional coherent vortices imbedded within this flow. For instance, the influence of the entrainment of the secondary streamwise vortices has been studied by Liepmann and Gharib [11]. On the numerical side, several simulations of two-dimensional or temporally evolving jets have been performed. Very few have however investigated the three-dimensional spatial development of the round jet. We here present a statistical and topological numerical study of the spatial growth of the round jet from the nozzle up to sixteen diameters downstream. The use of large-eddy simulations (LES) techniques allow us to reach high values of the Reynolds number: here, Re is 25000. The advantage of these techniques is that high Reynolds number flows can be simulated with a moderate number of resolution points and a reasonable computer cost. Our goal here is to demonstrate the ability of the LES to properly reproduce the coherent vortex dynamics in the transitional region of the jet. We also show the possibility of controlling the jet behaviour by manipulating the inflow conditions. The detailed results are presented in Urbin (1997) [21], Urbin and Métais [22] and Urbin *et al.* [23].

2 Subgrid-scale model and numerical methods

We have here chosen the *structure function model* proposed by Métais and Lesieur [14] in its selective version (see Lesieur and Métais [10]), which is well adapted for transitional flows and accepts non uniform grids. It is based upon an eddy viscosity that is supposed to represent a three-dimensional turbulence with nearly a Kolmogorov cascade in the subgrid-scales. One of its interesting characteristic is that it switches off the eddy-viscosity within the flow regions where the motion is not three-dimensional enough (see Lesieur and Métais, [10], for details). The LES filtered Navier-Stokes equations are solved using the TRIO-VF code. This is an industrial software developed for thermal-hydraulics applications at the Commissariat a l'Energie Atomique de Grenoble. It has been thoroughly validated in many LES of various flows (see e.g. Silveira *et al.* [20], for the backward facing step). It uses the finite volume element method on a structured mesh.

3 Numerical experiments

The experimental studies by Michalke and Hermann [15] have clearly pointed out the capital effect of the inflow momentum boundary layer thickness θ and of the ratio R/θ (R : jet radius) on the jet downstream development. It was shown that the detailed shape of the mean velocity profile strongly influences the nature of the coherent vortices appearing near the nozzle: either axisymmetric structures (vortex rings) or helical structure can indeed develop. Here, we did not simulate the flow inside the nozzle but we imposed a mean axial velocity profile in accordance with the experimental measurements:

$$W(r) = \frac{1}{2} W_o \left[1 - tanh \left(\frac{1}{4} \frac{R}{\theta} \left(\frac{r}{R} - \frac{R}{r} \right) \right) \right] \tag{1}$$

where W_o is the velocity on the axis. We have restrained ourself to a relatively small value of R/θ with $R/\theta = 10$ since a correct resolution of the shear zone at the border of the nozzle is crucial to correctly reproduce the initial development of the instabilities. We consider a computational domain starting at the nozzle and extending up to $16D$ downstream. The section perpendicular to the jet axis consists of a square $10D * 10D$, which has been shown to be sufficient to avoid jet confinement. The computational mesh is refined at the jet shear-layer (stretched mesh). The "natural" jet is forced upstream by the top-hat profile given by (1) to which is superposed a weak 3D white noise. The "excited" jet development is controlled with the aid of various deterministic inflow forcing (plus a white noise) designed to trigger specific types of three-dimensional coherent structures.

3.1 The natural jet

We have thoroughly validated our numerical approach by comparing the computed statistics with experimental results for the mean and for the r.m.s. fluctuating quantities. The frequency spectra have furthermore revealed the emergence of a predominant vortex-shedding Strouhal number, $Str_D = 0.35$ in good correspondance with the experimental value.

Temporal linear stability analysis performed on the inlet jet profile given by (1) (with $R/\theta = 10$) predicts a slightly higher amplification rate for the axisymmetric (varicose) mode than for the helical mode (see Michalke and Hermann [15]). For this reason, we have checked that the Kelvin-Helmholtz instability along the border of the jet yields, further downstream, vortex structures mainly consisting in axisymmetric toroidal shape. However, the simulations reveal that these structures are not always present and alternate with vortices of helical shape (see Urbin and Métais [22]).

The 3D visualization (Figure 1) exhibits an original vortex arrangement subsequent to the varicose mode growth: the "alternate pairing". Such a structure was previously observed by Fouillet [6] and Comte *et al.* [3] in a direct simulation of a temporally evolving round jet at low Reynolds number ($Re = 2000$). The direction normal to the toroidal vortices symmetry plane, during their advection downstream, tends to differ from the jet axis. The inclination angle of two consecutive vortices appears to be of opposite sign eventually leading to a local pairing with an alternate arrangement. Note that vortex loop's inclination at the end of the potential core was experimentally observed by Petersen [17]. Experimental evidence of "alternate pairing" was recently shown by Broze and Hussain [2]. This alternate-pairing mode corresponds to the growth of a subharmonic perturbation (of wavelength double of the one corresponding to the rings) developing after the formation of the primary rings. It therefore presents strong analogies with the helical pairing mode observed in plane mixing layers (see Pierrehumbert and Widnall [19]; Comte *et al.* [4]).

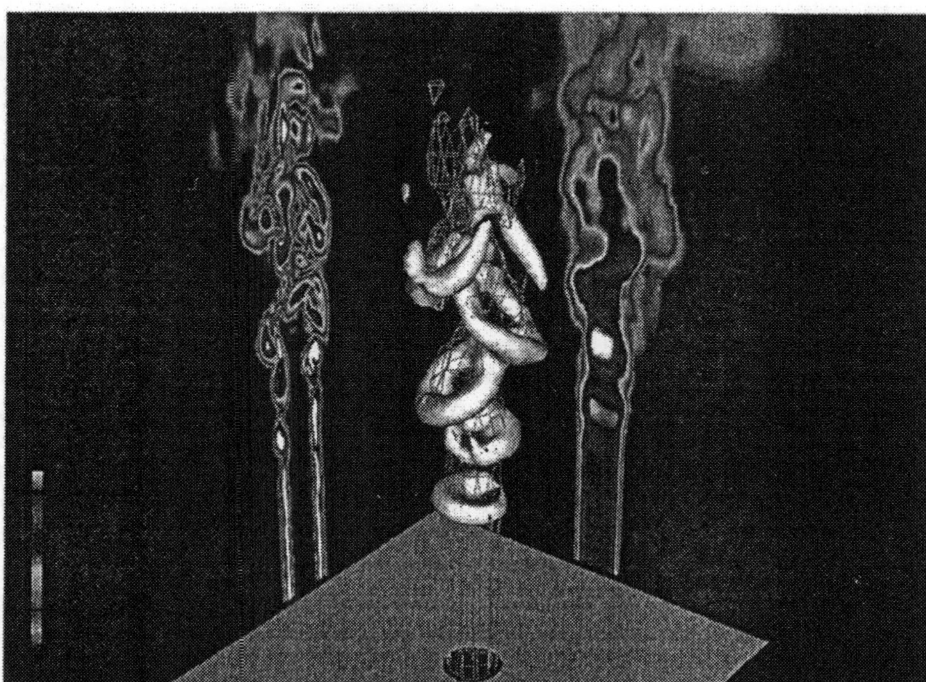

Figure 1: Natural jet: instantaneous visualization. Light gray: low pressure isosurface; wired isosurface of the axial velocity $W = W_o/2$; YZ cross-section (through the jet axis) of the vorticity modulus; XZ cross-section of the velocity modulus.

3.2 The forced jet

Our natural jet simulations have revealed three different types of vortical organization: the toroidal vortices (rings), the helical structure and the alternate-pairing. We now apply deterministic inflow perturbations to trigger one of these three particular flow organization and study the influence of the forcing on the statistics. Crow and Champagne [5] first noticed that the jet response is maximal with a preferred mode frequency corresponding to Str_D between 0.3 and 0.5. We then applied a periodic fluctuation associated with a frequency corresponding to $Str_D = 0.35$ in superposition to the white noise.

3.2.1 Varicose excitation

We first excitate the varicose mode by imposing a periodic perturbation (alternatively low-speed and high speed) to the axial velocity at the nozzle:

$$W(r) + \epsilon\, W_o\, sin\left(\frac{Str_D W_o}{D}t\right) \tag{2}$$

where $W(r)$ is given by equation (1) and $\epsilon = 1\%$.

Comparisons of the velocity fluctuations with experimental results, show that, as opposed to the unexcited jet, a strong and fast amplification of the instability appears.

The visualizations show that the varicose mode is now present at every instant at the beginning of the jet $Z/D < 6$. The vortex structures are more intense than in the natural case with well marked and organized pressure trough (Figure 2). The rings resulting from the varicose mode are linked together with longitudinal vortices. The maximum vorticity within these structures is about 40% of the vorticity of the associated rings. These have already been observed experimentally at moderate Reynolds number flow (see e.g. Lasheras, Lecuona and Rodriguez [8]; Monkewitz and Pfitzenmaier [16]; Liepmann and Gharib [11]). The present simulation indicates that they are also present at high Reynolds number. These longitudinal vortices are known to entrain and eject fluid outside, thus creating transverse side jets and "branches". The latter were studied numerically, in temporal simulations of Martin and Meiburg [13] and Abid and Brachet [1]. One possible explanation for the origin of these longitudinal vortices is an azimuthal oscillation of the vortex lines at the stagnation points between consecutive primary rings followed by a strong stretching mechanism by the latter (Lasheras, Lecuona and Rodriguez [8]). We have numerically checked that the number of longitudinal structures is directly linked with the most unstable azimuthal mode of the primary rings predicted by Widnall *et al.* [24].

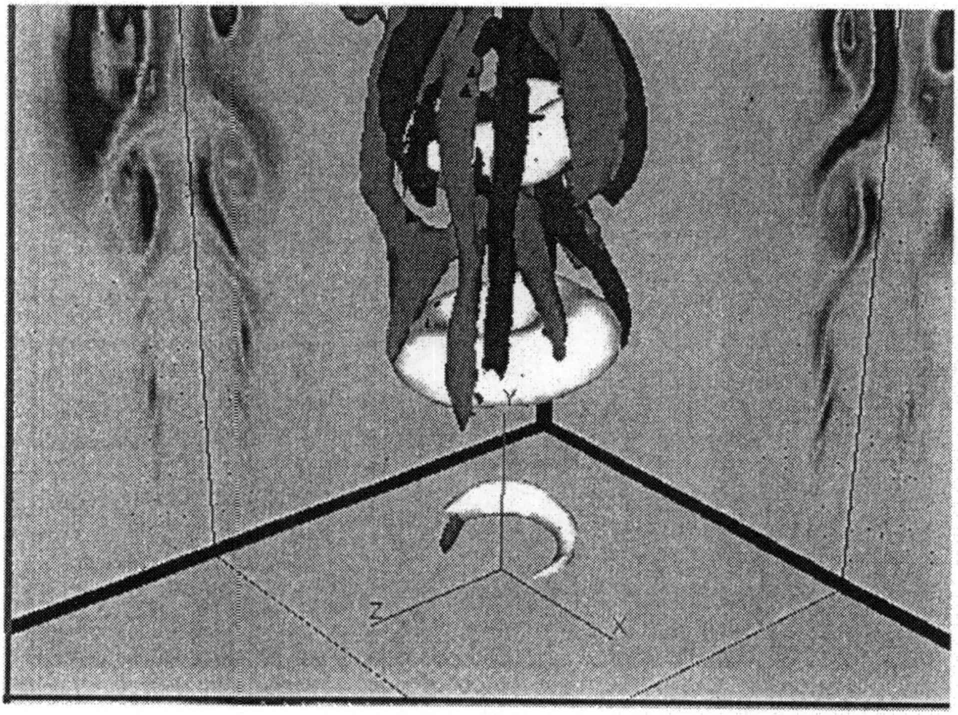

Figure 2: Jet with varicose mode excitation. Black and grey: positive and negative longitudinal vorticity isosurfaces corresponding to $\omega_z = \pm 1.2 W_o/D$. YZ and XZ cross-sections (through the jet axis) of the longitudinal vorticity component ($min = -4.W_o/D$; $max = +4.W_o/D$).

3.2.2 Helical excitation

The next excitation is designed to trigger the first helical mode by imposing the following inflow velocity profile:

$$W(r) + \epsilon \, W_o \, sin\left(\Theta - 2\pi \frac{Str_D W_o}{D} t\right) \frac{r}{D/2} \tag{3}$$

where Θ stands for the azimuthal angle.

The response of the jet indeed consists in the development of an helical coherent vortex structure (Figure 3). This is in concordance with Kusek *et al.* [7] who experimentally observed the helical mode development with an appropriate inflow excitation. The signature of the helical excitation on the statistics consists in an increase of the potential core length as compared with the natural case, and a reduction of the spreading rate. For the present jet (no swirl), the velocity circulation on a circular contour of large radius, contained in a plane perpendicular to the

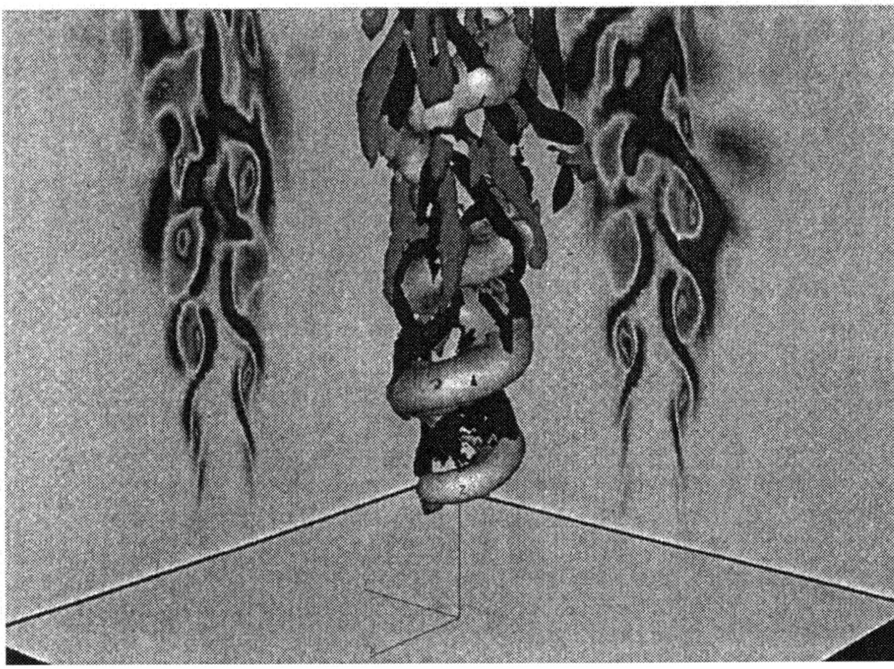

Figure 3: Jet with helical excitation: same vizualisation as figure 2.

jet axis and centered on the later remains zero. This implies that the longitudinal vorticity flux through the surface limited by this contour is also zero. The present excitation gives rise to an helix structure which rotates in the anti-clockwise direction, when moving away from the nozzle: it is therefore associated with a negative longitudinal vorticity component. This generation of negative longitudinal vorticity has necessarily to be compensated by regions of positive longitudinal vorticity. Indeed, in the vicinity of the nozzle, Figure 3 shows the appearance of positive longitudinal vorticity on the helix border. However, further downstream both positive and negative longitudinal vortices do appear, but the former are more intense than the latter. At $Z = 4.5D$, the vorticity maximum within the positive vortices is $\approx 50\%$ of the vorticity (modulus) maximum within the helix, while it is only $\approx 25\%$ in the negative ones. Martin and Meiburg's [13] results, using vortex filament numerical techniques, display the same trend.

3.2.3 Alternate-pairing excitation

The excitation method is here based upon the same principle as previously described. The perturbation intensity is here $\epsilon = 5\%$. Its frequency is the same as before except that now half of the jet presents a speed excess, while a speed defect is imposed on the other half, and this alternatively. Note that this perturbation has a preferential direction, chosen along the Y axis. The resulting structures are analogous to Figure 1 except that the alternatively inclined vortex rings now appear from the nozzle (see Figure 4). These inclined rings exhibit localized pairing and persist far downstream till $Z/D = 10$.

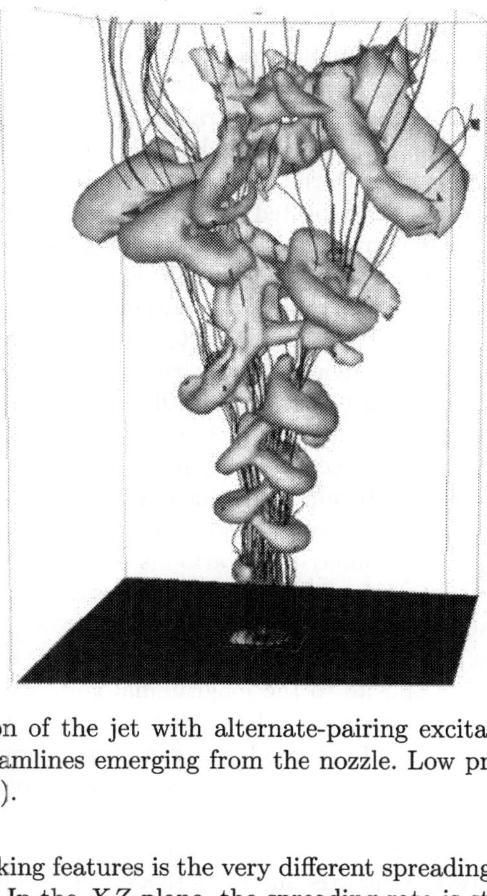

Figure 4: Bifurcation of the jet with alternate-pairing excitation. Instantaneous vizualisation of streamlines emerging from the nozzle. Low pressure isosurface in grey ($P = 25\% P_{min}$).

One of the striking features is the very different spreading rates in the X and in the Y directions. In the XZ plane, the spreading rate is strongly increased as compared to the natural jet case: it reaches $50^o - 55^o$ at $Z/D = 6$. Conversely, it is close to zero (3° at $Z/D = 6$) in the YZ plane. Note that the present jet exhibits strong similarities with the "bifurcating" jet of Lee and Reynolds [9] (see also Parekh, Leonard and Reynolds [18]). They have experimentally showed, that a properly-combined axial and helical excitations can cause a turbulent round jet to split into two distinct jets. Such a bifurcation is indeed observed here (see Figure 4). The streamlines originally concentrated close to the nozzle tend to clearly separate for $Z/D > 4$. Furthermore, the alternatively inclined vortex-rings seem to separate and move away from the jet centerline to form a Y-shaped pattern. More recently, the experiment performed by Longmire and Duong [12] has displayed similar vortex topology by using a specially designed nozzle made of two half nozzles. One of the important technological application of this peculiar excitation resides in the ability to polarize the jet in a preferential direction.

4 Conclusion

With the aid of Large-Eddy Simulations based upon the *selective structure function model*, the statistical and topological behaviour of the round jet spatial evolution has been studied. For the simulated jet, we have chosen a velocity profile close to a top-hat shape. Due to numerical constraints, the simulated profile is however shallower than the experimentally measured profiles: here $R/\theta = 10$. Despite this difference, a good agreement with the experimental data is found for the mean and fluctuating axial velocities. As expected, the growth of the axisymmetric instability mode leads to axisymmetric vortices (rings). This preferential mode appears at a Strouhal number $Str_D = 0.35$ in good correspondance with the experimental predictions. In alternance with rings, we also observe helical vortices. Further downstream, the rings also give rise to a vortex structure consisting in alternate localized pairings between two consecutive vortex rings. We have designed a specific excitation of the jet aimed at triggering this "alternate-pairing" mode. The jet behaviour is dramatically modified in that case: in one radial plane, the jet expansion rate is greatly enhanced as compared to the natural jet, while in the perpendicular plane it is totally inhibited. One important technological application of this excitation lies in the ability to control the jet by polarization in a preferential direction. When axisymmetric excitation is applied, after vortex rings have formed, pairs of counter-rotating longitudinal vortices appear linked with primary rings and these create horizontal side jets. Counter-rotating longitudinal vortices are still present when a helicoidal excitation is imposed. As a consequence of Stokes theorem, those of sign opposite to the longitudinal vorticity associated with the helix vortex are however more intense. We have therefore showed that Large-Eddy Simulations techniques allow for a good representation of the great variety of coherent structures present in the near region of a turbulent round jet. In agreement with several experimental results, we have showed how a specific excitation (axisymmetric, alternate, helicoidal, etc) of the jet at the nozzle renders possible the control of the jet vortical structures and leads to a strong modification of its statistics (spreading rate, turbulence level, etc). The present work clearly demonstrates that a good knowledge of the vorticity topology is compulsory to design efficient turbulent flow control.

Acknowlegments : This work is supported by the CEA, EDF, FRAMATOME (fiche 4443).

References

[1] Abid, M. and Brachet, M.E. (1992) Mécanisme de génération des jets latéraux dans les jets axisymetriques forcés, *C.R. Acad. Sci. Paris.*

[2] Broze, G. and Hussain, F. (1996) Transition to chaos in a forced jet: intermittency, tangent bifurcations and hysteresis. *J. Fluid Mech.*, **311**, pp. 37–71.

[3] Comte, P., Fouillet, Y. and Lesieur, M., (1992) Simulation numérique des zones de mélange compressibles. *Revue scientifique et technique de la defense*, 3ème trimestre. *Phys. Fluids*, **6** (2), pp. 751–759.

[4] Comte, P., Lesieur, M., and Lamballais, E. (1992) Large- and small-scale stirring of vorticity an a passive scalar in a 3-D temporal mixing layer, *Phys Fluids A*, **4**, pp. 2761–2778.

[5] Crow, S.C. and Champagne, F.H. (1971) Orderly structure in jet turbulence. *J.Fluid Mech.*, **48**, pp. 547–591.

[6] Fouillet, Y. (1992) *Contribution à l'étude par expérimentation numérique des écoulements cisaillés libres. Effets de compressibilité.* PhD thesis. National Polytechnic Institute, Grenoble.

[7] Kusek, Corke and Reisenthel, Seeding of helical modes in the initial region of an axisymmetric jet (1990), *Experiments in Fluids*, **10**, pp. 116–124.

[8] Lasheras, J.C., Lecuona A. and Rodriguez, P. (1991) Three dimensionnal structure of the vorticity field in the near region of laminar co-flowing forced jets. In *The Global Geometry of Turbulence*, edited by J. Jimenez (Plenum Press, New-York).

[9] Lee, M., Reynolds, W.C. (1985) Bifurcating and blooming jets at high Reynolds number. *Fifth Symp. on Turbulent Shear Flows*, Ithaca, New York 1.7–1.12.

[10] Lesieur, M. and Métais, O. (1996) New trends in large-eddy simulations of turbulence. *Annu. Rev. Fluid. Mech.*, **28**, pp. 45–82.

[11] Liepmann, D. and Gharib, M. (1992) The role of streamwise vorticity in the near-field entrainment of round jets. *J. Fluid Mech.*, **245**, pp. 643–668.

[12] Longmire, E.K. and Duong, L.H. (1996) Bifurcating jets generated with stepped and sawtooth nozzles. *Phys. Fluids*, **8** (4), pp. 978–992.

[13] Martin, J.E. and Meiburg, E. (1991) Numerical investigation of three-dimensionally evolving jets subject to axisymmetric and azimuthal perturbations. *J. Fluid Mech.*, **230**, pp. 271–318.

[14] Métais, O. and Lesieur, M. (1992) Spectral large-eddy simulation of isotropic and stably stratified turbulence. *J. Fluid Mech.*, **239**, pp. 157–194.

[15] Michalke, A. and Hermann, G. (1982) On the inviscid instability of a circular jet with external flow. *J. Fluid Mech.*, **114**, pp. 343–359.

[16] Monkewitz, P.A. and Pfizenmaier, E. (1991) Mixing by 'side jets' in strongly forced and self-excited round jets. *Phys. Fluids A*, **3** (5), pp. 1356–1361.

[17] Petersen, R.A. (1978) Influence of wave dispersion on vortex pairing in a jet. *J. Fluid Mech.*, **89**, pp. 469–495.

[18] Parekh, D.E., Leonard, A. and Reynolds, W.C. (1988) Bifurcating jets at high Reynolds number. Air Force Office of Scientific Research contractor Rep n^{o}AF-F49620-84-K-0005 and n^{o}AF-F49620-86-K-0020.

[19] Pierrehumbert, R.T., and Widnall, S. E. (1982) The two- and three-dimensional instabilities of a spatially periodic shear layer, *J. Fluid Mech.*, **114**, pp. 59–82.

[20] Silveira-Neto, A., Grand, D. Métais, O. Lesieur, M. (1993) A numerical investigation of the coherent structures of turbulence behind a backward-facing step. *J. Fluid Mech.*, **256**, pp. 1–55.

[21] Urbin, G., (1998) Etude numérique par simulation des grandes échelles de la transition à la turbulence dans les jets. PhD thesis. National Polytechnic Institute, Grenoble.

[22] Urbin, G. and Métais, O. (1997) Large-eddy simulation of three-dimensional spatially-developing round jets, in *Direct and Large-Eddy Simulation II*, J.P. Chollet, L. Kleiser and P.R. Voke eds., Kluwer Academic Publishers, pp. 35–46.

[23] Urbin, G., Brun, C. and Métais, O. (1997) Large-eddy simulations of three-dimensional spatially evolving roud jets, 11th symposium on Turbulent Shear Flows, Grenoble, September 8-11, pp. 25–23/25–28.

[24] Widnall, S.E., Bliss, D.B. and Tsai, C. (1974) The instability of short waves on vortex ring. *J. Fluid Mech.*, **66** (1), pp. 35–47.

[25] Zaman, K.B.M.Q., Reeder, M.F. and Samimy, M. (1994) Control of an axisymmetric jet using vortex generators. *Phys. Fluids*, **6** (2), pp. 778–794.

3 Observational aspects

Trends in Mathematics, © 1999 Birkhäuser Verlag Basel/Switzerland

The lack of a simple paradigm in fully developed turbulence: characteristics of local concentrations of vorticity and Reynolds stress in turbulent shear flows[1]

J. G. Brasseur

I. Issues

Fully developed turbulence is characterized by regions of concentrated temporally coherent activity in which dynamically relevant turbulence variables are concentrated. Concentrations can be identified in all turbulence variables at all turbulence scales, and are often perceived to play a major, even dominant, role in the kinematic and dynamic description of turbulent flows[2]. The highest intensity values of enstrophy, for example, are observed to lie within structures that resemble classical models of vortex tubes and sheets (many studies). Implicitly, and sometimes explicitly, we assume that these vortical structures dominate the dynamics of turbulence evolution, at least at the small scales. Similarly, the dynamical role of short-lived quadrant two (Q2) Reynolds shear stress events in the near-wall region of turbulent boundary layers, as associated with the instability and breakdown of horseshoe-shaped vortices and internal shear layers, is often implicitly transferred to the description of fully turbulent shear flows in general. Underlying this discussion is the question "to what extent do paradigms such as vortex sheets, vortex tubes and Q2 Reynolds shear stress events characterize the dynamical evolution of fully developed turbulent flows?"

Dr. Lin and I (Lin 1993; Brasseur & Lin 1991, 1995) explored the relative importance of concentrations of dynamically relevant turbulence variables to the kinematics and dynamics of fully developed turbulence under the influence of mean shear, focusing on vorticity, strain-rate and Reynolds shear stress fluctuations. To avoid the complications arising from inhomogeneity and the local instability and breakdown of friction-dominated layers near a surface, we examined homogeneous turbulent shear flow at nondimensional mean shear-rates comparable to those found in free shear layers and inertial law-of-the-wall layers of wall-bounded shear flows ($S^* = Sq^2/\epsilon \sim 6-9$, where S is mean shear, $1/2q^2$ is the turbulent kinetic energy and ϵ is the dissipation-rate). Limitations in the length of this dis-

[1]The analysis underlying this discussion is mostly contained in the Ph.D. thesis of Dr. Wenquei Lin (1993), currently of Delphi Harrison Thermal Systems, 200 Upper Mountain Road, Lockport, NY 14094 USA.

[2]"Dynamics" implies a significant contribution to the forces, accelerations and time evolution of the turbulence from Newton's second law, while "kinematics" implies a description of structure without explicit reference to Newton's law.

cussion allow only a brief description of primary results which are yet to be fully presented in the open literature.

A question similar to the one we ask above was posed also by Tsinober at a Euromech Colloquium in June 1997 in a discussion entitled "Is concentrated vorticity that important?" (Tsinober 1997 and references therein). Tsinober makes the argument that the interstitial regions outside concentrations of vorticity play at least as important a dynamical role in turbulence evolution as the concentrations themselves. We are lead in our work to the same conclusion, but from a somewhat different perspective. We recognize, for example, the important distinction between a fully developed turbulence with a wide range of interacting turbulence scales, and a transitional turbulence which has a very different dynamics and which fits better the simplified view of turbulence dynamics dominated by coherent concentrations of turbulence variables.

A. Transitioning Turbulence vs. Fully Developed Turbulence

To understand the question we pose above, it is necessary to appreciate the role that observations of transitional turbulence have played in the description of the dynamics and evolution of fully developed turbulence. "Fully developed" here implies a turbulence which has progressed through transition from a nonsteady unstable laminar state, through the growth of harmonics and three dimensional instability, and into a chaotic state of approximate equilibrium where the full range of turbulence length and time scales which can be supported at the Reynolds number of the flow exist and continuously interact. In this fully turbulent state, the term "large scales" applies to the range of turbulence length scales which contain most of the turbulent kinetic energy and Reynolds stress fluctuations, and "small scales" to the range of scales which contain most of the enstrophy (vorticity) and dissipation-rate (strain-rate) fluctuations[3]. In a Fourier spectral description of 3-D quasi-homogeneous turbulence, the large-scale motions center on a peak in the energy spectrum $E(k)$ with wavenumber $k_L \sim 1/L$, where L is an integral scale, and the small-scale motions center on a peak in the dissipation-rate spectrum $D(k)$ with wavenumber $k_D \sim 1/\eta$, where η is the Kolmogorov microscale. Fully developed turbulence is characterized by a separation in scale in which vorticity and strain-rate variance is concentrated as length scales smaller than the length scales which contain most of the turbulent energy and Reynolds stress. At high Reynolds numbers these two scales can be orders of magnitude apart.

By contrast, a transitional turbulence has no clear separation between energy-dominated and vorticity-dominated scales. The structures in which vorticity is concentrated are also the structures which contain most of the turbulent energy. The nonlinear dynamics of instability and breakdown is a process that creates *new* structure at scales different from existing structure. Visually, larger-scale concentrations of vorticity (tubes and sheets) are observed break into smaller-scale vortical structures, creating new structure at smaller scales. Spectrally, one ob-

[3] "most" might mean 85–90%, for example.

serves the formation of *new* discrete higher-wavenumber Fourier modes from nonlinear triadic interactions between pairs of existing (i.e. energetic) Fourier modes (Brasseur & Wei 1994). As nonlinear instability dominates the transitional process, discrete spectral modes and a relatively narrow range of fluctuations give way to a broad-band spectrum with a wide distribution of turbulence fluctuations.

An important point is that, whereas in a transitioning turbulence "large" and "small" scales overlap, in fully developed turbulence at higher Reynolds numbers these motions separate in scale; consequently, both the physical space structure and the dynamical interactions among those structures are very different. In the transitional state the vortex elements are also the energy-containing motions, and the breakdown of these energetic vortex structures is a non-equilibrium process leading to energy being deposited into new smaller-scale structures. By contrast, whereas the fully developed state contains small-scale structures with high levels of vorticity fluctuation, the highest levels of vorticity contain the lowest levels of kinetic energy. Thus, whereas the popular description of the turbulent "energy cascade" draws from observations of transitioning turbulence with larger structures breaking into new smaller structures, in fully developed turbulence the energy cascade takes place via quasi-equilibrium energy exchanges among large numbers of interacting spectral modes, and therefore a wide range of interacting motions, of roughly the same scale and already populated with energy. Given that the vortex tubes observed with isosurfaces live at the smallest and least energetic dynamical scales in fully developed turbulence, and therefore do not directly participate in transfer of energy out of the large scales, describing turbulence evolution in terms of dynamical interactions among vortex tubes and sheets is akin to a describing a happy chubby dog as being wagged by its skinny vibrating tail.

B. Visual Description of Turbulence Structure

All visual techniques for evaluating turbulence structure are necessarily subjective in nature. Perhaps the most useful visual technique applied to three dimensional databases is the imaging of single isosurface levels. With this approach, however, the subjectivity inherent in conclusions drawn from visual observation is compounded by the use of single solid isosurfaces to analyze the structure of a space-filling turbulence variable which fluctuates over a wide range of concentration levels. Because the isosurface level is chosen to minimize visual clutter, the resulting threshold is generally several standard deviations above the mean. Consequently, single isosurfaces select only the highest peaks in local concentrations of activity, with little information on the spatial extent of the concentration or the structure of the lower level fluctuations surrounding the isosurface.

There is a natural tendency, when examining the topology of the highest concentration levels with an isosurface, to assign to the observed structure a high level of kinematical and dynamical significance and, implicitly, to regard the lower-level fluctuations as of lesser importance to turbulence evolution. This tendency is especially strong with the vorticity field which, at high intensity levels, is observed to contain relatively well-defined tube- and sheet-like structures. The temptation to apply the dynamics of instability and breakdown of well-defined interacting vortex structures in transitioning turbulence as paradigms for the dynamical description of fully developed turbulence seems irresistible. As discussed above, however, there are fundamental differences between transitioning and fully developed turbulence which make such comparisons highly questionable.

C. Statistical Description of Turbulence Structure

Distribution functions emphasize very different aspects of turbulence structure than do visual observations of isosurfaces. Pdfs show that turbulence variables fluctuate over a wide range of fluctuation levels in fully developed turbulence. Isosurface-defined structures live at several standard deviations above the mean – in the tails of pdfs. In contrast, weighted pdfs (e.g., $\omega^2 P(\omega^2)$ and $uvP(uv)$) generally peak at about 1 standard deviation, indicating that the greatest contributions to the variance come from relatively low-level turbulence fluctuations – those which live outside the commonly visualized concentrations of activity (many experimental and numerical studies; explicitly discussed in Brasseur & Lin 1991). Thus, whereas isosurfaces visually imply dynamical relevance of the highest magnitude fluctuations in vorticity, for example, distribution functions suggest that the vast regions surrounding the vortex tubes and sheets contain most of the vorticity variance, and therefore may contribute as much or more to the dynamical evolution of fully developed turbulence (as argued by Tsinober 1997).

Scale decompositions such as Fourier demonstrate that quasi-homogeneous fully developed turbulence lives over a very wide range of spatial and temporal scales characterized roughly by the separation between the integral, Taylor and Kolmogorov scales. As discussed above, statistically most of the energy lives within a narrow band of Fourier spectral modes with peak wavenumber that scales on an inverse integral scale (L), whereas the vorticity and strain-rate/dissipation-rate fluctuations are over a broad range of spectral scales with a peak that scales on the inverse Kolmogorov scale (η). Thus, one expects the highest level vorticity and strain-rate fluctuations, those that live at the tails of the pdfs and visualized within enstrophy isosurfaces, to be found predominantly near the peak in the dissipation spectrum, and therefore to live within structures with a dimension that scales on η (as suggested by Jimenez et al. 1993 and several studies since). The highest level Reynolds stress fluctuations, on the other hand, likely concentrate within structures that scale on L in size. Filtered direct numerical simulations support this expectation (e.g., Brasseur & Wang 1992, 1995; She et al. 1993, Yeung et al. 1995).

It follows from this reasoning that, in high Reynolds number fully developed turbulence, the large integral-scale structures contain much of the lowest-level vorticity/dissipation-rate fluctuations, whereas the small Kolmogorov-scale structures, the vortex tubes and sheets, contain the lowest levels of energy and Reynolds stress. Where do the fluctuations that contribute mostly to the variance, those around 1 standard deviation, live, and what dynamics are associated with these low-level volume-filling fluctuations?

D. Analysis

To address some of the issues raised above, we identify kinematic and dynamic characteristics of volumes in which turbulent fluctuations are concentrated. Unlike isosurfacing, these "structures" are extracted objectively to include the lower level fluctuations surrounding the peaks which may be defined as part of the region of concentration. Of particular interest are the relative roles of low, medium and high-level concentrations in the kinematics and dynamics of fully turbulent shear flow. We use fluctuating Reynolds shear stress to describe large scale structure in this shear flow, fluctuating enstrophy and strain-rate as primary variables at the small scales, and the vectorial relationships between vorticity and strain-rate in the vortex stretching term as the primary dynamical interaction.

II. Methodology and approach

A. "Structures" of Concentrated Turbulence Fluctuations

Whereas traditional distribution functions and isosurfaces are applied in our studies, we also depart from standard pdf-based analysis by systematically extracting individual concentrations in given turbulence variable from our data and identifying their individual spatial locations for subsequent statistical and visual analysis. We define "a structure" as the entire volume surrounding a peak in magnitude of a fluctuating turbulence variable which is concentrated in the vicinity of the peak, at fixed time. To understand the approach, imagine a three-dimensional field of turbulence fluctuations as qualitatively similar to two-dimensional mountains of varying height, expanse and connectivity rising from a rough terrain of lower level crags and cervices. "Mountains" might be localized 3-D concentrations of vorticity within an array on interconnected vortex tubes and sheets surrounded by 3-D "crags and crevices" of lower-level vorticity fluctuation. Pdfs select from pieces of each mountain to quantify global characteristics of "mountain segments" together as a group (as applied, for example, by Jimenez et al.(1993) to analyze high intensity portions of vortex tubes). Here we analyze entire mountains (e.g., an entire vortex), including both the high intensity peak and the lower intensity values surrounding the peak which are part of the contiguous, presumably coherent, structure. Furthermore, the structures are extracted individually from the dataset, beginning with the "mountain" surrounding the highest peak, then moving progressively to the next lowest peak, and so on. The beauty of the method is the flexibility it provides to both visualize and quantify the characteristics of either individual structures or groups of structures as desired. Once the 3-D locations of a sufficiently large number of individual structures have been recorded, the collection may be rearranged at will using whatever conditions one wishes to impose for visual and/or quantitative analysis.

A robust algorithm was developed for systematic extraction of 3-D structures within an arbitrary fluctuating variable, ordered initially by peak value in each structure (Lin 1993). Like all structure-based studies, subjectivity arises in the definition of the "boundary" of a "structure." Here the boundary is defined by identifying the surface that defines the "base" of each 3-D "mountain" of high concentration of the turbulence variable. Because the "mountain base" is not necessarily at the same height all around the base, the boundary of an extracted structure is *not* an isosurface, in general. Our method extracts the boundary of a structure according to the topology of concentration within the structure rather than an via an arbitrary threshold value.

A difficulty is that one often encounters (especially in the vorticity field) what appear to be individual structures that interconnect. Thus, to define the boundary of concentration in 3D space, one must distinguish lower level concentrations of activity *within* a primary structure from concentrations of activity that are either completely new structures, or are "substructures" of a larger interconnected structure. Furthermore, algorithm must be automated and be successful a large percentage of the time. All requirements were successfully met (after much effort).

B. The Shear Flow Calculations

To learn how mean shear changes the structure and dynamics of the more intense structures within the vorticity, strain-rate and Reynolds shear stress fields we analyzed two DNS datasets of homogeneous turbulent shear flow, one created by Rogers (1986) and one by Lin (1993). Both used the Rogallo pseudo-spectral algorithm in a computational box with aspect ratio $2:1:1$ in $x:y:z$ to accommodate elongation in the mean flow direction (x), and both simulations ultimately were carried out on a 128^3 grid. The two data sets differed primarily in the initial conditions and corresponding evolution. Rogers switched on shear at the beginning of the simulation with Gaussian initial conditions so that shear acted during the transition from a random field of velocity fluctuations to fully developed shear turbulence. In the Lin simulation isotropic decaying turbulence was first created (on a $256 \times 128 \times 128$ grid) before shear was switched on (on a 128^3 grid) so as to study the transition to a shear-dominated anisotropic structure. During the Lin simulation period, $St = 0^+ - 4$, S^* increased from 2.6 to 10. Mean shear causes the vortex elements to rotate and align, with average orientation angle (from x) decreasing with time (Brasseur & Wang 1995). The turbulence is found to move towards a state of asymptotic growth after an initial transition period between $St \sim 0-1$. The data were analyzed in the isotropic and asymptotic states at $S^* = 10, R_\lambda = 65$ with the Lin calculation, and at $S* = 9, R_\lambda = 74$ with the Rogers calculation ($R_\lambda = \lambda_{11,1}q/\nu$).

C. Analyses

Four classes of study were carried out using the DNS shear-flow data described above. Firstly, the global statistical characteristics of different turbulence variables were quantified using classical statistical measures such as lower order moments, pdfs and conditional pdfs as the reference for a second class of statistical studies based on groups of structures extracted from the datasets. The third class of studies was more subjective, contrasting visually the topologies and spatial relationships among individual structures, and contrasting our extracted structures with traditional isosurfaces. The three principle dimensions of extracted structures were quantified. In the fourth study visual analysis was combined with quantitative analysis by following in time the development of a single hairpin vortex (the paradigm vortex structure in shear flow turbulence) from the initial isotropic state while analyzing concurrently the statistical structure of the vortex as it developed and evolved.

A number of variables were analyzed to obtain information both on the kinematic structure and on the dynamic evolution of shear turbulence. Kinematic analysis was carried out with fluctuating vorticity and strain-rate (small-scale variables) and Reynolds shear stress (large scales), including the contributions to the variances, and (at the small scales) contributions to anisotropy and non-Gaussianity. Study of dynamics was confined to the production terms in the fluctuating enstrophy and strain-rate variance equations, $\mathbf{P}_{\omega^2} = 2\omega_i s_{ij} \omega_j$ and $\mathbf{P}_{s^2} = -s_{ij}s_{jk}s_{ki} - 1/4\omega_i s_{ij}\omega_j$, respectively.

Of particular dynamical interest are the vectorial relationships between the vorticity vector and strain-rate tensor in principle coordinates. A conclusion from this study is that two classes of strain-rate fluctuations are relevant to the production of vorticity. We call these "passive" and "active" strain-rate fluctuations. Passive strain-rate is the contribution to s_{ij} that is kinematically induced by the vorticity distribution within vortex tubes and sheets. The structure of passive strain-rate is such that the second principle eigenvector of s_{ij} is nearly aligned with the local vorticity vector, but does not contribute significantly to enstrophy production. Active s_{ij} are those fluctuations which do contribute significantly to $\mathbf{P}_{\omega^2}$. A detailed statistical analysis of the alignment between principle strain-rate components and the fluctuating vorticity vector and relative active to passive strain-rate.

To quantify the principle dimensions of the extracted structures we applied a technique based on moment invariants of a density function defined as 1 and 0 interior and exterior to the structure, respectively. The method, borrowed from the pattern-recognition literature, optimally surrounds a structure with an ellipsoid of minimum volume. The three principle diameters of the ellipsoid then define the principle dimensions of the structure. The ratios of these dimensions were used to evaluate tube- vs. sheet-like topology of vorticity, strain-rate and Reynolds shear stress concentrations.

III. Brief review of results

A. Relative Contribution of Structures to Variances and Pdfs

Higher intensity structures of the enstrophy field were found to contain 3-5 substructures, on average, whereas lower intensity enstrophy structures and Reynolds stress structures contained 1-2 substructures, on average. Whereas the high magnitude fluctuations which live in the tails of vorticity and strain-rate pdfs contribute very little both to the global volume and to the variance, because the extracted structures extend to lower level fluctuations, when ordered by intensity they contribute a much larger percentage to the global variance relative to the global volume (by a factor of 3-4). This was found to be a general result for vorticity, strain-rate and Reynolds shear stress structures. Both vorticity and strain-rate structures grew in size as a result of shear but contribute the same amount to the variance relative to volume.

B. Internal Structure of Concentrations

Defining $P(\omega^2)$ as the pdf of enstrophy fluctuations, we find that $\omega^2 P(\omega^2)$ peaks at about $1\times$ the global rms, indicating that the variance is dominated by relatively low-level vorticity fluctuations. The extracted high level structures may be thought of as "complete" concentrations of vorticity surrounding the high-level vortex tubes Jimenez et al. (1993) called "worms." Because of the lower-level fluctuations within the extracted structures, the structures contribute a great deal more to the variance than do the "worms.". However, although significant contribution from fluctuations in the range 1 rms are included in the extracted vorticity structures, a large percentage ($> 50\%$) of the variance is still not accounted for, implying that most of the vorticity resides in the spaces between the vortex structures.

The internal vorticity vector field within many of the higher magnitude vorticity structures were visually analyzed. It was observed subjectively that most structures which topologically are tube-like when observed using isosurfaces are, in reality, the legs of hairpin vortices. Whereas well-defined tube-like topology invariably implied straight or curved vortex tube internal structure, vortex sheets were found to be less well defined. A "vortex sheet" must, in principle, contain overlapping vorticity and strain-rate sheet-like structures. We found that sheet-like topology in the vorticity field often does not coincide with sheet-like topology in strain-rate and therefore is more akin to a "squashed vortex tube" than a vortex sheet. Whereas an axisymmetric vortex tube is surrounded symmetrically by high levels of passive strain-rate, the "squashed" vortex tube is surrounded by a non-axisymmetric distribution of passive strain-rate. Only high aspect ratio sheet-like topology (> 10) in concentrated vorticity was observed visually to approximate a local vortex sheet structure.

Reynolds shear stress structures were analyzed conditioned on the quadrant of the average uv fluctuations within the structure. The higher intensity uv structures where entirely dominated by negative uv fluctuations and were found to be the largest structures, suggesting major contributions to the production of turbulent kinetic energy. Although in this homogenous turbulence there was an even split between Q2 and Q4 structures, the smaller $-uv$ structures tended to be Q2 dominant. Interestingly, there were significantly less interstitial low-level fluctuations in Reynolds stress than in vorticity.

Not surprisingly, we find that the highest level vorticity structures are the strongest contributors to non-Gaussianity in the vorticity field. More interestingly, by removing the higher intensity vortex structures within groups and analyzing the deviatoric part of the vorticity-vorticity correlation tensor in the remaining field, we find that the anisotropy level of vorticity fluctuations progressively decreases with intensity level of the structure.

Ejection events with strong concentrations of negative uv fluctuations observed in the near-wall turbulent boundary layer have been shown to be kinematically associated with hairpin-shaped vortices created by the lifting and bursting of near-wall streaks and shear layers (see Robinson 1991 and references therein). In the low Reynolds number shear flow simulations we observe visually no apparent spatial correlation between our extracted uv and ω^2 events. We have argued that in fully developed turbulence this lack of kinematic, and therefore dynamic, relationship between small-scale vortex structures and large-scale turbulence-producing events is to be expected, especially at high Reynolds numbers. Our observations support the caution stated above, that interpretations obtained from transitional turbulence should not be transferred directly to the description of fully developed turbulence.

C. Topology of Enstrophy, Strain-rate and Reynolds Stress Structures

Using the density matrix technique mentioned above, the three principle dimensions L_1, L_2 and L_3 of each structure were quantified and evaluated. Joint pdfs of the ratios L_1/L_2 vs. L_2/L_3 yielded a general description of structure as "tube-like," "sheet-like," and "slab-like." Tube- and sheet-like topology was concentrated in the highest magnitude enstrophy structures. Low intensity enstrophy concentrations tended to "thick slabs," although aspect ratios as high as 4 were found. Although many structures within the highest intensity ω^2 group could be classified as "tube-like," with aspect ratios as high as 7, it was also the case that many could be classified as "thick slabs" with aspect ratios of order 2. Horseshoe-shaped vortex structures, however, likely accounted for many of the apparent slab-like structure found in the high-intensity group. The strain-rate structures tended to be lower in aspect ratio (< 4), equally split between "fat tubes" and "thick sheets." Lower intensity strain-rate structures and Reynolds shear stress structures have a decidedly "slab-like" topology with aspect ratios between 1.5 and 2.

D. Relative Contributions of "Active" and "Passive" Strain-rate

We studied small-scale dynamics of turbulent shear flows by analyzing the overlaps between vorticity and strain-rate structures, the structure of the fluctuating strain-rate tensor and its alignment with the vorticity field, and shear-induced changes in the evolution of vorticity and strain-rate production through turbulence-turbulence interactions. We found that mean shear creates a realignment of both the vorticity and strain-rate fields relative to mean strain-rate tensor so as to enhance the global alignment between the second eigenvalue of the fluctuating strain-rate with the vorticity vector while reducing the relative magnitude of the second eigenvalue itself. At the same time, shear reduces the relative numbers of concentrated strain-rate structures in irrotational regions while retaining the overlap between vorticity and strain-rate that existed in the isotropic state. From these and related results we argue that shear enhances "passive" strain-rate fluctuations. By contrast, "active" (vorticity producing) strain-rate fluctuations retain similar structure before and after shear.

We find that, although the second strain-rate eigenvalue has a tendency to aligned with the vorticity vector, both alignment with the second strain-rate eigenvalue and alignment with the first (extensional) eigenvalue contribute significantly to enstrophy production. The analysis indicates that a wide range of relative orientations between s_{ij} and ω_i contribute to the production of enstrophy fluctuations by vortex stretching and that the strong tendency for alignment with the second principle eigenvalue of s_{ij} is likely a consequence of kinematically induced "passive" strain-rate which is enhanced under the action of mean shear.

IV. Discussion

The use of DNS data has enhanced our understanding of turbulence structure and dynamics immeasurably. However, one thought that continually enters my mind as we sift through different DNS datasets using these and other analysis techniques, is that turbulence structure and dynamics is not simple. Whereas the search for simplification underlies much scientific research, time and time again we are faced with a complexity in turbulence that cannot be ignored, and we must recognize that to reduce turbulence to a few relatively simple paradigms is to deny turbulence its true character. The analyses described above suggest a structure and dynamics that defies a description based primarily on such simple paradigms as vortex tubes, vortex sheets and uv ejection events.

It appears that turbulent flows are simply *not* low-degree-of-freedom systems that respond in simple predictable ways to simple inputs. A forced turbulence does not respond as a semi-rigid body, in relatively straightforward and predictable ways that can be represented with relatively few independent degrees of freedom. Turbulence appears to respond to forcing more like a soft water-filled sponge with a very wide range of hole sizes, hole topologies, and varying material properties – pushes and pulls are absorbed within a tangled web of nonlinear interactions among a sea of structural elements embedded within a wide range of fluctuation levels and spatial scales. The sponge does, of course, respond, but its response is necessarily complex and involves strong couplings between the higher concentrations of material within the sponge and in the lower concentration levels within interstitial spaces. Both the higher and lower concentration levels live over a wide range of length scales and respond over a wide range of time scales, in a peculiar way so that the dynamical response at the largest scale motions is only weakly coupled to the dynamical interactions within the small scale motions.

This and many other studies point to the difficulty inherent in simplifying the structure and dynamics of fully developed turbulence to relatively simple structural paradigms which can be used to describe and predict the primary features of turbulence evolution. Whereas the highest intensity concentrations of vorticity and other fluctuating turbulence variables may have reasonably well-defined structure, it is perhaps too much of an oversimplification to build models of turbulence structure and dynamics based primarily on these paradigms. The lower level fluctuations surrounding the vortex tube- and sheet-like structures likely contribute at least as much to the structure and dynamics of turbulence evolution as do the structures themselves.

References

[1] Brasseur, J.G., Lin, W.-Q. 1991 Structure and statistics of intermittency in homogeneous turbulent shear flow. *Advances in Turbulence 3*, Springer-Verlag, Heidelberg: 3–12.

[2] Brasseur, J.G., Lin, Q. 1995 Dynamics of small-scale vorticity and strain-rate structures in homogeneous shear turbulence. *Proc. Tenth Symposium on Turbulent Shear Flows*: 3-19 P 3–24.

[3] Brasseur, J.G. Wang, Q. 1992 Structural evolution of homogeneous turbulence at different scales analyzed using 3D wavelet transforms. *Phys. Fluids A* **5**: 2538–2554.

[4] Brasseur, J.G., Wang, Q. 1995 Relationship between structure and scale in homogeneous isotropic and shear turbulence. *Proc. Tenth Symposium on Turbulent Shear Flows*: P1-91 P P1-96.

[5] Brasseur, J.G., Wei, C-H. 1994 Interscale dynamics and local isotropy in high Reynolds Number turbulence. *Phys. Fluids* **6**: 842–870.

[6] Jimenez, J., Wray, A.A., Saffman, P.G., Rogallo, R.S. 1993 The structure of intense vorticity in homogeneous isotropic turbulence. *J. Fluid Mech.***255**: 65–90.

[7] Lin, W-Q. 1993 Structural and dynamical characteristics of intermittent structures in homogeneous turbulent shear flow. Ph.D. Thesis, Department of Mechanical Engineering, Pennsylvania State University, University Park, PA.

[8] Robinson, S. 1991 Coherent motions in the turbulent boundary layer. *Ann. Rev. Fluid Mech.* **23**: 601–639.

[9] Rogers, M.M. 1986 The structure and modeling of the hydrodynamic and passive scalar fields in homogeneous turbulent shear flow. Ph.D. Thesis, Department of Mechanical Engineering, Stanford University, Stanford, CA.

[10] She, Z-S., Chen, S., Doolen, G., Kraichnan, R.H., Orszag, S.A. 1993 Reynolds number dependence of isotropic Navier-Stokes turbulence. *Phys. Rev. Lett.* **70**: 3251.

[11] Tsinober, A. 1997 Is concentrated vorticity that important? (or is it possible to represent turbulence as a collection of 'simple' objects?) Presented at Euromech Colloquium 364, Carry-le-Rouet, France, June 24–27.

[12] Yeung, P.K., Brasseur, J.G., Wang, Q. 1995 Dynamics of large-to-small scale couplings in forced turbulence: physical and Fourier-space views. *J. Fluid Mech.* **283**: 43–95.

Trends in Mathematics, © 1999 Birkhäuser Verlag Basel/Switzerland

Near-singular flow structure in small-scale turbulence

J. C. Vassilicos

1 Introduction

Two aspects of turbulence are of pivotal importance to both the fundamentals and the applications of turbulence: (i) the relation of dissipation rates to molecular viscosity and diffusivities and (ii) the dispersive properties of turbulence. Both these aspects involve averages over an ensemble of ordered and disordered flow patterns. A future theory of such averages may presumably require some understanding of the dissipative and dispersive properties of the particular classes of flow patterns that we may expect to find in turbulent flows. It is the object of this paper to give a summary overview of some research carried out in this preliminary direction: the understanding of the dissipative and dispersive properties of local flow structures, and by recourse to a choice of assumptions at this stage, the relation of these local properties to global statistics.

2 The physics of fractals and spirals

Various universality and self-similarity arguments lead to $k^{-5/3}$ high-wavenumber power spectra for turbulent scalar and velocity fields and experiments indeed confirm, at the very least, that these spectra scale as k^{-p} where $p < 2$ (see, for example, Maurer *et al.* 1994 and Sreenivasan 1991). As noted by Hunt & Vassilicos (1991), power spectra with a k^{-p} dependence on wavenumber and $p < 2$ imply the existence of near-singularities in the turbulence that are either cusp-like, spiral-like or fractal-like. In particular, black-or-white tracer fields with a well-defined grey interface between black and white regions and a high-wavenumber spectrum of the form k^{-p} with $p < 2$ must have near-singularities in the geometry of the interface that are either spiral-like or fractal-like. The dissipative properties of such scalar interfacial patterns are examples of dissipative properties of generic scalar near-singular flow structures that may be generated by either steady vortices in the case of spirals, or chaotic advection in the case of fractals (see Fung & Vassilicos 1991, Vassilicos & Fung 1995). Other examples are the dissipative properties of spiral vortex sheets which also have a well-defined interfacial structure corresponding to high gradients of velocity and can also have well-defined k^{-p} power-spectra over a large range of wavenumbers k. The study of the dissipative properties of spiral and fractal fields seems to be still in its infancy and may be a necessary prerequisite for the understanding of turbulence dissipative properties. Such a study is but one aspect of a broader field of study, the physics of fractals and spirals (Berry 1979, 1981, 1989, van den Berg 1994, Fleckinger *et al.* 1995, Gurbatov & Crighton 1995, Lapidus *et al.* 1996), also currently in its infancy.

Concerning the dissipative properties of spiral and fractal structures, the following broad principles can be drawn from the calculations of Vassilicos & Hunt (1991), Vassilicos (1995), Flohr & Vassilicos (1997), Angilella & Vassilicos (1998) and Angilella & Vassilicos (1999).

1. Dissipation is faster where the field is less autocorrelated in space. It is slower where the field is more autocorrelated in space.

2. On-off (black-or-white) fields with fractal or spiral interfacial structure are more autocorrelated in space for lower values of the fractal dimension (Kolmogorov capacity) and less autocorrelated in space for higher values of the fractal dimension of the interface. An increase in the degree to which an interface, whether spiral or fractal, is space-filling leads to a faster spatial decorrelation of the on-off scalar field.

3. A non-alternating field of delta functions on a fractal or spiral set of points on the 1-D axis is more autocorrelated in space for *higher* values of the fractal dimension of the set of points. It is less autocorrelated for *lower* values of this fractal dimension. This property of delta functions is in stark contrast with the corresponding property of on-off functions. The autocorrelation properties of non-alternating delta functions on a fractal or spiral curve in the 2-D plane or on a fractal or spiral curve or surface in 3-D space are not a trivial extension of the corresponding 1-D properties and remain the object of current research in the turbulence and mixing group in DAMTP.

4. From points 1 and 2 above it follows that on-off fields with well-defined fractal or spiral interfaces dissipate faster for more space-filling and slower for less space-filling interfaces. This qualitative conclusion can be made quantitative as follows: letting $\theta(\mathbf{x}, t)$ be the scalar field, initially sharply on-off (i.e. $\theta_0(\mathbf{x}) = \theta(\mathbf{x}, 0) = 0$ or 1) across a sharp fractal or spiral interface and taking averages over space (overbar),

$$\frac{\overline{\theta_0^2} - \overline{\theta^2}(t)}{\overline{\theta_0^2}} \sim (\eta(t)/L)^{E-D} \tag{1}$$

where $\eta(t)$ is the diffusive microscale at time t, L is an integral length-scale, E is the Euclidian dimension of the embedding space and D is the fractal dimension of the interface ($E - 1 \leq D < E$). The molecular diffusivity of the scalar θ being κ, the microscale $\eta(t)$ is equal to $\sqrt{\kappa t}$ when θ evolves under the action of diffusion alone,

$$\frac{\partial}{\partial t}\theta = \kappa \nabla^2 \theta, \tag{2}$$

but $\eta(t)$ can have a different dependence on time t when θ evolves under the combined action of both advection and diffusion,

$$\frac{\partial}{\partial t}\theta + \mathbf{u} \cdot \nabla \theta = \kappa \nabla^2 \theta. \tag{3}$$

In the case of equation (2) the interface is initially given a fractal or spiral structure with dimension D whereas in the case of equation (3) it is the velocity field $\mathbf{u}(\mathbf{x},t)$ that generates a fractal or spiral interface from an initially regular interfacial geometry. When $\mathbf{u}(\mathbf{x},t)$ is induced by a steady 2-D vortex and is purely azimuthal with a differential angular frequency $\Omega(r) = \Omega_0(r/L)^{-\alpha}$ where $\alpha > 1$, r is the radial distance to the centre of the vortex and Ω_0 is the angular frequency at $r = L$, then the interface adopts a spiral structure of dimension $D = \frac{2\alpha}{1+\alpha}$ at times $t \gg \Omega_0^{-1}$ and

$$\eta(t) = \Omega_0 t\sqrt{\kappa t} = \Omega_0 \kappa^{1/2} t^{3/2}. \tag{4}$$

In both cases of equations (2) and (3), the scaling (1) is valid as long as the fractal or spiral structure of the interface remains well-defined over a sufficient range of length-scales, that is for times t such that $\eta(t) \ll L$. When molecular diffusion acts alone without advection, $\eta(t) \ll L$ implies $t \ll \frac{L^2}{\kappa}$, but when molecular diffusion acts concurrently with the azimuthal advection of the steady vortex, $\eta(t) \ll L$ implies $t \ll \Omega_0^{-1} Pe^{1/3}$ where $Pe \equiv \frac{\Omega_0 L^2}{\kappa}$. At this stage we should specify that (1) with (4) is valid for $Pe \gg 1$.

The dissipation of scalar variance in a steady vortex is therefore accelerated by two different effects, one local and one global. The local effect is the effect of the local shear caused by the differential rotation which acts locally on the interface to enhance gradients and thereby accelerate dissipation (Moffatt & Kamkar 1983, Rhines & Young 1983). This effect leads to an increase of the local diffusive microscale $\eta(t)$ from $\sqrt{\kappa t}$ to $\Omega_0 t\sqrt{\kappa t}$ and thereby to a reduction of the length of time for which the spiral interface is well defined from $\Omega_0^{-1} Pe$ to $\Omega_0^{-1} Pe^{1/3}$. The global effect is that of the space-filling geometry of the spiral interface and is manifest in (1). The more space-filling the spiral interface, the larger the value of D and the faster the decrease of the scalar variance in time. The dissipation rate of scalar variance is $\chi \equiv -\frac{d}{dt}\overline{\theta^2}(t)$ and (1) implies

$$\chi \sim \frac{\overline{\theta_0^2}}{t}(\eta(t)/L)^{E-D} \tag{5}$$

in the limit $Pe \gg 1$, which means that the dependence of χ on κ is progressively weaker as $D \to E$ and χ does not depend on κ in the limit $Pe \gg 1$ only if $D = E$, i.e. only if the spiral is completely space-filling.

5. From points 1 and 3 above, it follows that non-alternating delta functions on a fractal or spiral set of points on the 1-D axis evolving under the action of diffusion alone (equation 2) dissipate *slower* for more space-filling geometries. Quantitatively, for an initial field $\theta_0(x) = \sum_i \delta(x - x_i)$ where the points x_i are those of a homogeneous fractal set with a fractal dimension D ($0 \le D < 1$) that is well-defined over a range of scales between η_0 and L ($\eta_0 << L$),

$$\frac{\overline{\theta^2}(t)}{\overline{\theta^2}(\eta_0^2/\nu)} \sim (\eta(t)/L)^{-1+D} \tag{6}$$

for $\eta_0^2/\nu << t << L^2/\nu$. The diffusive microscale $\eta(t)$ is equal to $\sqrt{\kappa t}$. (See Angilella & Vassilicos (1998) for a definition of 'homogeneous fractal' and for the 1-D spiral case which is qualitatively similar but quantitatively different.) It is instructive to contrast (6) with the corresponding result for *alternating* delta functions on a fractal or spiral (e.g. $x_i \sim i^{-a}$ with $a > 0$) set, i.e. $\theta_0(x) = \sum_i (-1)^i \delta(x - x_i)$, where dissipation of scalar variance is *faster* for more space-filling geometries. Quantitatively,

$$\frac{\overline{\theta^2}(t)}{\overline{\theta^2}(\eta_0^2/\nu)} \sim (\eta(t)/L)^{-1-D} \tag{7}$$

for $\eta_0^2/\nu << t << L^2/\nu$.

6. However, the dissipation of 2-D spiral vortex sheets is different. Lundgren (1982) has shown that $\omega(r, \phi, t) = \gamma(r)\delta(\phi - \Omega(r)t)$ is a long-time asymptotic solution of the 2-D Helmholtz equation

$$\frac{\partial}{\partial t}\omega + \mathbf{u} \cdot \nabla\omega = \nu\nabla^2\omega \tag{8}$$

when $\nu = 0$ and provided that $\Omega(r)$ is a monotonic decreasing function of r and $r\gamma(r) = 2\pi \frac{d}{dr} r^2 \Omega(r)$. The latter condition is a result of the coupling $(0, 0, \omega) = \nabla \times \mathbf{u}$ wich makes the difference between (8) and (3). Setting $\nu \neq 0$, and chosing $\Omega(r) = \Omega_0 (r/L)^{-\alpha}$ with $\alpha > 0$ so that the spiral's co-dimension is $D_0 = \frac{\alpha}{1+\alpha}$, the average enstrophy of this solution decays as follows:

$$\frac{\overline{\omega^2}(t)}{\Omega_0^2} \sim (\eta(t)/L)^{2-4D_0} \tag{9a}$$

for $D_0 \geq \frac{3}{4}$ and

$$\frac{\overline{\omega^2}(t)}{\Omega_0^2} \sim (\eta(t)/L)^{-1} \tag{9b}$$

for $D_0 \leq \frac{3}{4}$. The viscous microscale is given by (4) where κ should be replaced by ν, and equations (9) are valid for as long as $\eta(t) \ll L$, that is $t \ll \Omega_0^{-1} Re^{1/3}$ where $Re \equiv \frac{\Omega_0 L^2}{\nu}$. Lundgren's spiral vortex sheet solution of (8) is valid for large times in the sense that $\Omega_0^{-1} \ll t$, and our conclusions (9) are therefore valid in the limit $Re \gg 1$.

The dissipation of enstrophy is neither decelerated nor accelerated by the space-filling property of the spiral when $D_0 \leq 3/4$ and is in fact accelerated rather than decelerated by this space-filling property when $D_0 > 3/4$. This may be a surprising conclusion because the dissipation of non-alternating delta functions on a 1-D spiral or fractal set of points is decelerated by the space-filling property of the spiral (Angilella & Vassilicos 1998). For some understanding of this conclusion, it must be noted that the high-wavenumber energy spectrum of the Lundgren spiral vortex sheet considered here is $E(k) \sim k^{-2}$ for $D_0 \leq 3/4$ but $E(k) \sim k^{-5+4D_0}$ for

$D_0 \geq 3/4$ (see Angilella & Vassilicos 1999). Hence, the autocorrelation property of the Lundgren spiral vortex sheet does not depend on D_0 if $D_0 \leq 3/4$ but does depend on D_0 otherwise, and in fact in a way that makes this spiral vortex sheet *less* autocorrelated as D_0 increases. Hence, conclusions (9) are consistent with point 1.

3 Turbulent diffusion

Small-scale flow structures can also have important effects on turbulent diffusion. Fung *et al.* (1992), Fung & Vassilicos (1998), Malik & Vassilicos (1998) and Flohr & Vassilicos (1998) put forward a non-Markovian mechanism of two-particle dispersion (particles and fluid elements are synonymous here) based on the separating effect of straining flow structures. The picture suggested in these works is one of two particles travelling side by side for some time untill they meet a straining flow structure where they separate violently. They then continue moving together but at an increased distance from each other till they meet a second straining region of a larger scale and suddenly separate again even further. This is a picture of two-particle dispersion operating by bursts, that is intermittently when straining regions are encountered by the particle-pairs, and it is a picture that is qualitatively very different from that of stochastic models of two-particle dispersion (Pedrizzetti & Novikov 1994, Heppe 1998 and references therein). Stochastic models of two-particle dispersion are based on the assumption that relative accelerations are Markovian in the inertial time-range. Indeed they do not fundamentally require this assumption, but they are effectively useless without it. The ensuing Lagrangian velocities vary with time in a way similar to a Brownian motion with drift and are therefore very different in character from the Lagrangian velocities assumed in the picture of two-particle dispersion suggested by Fung *et al.* (1992) and Fung & Vassilicos (1998). Even though statistics are obtained by averaging over many particle trajectories in many realisations of the turbulence, the qualitative difference in the nature of the Lagrangian velocity processes and consequently in the nature of particle trajectories can have a significant impact on some of these statistics.

Two-particle statistics are important for Lagrangian calculations of concentration fluctuations and concentration covariances (Durbin 1980). One-particle statistics determine average concentrations but they do not interest us here because they are mostly affected by the large scales and not by small-scale flow structure. Here we focus indirectly on concentration fluctuations by effectively focusing on two-particle dispersion and in fact on the non-Markovian effect of realistic Eulerian turbulence fields which incorporate turbulent-like eddying, streaming and straining flow structures. How can we model the non-Markovian aspect of turbulent diffusion in a workable and realistic way including the Lagrangian effects of small-scale flow structures and incompressibility?

It is an answer to this question that Kinematic Simulation (KS) provides. KS is a Lagrangian model of turbulent diffusion that incorporates turbulent-like small-scale flow structures. Individual particle trajectories are calculated by integrating

$$\frac{d}{dt}\mathbf{x}(t) = \mathbf{u}_E(\mathbf{x}(t), t) \tag{10}$$

in individual realisations of an Eulerian turbulent-like velocity field $\mathbf{u}_E(\mathbf{x}, t)$ generated as follows:

$$\mathbf{u}_E(\mathbf{x}, t) = \sum_{n=1}^{N_k} [\mathbf{A}_n \times \hat{\mathbf{k}}_n \cos(\mathbf{k}_n \cdot \mathbf{x} + \omega_n t) + \mathbf{B}_n \times \hat{\mathbf{k}}_n \sin(\mathbf{k}_n \cdot \mathbf{x} + \omega_n t)], \tag{11}$$

where N_k is the number of modes in the simulation, $\hat{\mathbf{k}}_n$ is a random unit vector ($\mathbf{k}_n = k_n \hat{\mathbf{k}}_n$), and the directions and orientations of $\mathbf{A}_n$ and $\mathbf{B}_n$ are chosen randomly under the constraint that they be normal to $\hat{\mathbf{k}}_n$. This random choice of directions for the nth wavemode is independent of the random choice of directions for all other wavemodes. Note that the velocity field (11) is incompressible by construction. The positive amplitudes A_n and B_n of the vectors $\mathbf{A}_n$ and $\mathbf{B}_n$ are determined by the energy spectrum $E(k)$ which is the major input of this model. The frequencies ω_n in (11) determine the unsteadiness associated with wavemode n. For example, $\omega_n = \lambda\sqrt{k_n^3 E(k_n)}$ where λ is a dimensionless constant that may be expected to be of order 1. This velocity field is turbulent-like because it is incompressible and unsteady by construction, has straining, streaming and eddying structures in every single realisation of the flow and the energy spectrum can be specified in accordance with turbulence theory, e.g. $E(k) \sim k^{-5/3}$ in the inertial range with appropriate shapes in the dissipation and the energy-containing ranges.

Two-particle dispersion in such turbulent-like flows can be local provided that the energy spectrum is not too steep and that extreme values of λ are avoided (see Malik 1996, Fung & Vassilicos 1998, Flohr & Vassilicos 1998). Locality is the property that, in the inertial range, the dominant contribution to the turbulent diffusivity $\frac{d}{dt}\overline{\Delta^2}(t)$ (where $\overline{\Delta^2}(t)$ is the mean square distance between two particles) comes from "eddies" of size $(\overline{\Delta^2})^{1/2}(t)$. An implication of this property is that if $E(k) \sim k^{-p}$, then $\overline{\Delta^2}(t) \sim t^{\frac{4}{3-p}}$. For 3-D homogenous and isotropic turbulence with a well-defined $-5/3$ spectrum in the inertial range, KS leads to $\overline{\Delta^2}(t) \approx G_\Delta \epsilon t^3$ where ϵ is the kinetic energy dissipation rate per unit mass and the Richardson constant $G_\Delta \approx 0.07$. It should be stressed that this value of G_Δ is in reasonable agreement with experiment and is obtained by KS without tuning of parameters, unlike stochastic models of two-particle dispersion where the energy spectrum is also effectively an input but with the addition of at least one extra free parameter (see Pedrizzetti & Novikov 1994 and Heppe 1998 and references therein). Fung & Vassilicos (1998) show how the value of G_Δ is a decreasing function of the spatial density of straining flow structures in the turbulent-like flow of KS. They also demonstrate that 2-D KS turbulent-like flows have a fractal structure consisting of cats' eyes whithin cats' eyes, or equivalently figures of 8 within figures of 8. The figure of 8 is a structure with a straining region between two eddying regions.

Davila & Vassilicos (1998) count the number of straining regions in turbulent-like flows and find that they have a power-law dependence on the Kolmogorov microscale η, i.e. $N_s(\eta) \sim \eta^{-D_s}$. By varying the power p of the energy spectrum $E(k) \sim k^{-p}$ between $p = 1$ and $p = 3$ they find that $3 = p + D_s$ in 2-D and $9 = 3p + 2D_s$ in 3-D KS turbulent-like flows. Hence, both the time-dependence and the overall magnitude of two-particle dispersion $\overline{\Delta^2}(t)$ are influenced by the straining structure of the turbulence. These works point to a sharper statement of the locality property: in the inertial range, the dominant contribution to the turbulent diffusivity $\frac{d}{dt}\overline{\Delta^2}(t)$ comes from straining regions of size $(\overline{\Delta^2})^{1/2}(t)$; the straining regions are embedded in an overall fractal structure of straining regions characterised by $N_s(\eta) \sim \eta^{-D_s}$ and therefore straining regions exist with a variety of length-scales over the entire inertial range.

The basic assumption of KS as a Lagrangian model of turbulent diffusion is that it should be sufficient, for the calculation of Lagrangian (*not* Eulerian) statistics, to have a qualitatively (but not necessarily quantitatively) realistic representation of the geometrical character of turbulent trajectories which are generated by the eddying, straining and streaming flow structures of the turbulence. The flow structures in KS are not quantitatively precise, but they are turbulent-like because they are qualitatively the type of structures expected and observed in turbulent flows. Malik & Vassilicos (1998) and Heppe (1998) have compared with DNS their results on Lagrangian two-particle statistics obtained, respectively, with KS and stochastic models of two-particle dispersion that have been adjusted to incorporate finite Reynolds number effects. The DNS results on two-particle dispersion for statistically isotropic and stationary turbulence are those of Yeung (1994). Both approaches lead to good comparisons with most Lagrangian two-particle statistics obtained from DNS except for the flatness factor of Lagrangian velocities which is underpredicted by stochastic models by up to one order of magnitude with the most favourably adjusted input tuning constant. However, KS reproduces the DNS flatness factor of Lagrangian velocities within factors of 1.3 to 1.8 without tuning constants.

4 Scalar intermittency

Concerning assumptions on how local flow structures are related to global statistics, it may be interesting to close this paper by recalling Lundgren's (1982) assumption that a time-average over the history of one flow structure is equivalent to a spatial average over many such flow structures at different stages of their history. Using Lundgren's assumption on averaging and Segel's (1995) method to calculate the scaling of structure functions of all orders, we find that if a scalar field is advected in a 2-D sea of identical non-interacting axisymmetric vortices characterised by the same differential angular frequency $\Omega(r) = \Omega_0(r/L)^{-\alpha}$, the scaling of the structure functions $< \delta\theta^n(r) >$ is

$$< \delta\theta^n(r) > \sim r^{2(4-3D_0)} \tag{12}$$

in a range of scales r bounded from below by a microscale that depends on both κ and n. The co-dimension D_0 ($0 \leq D_0 < 1$) is that of the spiral structures generated by each one of these identical vortices. The scaling (12) is obtained by solving the advection-diffusion equation (3) but is reminiscent of the structure function results obtained by Vassilicos (1992) without solving this equation. Indeed, Vassilicos (1992) calculated the structure functions of on-off fields characterised by either spiral or fractal interfaces and found, as in (12), that the scaling exponents of these structure functions do not depend on the order n and are a function of D_0. The scaling (12) must therefore be seen as a consequence of the spiral interfacial structure generated by the action of the vortices on the scalar field.

Furthermore, it should be noted, with reference to equation (5), that the dissipation rate of this scalar field tends to 0 as $\kappa \to 0$ but at an increasingly slow rate as $D_0 \to 1$, that is as the spiral structures tend to their space-filling limit, in which limit the dissipation rate tends in fact to be independent of κ.

Acknowledgements: I am grateful for financial support of the work described here to the Royal Society, EPSRC, NERC, European Commission, British Council and Hong Kong Research Grant Council.

References

Angilella, J.R. & Vassilicos, J.C. (1998) Spectral, diffusive and convective properties of fractal and spiral fields. Physica D (to appear).

Angilella, J.R. & Vassilicos, J.C. (1999) Time-dependent geometry and energy distribution in a spiral vortex layer. In preparation.

van den Berg, M. 1994 Heat content and Brownian motion for some regions with a fractal boundary. Probab. Theory Related Fields **100**, 439.

Berry, M. V. (1979) Diffractals. Phys. A: Math. Gen. **12** (6), 781.

Berry, M. V. (1981) Diffractal echoes. Phys. A: Math. Gen. **14**, 3101.

Berry, M. V. (1989) Falling fractal flakes. Physica D **38**, 29.

Davila, J. & Vassilicos, J.C. (1998) Does Kolmogorov's $-5/3$ spectrum reflect a number of straining flow structures in the turbulence that scales like η^{-2}? In preparation.

Durbin, P.A. (1980) A stochastic model of two-particle dispersion and concentration fluctuations in homogeneous turbulence. J. Fluid Mech. **100**, 279.

Fleckinger, J., Levitin, M. & Vassiliev, D. 1995 Heat equation on the triadic von Koch snowflake: asymptotic and numerical analysis. Proc. London Math. Soc. (3) **71**, 372.

Flohr, P. & Vassilicos, J.C. 1997 Accelerated scalar dissipation in a vortex. J. Fluid Mech. **348**, 295.

Flohr, P. & Vassilicos, J.C. 1998 Scalar subgrid model with flow structure for large-eddy simulations of scalar variances. In preparation.

Fung, J.C.H., Hunt, J.C.R., Malik, N.A. & Perkins, R.J. (1992). 'Kinematic Simulations in turbulent flow generated by unsteady random Fourier modes'. J. Fluid Mech. **236**, 281.

Fung, J.C.H & Vassilicos, J.C. (1991) Fractal dimensions of lines in chaotic advection. Phys. Fluids **11**, 2725.

Fung, J.C.H. & Vassilicos, J.C. (1998) Two-particle dispersion in turbulent-like flow. Phys. Rev. E **57**, no. 2, 1677.

Gurbatov, S.N. & Crighton, D.G. 1995 The nonlinear decay of complex signals in dissipative media. Chaos **5** (3), 524.

Heppe, B.M.O. 1998 Generalized Langevin equations for relative turbulent dispersion. J. Fluid Mech. **357**, 167.

Hunt, J.C.R. & Vassilicos, J.C. (1991) Kolmogorov's contributions to the physical understanding of small-scale turbulence and recent developments. Proc. R. Soc. Lond. A **434**, 183.

Lapidus, M., Neuberger, J., Renka, R. & Griffith, C. (1996) Snowflake harmonics and computer graphics: numerical computation of spectra on fractal drums. Int. J. Bifurcation Chaos **6**, 1185.

Lundgren. T.S. 1982 Strained spiral vortex model for turbulent fine structure. Phys. Fluids **25**, 2193.

Malik, N.A. (1996) Structural diffusion in 2-D and 3-D random flows. In *Advances in Turbulence VI* (ed. S. Gavrilakis *et al.*), 619–620. Kluwer.

Malik, N.A. & Vassilicos, J.C. (1998) A Lagrangian model for turbulent disperion with turbulent-like flow structure: comparison with DNS for two-particle statistics. Submitted.

Maurer, J., Tabeling, P. & Zocchi, G. 1994 Statistics of turbulence between two counter-rotating disks in low temperature helium gas. Europhys. Lett. **26**, 31.

Moffatt, H.K. & Kamkar, H. (1983) The time-scale associated with flux expulsion. In *Stellar and Planetary Magnetism* (ed. A.M. Soward), pp. 91–97. Gordon & Breach.

Pedrizzetti, G. & Novikov, E.A. (1994). On Markov modelling of turbulence. J. Fluid Mech. **280**, 69.

Rhines, P.B. & Young, W.R. (1983) How rapidly is a passive scalar mixed within closed streamlines? J. Fluid Mech. **133**, 133.

Segel, D. 1995 The higher moments in the Lundgren model conform with Kolmogorov scaling. Phys. Fluids **7** (12), 3072.

Sreenivasan, K.R. (1991) On local isotropy of passive scalars in turbulent shear flows. Proc. R. Soc. Lond. A **434**, 165.

Vassilicos, J.C. (1992) The multispiral model of turbulence and intermittency. In *Topological Aspects of the Dynamics of Fluids and Plasmas* (ed. H.K. Moffatt *et al.*), 427-442. Kluwer.

Vassilicos, J.C. (1995) Anomalous diffusion of isolated flow singularities and of fractal or spiral structures. Phys. Rev. E **52**, R5753.

Vassilicos, J.C. & Fung, J.C.H. (1995) The self-similar topology of passive interfaces advected by two-dimensional turbulent-like flows. Phys. Fluids **7** (8), 1970.

Vassilicos, J.C. & Hunt, J.C.R. (1991) Fractal dimensions and spectra of interfaces with application to turbulence. Proc. R. Soc. Lond. A **435**, 505.

Yeung, P.K. (1994). Direct numerical simulation of two-particle relative diffusion in isotropic turbulence. Phys. Fluids **6** (10), 3416.

Trends in Mathematics, © 1999 Birkhäuser Verlag Basel/Switzerland

Some thoughts on similarity, the POD, and finite boundaries

W. K. George

1 Introduction

This paper will take several seemingly disconnected ideas and show how they might be related. In keeping with the spirit of the Monte Verita Colloquium, I have chosen to talk about the way in which I believe the world might work in the hope that it will stimulate my own thinking and that of others. Let me begin with the physical problem which has troubled me for the two past decades: *When does a laboratory experiment or computer simulation truly represent an acceptable approximation to a flow of infinite extent?*

One can, of course, quite reasonably ask: *Who cares?* Or equally: *Why should we care?* Both questions would presumably be followed by the observation that all real flows have finite boundaries. But the simple fact is that with the exception of a few confined flows (like channel and pipe flows), almost every bit of knowledge we have about the behavior of turbulence solutions to the Navier-Stokes equations *from the equations themselves* comes from flows in which there are no boundaries. Examples include the homogeneous flows of such interest to modellers, as well as the familiar jet, wake and plume flows which are very close to flows which occur naturally. These solutions, with their simple scaling laws and similarity solutions, can be of immense value in both validating and understanding experiments, numerical simulations and models. And the point of this paper: *departures from similarity help us understand when boundaries matter.* Without this understanding, we might build models and theories for turbulence based on phenomena which have nothing to do with the dynamics of the flow, but instead are imposed by us and the domain we have chosen – and, in fact, we may have done so already. The consequences for engineering applications are of enormous importance since we seldom have the luxury of modelling (in lab or computer) a flow of realistic extent.

So when do finite boundaries matter? Now with experiments one might think the answer is obvious, but it seldom is. I first confronted this problem in 1979 when I realized that accepted measured profiles did not satisfy even the simplest momentum conservation requirements for a jet in a quiescent environment. In George (1990a) I documented four of the failures of me and my students to properly understand the subtle effects of finite boundaries – the round buoyant plume, our own round jet, grid turbulence through a contraction and free shear layers in a co-flowing stream. All of the experiments described deviated from the expected similarity results in subtle, but important, ways because of the unsuspected influence of boundaries. Since writing that article I have encountered a host of other flows where the finite boundary limits of the flow have made their presence felt, some of which I shall discuss below.

How does one *prove* the finite boundaries are important? The usual approach for the more difficult engineering problems, either computationally or experimentally, is to pretend the boundary conditions are not too important since the limitations of memory or space may leave little alternative. Sometimes an attempt is made to double the box size or move out the experimental boundaries, but more often not. Numerical studies may seek an experiment to "validate" the computation, and it is usually possible to find an experiment which does so. One surprising thing about the afore-mentioned jet measurements was that most of the experiments agreed with each other, but almost all of them were carried out in the same size room with even the same size jet. So agreement with experiment, in and of itself, should not be too satisfying. In fact, the whole point of the George (1990a) paper was that an experiment may be assumed valid *ONLY* when the data satisfy the governing equations and boundary conditions which are *ASSUMED* to apply. If they do not, then either the measurements are wrong, or the flow is not the flow it is believed to be.

The problem is a little more obscure for the the DNS and many theoreticians: Their results *always* satisfy the governing equations and imposed boundary conditions since they obtain solutions by solving them. And they have avoided the problems of the experimentalist by simply *defining* their world to be periodic (*cf* Frisch 1995, Doering and Gibbon 1996). Now while this may self-consistent, there is no assurance that these periodic results bear any resemblence to the non-periodic world around us. So how do they decide whether these results mimic reality? They can, of course, compare to experiments which are themselves done in finite boxes. Even so, the results are not always comforting. For example, I was surprised to learn some years ago that the DNS isotropic turbulence simulators usually throw away all the data prior to the peak in the derivative skewness since the rapid rise is believed to be part of the starting transient. Now if this is true, then all of the wind tunnel experiments (with a single exception) are invalid, since in all of the classic experiments the skewness continues to rise down the tunnel. In fact, if one looks at the eddy turn-over times and calculates an equivalent tunnel length for when the simulators think the data is good, this is 2 to 4 times the longest tunnels in the world. Now interestingly, the main reason experimenters don't build longer tunnels is because they believe the turbulence to be influenced by the walls when the integral scale grows to more than about a tenth of the tunnel width. Yet this is the ratio of integral scale to box size (or its spectral equivalent) at which many simulators *begin* their calculations.

Now let me make it clear that I am not questioning the validity of the DNS and LES results for periodic domains, only the degree to which they model turbulence in an infinite domain. Or put another way, my concern is about how much the results are determined by the non-linear dynamics as opposed to being dictated by the boundary conditions. And lest the experimentalists feel left out, I have exactly the same concern about experiments. Unfortunately, unlike the jet experiments cited earlier, none of the above examples provide indisputable proof (at least to some) that the finite domain adversely affects experiments and the simulations. At most they offer a clue. Nonetheless, our knowledge of even potential theory where the entire solution everywhere is determined by the boundaries should at least make us cautious. We have no corresponding theorems for turbulent flows, but certainly have no reason to discard boundaries as being important.

2 The role of similarity solutions

The question then remains: How does one ever prove that boundaries are not important in a particular simulation or experiment? This is where similarity solutions are the most useful, since they prescribe exactly how the flow must evolve (at least statistically) for a particular set of boundary conditions (usually homogeneous in some sense). If the flow that is being modelled in the computer or laboratory is supposed to be similar, then it is relatively easy to conclude whether or not the boundary conditions are affecting it. So the big question is whether the flow is supposed to be similar?

Of course, it is important to build the similarity theory correctly. I addressed this subject in some detail in George (1989a) where I argued that the traditional self-preservation approach to turbulence used in most texts was fundamentally wrong. I also argued for a general approach, applicable to all equations, in which each dependent variable is allowed to have its own scale. These scales, and the relations among them, are then in turn determined by asking whether the equations admit to solutions in which all relevant terms evolve together. Sometimes they do, and sometimes they do not. This is, of course, the way classical *non-turbulence* similarity theory has always been done, but somewhere along the way the turbulence community got lost. Two surprising results of doing things correctly are that most flows require more than one velocity scale *and* retain a dependence on source conditions.

In that same article (George 1989a) I tried to codify my own *beliefs* (which certainly are not unique) into two formal conjectures:

- *Conjecture I*: If the equations of motion, boundary and initial conditions admit to similarity solutions, then the flow will *always* asymptotically behave in this manner.

- *Conjecture II*: If the equations, boundary and initial conditions governing the flow do not admit to similarity solutions, the flow will adjust itself as closely as possible to a state of full similarity.

Note that Conjecture I, if ever proven, is at least one manifestation of the long-sought-after uniqueness theorem for turbulent flows. And Conjecture II is the basis for all of the local similarity theories we hold so dear (like K41, etc.).

There are numerous recent examples that are consistent with Conjecture I; *eg* George 1992 (isotropic decay), George 1990 (isotropic scalar decay), George and

Gibson 1992 (homogeneous shear flow), Moser *et al* 1996 (time-dependent wake), Ewing 1995 (axisymmetric and plane jets), Chasnov 1996 (two-dimensional turbulence), Boersma *et al* 1998 (round jets), Wosnik and George 1995 (natural convection boundary layers), and George and Castillo 1997 (boundary layers). All of these are consistent with the possibility of a similarity state which retains a dependence on initial (or upstream) conditions, contrary to the conventional wisdom. The last two show that the same considerations even apply to wall-bounded flows. Most importantly, *there still are no known exceptions* which would disprove the conjectures. In fact, one could argue that if Conjecture II were false, attempts to study canonical flows by experiments and simulations would be impossible because of the finite boundaries.

Now let's be very precise here: these remain *conjectures, and not theorems* until formally proven. And they can easily be disproven by counter-example. In fact some would argue that the failure of experiments and simulations to *always* conform disproves them. But this ignores the important question raised above about the importance of boundary conditions, which are crucial to the very existence of a similarity solution. Since for similarity theory they generally must be imposed at infinity, there can be no experiments or simulations which exactly satisfy the essential requirement to test the conjecture, *except possibly over a limited domain and/or for a limited time.* In fact I argue that the only place one could expect an experiment or simulation to provide a reasonable test of similarity theory is when the scales of the motion of importance are much smaller than the distance to the boundaries.[1] When the scales are growing in time or with streamwise distance, this means the solution is limited in its domain of applicability. OR put another way, the experiment ceases to be a valid test of the theory.

For example, in George (1992) I was able to show that a single length scale similarity theory was able to account for most of the observations in long wind tunnels (including about eight decades for spectral data), but it was not able to account for the DNS simulation results beyond what was generally believed (by the DNS community at least) to be the initial transient. Most troubling was that in the DNS results, the velocity derivative peaked whereas the similarity theory demanded that SR_λ = *constant for fixed initial conditions.* For the wind tunnel experiments, almost all the data satisfied the similarity constraint, and the constant increased with *source* Reynolds number. (Note that R_λ decreases during decay and the derivative skewness *increases*, contrary to the popular view that it decreases, at least for finite grid Reynolds numbers and *fixed initial conditions*). Subsequently, Huang and Leonard (1994) found that by introducing another length scale, they could account for the peak in the DNS results. The question they left unanswered is: Where does the other length scale come from[2]? A possible answer is: from the tunnel walls or the limits imposed by the computational box (or equivalently the lowest wavenumber allowed). If so then all of the departures

[1] Curiously, for grid and homogeneous shear flow turbulence this does not seem to have been pointed out before George 1992.

[2] This question was first posed to me in this form by D. Ewing.

from the simple similarity solution can be attributed to the effect of the finite boundaries.

Consider as a second example the time-dependent wake of Moser *et al* 1997 (see also Ewing 1995). The data were computed in a large DNS simulation and quickly settled into a similar state, both single point *and two-point* statistical properties. The similarity state was *not* the *self-preserving* state of classical theory where all statistical quantities are characterized by single length and velocity scales (*cf* Tennekes and Lumley 1972, Chapter 4). Instead it was the more general similarity solution of the averaged equations obtained by following the methodology of George 1989a, the most important difference being that the Reynolds shear stress scaled as $\Delta U d\delta/dt$ instead of $(\Delta U)^2$. (Note that these are the counterparts of $U_\infty \Delta U d\delta/dx$ and $(\Delta U)^2$ for the more familiar spatially developing wake.) The most striking feature, however, was the normalized dissipation which rapidly achieved a near constant value, but then started to increase, slowly at first, then ever more rapidly. Coincident with the slow rise of the dissipation was the beginning of the breakdown of the similarity spectral scaling for the very largest wavenumbers. Prior to this breakdown, visualizations of the flow showed no discernable large scale structure, but subsequent to the breakdown of similarity the flow began to show large scale roller eddies which were clearly visible. As time evolved and the shear layer grew, the number of eddies continued to decrease until only two such eddies were present in the computational box.

Now which of these flow states was the transient and which was the asymptotic flow? Obviously the rolls were the asymptotic flow for the box in which the computation was performed. But if the purpose of the computation was to study turbulence in the absence of artificial boundary conditions, then I would argue (as did the authors) that the near similarity intermediate state was the best approximation to boundary free flow. Thus, here similarity theory is being used to bound the validity of the "experiment", and hence the boundary-independent part of the flow. Interestingly, this work was carried out using spectral techniques and periodic boundary conditions for two of the three directions, just like many other simulations on which we base much of our new understanding.

3 The POD and Galerkin expansions

Some insight into the role of boundary conditions can be obtained using the POD (Proper Orthogonal Decomposition). These techniques are currently in vogue to generate appropriate bases for dynamical systems models of turbulence (*v* Holmes *et al* 1996), but they have been used for more than 30 years to investigate coherent structures in turbulence (*eg* Lumley 1967, George 1989b, Moin and Moser 1989). The problem was originally posed for turbulence by Lumley in the following manner: Suppose we have a random velocity field, $u_i(\cdot)$ where "$\cdot$" represents x_i, t or some subset of them. We seek to find a deterministic vector, say $\phi_i(\cdot)$ which has the maximum projection on u_i in a mean square sense; *ie* $\phi_i(\cdot)$ is chosen so that $< |u_i(\cdot)\phi_i(\cdot)|^2 >$ is maximized. The appropriate choice of $\phi_i(\cdot)$ can be shown by the calculus of variations to be given by

$$\int_{region} R_{ij}(\cdot,\cdot')\phi_j(\cdot')d(\cdot') = \lambda\phi_i(\cdot) \tag{1}$$

This is an integral equation for $\phi_i(\cdot)$ in which the kernel is given by the two-point correlation function, $R_{ij} =< u_i(\cdot)u_j(\cdot') >$. In general, equation 1 does not have a single solution but many, and their character depends on both the kernel *and* the region over which the integral is taken.

3.1 Fields of finite extent

The most familiar application of the POD is to flows in which the region is of finite extent in one or more directions (or time), either naturally or because of artificially imposed boundaries. It is well-known that when the POD is applied to flows which are of finite total energy, then the classical Hilbert-Schmidt theory applies. In this case there are denumerably infinite POD modes (or eigenfunctions), and they are orthogonal. Thus the original velocity field can be reconstructed from them as

$$u_i(\cdot) = \sum_{n=1}^{\infty} a_n \phi_i^n(\cdot) \tag{2}$$

The *random* coefficients a_n are functions of the variables not used in the integral, and must be determined by projection back onto the velocity field; *ie*

$$a_n = \int_{region} u_i(\cdot)\phi^{*n}_i(\cdot)d(\cdot) \tag{3}$$

The eigenfunctions are ordered (meaning that the lowest order eigenvalue is bigger that the next, and so on) so the representation is optimal in the sense that the fewest number of terms is required to capture the energy.

Thus the POD has provided several insights and possibilities: First, because of the finite boundaries it has produced a *denumerably infinite* set of orthogonal functions which optimally (in a mean square sense) describe the flow. Second a finite subset of these functions can be used to produce a finite number of equations for analysis. This is accomplished by using them in a Galerkin projection of the governing equations (in our case the Navier-Stokes equations). Thus by truncating after a specified number of modes, the infinitely dimensional governing equations are reduced to a finite set (v Holmes *et al* 1996 for details).

3.2 Are homogeneous fields and periodic fields the same?

Really interesting things happen to the POD if the flow is homogeneous or periodic. Note that, contrary to popular assumption (especially in the DNS and LES communities), these are *not* the same thing. The velocity field is said to be *periodic* in the variable x if $u(x) = u(x + L)$ where L is the period and the dependence on the other variables has been suppressed for now, as has the fact that the field is a vector. *Homogeniety*, on the other hand, means the statistics are independent of origin. For example, if a flow is homogeneous in a single variable, say x, then the two point correlation with separations in x reduces to $R(x, x') = \tilde{R}(r)$ where $r = x' - x$ is the separation. Note that by definition, homogeneous flows are not of finite total energy since they are of infinite extent, so the Hilbert-Schmidt theory cannot apply to them. Moreover, periodic fields are of finite total energy only if a single period is considered, since otherwise they repeat to infinity.

Now if periodicity and homogeniety are so different, why does the confusion arise? The POD provides the answer. For fields homogeneous in x, equation 1 can be shown to transform to

$$\int_{-\infty}^{\infty} \tilde{R}(r)\tilde{\phi}(x+r)dr = \tilde{\lambda}\tilde{\phi}(x) \tag{4}$$

Since the $\phi(x)$ on the right-hand side is a function of x only, it can be included in the integral on the left. Since there is now no x-dependence left on the right-hand side, it is immediately obvious that solution itself must eliminate the x-dependence on the left-hand side. Therefore the eigenfunctions must be of the form $\tilde{\phi}(x) \sim \exp{(ikx)}$ where k is a wavenumber and *all* values of k are possible; *ie* $-\infty < k < \infty$. The coefficients, $\hat{u}(k)$, can be shown to be given by

$$\hat{u}(k) = \frac{1}{2\pi}\int_{-\infty}^{\infty} u(x)e^{-ikx}dx \tag{5}$$

and the velocity field can be reconstructed from them by

$$u(x) = \int_{-\infty}^{\infty} \hat{u}(x)e^{ikx}dk \tag{6}$$

Thus the POD for homogeneous fields reduces to the familiar *Fourier transform* which depends on the continuous variable k, so the number of eigenfunctions is *non-denumerable.*

The situation for periodic fields is almost the same, but not quite – and that little difference is at the root of the problems being addressed in this paper. Any periodic field, even a random one, can be represented by a *Fourier series*; *ie*

$$u(x) = \sum_{n=-\infty}^{\infty} a_n e^{i2\pi nx/L} \tag{7}$$

where the a_n are random and are determined in the usual manner. Using the orthogonality, the two-point correlation function can be written as

$$R(x, x') = \sum_{n=-\infty}^{\infty} < |a_n|^2 > e^{i2\pi n(x'-x)/L} \tag{8}$$

Thus the two-point correlation for periodic flows, like homogeneous flows, depends only on the difference variable $r = x' - x$. Hence the eigenvalue problem of the

POD reduces to exactly the form of equation 4, except now the limits of integration are $(L/2, -L/2)$. It is easy to see that the POD modes must also be harmonic functions, like those for homogeneous flows. But there is a very important diffence which is obvious from the integral: for periodic flows the wavenumber must be given by $k = 2\pi n/L$ *and* n can only take integer values! Moreover, the number of POD modes is now *denumerably infinite* instead of being *non-denumerable* (*ie* continuous in k). Moreover, *the POD modes and the Fourier modes are identical.* Thus the use of *Fourier series* to represent periodic fields is indeed optimal, at least in a mean square sense.

Now the relation between a boxed homogeneous field and a periodic field can be readily determined by noting that because the energy is made finite by the box, the Hilbert-Schmidt theory again applies; hence the number of eigenfunctions becomes denumerable. If the kernel of boxed field is now *in addition assumed to be periodic*, the Fourier series representation above follows immediately. Thus the periodic fields usually assumed for calculation are dynamically equivalent to a boxed homogeneous field with the additional assumption of periodicity of the *instantaneous* fields. The assumption of periodicity has not only made the eigenfunctions denumerable, but it has forced the phase relations of all the scales, and this must also be of particular concern for the largest ones.

Such calculations of bounded fields, like their experimental counterparts, can only be representative of homogeneous fields for scales of motion much smaller than the computational box (or lowest wavenumber) and for limited times. Whether current computations are acceptable is open to debate, but the departures from similarity theory of the three-dimensional calculations would suggest not. In fact, the success of the two length scale similarity analysis of Huang and Leonard (1994) in accounting for the DNS results is probably decisive, since the additional length scale must be externally imposed, consistent with the effect of confinement. In addition, the success of the two-dimensional simulations of Chasnov (1996) in producing similarity spectra (analogous to those suggested by George 1992 for three-dimensional turbulence) provides additional support since the range of scales in the calculation is substantially larger than is possible in three-dimensions.

3.3 Inhomogeneous fields of infinite extent

None of the approaches above applies to flows which are inhomogeneous, but of infinite extent (like most shear flows in the streamwise direction). In fact, it has not been at all clear until recently whether the POD integral even exists in all cases. All attempts to-date to apply the POD to the flow in these inhomogeneous directions have ended up applying the Hilbert-Schmidt theory to finite regions of the flow. And as a result, the eigenfunctions and eigenvalues found are dependent on the particular domain included in the decompostion. Clearly this is because it is the finite domain itself which is making the energy finite.

Recently, however, Ewing 1995 (see also Ewing and George 1995) was able to show that if similarity solutions of the *two-point* Reynolds stress equations were possible, then the POD could be applied *in similarity coordinates* and the eigenfunctions were harmonic functions in it. By using a logarithmic coordinate transformation he was able to identify a number of flows for which two-point similarity was possible, thus for these flows the POD modes were known analytically. Most importantly, the eigenfunctions were independent of the domain, at least in principle. For the far axisymmetric jet, the appropriate modes were

$$u(\kappa, x) \sim x^{-1} \exp(-i\kappa\xi) \tag{9}$$

where

$$\xi \equiv \ln x / L_o \tag{10}$$

and L_o is prescribed by the initial conditions.[3] Thus two-point similarity and the POD have yielded an optimal set of eigenfunctions into which the flow can be decomposed. The two point correlations, $R_{ij}(x, x') =< u_i(x)u_j(x') >$, could all be expressed in the form,

$$R_{ij}(x, x') = Q(x, x')exp[i\kappa(\xi' - \xi)] = Q(x, x')exp[i\kappa \ln x'/x] \tag{11}$$

where $Q(x, x') = U_s(x)U_s(x')d\delta/dx$ and for this flow $U_s(x) \sim 1/x$ and $d\delta(x)/dx = constant$. Note the dependence of the correlation in similarity variables on $\xi' - \xi$, an obvious counterpart to the $x' - x$ dependence of homogeneous flows.

Now these functional forms are interesting for a couple of reasons. First, because they settle the question of whether the POD can be applied to a flow of infinite extent that is not homogeneous: It can! Second, for similarity flows of infinite extent, the optimal basis functions are analytical functions, and they are harmonic functions *in the similarity variable* $\xi = \ln x/L_o$. Third, there is a continuum of eigenfunctions since all values of the reduced wavenumber, κ, are possible; *ie* $-\infty < \kappa < \infty$. This last fact is the most interesting of all since it represents the counterpart of the homogeneous analysis above. Hence the denumerable POD modes of the Hilbert- Schmidt theory for an inhomogeneous finite energy flow have given way to the non-denumerable modes of Ewing. Thus, once again, the POD suggests that confining a flow changes the fundamental nature of it, consistent with observation.

There is at least one more interesting aspect of the these inhomogeneous eigenfunctions. It is easy to show by expanding the logarithm of equation 11 that the limiting forms of at least these inhomogeneous eigenfunctions are ordinary Fourier modes. From its Taylor expansion about $x = x'$, $\ln x'/x = (x' - x)/x + \cdots$. It follows for small values of $(x' - x)/x$ that $R_{ij} \sim \exp[ik(x' - x]$ where k is the ordinary, but local, wavenumber defined by $k = \kappa x$. Thus the usual *assumptions* of *local* homogeniety and the use of spectral analysis for the small scale motions are justified, at least in this case. Whether this is a general property of the POD is still very much the subject of debate (*cf* Holmes *et al* 1996).

[3]Interestingly, no length scale can be formed for a point source jet from the two parameters available, the kinematic viscosity and the rate at which momentum is added per unit mass. Hence L_o must depend on 'finite' source effects, like perhaps $(B_o^2/M_o)^{1/2}$ where B_o is the rate of mass addition per unit mass (*v* George 1989a).

4 Other possible implications

In 1995 at a Stanford/Ames CTR 'tea' I asked, "Is there a need for a *SUPER-grid-scale* model?" My basic hypothesis was that the finite boundaries of experiments and computational domains prohibit a necessary flux of energy to scales larger than the size of the domain. I argued that the energy which should have left "the box" for larger scales, shows up at the largest scale available (or lowest wavenumber) and saturates it. Worse, since it ultimately must be transferred by non-linear interactions (possibly non-local) to the dissipative scales, it causes a hyper-dissipation.

The phenomenon of energy accumulation in the lowest wavenumber defining the boundary of the system is well-known in two-dimensional turbulence and is refered to as Bose condensation (Hossain *et al* 1983). Smith and Yakhot 1984 were even able to show that it was responsible for the emergence of coherent structures in two-dimensional turbulence. While an inverse energy cascade is a necessary consequence of the unique nature of two-dimensional turbulence, a continuing flux of energy to *larger* scales is also a necessary feature of any three-dimensional developing turbulent flow as well. Therefore it is not a big intellectual leap to suggest that similar effects might be important in all flows constained by boundaries. Certainly all of the phenomenon suggested above have been observed in three-dimensional flow, as described earlier, and the non-local spectral energy transfer as well (*v* Yeung *et al.* So at least an inferential case has been made that the finite box is responsible.

If the above hypothesis is correct, then it is possible that the problem imposed by the finite boundaries might be tractable, since models could conceivably be constructed to let the large scale energy leave. If so, this could have important practical implications for the engineering turbulence models. At least one test of such a super-grid scale model would be whether similarity can be maintained longer and farther than without it.

There are other possible implications for the future. First the proof of Conjecture I may follow directly from a requirement for a boundary-independent POD representation of an inhomogeneous, unbounded flow. Thus the POD may itself provide the required variational principle (George 1990), or at least be a consequence of whatever is (say a miminum entropy requirement). Second, the differences between the generalized Fourier modes of the unbounded (but inhomogeneous) similarity flows and the POD modes from the Hilbert-Schmidt theory for bounded flows may offer clues about how boundary limits affect the flow. This may lead to a proof of Conjecture II. Third, these same differences may provide general specific criteria for evaluating *when* the finite domains of experiments and simulations are too small to capture the essential dynamics of the flow.

Acknowledgements I am grateful to Prof. Dan Ewing (McMaster U.) and Dr. Farhad Jaberi (SUNY/Buffalo) for many helpful discussions on these topics. Their insights and comments have played an important part in the development of these ideas, as have the opportunities to visit the Stanford/Ames CTR and interact with its staff.

References

[1] Boersma BJ, Brethouwer G and Nieuwstadt FTM (1998) A Numerical Investigation of the Effect of the Inflow Conditions on the Self-similar Region of a Round Jet. *Phys. Fluids, 10*, 4, 899–909.

[2] Chasnov JR (1996) On the decay of two-dimensional turbulence. *Phys.Fluids, 9*, 171–180.

[3] Ewing D (1995) On Multi-point Simlarity Solutions in Turbulent Free-Shear Flows. PhD diss., Dept Mech Engr, SUNY/Buffalo, Buffalo, NY.

[4] Ewing D and George WK (1995) Similarity Analysis of the Two-Point Velocity Correlation Tensor in the Turbulent Axisymmetry Jet. *Turbulence, Heat and Mass Transfer, Lisbon 1994* (Hanjalic and Pereira, eds.), Begell House Inc., NY, 49–56.

[5] Doering CR and Gibbon JD (1996) *Applied Analysis of the Navier-Stokes Equations* CUP. Cambridge, UK.

[6] Frisch U (1995) *Turbulence: the Legacy of AN Kolmogorov*CUP, Cambridge, UK.

[7] George WK (1989a) The Self-Preservation of Turbulent Flows and Its Relation to Coherent Structures and Initial Conditions. in *Advances in Turbulence* (George and Arndt, eds), Hemisphere (now Taylor and Francis), NY, 39–73.

[8] George WK (1989b) Insight into the Dynamics of Coherent Structures from a Proper Orthogonal Decomposition. in *Zorin Zaric Symp on Near-wall Turbulence, Dubrovnik, Yug* (S.Kline ed), Hemisphere, NY.

[9] George, W.K., (1990a) Governing Equations, Experiments, and the Experimentalist. *J. Exper. Thermal and Fluid Sci.*, **3**, 557–566.

[10] George, W.K. (1990b) Self-Preservation of Temperature Fluctuations in Isotropic Turbulence. *Studies in Turbulence*, (T.B. Gatski, Sutanu Sarkar, Charles G. Speziale, eds.), Springer Verlag, Berlin, 514–527.

[11] George WK (1992) The Decay of Isotropic Homogeneous Turbulence. *Phys Fluids A, 4*, 1492–1508.

[12] George WK (1994) Some New Ideas for the Similority of Turbulent Shear Flows. *Turbulence, Heat and Mass Transfer, Lisbon 1994* (Hanjalic and Pereira, eds.), Begell House Inc., NY, 13–24.

[13] George WK and Castillo L (1997) The Zero-Pressure-Gradient Turbulent Boundary Layer. *Appl Mech Rev, 50*, 680–729.

[14] George WK and Gibson MM (1992) A Similarity Theory for Homogeneous Shear Flow Turbulence. *Expts in Fluids, 13*, 229.

[15] Holmes P, Lumley JL and Berkooz G (1996) *Turbulence, Coherent Structures, Dynamical Systems and Symmetry* CUP, Cambridge, UK.

[16] Hossain M, Matthaeus W and Montgomery D (1983) Long-time state of inverse cascades in the presence of a maximum length scale. *J Plasma Phys, 30*, 479–493.

[17] Huang MJ and Leonard A (1994) Power-law decay of homogeneous turbulence at low Reynolds number. *Phys. Fluids, 6*, 3765.

[18] Moin P and Moser RD (1989) Characteristic-eddy decomposition of turbulence in a channel. *JFM, 200*, 471–509.

[19] Moser R, Rogers M and Ewing D (1996) Self-similarity of time-evolving plane wakes. TAM Rept 829, Univ Ill, Urbana, Ill (to appear in *Phys Fluids*).

[20] Lumley JL (1967) The Structure of Inhomogeneous Turbulent Flows. in *Atm Turb and Radio Wave Propag*, Nauka, Moscow.

[21] Smith LM and Yakhot V (1994) Finite-size effects in forced two-dimensional turbulence. *J Fluid Mech, 274*, 115–138.

[22] Wosnik M and George WK (1995) Another Look at the Turbulent Natural Convectioin Boundary Layer Next to Heated Vertical Surfaces. *Turbulence, Heat and Mass Transfer, Lisbon 1994* (Hanjalic and Pereira, eds.), Begell House Inc., NY, 346–352.

[23] Yeung PK, Brasseur JG, and Wang Q Dynamics of direct large-small coupling in coherently forced turbulence: concurrent physical and Fourier space views. *J. Fluid Mech., 283*, 43–95.

Trends in Mathematics, © 1999 Birkhäuser Verlag Basel/Switzerland

Worms in a turbulence experiment, from hot wire time series

F. Belin, F. Moisy, P. Tabeling, H. Willaime

Abstract

We analyze the structure of the events associated to the high velocity gradients in a series of experiments performed in low temperature helium, between counter-rotating disks, and in which the local velocity is measured by using hot wire anemometry. Such events are interpreted as resulting from the sweeping of vorticity filaments (identified as worms in a previous study [1]) onto the hot wire. Several characteristics of the worms (mean shape, size, internal increment, and density) are analyzed, in a range of microscale Reynolds numbers R_λ comprised between 150 and 2300.

1 Introduction

An issue in turbulence is whether there exist well-defined structures at small scales, i.e. in the inertial and dissipative ranges. Several observations suggest to respond positively: numerical studies [2] have shown that the intense vorticity levels are organized in Burger type filaments, with radius on the order of 4-5 Kolmogorov scale, and with velocity increments on the order of the standard deviation of the turbulent fluctuation. The same objects have been identified by inspecting the local minima of the pressure field in numerical experiments [1]. These filaments are now currently termed as "worms" [3]. Similar objects have also been detected (albeit indirectly) in flows between counter rotating disks in low temperature helium [1], by inspecting the high velocity gradients on local velocity time series. Additionally, long vorticity filaments have been observed in water [4], between counter-rotating disks, by using a bubble technique and further behind an oscillating grid [5]. There are indications that these objects are responsible for the intense, localized, low pressure events observed on local pressure probes [6]. Although often implicitly assumed, it is not ascertained that such filaments and the worms are the same objects, nor that they form elements of the same family of objects. The internal structure of the long vorticity filaments appears as complex, whereas the worm structure does not seem more complicated than a Burger vortex. There are strong indications that the length of the vorticity filaments is comparable to the integral scale [4] whereas the worms seem much shorter [7]. In Fourier space, the wave-number range in which the long filaments operate seems to correspond to large scales [6] whereas, in the same space, the worm contribution appear to be confined to the dissipative range [2] [1]. In the present state of knowledge, one should perhaps consider the existence of a single family of vorticity filamentary structures as an open issue, and it might be preferable to treat them as two distinct objects. Note also that non filamentary vorticity structures have been detected in turbulent flows, suggesting the turbulent field anyhow contains an abundant variety of objects. In particular, the existence of vorticity layers at small Reynolds numbers seems well supported by numerical studies [8].

The aim of the paper is to convey new information on the objects associated to the large gradients; throughout the paper, we assume they have a filamentary structure, and we analyze their geometrical characteristics. New results are reported on their mean shape, and their density (i.e the number of filaments per unit of volume); in this paper, the range of the Reynolds number we consider is comparable to earlier studies, but we explore it in more detail.

The working fluid is helium, and the flow is driven between counter-rotating disks; we refer the reader to [10] [11] [12]; its main advantage is to span an unconventionally large range of microscale Reynolds number – from 150 to 2300 in the present case This situation is particularly suitable to determine power laws, in terms of the Reynolds number, concerning the objects we detect; therefore, we are in position to identify their scales, which is one of the central issues of any decomposition of the signal into elementary structures.

2 Definition of the objects associated to the large gradients

Defining objects associated to large gradients, at various Reynolds numbers, raises the delicate question of introducing thresholds which depend on this parameter. To "solve" this difficulty, we consider the pdfs of the velocity derivatives for various Reynolds numbers, and seek a transform, in terms of this parameter, which allows to collapse the tails of such pdfs onto a single curve. Such a transform has already been presented in [10]. It consists in introducing a set of reduced variables – s* and p* – defined by the following relations:

$$s* = \alpha(R_\lambda)s \tag{1}$$

$$p* = \beta(R_\lambda)p \tag{2}$$

where R_λ is the microscale Reynolds number defined by:

$$R_\lambda = \frac{u\lambda}{\nu} \tag{3}$$

in which u is the rms of the velocity fluctuation, ν is the kinematic viscosity, and λ is the Taylor microscale. Variable s is defined, as usual, by:

$$s = \frac{\frac{\partial u}{\partial x}}{< (\frac{\partial u}{\partial x})^2 >^{1/2}} \tag{4}$$

u(t) is the velocity fluctuation, x = Ut (where U is the local mean velocity and t is time) and the brackets mean time average (we thus use here Taylor hypothesis

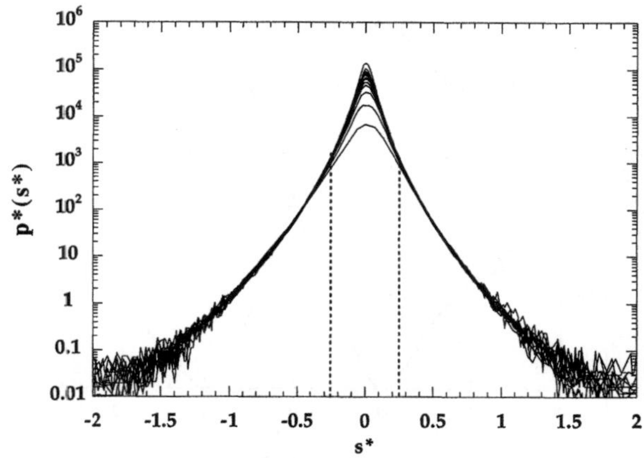

Figure 1: Pdf of the velocity derivatives, using s* and p* as the coordinates. The plot is a compilation of eleven pdfs, obtained for the following values of R_λ: 159, 255, 345, 489, 550, 745, 992, 1100, 1731, 1870, 2313. The dashed line represents the threshold s*=0.24.

to convert time into space); p(s), which is the density probability function for the variable s, has a variance equal to one, by construction; $\alpha(R_\lambda)$ and $\beta(R_\lambda)$ have been introduced previously [10]. Below $R_\lambda = 700$, they follow power laws in the form:

$$\alpha(R_\lambda) = R_\lambda^{-1/2} \quad \textit{and} \quad \beta(R_\lambda) = R_\lambda^2 \tag{5}$$

Above 700, the evolution is more complicated (it is displayed on Fig. 4 of [10]).

A set of pdf of velocity derivatives, obtained for different R_λ, and plotted by using s* and p*, is shown in Figure 1. One sees that the tails collapse pretty well for s*, say, larger than 0.24 (although the limit is not so sharp indeed), whereas below this value, the curves are dispersed. This suggests to define the large gradients by the condition:

$$s* > 0.24 \tag{6}$$

for all Reynolds numbers. The threshold used here is a little above the one of [1]. It is consistent with other proposals made in a lower range of R_λ [2]. The interest of using condition 6 for defining the large gradients is that the threshold stays at the same place on the tail as R_λ is varied. We had the opportunity to check that the results do not critically depend on the threshold definition.

3 Mean shape of the objects composing the tails of the pdf (defined as s* $\geq$ 0.24)

The procedure to determine the mean shape of the objects associated to the largest gradients is the following: (i) sort out the positive and negative gradients above

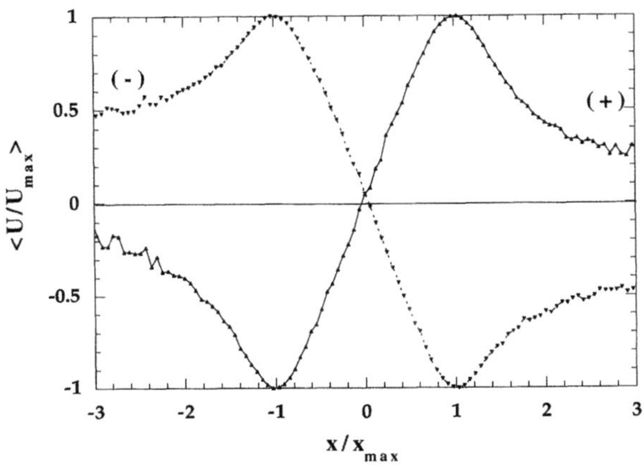

Figure 2: Mean shapes of the objects associated to the largest gradients of the system for $R_\lambda = 371$, averaged after amplitude and scale normalization is performed; the rescaling factors are the maximum velocity U_{max} and its corresponding position x_{max}. The positive and negative gradients are signalled by (+) and (-) symbols. The brakets represent statistical averaging.

the threshold defined by $s* > 0.24$ (ii) determine the two local extrema around the gradient, their distance and the peak to peak amplitude, (iii) normalize to unity these two quantities, and (iv) calculate the statistical average of the corresponding normalized curves, for each sign.

The result is shown in Fig. 2, for the positive and negative signs, and for $R_\lambda = 371$. The averages are performed on a population of 5000 events. The mean shapes look S shaped and the positive and negative gradients are different. Such an asymmetry presumably signals the action of vortex stretching, which tends to steepen the gradients, at a rate depending on the amplitude of the local velocity fluctuation. The standard type deviation around the mean shape is typically 40%, which is sizeable. We have not seen any measurable evolution of the shape with R_λ throughout the range investigated. A similar statement holds for the standard type deviation around the mean shape. This reassesses a statement made in [1], which was based on a poor statistics.

We have attempted to apply procedure (i)–(iv) on artificially generated signals, so as to appreciate to what extent the shapes of Figure 2 are specific of a given process. Clear differences appear with pure random noise, but shapes similar to those of Figure 2 are obtained with a low pass-band random noise, filtered in a way mimicking the energy spectrum. We concluded, without surprise, that the shape, although conveying interesting and non trivial information, is not accurately specific of a given process.

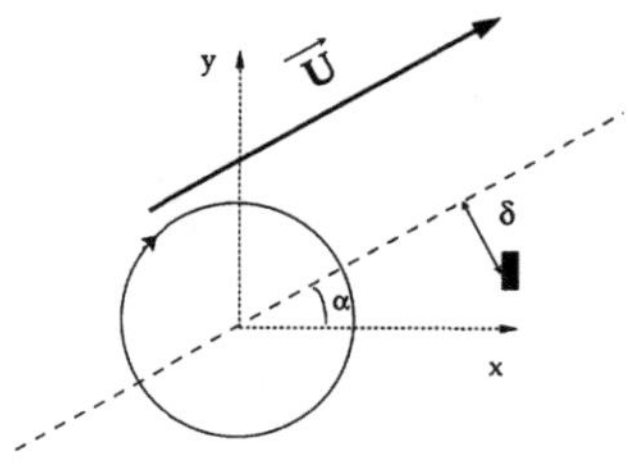

Figure 3: Representation of an isolated vortex filament, of axis normal to the plane (x,y), crossing the hot wire (black rectangle), at a speed U, with an angle a and a distance vortex center–probe equal to δ.

4 Geometrical interpretation of the events associated to the large gradients

We use here a geometrical interpretation previously advanced for the events associated to the large gradients [1]. The interpretation is based on the hypothesis that such events correspond to the sweeping of isolated Burger filaments onto the probe. As a starting point, the filament is taken normal to the plane defined by the mean flow and the sensor axis, as shown below.

Its vorticity field is:

$$\omega(r) = \omega_0 exp(-\frac{r^2}{2R^2}) \tag{7}$$

in which ω_0 is the core vorticity, and R its radius. In this notation, the worm radius (unfortunately) differs by a factor $\sqrt{2}$ with the usual convention. A useful quantity to be introduced here is the circulation Γ defined by

$$\Gamma = 2\pi R^2 \omega_0 \tag{8}$$

The calculation performed in [1] led to the following probe response when the vortex center hits the probe:

$$S(\frac{Ut}{r}) = U_\theta(r) sin\alpha \tag{9}$$

in which $U_\theta(r)$ is the velocity induced by the worm, i.e:

$$U_\theta(r) = \frac{\Gamma}{2\pi R}(1 - exp(-\frac{r^2}{2R^2})) \tag{10}$$

and α is the impact angle, i.e the angle between the trajectory of the worm and the axis of the sensor. The profile S(t) is similar to the mean shapes obtained experimentally, but there are differences, as shown in Figure 4. In particular, the velocity decreases much faster in the experiment, far from the core, than for an individual Burger vortex.

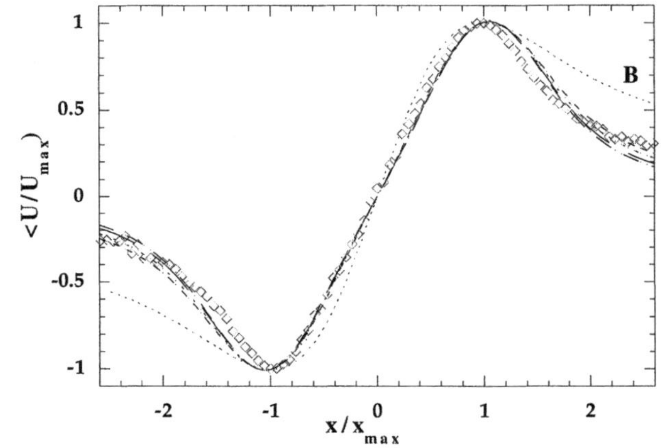

Figure 4: Comparison between the average of the mean shapes of the largest gradients for $R_\lambda = 371$, and a statistical average of events produced by Burger type filaments hitting the probe at different impact parameters α and δ, for five fluctuation rates u/U respectively equal to 5, 10, 20 and 30 % (they correspond to the set of unlabelled dashed lines). The dashed line labelled by a B corresponds to the response generated by an individual Burger vortex, crossing the probe at $\delta = 0$ and extremely small angle.

We have examined whether this discrepancy originates in the fact that the impact parameters fluctuate. To address this issue, we have calculated, for a fixed R and Γ, a mean profile $< S(t) >$ by conditionally averaging S(t) over a range of impact distances δ and impact angles α. The distances are distributed uniformly between −3R and 3R. The angles are distributed normally around zero, and their standard deviation is given by the following formula:

$$\alpha_{rms} = Arctg(\frac{U}{U+u}) \tag{11}$$

According to this formula, the higher the fluctuation rate, the larger the impact angle deviates from its average value. We further select the realizations S(t) giving rise to the largest gradients (this leads to reject more than 90 % of the realizations). Eventually we calculate the statistical average over the corresponding subensemble and renormalize.

The result is shown on Figure 4, for four different values of the ratio $\frac{u}{U}$, ranging between 5 and 30 %. It turns out that the average Burger profile does not sensitively depend on the precise values of the turbulent fluctuation rate $\frac{u}{U}$. It moreover agrees fairly well with the experimental curve obtained by averaging the mean profiles of the intense positive gradients. We may then conclude that the average shapes displayed on Figure 4 are likely to correspond to the sweeping of Burger like vortices onto the probe.

To be complete, one must also consider worms which are not normal to the plane defined by the mean flow and the sensor axis. This is a delicate problem which has been tackled long ago by Kuo and Corrsin [13] and more recently was addressed in [1]; one must introduce two additional angles (the pitch and the roll angles) and estimate their influence on the probe response. In practice, it seems enough to introduce correction factors on the distance measurements to account for the fluctuations of those two angles. Dividing by a factor of 1.40 seems a reasonable rule to infer the "true" worm radius (i.e. the one obtained if the worm axis was normal to the plane), from the "apparent" one (i.e. the one we measure). This rough procedure yields figures consistent with the Kuo and Corrsin estimates [13]. Nonetheless, a full analysis of the problem remains to be done.

Eventually, the calculation presented above (along with the incorporation of the factor 1.40) allows to determine the relation between the worm radius R and the temporal distance Δt between the two extrema located around an intense gradient, and that between the maximal orthoradial velocity U_{mw} and the peak to peak amplitude ΔS associated to such extrema. We find the following formula:

$$R = 0.15\,U\Delta t \;\; and \;\; U_{mw} = 1.1\Delta S \tag{12}$$

from which the worm Reynolds number is:

$$Re_w = \frac{\Gamma}{\nu} = 2.3U\frac{\Delta t \Delta S}{\nu} \tag{13}$$

The prefactors in Equation 12 update those previously published in [1], in which the impact parameters were assumed to be fixed, the impact angle vanishingly small, whereas here, those parameters are free to fluctuate in an appreciably wide range. Both analyses anyhow give essentially the same prefactor for the radius; actually, that found for the velocity increments is lower than the one of [2]. Let us emphasize again that a complete analysis of the problem remains to be done so as to propose accurate prefactors.

5 Geometrical characteristics of the worms

The distributions of the worm size $p(R/\eta)$ (where η is the Kolmogorov scale), along with that of the internal maximum velocity $p(U_{mw}/u)$, are shown in the figures below, in linear and semilogarithmic coordinates.

The figures 5 (a),(c) are close to those published previously [1], obtained with a different threshold. They confirm the scaling already proposed, i.e. the mean size is on the Kolmogorov scale, and the mean velocity increment across them is on the order of the standard type deviation of the large scale velocity. Moreover, we have enough statistics here to note that the distributions are extremely close to lognormal distribution, as seen on Fig. 5 (b),(d).

From those distributions, one can extract several quantities of interest. The evolution of the mean radius, the maximum orthoradial velocity and the worm Reynolds with R_λ are displayed on 6.

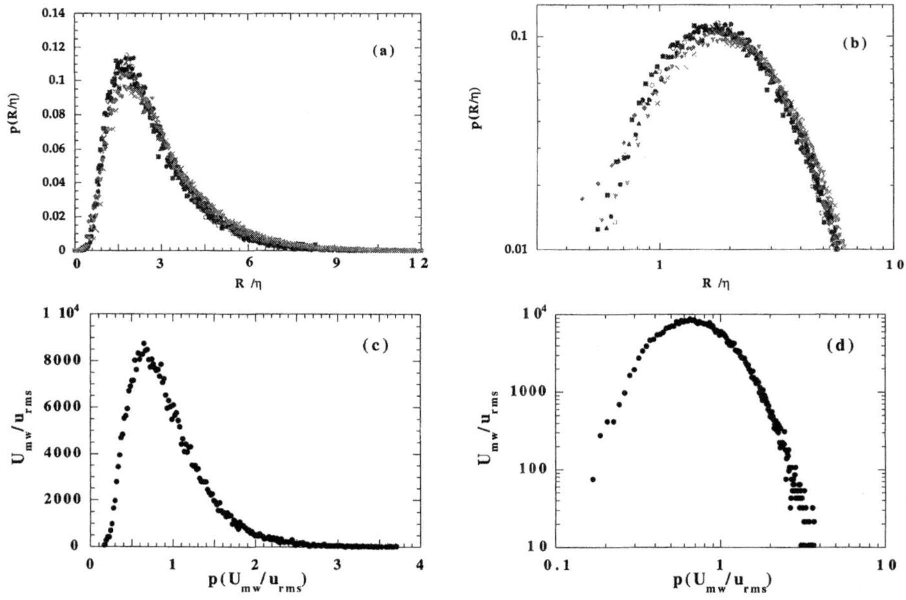

Figure 5: Distribution of worm radius R and maximum worm velocity UmW, (a) by using linear scales, and Kolmogorov units, for various R_λ: 692, 465, 649, 273, 371, 151, 225, 302, 255, 416, 359, 438, 514; (b) The same as (a), but using logarithmic scales; (c) internal velocity extrema distribution, on linear scales, for $R_\lambda = 438$; (d) the same as Fig. 6c, but using logarithmic scales. Note that in the case of Figures (a) and (b), the ordinate is rescaled so as to optimize collapse.

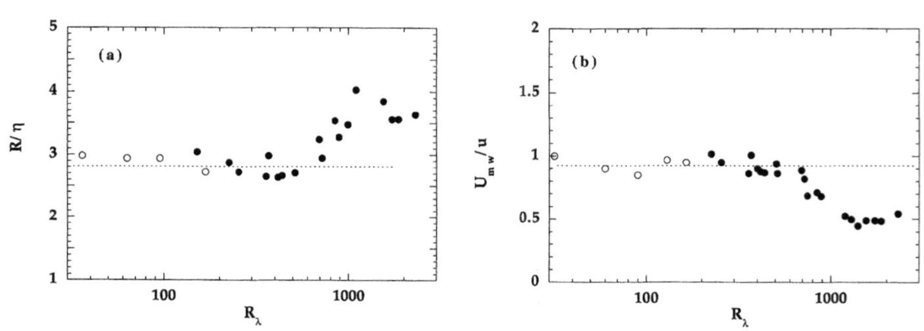

Figure 6: (a) Variation of the ratio $\frac{R}{\eta}$ with R_λ; (b) Variations of the ratio U_{mw}/u with R_λ; The circles are a numerical compilation made in Ref 9 and the black disks correspond to the present work; the dashed lines helps to localize the region where no evolution is visible

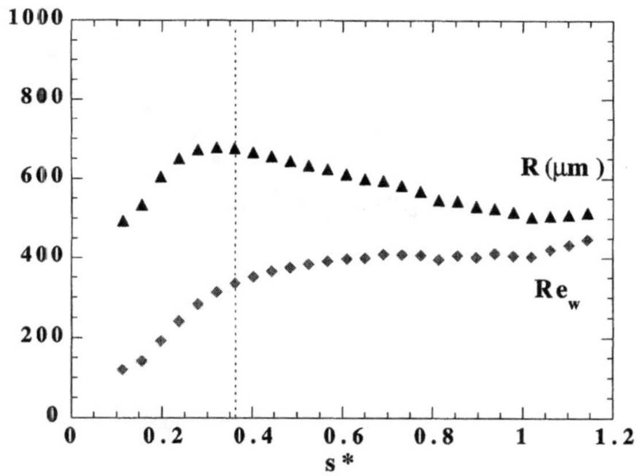

Figure 7: evolutions of R and Re_w with the threshold s* used to defined the intense gradients, for R_λ=225

Data obtained in numerical studies [9], and reconverted in our notations, are displayed on the same graph. Both are consistent. One sees that, in our units, the mean radius is three times the Kolmogorov scale up to R_λ, above which a transitional behavior takes place; the maximum orthoradial velocity is about 0.9 the standard deviation of the turbulent fluctuation below 700, then decreases and perhaps saturates as R_λ is further increased.

An interesting plot for checking the consistency of the analysis is the sensitivity of the mean radius R, and the mean worm Reynolds number Re_w to the threshold value of s* taken for the worm definition.

The evolution of R, and Re_w with s* are shown on Figure 7, for R_λ=225. Both quantities tend (albeit roughly) to saturate as s* is increased above 0.2-0.3. It is thus reasonable to define a region in the distributions of the velocity derivatives, where intrinsic characteristics can be sought. By contrast, below 0.20, the measurements sensitively depends on the threshold.

6 Density of the worms

We now turn to the spatial organization of the worms. We restrict ourselves to determining a quantity proportional to the worm density (i.e the number of worm per unit of volume), which is a global information; We thus have determined the quantity $N_w = fN$ where N is the number of worms per unit of time, detected by the probe, and f is the disk frequency rotation. It corresponds to the number of worms hitting the probe in one circumferential length (i.e 40 cm in the present case). The evolution of N_w versus R_λ is represented on Figure 8.

One can see that it is a constant, in the range lying between 150 and 700; to fix the ideas, one typically has, in this range, nine worms per disk revolution.

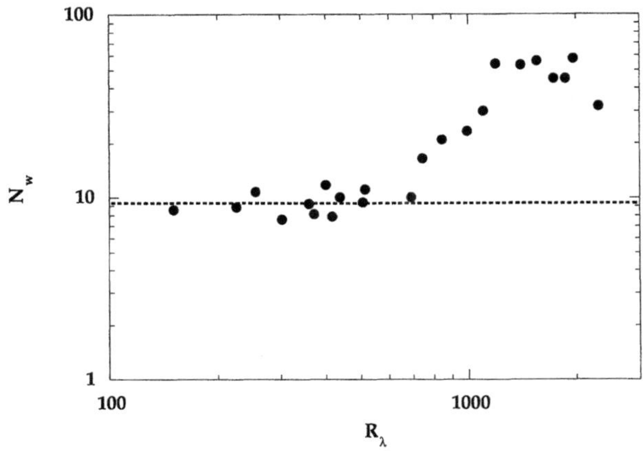

Figure 8: Evolution of the number of worms hitting the probe in one disk rotation, with the Reynolds number. The dashed line helps to guide the eyes.

This number increases with R_λ above 700. The fact that N_w (and hence the worm density) is independent of R_λ in the pretransitionnal range may seem surprising: it is at variance with previous numerical studies [2], which show the density decreases with R_λ, in the range $30 < R_\lambda < 180$. There is nonetheless no contradiction between the two results, since we address different ranges of R_λ. We unfortunately did not succeed to decrease the Reynolds number below 150, so as to obtain a sizeable overlap with numerical studies. Although unlikely, it would be interesting to confirm that the difference in the results does not originate from the procedures of worm detection, but is effectively due to the fact that the ranges of Reynolds numbers are different.

7 Conclusion

Before summarizing the findings, let us recall that all the results we obtain here are deduced from the analysis of hot wire signals. Throughout the paper, we have assumed that objects, of filamentary structure, are responsible for the intense gradients observed on the time series. If this assumption is rejected, the paper can nonetheless be viewed as presenting a particular way of processing the data, which, to the best of our knowledge, has not been developed so far. If the assumption is accepted, then we are left with technical issues to infer, from the signal, plausible geometrical characteristics of the underlying filamentary objects. To summarize the paper, we thus have found the following results:

- For a range of R_λ comprised between 150 and 700, worm radii equal three Kolmogorov scales (in our notation), and maximum orthoradial velocity inside them are close to the standard deviation of the turbulent fluctuation. Although such estimates rely on an approximate analysis of the impact of the worms onto the probe, it is remarkable they agree rather well with available numerical findings [2]. Nonetheless, a detailed analysis of the impact might be worth being performed.

- The mean shape of the large gradient events is consistent with that generated by Burger like vortices hitting the probe.
- The worm density is a constant in a range of R_λ comprised between 150 and 700.
- A transitional behavior on some of the quantities we have examined; such a transition has already been reported(13), and its observation here is not the main point of the paper.

Acknowledgments

The authors have benefited of discussions with J. Jimenez, Y. Couder, S. Douady, B. Andreotti, U. Frisch, A. Pumir, Y. Gagne, M.E. Brachet, S. Kambe. This work has been supported by Ecole Normale Supérieure, CNRS, the Universities Paris 6 and Paris 7.

References

[1] F. Belin, J. Maurer, P. Tabeling, H. Willaime, Journal Phys II, 6, 1 (1996).

[2] J. Jimenez , A. A. Wray , P. G. Saffman and R. S. Rogallo, J. Fluid Mech., 255, 65 (1993), M. Tanahashi, T. Miyauchi, K. Matsuoka, Proc IUTAM on Simulation and Identification of Organised Structures in Flows, Lyngby, Denmark, May 1997.

[3] It has been proposed to use the word "sinews" (H. K. Moffatt, S. Kida, K. Okhitani, J. Fluid Mech., 259, 241 (1993)), but this convention does not seem to prevail.

[4] S. Douady, Y. Couder, M. E. Brachet, Phys. Rev. Lett., 67, 983 (1991), O. Cadot, S. Douady, Y. Couder, Phys. Fluids, 7, 3 (1995)

[5] E. Villermeaux, B. Sixou, Y. Gagne, Phys. Fluids, 7, 8 (1995).

[6] P. Abry, S. Fauve, P. Flandrin, C. Laroche, J. Phys. II France, 4, 725 (1994).

[7] Worm length estimates provide values comprised between the Taylor and the integral scale (see Ref below for instance) ; from the inspection of the vorticity field of Jimenez et al(2), the distribution seems centered around a length well below the integral scale.

[8] Z. S. She, E. Jackson, S. A. Orszag, Nature 344, 226 (1990), M. Tanaka, S. Kida, Phys. Fluids A5 (9).

[9] J. Jimenez, to appear in the Proceedings of the Euromech colloquium 384, Dynamics and Statistics of Concentrated Vortices in Turbulent Flows, Carry le Rouet, June 24–27 (1997).

[10] F. Belin, J. Maurer, H. Willaime, P. Tabeling, Phys. Fluids, 9 (12) (1997).

[11] P. Tabeling, G. Zocchi, F. Belin, J. Maurer, H. Willaime, Phys. Rev. E, 53, 1613 (1996).

[12] G. Zocchi, P. Tabeling, J. Maurer, H. Willaime, Phys. Rev. E, 50, 5 3693 (1994).

[13] A. Kuo, S. Corrsin, J. Fluid Mech., 56, 3, 447 (1972).

Trends in Mathematics, © 1999 Birkhäuser Verlag Basel/Switzerland

Intermittency in the closed flow between coaxial disks

J.-F. Pinton, F. Chillá, S. Fauve, R. Labbé

Abstract

The flow created in the gap between coaxial, fast rotating disks provides a rich variety of dynamical regimes at high Reynolds numbers. In the contrarotating geometry it provides a homogeneous turbulent flow in a compact volume of space while the corotating geometry adds a large scale vortex motion to the turbulent background. These regimes are used to address several issues in turbulence related to the influence of large scale dynamics: the fluctuations of global (i.e. spatially averaged) quantities, the geometrical characteristics of vorticity filaments and the statistical description of inhomogeneous turbulent flows.

1 Power fluctuations

The contrarotating flow has many characteristics of the turbulence observed in large wind tunnels [1, 2]. In addition it has allowed the measurement of new quantities such as pressure fluctuations at the flow wall [3, 4, 5] and power fluctuations [6]. This last measurement raises the question of the fluctuations of global quantities; it has implications on the statistical treatment of turbulence.

In a stationary turbulence experiment, one parameter has to be fixed. If it is the integral Reynolds number then the power consumption of the flow fluctuates in time. Indeed it can be conceptually expressed as $\vec{f} \cdot \vec{u}$, where the force $\vec{f}$ and the velocity $\vec{u}$, being conjugate parameters, cannot be both kept at constant values. The variations are related to the kinetic energy and dissipation through:

$$\frac{dE}{dt} = \epsilon_I - \epsilon_D \quad .$$

If time averages are taken the above equation is trivially verified. Instantaneously all quantities fluctuate; of the three, only the power input ϵ_I can be measured experimentally by monitoring the power consumption of the motors driving the disks. Then two questions may be asked: *(Q1)* what are the magnitude and probability density function (PDF) of its fluctuations in time?; and *(Q2)* how do they change with the Reynolds numbers?

The mean value $\overline{\epsilon_I}$ can be obtained by a dimensional argument: $\overline{\epsilon_I} = \rho R^5 f^3$, if ρ is the fluid density, R the disks radius and f their rotation rate. The variance ϵ_I^{rms} of the time fluctuations is about 6% of the mean, with peak-to-peak fluctuations as high as 30%. Despite the fact the ϵ_I is a large scale quantity, we observe that its PDF is strongly non-gaussian, with the development of an exponential tail towards low values – see Figure 1.

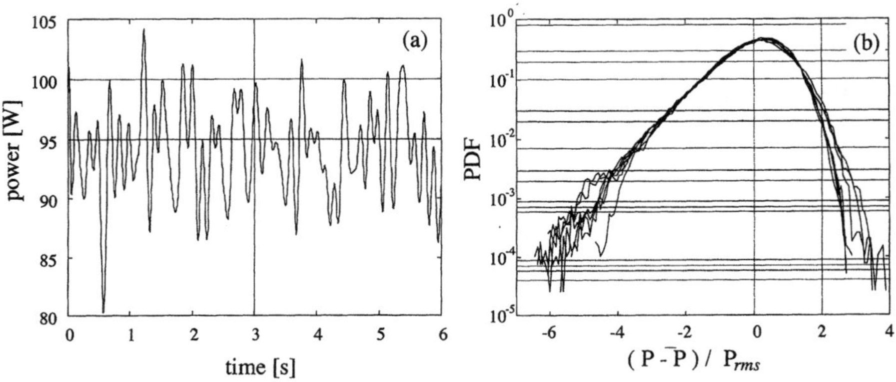

Figure 1: Power fluctuations in the closed von Kármán flow. Rotation frequencies $f = 25, 30, 35, 40, 45$Hz, with disks of radius $R = 10$cm and a vertical separation of 27.5cm. When the reduced variable $\Pi = (P - \overline{P})/P_{rms}$ is used, the PDFs obtained for different rotation rates all collapse on a single curve.

We believe that the asymmetry of the PDF is related to the formation of large scale vortical structures in the flow. Although the range of explored Reynolds number is too small to give a definite answer to question *Q2*, we do not observe a significant decrease of $\epsilon_I^{rms}/\overline{\epsilon_I}$, certainly not as one over the square root of the number of degrees of freedom , i.e. as $Re^{-9/8}$ using Landau's argument.

We note that non-gaussian fluctuations of a global quantity have been observed with the same law PDF curve in (*non-dissipative*) critical systems such as the 3D Ising model at the critical point and the 2D-XY model below the Kosterlitz-Thouless temperature [7].

2 Vorticity filaments

The flow between the contrarotating disks has also lead to the observation and to the experimental study of vorticity filaments [8, 3, 4, 5, 9], as the strong elongated vortices that are formed at rare intervals have been termed. These structures have raised much interest as they are reminiscent of the intense vortices observed in numerical simulations [10, 11] which may be connected to the intermittency. The characteristics of these filaments are the following: *(i)* their length is of the order of the integral scale, as indicated by the visualization; *(ii)* their velocity difference is of the order of the *rms* velocity fluctuations in the flow, as shown by the pressure measurements at the flow wall; *(iii)* their core size corresponds to a small spatial scale, probably between the Kolmogorov length and Taylor microscale.

We have used the sound scattering technique developed in our laboratory to obtain a direct and non intrusive measurement of the filaments in the bulk of the flow. When a plane sound wave is incoming on a flow, the amplitude of the

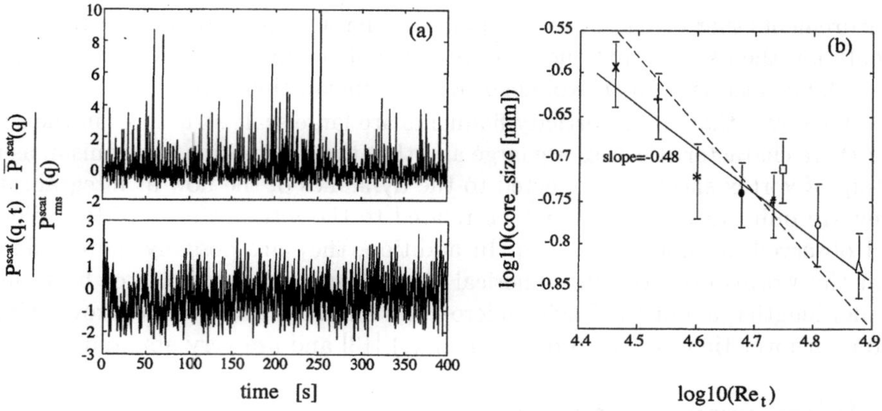

Figure 2: (a) Time evolution of the scattered sound amplitude $P^{scat}(q,t)$ for an incoming sound frequency equal to 10 MHz ($b = 150\mu$m) in the upper curve, and to an incoming sound frequency at 6 MHz ($b = 250\mu$m in the lower curve, sat a rotation rate of 1000 rpm. (b) Filaments core size as a function of the turbulent Reynolds number; the solid line is a $Re_t^{-\alpha}$ fit, with $\alpha = 0.48$. The dashed line materializes a $Re_t^{-3/4}$ scaling law.

scattered wave is proportional to the Fourier transform of the vorticity component perpendicular to the scattering plane:

$$P_{scat}(\vec{r},t) = P^{\circ}_{\text{inc}} \, \frac{-\cos\theta\sin\theta}{1-\cos\theta} \, \frac{i\pi^2\nu}{c^2} \, \frac{e^{2\pi i(\nu r/c-\nu_0 t)}}{r} \, \tilde{\Omega}_{\perp}(\vec{q},t) \quad ,$$

where P°_{inc} is the amplitude of the incoming wave of frequency ν_0, c the speed of sound, θ the scattering angle and $\vec{q}$ the scattering wave vector:

$$q = \frac{4\pi\nu}{c} \, \sin\left(\frac{\theta}{2}\right)$$

which defines the scale of the flow under analysis $b = 2\pi/q$.

When b is of the order of the size of the filaments'core, the Fourier component of the vorticity field shows a very intermittent behavior with the rare occurrence of very high events (as large as 10 standard deviations) – see Figure 2(a). They obey a Poisson statistics with a mean waiting time of about a hundred integral scale turnover time and correspond to the passing of a vorticity filament across the measurement volume. Varying the Reynolds number of the flow, it is observed that this core size is of the order of and scales like the Taylor microscale of the flow – Figure 2(b).

Although the scaling range is limited, this establishes the relevance of the Taylor microscale as the filaments's characteristic core size. Relying on pressure measurements that give u_{rms} as the filaments characteristic rotation rate, a simple calculation then shows that the turnover time of the filament is $\sqrt{\nu_k/\epsilon}$ (ν_k is the fluid's kinematic viscosity), Kolmogorov's lagrangian time scale.

We stress that these vorticity filaments are large scale structures in the sense that there characteristic sizes are large and that their formation mechanism, as the roll-up of vortex sheet, is connected to the dynamics of the flow at integral scale. They are quite rare and may not be related to the intermittency of the velocity field observed at small scales [12]. In addition, they are possibly quite different from the worms observed in numerical simulations. Indeed the later have much smaller lengths (about the Taylor microscale) and core size (Kolmogorov scale), a faster turnover time (of the order of η/u_{rms}) [13] and are very abundant [14].

3 Inhomogeneous turbulence

Many studies in turbulence have been devoted to the homogeneous and isotropic (HI) case, and certain universal feature have emerged, as the Kolmogorov $k^{-5/3}$ spectrum or the values of the structure function exponents [15, 16]. In particular, most of the turbulence phenomenology is based on the existence of a "universal" properties in an inertial range of scales whose dynamics is independant of the characteritics of the large scale motion and of the dissipative zone. However, real flows at high Reynolds numbers are strongly inhomogeneous, anisotropic and it is worthwhile to study the relevance and usefulness of HI concepts in these situations.

We analyze here the turbulence features in the neighborhood of a strong vortex which may be stable or unstable. One motivation is that in inhomogenenous flows coherent structures in the form of vortex tubes are commonly observed [17].

The flow is produced in the gap between two coaxial disks of radius R rotating in the same direction. Air is the working fluid; the flow is enclosed in a cylindrical vessel. The rotation rates of each disk is set independantly and controled with a feedback loop. In this situation one observes a large scale dynamics that depends on the disks rotation rates [6, 18, 19], superimposed to turbulent fluctuations (the integral Reynolds number of the flow is $Re > 5\ 10^4$). For equal rotation rates ($f_1 = f_2 = 30$Hz), the vertical vorticity concentrates to build a strong stable axial vortex. The flow has a large core in solid body rotation. When the disks rotate at quite different rates ($f_1 = 12$Hz, $f_2 = 40$Hz) the axial vortex is unstable. One observes an intermittent sequences of formation and breakdown of a strong and thin axial vortex. We consider here the influence of the large scales dynamics on the small scale turbulence features. We use local hot-wire anemometry to study the velocity field characteristics as a function of the distance r to the rotation axis. The component $v_{r\theta} = \sqrt{v_r^2 + v_\theta^2}$ is recorded; in the region discussed here the azimuthal component dominates. The velocity profiles for both flow configurations are shown in Figure 3 (top curves).

One issue is the shape of the velocity spectrum. We observe that, at each distance r, it displays a power law behavior indicating the existence of a self similar region. Sufficiently far away from the rotation axis one gets a Kolmogorov-like

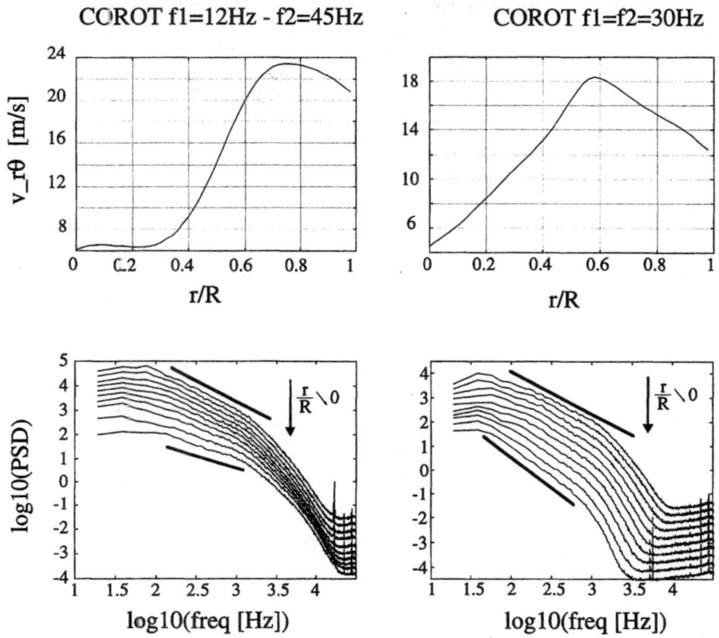

Figure 3: (Top): mean velocity profile; (Bottom): power spectrum. In each case the left column pertains to the unstable vortex dynamics and the arrow indicates the direction of variation as one gets closer to the rotation axis.

$k^{-5/3}$ spetrum. It is no longer the case as one gets nearer to the vortex structure; the slopes of velocity spectra are not constant and change continuously (see Figure 3 – bottom curves):

– in the case of equal rotation rates (stable vortex) the slope of the spectrum goes from -1.5 at $r/R \sim 0.5$ to -2.2 at the center.

– in the case of different rotation rates (unstable vortex), the spectrum slope varies from -1.65 to -1.1 as r/R again ranges from 0.5 to 0.

As can been seen the large scale dynamics influences the energy transfers in the neighborhood of a coherent structure. In the case of a stable vortex, the spectra exhibit a steeper slope; this may indicate a tendency to bidimensionalization. On the opposite the spectra have a slower decrease as one approaches the core of an intense, unstable vortex – this again may be expected: sharp discontinuous ‘spikes’ would yield a k^{-1} spectrum. We observe that this behavior is at odds compared to our study of the open flow [6] (i.e. with the lateral walls removed). In that situation the vortex structure is weaker and has a slow precession motion; the velocity spectra at every distance from the vortex have a Kolmogorov scaling region.

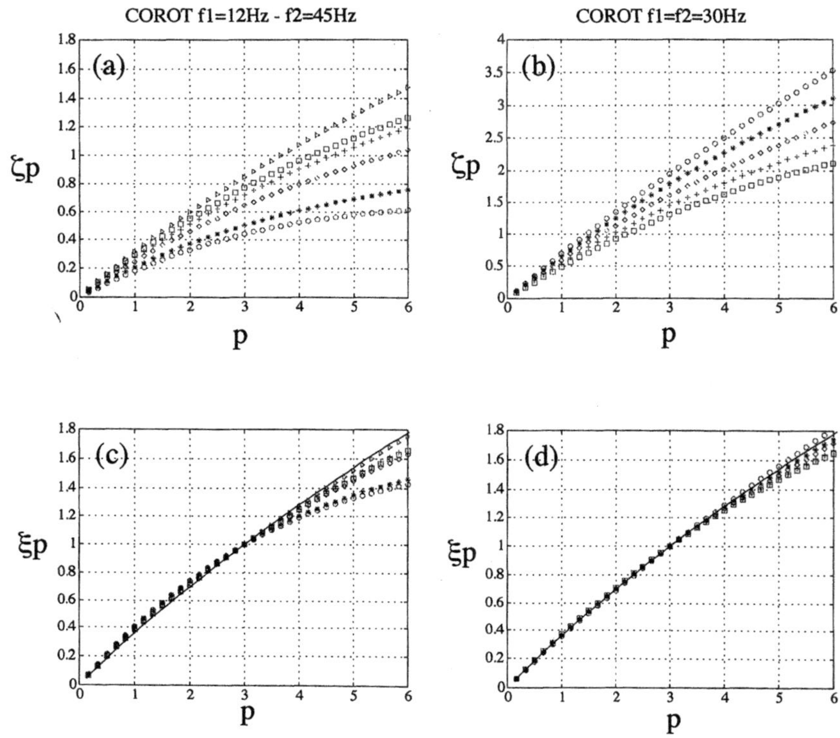

Figure 4: Structure function exponents at several distances near the vortex core: unstable vortex case on the left (a&c) $\{(\triangle 3.5\mathrm{cm}), (\square 3\mathrm{cm}), (+2.5\mathrm{cm}), (\diamond 2\mathrm{cm}), (*1\mathrm{cm}), (\circ 0\mathrm{cm})\}$ and stable vortex case on the right (b&d) $\{(\square 2\mathrm{cm}), (+1.5\mathrm{cm}), (\diamond 1\mathrm{cm}), (*0.5\mathrm{cm}), (\circ 0\mathrm{cm})\}$.

Another issue is the intermittency behavior, i.e. the change in the shape of the PDFs of the velocity increments through the scales. It can be measured *via* the structure function exponents:

$$\zeta(p) \equiv \frac{d \ln < \delta|v(\ell)|^p >}{d \ln \ell} \quad ,$$

calculated in the inertial domain evidenced in the spectra. The results are displayed in Figure 4 – top curves. Since the 2nd order exponent is related to the scaling of the velocity spectrum, the $\zeta(p)$ functions at each location are ordered as the spectra slopes. In particular, the Kármán-Howarth value $\zeta(3) = 1$ is no longer a fixed point. However, in each flow geometry and for every location of the probe, the $\zeta(p)$ functions are non linear and the magnitude of intermittency effects is related to their curvature. To study this and to offset the variations of $\zeta(3)$, we calculate the relative exponents:

$$\xi(p) \equiv \frac{d \ln < \delta|v(\ell)|^p >}{d \ln < \delta|v(\ell)|^3 >} ,$$

again in the 'inertial range' detected on the spectra – although the scaling range is somewhat extended using this scheme. As can be observed in the bottom part of Figure 4, the relative exponents are similar. Deviations from the isotropic ('She-Lévêque'[7]) values are marked in the neighborhood of the instable vortex and otherwise smaller (possibly within the error bars – this present work uses 10^6 data points in the calculations of the structure functions). However we note a trend: the $\xi(p)$ curvature increases as one gets closer to the axis (vortex core) in the unstable vortex configuration whereas it decreases in the stable vortex geometry.

These results show that the velocity cascade is inhomogeneous in space. Its characteristics depend on the distance to the coherent vortex and are quite different from that of homogeneous, isotropic turbulence. The flow is also anisotropic at all scales, as indicated by analysis of other velocity components. The influence of large scale coherent motions on the turbulent cascade has also been observed in a similar geometry [8] and in the neighborhood of large shearing zone [9]. It remains a goal of further studies to relate the observed changes to the global structure of the flow.

References

[1] Pinton J.-F. and Labbé R., *J. Phys. II*, **4**, 1461, (1994).

[2] Maurer J., Tabeling P. and Zocchi G., *Europhys. Lett.*, **26**, 31, (1994).

[3] Fauve S., Laroche C., Castaing B., *J. Phys. II France*, **3**, 271–278 (1993).

[4] Abry P. , Fauve S., Flandrin, P., Laroche C., *J. Phys. II France*, **4**, 725–733 (1994).

[5] Cadot O., Douady S. and Couder Y., *Phys. Fluids* A**7**, 630–646 (1995).

[6] Labbé R., Pinton J.-F. and Fauve S., *J. Phys. II France*, **6**, 1099, (1996).

[7] Archambault P., Bramwell S., Holdsworth P.C.W., *J. Phys. A*, **30**, (1997).

[8] Douady S., Couder Y. and Brachet, M.-E., *Phys.Rev. Lett.* **67**, 983–986 (1991).

[9] Dernoncourt B., Pinton J.-F. and Fauve S., *Physica D*, in review.

[10] Siggia E. D., *J. Fluid Mech.* **107**, 375–406 (1981).

[11] Jiménez J., Wray A. A., Saffmann P. G. and Rogallo R., *J. Fluid Mech.* **255**, 65–90 (1993).

[12] Chainais P., Graduate Report, École Normale Supérieure de Lyon, (1997).

[13] Jiménez J., *Proc. Euromech Coll. 364*, Carry-le-Rouet, June 1997.

[14] Kida S., Miura H.,*Proc. Euromech Coll. 364*, Carry-le-Rouet, June 1997.

[15] Nelkin M., *Adv. in Physics*, **43**(2), 143–181, (1994).

[16] Arnéodo A. *at al.*, *Europhys. Lett.*, **34**(6), 411–416 (1996) .

[17] 'Eddy structure identification in free turbulent shear flows', Bonnet J.-P., Glauser M.N. eds., Kluwer Acad. Press, (1995).

[18] Pinton J.-F., Chillà F., Morant N., "Intermittency in the closed flow between coaxial corotating disks", *European J. Mech./B Fluids*, **17**(4), 535–547, (1998).

[19] Chillà F., Pinton J.-F., Labbé R., *Europhys. Lett.*, **35**, 271, (1996).

4 Two- and quasi-two-dimensional flows

Trends in Mathematics, © 1999 Birkhäuser Verlag Basel/Switzerland

Turbulent statistical dynamics of a system of point vortices

H. Aref, P. L. Boyland, M. A. Stremler, D. L. Vainchtein

1 Introduction

Turbulence is the name usually given to an imperfectly understood class of chaotic solutions to the Navier–Stokes equation in which many degrees of freedom are excited. The main premise of our work is that related or even similar solutions exist for a much simpler dynamical system that is closely related to the 2D Euler and Navier–Stokes equations: the Hamiltonian system defined by N interacting point vortices. An understanding of such solutions to point vortex dynamics should allow one to approach the phenomena of 2D turbulence without the use of *a priori* statistical postulates.

Some workers in turbulence are of the opinion that even for the problem of 2D turbulence point vortex dynamics is an inadequate model. This opinion is based on the singular nature of the vorticity distribution and on the inviscid nature of the dynamics of point vortices. In support of this opinion the existence of 'vortex collapse' solutions (Gröbli 1877; Aref 1979), in which the point vortex system develops a singularity after a finite time, is sometimes cited, as it is well known that the 2D Euler equation with smooth initial vorticity data does not permit such finite-time singularities.

In spite of such differences between the 2D Euler equation and the N-vortex problem for large N, the latter does have solutions with qualitative features strongly analogous to 2D turbulent flows, as is seen, for example, by comparing numerical solutions of the 2D Euler or Navier–Stokes equation with numerical simulations of a large system of interacting vortices. Indeed, one of the key insights in 2D turbulence has been the observation that the flow evolves into an assembly of isolated, strong vortices, and that this state of affairs persists for a considerable amount of time.

Benefits of studying the N-vortex problem from the outset, then, are that the dynamics are considerably simpler for this discrete problem than for the continuum PDEs; that the degrees of freedom, i.e., the vortices, are physically intuitive; and that the approach via a multi-particle system suggests analogies, approaches and tools from areas of physical theory not commonly associated with turbulence, such as kinetic theory and the theory of dynamical systems. This suggestion to approach 2D turbulence dynamics via point vortices is not new. It goes back some 50 years to a seminal paper by Onsager (1949) in which equilibrium statistical mechanics was used to argue the formation of large-scale vortices, a phenomenon we would today label the 'inverse energy cascade'.

In the following sections we comment on several results obtained while pursuing solutions of the N-vortex problem that appear, in one way or another, to shed light on the problem of turbulent flows. Our intent here is not to discuss any one approach or set of results in detail. For that we direct the reader to the literature cited. (Manuscripts under review or in press may be requested directly from the first author).

This work was supported by NSF grant CTS-9311545. MAS was also supported by a fellowship from the Office of Naval Research.

2 Specification of the dynamics

Under consideration, then, is the dynamical system defined by N interacting point vortices. These equations are well known for vortices on the unbounded plane. It is also well known that the equations can be cast in Hamiltonian form, so that ideas of integrability and chaos in Hamiltonian dynamical systems can be immediately applied.

On the unbounded plane the N-vortex problem has four general integrals: the Hamiltonian itself, the two components of the linear impulse, and the angular impulse. These integrals arise in the usual way from the time-independence of the Hamiltonian and from its invariance to translation and rotation of the coordinates. It may be shown that the four integrals always yield three that are in involution for arbitrary choice of the vortex circulations. Hence, the dynamical problems of one, two, and three vortices are integrable. Numerical experiments, in some cases backed up by analysis, suggest that the four-vortex problem is non-integrable for general choices of the vortex circulations.

For vortices in a periodic strip, a model problem of relevance to the evolution of plane wakes and jets, rotational invariance of the Hamiltonian is forfeited, and the angular impulse is no longer an integral. The problem still has obvious translational invariance along the strip, so the corresponding component of the impulse is still an integral. The other component of impulse turns out also still to be an integral, as was apparently first observed by Birkhoff & Fisher (1959). The Poisson bracket algebra of the integrals now shows that the three-vortex problem in the periodic strip is integrable when the sum of the three circulations vanishes. This has been recognized for many years, but only recently have we succeeded in elucidating the details of the motion and in providing a solution method applicable to arbitrary vortex triples with vanishing sum of circulations (Aref & Stremler 1996).

For the periodic parallelogram the periodicity of the flow guarantees that the circulation integral taken around the parallelogram vanishes, and so the sum of the circulations of the vortices must be zero by periodicity. Various versions of the equations of motion have appeared in the literature (Kraichnan 1975; O'Neil 1989), but they are not particularly convenient since they involve infinite sums. We have recast the equations of motion into a compact, explicit form (Stremler & Aref 1998). The main interaction involves the Weierstrass ζ-function. By similar logic as for the periodic strip case one finds that both components of linear impulse are integrals, and that they are in involution since the sum of the vortex strengths is zero. One concludes that the three-vortex problem in the periodic parallelogram is integrable. In the paper just cited we provide a solution method that is, in principle, capable of reducing the problem to quadratures for all vortex triples. Although the problem is integrable, the solutions can become very complicated as we shall discuss in §4.

3 An aside on steadily rotating states with N identical vortices

The equations of motion for N identical vortices become non-integrable for N=4 (and, thus, presumably for any $N > 4$). One can think of these solutions as vortex flows with simple spatial structure at any instant, but complicated temporal evolution. Here we mention some recently discovered solutions that have very simple temporal evolution (they rotate like a rigid body) but 'complex' spatial structure (in the sense that they lack any symmetry).

Solutions of the point vortex equations that describe configurations moving without change of shape or size have been sought for more than a century. In the late 1870's Kelvin believed that such configurations were models of atoms with the vortices as some kind of excitation of the 'ether', and since then various other 'stimuli' have occurred, from atmospheric science to vortices in superfluid Helium, to keep the quest for such solutions alive. One is led to an eigenvalue problem for a system of coupled algebraic equations. Analytical solutions are known in which the vortices form a system of concentric, nested polygons each with the same number of vertices with or without a vortex at the center. Numerical calculations reveal additional solutions with the vortices arranged approximately, but not exactly, on a system of concentric circles. All known configurations have shown at least reflection symmetry about an axis, but may not have had more symmetry than that. These vortex flows have both a simple spatial structure (because of the symmetry) and simple temporal evolution.

Recently we found several 'families' of qualitatively new solutions, including configurations that are spatially asymmetric (Aref & Vainchtein 1998). These vortex flows, then, have simple temporal evolution but complex spatial structure. One methodological aspect of this investigation is worth highlighting. In searching for new equilibrium configurations we found the co-rotating points, i.e., points in the fluid that rotate as if attached to the rigidly rotating vortex configuration, to be of great utility. Such points can be thought of as 'embryonic' vortices. Starting from a co-rotating point in an N-vortex configuration one can 'grow' an $(N+1)$-vortex configuration rotating with the same angular frequency by gradually increasing the vortex strength of the co-rotating point up to the strength of the other vortices. This algorithm, with no basis in physical processes, proved surprisingly versatile in generating new states. The number of co-rotating points for an N-vortex equilibrium can be quite large – for vortices placed in a regular N-gon, for example, there are $3N - 1$ co-rotating points. It seems to be a recurring theme that dynamically defined points, such as instantaneous stagnation points (Aref & Brøns 1998) or co-rotating points, play significant roles in understanding the flow field.

While these results may not directly have a bearing on the problem of 2D turbulence, the asymmetric nature of the states found has a 'flavor' of spatial irregularity usually associated with turbulent flows. We shall encounter further examples of spatially irregular vortex equilibria below.

4 The role of the three-vortex problem

The generation of new inter-vortex distances in an N-vortex system hinges on three-vortex interactions. The one- and two-vortex problems, whether on the unbounded plane or in a periodic strip or parallelogram, do not generate new scales. From the equations of motion one can derive a formula for the rate of change of the inter-vortex distance in which it is expressed as a sum over contributions from all 'third' vortices. This formula, thus, gives a 'decomposition' of the rate of change of a 'flow scale' (i.e., separation) associated with any particular pair of vortices into contributions from all vortex triples of which the pair in question is a part. On the other hand, it is clear that the two-vortex problem, with sum of circulations equal to zero and with the linear impulse conserved, will lead to the two vortices having the same velocity, regardless of the explicit form of the interaction. Hence, on the unbounded plane or for the periodic parallelogram, or its limiting form the periodic strip, the vector connecting the two vortices will be constant and, *a fortiori*, its length will be constant. On the unbounded plane the formula for the rate of change of inter-vortex distances has a particularly simple geometrical interpretation (Gröbli 1877; Novikov 1975; Aref 1979). In summary, if we wish to generate new scales of motion, in the sense of new inter-vortex separations, we must at least include three-vortex interactions. Conversely, we shall see below that once three-vortex interactions are included, at least in the periodic problems, the advection produced by three interacting vortices is, in a precise sense, as complicated as we may expect over a subdomain of the flow, and cannot be made 'more complicated' by adding further vortices.

4.1 Kinetic theory

One obvious idea in a particle model in the limit of many particles is to try a kinetic theory. As we have just argued, the interaction on which to base such a theory is not the usual two-body problem, as in Boltzmann's theory, but rather the three-vortex interaction. This insight was already provided by Novikov (1975), who pursued a kinetic theory for identical vortices. When circulations of both signs are allowed, the formalism becomes more complex. Marmanis (1995) has provided details in his thesis (and in a recent paper, Marmanis 1998) for the case when all vortices have the same absolute value of circulation and so can be distinguished simply by the sign of that circulation. It is not terribly difficult to write down the BBGKY hierarchy of kinetic equations for the various distribution functions in this case.

A convincing closure of the kinetic equations for three vortices has not been achieved thus far. If one tries to achieve such a closure by thinking of a pair of vortices (of opposite circulations) scattering off a stationary single vortex, one problem that arises is that this scattering cross-section is itself quite complicated. The scattering angle as a function of an impact parameter has singularities, since the three vortices may become trapped in a uniformly rotating state. One also has to contend with the fact that the 'target' vortex may become part of the outgoing, scattered vortex pair (a process that we have called 'exchange' scattering; Aref 1979).

A very interesting feature of this approach is that the vortices play the role of the 'atoms' in conventional kinetic theory. Their interactions are what define the turbulence which, thus, becomes a larger-scale 'meta-fluid'. The transport coefficients of this 'meta-fluid' are then to be calculated as averages (in the sense of kinetic theory) over this 'gas' of interacting vortices. In particular, the 'meta-fluid' may acquire a viscosity, in the same way that kinetic arguments applied to the reversible mechanical interactions of massive particles lead to an expression for the viscosity of a gas. This line of reasoning appears as a formalized version, albeit only for 2D flows, of classical ideas in turbulence related to such concepts as 'eddy viscosity'.

4.2 Solutions of the three-vortex problem

In a periodic parallelogram the three-vortex problem is still integrable. The solutions are, however, sufficiently complicated that we have been led to believe that they will directly pertain to the problem of 2D turbulence even without the introduction of kinetic arguments. We shall not provide full details, which may be found in our papers (Aref & Stremler 1996; Stremler & Aref 1998). We simply highlight three points, two methodological, the last more phenomenological.

On the methodological side, we recall that the method of solution consists in a mapping of the separation between two vortices in the original three-vortex problem onto a problem of advection of a passive particle in the flow field of a certain system of fixed vortices. This 'symplectic reduction' has the pleasing interpretation in terms of a closely related physical problem. If the original three-vortex problem is posed for a square domain of side L, the advection problem takes place on a larger domain depending on the vortex strengths. (Similar results are obtained for a general parallelogram, but it is simpler to think in terms of a square.) Interestingly, this domain is infinite if the ratio of two of the vortex circulations is irrational (since the circulations sum to zero, all ratios of circulations of different vortices are either rational or irrational). If the ratio is rational, the advection problem 'lives' in a square of larger size. The number of fixed vortices, and the size of the periodic square to be used for the advection problem, grows rapidly with the 'degree of irrationality' of the circulation ratios.

As a second methodological point we note that the shape of the periodic parallelogram plays an important role in the nature of the solutions of the three-vortex problem. For example, even in the simplest case of three vortices with circulations in the proportion 1:1:(-2) we find that there are solutions that appear when the parallelogram is a rhombus of general angle that are 'forbidden' when the parallelogram is a square. This raises the possibility that a numerical simulation of 2D turbulence with many discrete vortex structures might unintentionally be biased by the canonical choice of a square shape for the periodic computational domain!

Phenomenologically, the nature of the solutions follows from the construction of the 'reduced' advection problem. The size of the square to which one is led by this reduction defines the overall size of motions of the three vortices in the original problem since the separation of any two of the vortices is directly related to the change in coordinates of the advected particle. Thus, if the advection problem requires a periodic square of side $10L$, which will happen for three vortices with strengths in the proportion 7:3:(-10), then for some initial conditions the vortices will roam over a domain of that size during one period of their relative motion, i.e., the length of individual vortex trajectories will be ten times the periodic square in which the three-vortex problem is defined. The resulting trajectories pass through the basic periodic square several times, and easily become as complex as the trajectories found by direct tracking of vortices in large-scale numerical simulations of 2D turbulence (cf. McWilliams 1990).

4.3 Advection by three vortices

It has been known for many years that the advection produced by three interacting vortices can be chaotic. We have recently shown that topological results embodied in so-called Thurston–Nielsen theory provide a further classification of the nature of the advection. Most significantly, these results show that for three vortices in a periodic strip or parallelogram the 'mapping' induced by the advection during one period of the relative motion is everywhere hyperbolic in a subdomain of the periodic domain, i.e., within this subdomain the advection is like the well known 'cat map'.

The application of Thurston–Nielsen theory to advection problems has been illustrated in the analysis of a low-Reynolds-number experiment using three stirring rods in a viscous fluid (Boyland, Aref & Stremler 1998). However, point vortices can play the role of 'rods', and because of the topological nature of the arguments, the inviscid flow induced by the point vortices 'works just as well' as the Stokes flow of the experiment. Upon (not inconsiderable) analysis, the conclusion is that three-vortex motion in a periodic strip or parallelogram can, in some regimes, stir the surrounding fluid over some subdomain as effectively as a 'cat map'. It may be that adding further vortices will enlarge the region over which this kind of stirring takes place, or may enhance the stretching rate (or both). Everywhere-hyperbolic advection is the most active process of advection that one can imagine, and in this sense the fluid stirring cannot become more chaotic. The theory seems to have a notion of 'intermittency' built in, in the sense that only a certain, probably quite complicated, subdomain supports this kind of everywhere-exponential stretching. Thurston–Nielsen theory says little about the size or shape of the subdomain. Laboratory and numerical experiments suggest that this well-stirred region is on a scale comparable to the scale of motion of the stirring agents.

The unbounded plane is apparently simpler than the periodic parallelogram case in the sense that Thurston–Nielsen theory applied to the stirring produced by the three vortices does not imply a subdomain with everywhere hyperbolic advection when the periodic boundary conditions are relaxed.

5 The energy spectrum

The historical evolution of the theory of turbulence has focussed attention on a quantity known as the energy spectrum, defined as the energy content of the flow per unit wave-number. This quantity is particularly well adapted to a representation of the flow in terms of Fourier modes. When thought of in terms of interacting vortices the energy spectrum becomes considerably less intuitive and appears more as an after-the-fact flow diagnostic than as a key dynamical entity.

Currently, there are at least three competing approaches to provide a physical interpretation of the energy spectrum and to attempt to relate it to real space structures and processes. The first of these is represented by the work of Berry *et al.* (1979), Moffatt (1984), and Gilbert (1988). One starts from essentially kinematic considerations. The instantaneous flow field produced by a system of vortices consists of a set of 'eddies' separated by 'dividing streamlines'. The 'eye' of each eddy is an elliptic fixed point of the flow. Since this 'eye' is situated within vortical fluid, the ensuing motion wraps vortical fluid into a spiral pattern along with irrotational fluid (or fluid of much less vorticity). If one now samples this spiral along a line intersecting its arms many (in principle, infinitely many) times, and if one assumes the vorticity jumps, essentially discontinuously, from one finite value to another every time the line crosses a surface between vortical and irrotational fluid, then it may be shown that a power-law form of the energy spectrum emerges. Different spiral models translate into different spectral exponents.

The second approach is represented by the work of Carnevale *et al.* (1991, 1992). A scaling theory is proposed that addresses the temporal evolution of the energy spectrum (and says very little about the power-law form in terms of wavenumber). The vortex merger process, observed for finite area vortices, is the main physical mechanism that is believed to scale. These authors demonstrate correspondence between full numerical simulations and a point vortex model in which like-signed vortices are merged by fiat subject to a certain criterion of closeness. Before and after such merger events the dynamics is the Hamiltonian dynamics of point vortices that we have been considering. The system with merging has been referred to as a 'punctuated' Hamiltonian system.

The third approach, with which we are particularly intrigued, stems from a formula for the energy spectrum of the flow due to an assembly of point vortices first derived by Novikov (1975). This formula gives the energy spectrum as a sum of zero order Bessel functions. The 'scales of motion' in this representation are the inter-vortex distances. Relatively little work has been done on this formulation.

Sedov (1976) performed numerical simulations on the interaction of 100 *identical* vortices, initially placed uniformly on a circle, and monitored the spectrum via Novikov's formula. The choice of identical vortices is somewhat unfortunate, since an appropriate vorticity distribution for 2D turbulence should have the sum of circulations equal to zero. Sedov (1976) claims to see familiar power-law exponents, k^{-3} and $k^{-5/3}$, emerge in his numerical data. However, the numerical resolution is low, and the fits to the data are not particularly convincing.

Regarding the first approach, we comment that besides the elliptic points, of which there are N in an N-vortex system, there are also $O(N)$ hyperbolic points. These instantaneous stagnation points are the nodes of the network of dividing streamlines. We mention in passing that their location has an intriguing geometrical interpretation in terms of the foci of a certain algebraic curve (Aref & Brøns 1998). This provides some information on the location of these stagnation points relative to the vortices as the flow evolves (and, for a fixed configuration of vortices, parametrically in terms of the vortex strengths). The spectral 'signature' of each hyperbolic point, then, is likely to be at least as important as those of the elliptic points. It is well known from dynamical systems theory that the kinematic nature of the flow in the vicinity of a hyperbolic point is very different from the flow close to an elliptic point, and generally consists of a 'lattice' of intersecting manifolds. A passively advected substance will be stretched into strips of geometrically increasing length and decreasing width. The spectral form of such a structure would be of interest.

Regarding the third approach, we mention that we have found, as a by-product of our solution of the three-vortex problem, a family of infinite, stationary vortex arrays in which the total circulation vanishes and the total impulse is zero. These arrays are simply the fixed, advecting vortices that emerge in the 'reduced' problem when solving for three-vortex motion in a periodic parallelogram. The nature of these states is as follows: They contain vortices of three different circulations in the proportion $-(1-\lambda^2) : 2(1-\lambda) : 2(1+\lambda)$, where λ can be any real number between 0 and 1. Vortices of the same circulation are situated on a regular lattice. The three lattices are similar with periods in the proportion $(1-\lambda^2) : 2(1-\lambda) : 2(1+\lambda)$. For given λ there is a two-parameter family of solutions parametrized by the offset of the three lattices from one another. If λ is rational and equal, say, to p/q in lowest terms, the three sublattices in the vortex configuration are commensurate and will all repeat periodically on the scale of $2q$ times the period of the first lattice. Hence, the configuration has at most this longer spatial period. When λ is irrational, on the other hand, the three lattices are incommensurate, and there is no overall periodicity to the configuration. In these cases we produce what might be called a 2D *vortex quasi-crystal.* This is an infinitely extended equilibrium configuration, i.e., a solution to the point vortex equations, without spatial periodicity. We are currently exploring the nature of the energy spectrum for such configurations, in the sense of Novikov's formula, and whether there are particular choices of the parameters that yield results of relevance to 2D turbulence.

Different power law regions of the spectrum of 2D turbulence may be associated with the different views described above. Thus, the k^{-3} spectrum may, indeed, be produced by the winding around individual strong vortices (and maybe even a hierarchical arrangement of such 'windings'). The distribution of strong vortices, treated as points, may lead to another scaling and power law region of the spectrum. A recurring issue as one approaches the problem of 2D turbulence from the viewpoint of interacting vortices is that the canonical spectral power laws appear as a rather artificial way of viewing the problem. Most of the results that one obtains in vortex dynamics have an immediate expression in terms of real space processes, phenomena and structures. Fourier transforming everything and looking at these same results in wave-number space is really only done in order to make connections with a large body of established practice in turbulence research and is in this context more of a 'diagnostic' than an introduction of obvious or necessary dynamical variables.

6 Conclusions

We may summarize our insights to date as follows: In the point vortex system inter-vortex separations play the role of 'scales of motion' of the turbulence. The three-vortex problem is the basic interaction responsible for generating new scales; the one- and two-vortex problems do not produce new scales. In a kinetic theory for point vortices, then, the key equation to be modelled, and the level at which closure may be attempted, is for the three-vortex distribution function.

The three-vortex problem itself is integrable in two important cases: on the unbounded plane for arbitrary vortex strengths, and for vortices in a parallelogram with periodic boundary conditions, where periodicity implies that the sum of vortex circulations is zero. The three-vortex problem on the unbounded plane was solved long ago. The problem of three vortices in a periodic parallelogram was solved recently, and we have found that individual vortex trajectories for this integrable problem may be considerably more complicated than expected and easily as complex as the trajectories of strong, isolated vortices traced in large-scale numerical simulations of high Reynolds number 2D Navier–Stokes equation turbulence.

We have studied the advection of a fluid particle in the field of three vortices and shown, using Thurston–Nielsen theory, that there is, in general, a subdomain of the periodic parallelogram in which the advection is of the nature of a 'cat map'. Since this map is everywhere hyperbolic, the fluid advection is 'as chaotic as possible' in this subdomain. Adding further vortices cannot make the advection more chaotic, although the size of the domain and the magnitude of the rate of stretching of material lines may increase. In this sense, for the periodic parallelogram the three-vortex problem is sufficient to produce 'turbulent mixing'.

We have used a formula for the energy spectrum, first given by Novikov, to study the spectral 'signature' of various point vortex motions and equilibrium stationary configurations.

References

Aref, H. 1979. Motion of three vortices. *Phys. Fluids* **22**, 393–400.

Aref, H. & Brøns, M. 1998. On stagnation points and streamline topology in vortex flows. *J. Fluid Mech.* (in press).

Aref, H. & Stremler, M. A. 1996. On the motion of three point vortices in a periodic strip. *J. Fluid Mech.* **314**, 1–25.

Aref, H. & Vainchtein, D. L. 1998. Asymmetric equilibrium patterns of point vortices. *Nature* (in press).

Berry, M. V., Balazs, N. L., Tabor, M. & Voros, A. 1979. Quantum maps. *Ann. Phys.* **122**, 26–63.

Birkhoff, G. & Fisher, J. 1959. Do vortex sheets roll up? *Rend. Circ. Mat. Palermo* **8**, 77–90.

Boyland, P. L., Aref, H. & Stremler, M. A. 1998. Topological fluid mechanics of stirring. *J. Fluid Mech.* (submitted).

Carnevale, G. F., McWilliams, J. C., Pomeau, Y., Weiss, J. B. & Young, W. R. 1991. Evolution of vortex statistics in two-dimensional turbulence. *Phys. Rev. Lett.* **66**, 2735–2737.

Carnevale, G. F., McWilliams, J. C., Pomeau, Y., Weiss, J. B. & Young, W. R. 1992. Rates, pathways, and end states of nonlinear evolution in decaying two-dimensional turbulence: Scaling theory versus selective decay. *Phys. Fluids A* **4**, 1314–1316.

Gilbert, A. D. 1988. Spiral structures and spectra in two-dimensional turbulence. *J. Fluid Mech.* **193**, 475–497.

Gröbli, W. 1877. *Specielle Probleme über die Bewegung geradliniger paralleler Wirbelfäden.* Zürcher und Furrer, Zürich, 86pp.

Kraichnan, R. H. 1975. Statistical dynamics of two-dimensional flow. *J. Fluid Mech.* **67**, 155–175.

McWilliams, J. C. 1990. The vortices of two-dimensional turbulence. *J. Fluid Mech.* **219**, 361–385.

Marmanis, H. 1995. *Statistical vortex dynamics and two-dimensional turbulence.* MS thesis, University of Illinois at Urbana–Champaign.

Marmanis, H. 1998. The kinetic theory of point vortices. *Proc. Roy. Soc. London A* **454**, 587–606.

Moffatt, H. K. 1984. Simple topological aspects of turbulent vorticity dynamics. In *Turbulence and Chaotic Phenomena in Fluids*, T. Tatsumi, ed., IUTAM/Elsevier, pp. 223–230.

Novikov, E. A. 1975. Dynamics and statistics of a system of vortices. *Sov. Phys. JETP* **41**, 937–943.

O'Neil, K. A. 1989. On the Hamiltonian dynamics of vortex lattices. *J. Math. Phys.* **30**, 1373–1379.

Onsager, L. 1949. Statistical hydrodynamics. *Nuovo Cim.* **6** (suppl), 279–287.

Sedov, Yu. B. 1976. Evolution of the energy spectrum of an eddy system. *Izv. Akad. Nauk SSSR, Mekh. Zhid. Gaza* **6**, 43–48.

Stremler, M. A. & Aref, H. 1998. Motion of three point vortices in a periodic parallelogram. *J. Fluid Mech.* (submitted).

Combining deterministic and statistical approaches to compute two-dimensional turbulent flows with walls

M. Farge, N. Kevlahan, C. Bardos, K. Schneider

1 Research programme

'Although this may seem a paradox, all exact science is dominated by the idea of approximation' (Bertrand Russell).

We are searching for the best approximation to compute fully-developed two-dimensional turbulent flows. Fluid mechanics is governed by the Navier–Stokes equations, which are entirely deterministic. Fully-developed turbulence corresponds to very large Reynolds number flows (for which the micro-scale Reynolds number Re is larger than 10^4) and is the regime where the nonlinear advective term of Navier-Stokes equations strongly dominates the linear dissipative term. In this limit, the solutions to the Navier–Stokes equations are highly chaotic and we are unable to integrate them. Therefore, in order to compute fully-developed turbulent flows we need to combine a deterministic numerical integration with a statistical model. In this paper we propose a possible solution to this problem, based on the wavelet representation.

The classical method of computing fully-developed turbulent flows is based on Reynolds averaging: in each realization the flow is separated into a mean part and a fluctuating part using a suitable averaging procedure. Such a separation is necessary because the huge number of degrees of freedom in a high Reynolds number turbulent flow prohibits a direct numerical simulation (DNS). The goal is to calculate the evolution of the mean part in detail using a deterministic equation and to design a statistical model which simulates the effect of the fluctuating part on the mean. However, because the Navier–Stokes equations are nonlinear, we must address the closure problem, i.e. the equations for the nth order moment depend on the n+1th order moment. Thus this method requires that the statistics of the fluctuating part be known completely. The closure problem can be solved if the statistics are Gaussian, since in this case all higher-order even moments can be expressed in terms of the second order moments and all odd moments are zero. Therefore the fundamental difficulty in turbulence is to find an averaging technique that produces a fluctuating part with Gaussian statistics. Unfortunately, the classical averaging techniques (e.g. separation into large scale and small scale eddies used in Large Eddy Simulation) do not guarantee this. We will propose one such averaging procedure that overcomes the closure problem, based on the wavelet-representation of the vorticity field.

In this paper we focus on two-dimensional turbulence, but our results could be extended to three-dimensional turbulence. From a physical point of view the two-dimensional approximation is relevant for studying large-scale geophysical flows, due to the combined effect of stable stratification and Earth's rotation. From a mathematical point of view there are existence, uniqueness and regularity theorems for the two-dimensional Navier–Stokes equations, which are not yet available in three dimensions. These theorems are necessary to validate the numerical procedure we use to solve Navier–Stokes equations. Therefore, from a numerical analysis point of view, numerical integration is better justified for two-dimensional turbulent flows than for three-dimensional turbulent flows. Moreover, the existence of an inertial manifold has been proven for two-dimensional turbulent flows, and upper bounds for the dimension of the attractor have been given, but this is still an open problem for three-dimensional turbulent flows. A final argument justifying our interest in two-dimensional turbulent flows is that, according to the usual estimation the minimal number of degrees of freedom N necessary to compute fully-developed turbulent flows without a model, namely by DNS, scales as Re in two dimensions and as $Re^{9/4}$ in three dimensions. Therefore using DNS we are able to compute much larger Reynolds number flows in two dimensions than in three dimensions. For two-dimensional turbulence we have already reached the fully-developed turbulent regime without resorting to any *ad hoc* turbulence models, but this is not yet the case for fully-developed three-dimensional turbulence.

2 Coherent vortex eduction

Since 1984 we have proposed using the wavelet representation to analyze, model and compute fully-developed turbulent flows. We have shown that the strong wavelet coefficients correspond to the coherent vortices, while the weak wavelet coefficients correspond to the incoherent background flow (for 2D turbulence see Farge and Rabreau 1988, Farge and Sadourny 1989, Farge, Holschneider and Colonna 1990, and for 3D see Farge, Meneveau, Guezennec and Ho 1990, Farge 1992). Both components are multiscale and therefore cannot be separated by Fourier filtering. We have developed a method, inspired by Donoho's denoising technique (wavelet shrinkage see Donoho 1993), to separate coherent vortices from the background flow (Farge, Schneider and Kevlahan 97). To extract the coherent vortices we reconstruct the vorticity field from its wavelet coefficients, retaining only those larger than a threshold value $\tilde{\omega}_T = (2Z \log_{10} N)^{-1/2}$, which depends on Z the total enstrophy and N the resolution without any *ad hoc* adjustable parameter. Figure 1 shows an example of this separation applied to a vorticity field which has been computed at a resolution a 256^2. Only 0.7% of the wavelet coefficients correspond to the coherent vortices, which have the same non-Gaussian Probability Distribution Function (PDF) as the total field, while the remaining 99.3% weaker wavelet coefficients correspond to the incoherent background flow, which has a Gaussian PDF. Using the wavelet representation higher resolutions produce stronger compression ratios.

We separate each turbulent flow into:

- A low-dimensional dynamical system out of statistical equilibrium, which we can compute with as little approximation as possible, using a deterministic equation.

- A high-dimensional dynamical system which has reached a statistical equilibrium state, for which we can compute averages that can be modelled by a Gaussian stochastic process.

In order to perform this separation we assume that there are two kinds of elementary motions characteristic of two-dimensional turbulent flows:

- *Self-organized vortices*, which are dynamically stable and each vortex belongs to an elliptic region and is characterized by the existence of a monotonic coherence function $\omega = F(\psi)$. These vortices are out of statistical equilibrium because the encounters with other vortices are too rare to produce enough mixing to generate an equilibrium state. The one-point PDFs of the coherent vortices are non-Gaussian and their entropy evolves over time, which means that we cannot rely on the Central Limit Theorem to define averages. This means that we cannot discard the phase information of the vortices (i.e. their positions) and we should therefore compute their motions with as little approximation as possible.

- *Incoherent background flow*, made of one-dimensional vorticity filaments which are dynamically unstable (they belong to hyperbolic regions and are stretched in one direction and compressed in the orthogonal direction) and have reached statistical equilibrium (because the strain imposed by the coherent vortices on the background flow inhibits the formation of new coherent vortices and mixes the filaments). Because the one-point PDFs of the background flow are Gaussian and entropy has reached its maximum we can rely on the Central Limit Theorem to define averages and discard the phase information (namely the spatial distribution of the vorticity filaments). Only its mean and variance are necessary to describe the stochastic effect the background exerts on the coherent vortices.

We have shown that the incoherent background flow is slaved to the coherent vortices, due to their straining which inhibits the development of any nonlinear instability in the background flow (Kevlahan and Farge 1997). From this result we have conjectured that the number of coherent vortices may saturate to a constant number when the Reynolds number is sufficiently large to produce enough vortices in boundary layers in order to inhibit any instability which would otherwise form new vortices in the bulk of the flow. We have shown that both coherent vortices and the incoherent background are multiscale, and therefore we propose the wavelet representation to compute their evolution. This is because the wavelet basis is scale invariant (being based on the affine group), and is thus better suited to compute turbulent flows than the Fourier basis which is wavenumber invariant but not scale invariant (being based on the Weyl-Heisenberg group). Similarly, the wavelet basis is preferable to a pure grid point representation since a grid point basis is translation invariant, but not scale invariant.

3 Coherent Vortex Simulation

We are presently developing a new method to compute fully-developed two-dimensional turbulent flows, which is based on a wavelet phase-space segmentation (Farge et al. 1992, Farge et al. 1996, Farge et al. 1997) and which uses the wavelet-based Navier–Stokes solver designed by Jorgen Frölich and Kai Schneider (Frölich and Schneider 1996). This method computes the dynamics of the coherent vortices with a limited number of wavelet modes, keeping only the most excited ones which correspond to coherent vortices, and re-mapping the wavelet basis at each time step. We have compared simulations using these wavelet techniques with standard spectral simulations and nonlinearly filtered spectral simulations (Schneider, Kevlahan and Farge 1997). The results showed that the wavelet method is very accurate and require fewer active modes than spectral methods. Moreover, the aliasing errors remain localized to regions of strong gradients, in contrast to the Fourier representation which spreads aliasing error over the entire solution. One can always reduce aliasing errors by locally adding more wavelet modes to improve the resolution where needed.

We have shown that the number of active wavelet modes is approximately constant in time, even during intense nonlinear interactions, whereas the number of active Fourier modes peaks when the interactions are more intense and strongly excite small scale. The discarded coefficients, which correspond to the incoherent background flow and have a Gaussian one-point PDF for velocity and vorticity, should be modelled statistically in order to take into account their effect on the coherent vortices. We can either model them by a stochastic forcing having the same statistical behaviour, compute the linear equation characterizing their motions, or design a one-point turbulence model (such as Boussinesq, Smagorinsky or k-ϵ). The justification for this procedure is that the coherent vortices are not numerous enough and their encounters are too rare events to have reached a statistical equilibrium state, and therefore we have to compute their dynamics with a deterministic method. On the contrary, for the well-mixed background flow we can assume stationarity, homogeneity and ergodicity in order to define a statistical equilibrium state from which we can design an appropriate statistical model.

4 Wall effect

4.1 Mathematical analysis

In two space dimensions and for any finite time both the solutions of the incompressible Navier Stokes equation and of the incompressible Euler equations are well controlled for smooth initial data. This result is true for a solution defined either in the whole space, in a periodic box, in the interior of a vessel or around an obstacle. In the last two cases the fluid domain is denoted by Ω, the boundary of this domain is denoted $\partial\Omega$ and the outward unit normal to this boundary is denoted by $\vec{n}$. Some boundary conditions have to be assumed. The natural boundary condition for the Navier–Stokes equation is the viscous boundary condition

$$u_{\nu\,\partial\Omega} = 0 \tag{1}$$

and for the Euler equation it turns out to be

$$u_\nu \cdot \vec{n}, \tag{2}$$

which means that the fluid does not penetrate or leave the domain.

The viscous boundary condition is a natural approximation of the effect of a 'rough boundary'. This can be seen as follows. The action of a rough boundary can be described at the level of the Boltzmann equation with the introduction of a scattering kernel $R(v, v')$. Then an asymptotic analysis, with a Knudsen number and a Mach number both of the order of $\epsilon \to 0$ gives a boundary condition for the macroscopic equation of the form $u = O(\epsilon)$ on the boundary. For the Euler equation the tangential component of the velocity is not usually equal to zero. Therefore when the viscosity goes to zero the quantity

$$u_\nu \wedge \vec{n} \tag{3}$$

goes to a non-zero value. The solution cannot remain uniformly smooth near the boundary and this implies that some boundary layer must appear. Since the problem is nonlinear, in many unstable cases this layer of strong vorticity does not remain confined near the boundary, but instead moves into the interior of the domain. Such a situation corresponds to detachment of the boundary layer. This implies that the solution u_ν of the Navier Stokes equation which remains uniformly bounded in energy norm may not converge to a solution of the Euler Equation (Bardos and Ghidaglia 1998). The fact that the u_ν converges in the weak sense in the space

$$L^\infty_{loc}(\Re_t; L^2(\Omega)) \tag{4}$$

to a function u does not implies the relation

$$\lim_{\nu\to 0}(u_\nu \otimes u_\nu) = (u \otimes u) \tag{5}$$

Hilbert space theory shows that when the above relation is not valid one has

$$\lim_{\nu\to 0}(u^i_\nu \otimes u^j_\nu) - (u^i \otimes u^j) = R^{ij} \tag{6}$$

with R^{ij} denoting a symmetric positive tensor, and thus the limit of the Navier Stokes equation may not be the Euler equation, but rather the equation:

$$\frac{\partial u}{\partial t} + \nabla(u \otimes u) + \nabla R = -\nabla p,\ \nabla \cdot u = 0 \tag{7}$$

R plays the role of the Reynolds tensor and it is interesting to notice that in this case its appearance is due only to the fact u is a "weak limit" of solutions of the Navier Stokes equation with viscous boundary condition. No separation of scale and no introduction of a family of solutions nor of ensemble averaging is needed. It is possible that the region where R is not zero (which is not necessarily confined near the boundary) is the turbulent region.

With the divergence free condition the tensor R, which is called the 'defect measure', can be chosen to be symmetric and with zero trace. Such a property is also true for the strain tensor $S(u)$ therefore one can write

$$R = \nu_S(x,t)S(u) + \nu_T(x,t)T(u) \tag{8}$$

with $\nu_S(x,t)$ and $\nu_T(x,t)$ scalar functions and T orthogonal to S (for the canonical scalar product). Even if no space invariance is present, it may be possible that the defect measure, which represents a fluctuation, is rotationally invariant. In this case, the formula (8) should reduce to the formula

$$R = \nu_S(x,t)S(u) \tag{9}$$

leading in (7) to a turbulent diffusion provided that the function $\nu_S(x,t)$ is positive. Such a property should be true in 'reasonable' situations; it is not a direct consequence of the positivity of the tensor R. In fact at present there is no direct way of constructing this 'enhanced diffusion'. The above consideration should only be used as guideline for the study of two dimensional turbulent flows generated by viscous boundaries. Following Chorin (Chorin 1996) one could introduce a fractional step method with a diffusion equation, which would generate vorticity from the boundary, and a treatment of the incompressible Euler equation. To study the propagation of the vorticity this may be similar to the decomposition between the coherent vortices and the incoherent background flow we have proposed in this paper.

4.2 Physical modelling

Although DNS are the only methods able to compute turbulent flows without resorting to *ad hoc* models, they are currently only of academic interest and have not proved directly useful to engineers. This is due to two drawbacks. First, as mentioned above, such simulations cannot reach the high Reynolds numbers typical of most flows of interest to engineers. We hope that the combination of wavelet-based numerical methods for the Coherent Structure evolution and statistical modelling of the Gaussian background will overcome this limitation. Secondly, DNS are usually limited to simple boundary conditions (e.g. periodic boundary conditions) and geometries (e.g. a rectangular prism). The most realistic DNS manage to compute flow in a rectangular channel with periodic boundary conditions in the streamwise and spanwise directions and no slip conditions in the perpendicular direction (using Chebyshev polynomials). Obviously, a DNS which computes flow in a box with periodic boundary conditions is not useful for calculating the flow over a wing!

The problems of treating no-slip boundary conditions and complicated geometries at high Reynolds numbers are both theoretical and numerical in nature. The problem is difficult theoretically because the only mathematical theory available for flow over a wall (e.g. Prandtl's law of the wall) fails as soon as the boundary layer detaches. Unfortunately, the boundary layer detaches as soon as the flow becomes turbulent, which severely limits the use of boundary layer theory for describing fully turbulent flow over a wall. There are two main classes of numerical difficulties, the first related to matching the correct boundary conditions at the wall (or edge of the computational domain) and the second related to calculating the flow around complicated shapes. The turbulent boundary layer also demands very high resolution since the boundary layer thickness decreases with Reynolds number according to $\delta \propto Re^{-1/2}$.

Matching boundary conditions requires that the basis function used to represent the solution have the correct behaviour near the wall (e.g. that they are zero at the wall, as is the case for the Chebyshev polynomials mentioned above). This limits the choice of basis functions and means that a set of basis functions that is very efficient and accurate numerically may not be able to be used (this is the reason Chebyshev polynomials were used for the channel flow instead of the more efficient Fourier basis functions). Wavelets can be constructed so as to have the desired properties at the boundaries, and recently an efficient algorithm for constructing wavelets over an interval on an irregular grid has been developed (Sweldens 1996). The fact that wavelets can be easily constructed on irregular grids allows them to increase resolution near the wall in order to fully resolve the thin boundary layer of a fully-developed turbulent flow.

Figure 1. (Next two pages.)
Wavelet compression of vorticity: (a) The vorticity. (b) The modulus of velocity. (c) The stream function. (d) The coherence scatter plot. (e) Cut of vorticity. (f) PDFs of velocity and vorticity. (g) Energy spectrum. The solid lines correspond to the total vorticity ω, the dashed lines to the coherent part $\omega_>$, and the dotted lines to the incoherent part $\omega_<$. We observe that only 0.7% of the total number of wavelet coefficients are sufficient to represent all coherent vortices, while the remaining 99.3% correspond to the incoherent background flow, which is much weaker and homogeneous. The coherent vorticity $\omega_>$ contains 94.3% of the total enstrophy. Moreover, the velocity associated with the coherent vortices is quasi-identical to the total velocity and contains 99.2% of the total energy. As for the coherent stream function, $\psi_>$, it is identical to the total stream function ψ. The fact that the scatter plot of the background, $F_<$ such that $\omega_< = F_<(\psi_<)$, is isotropic proves that our method has extracted all coherent vortices. The PDFs of velocity and vorticity show that only 0.7% of the wavelet coefficients are sufficient to capture the non-Gaussian one-point statistical distributions of vorticity and velocity, while the remaining 99.3% correspond to a Gaussian distribution. The energy spectrum, on the contrary, is dominated at small scales by the incoherent background flow and therefore is insensitive to coherent vortices because they are too rare to affect the energy spectrum (which is the Fourier transform of the two-point correlation function).

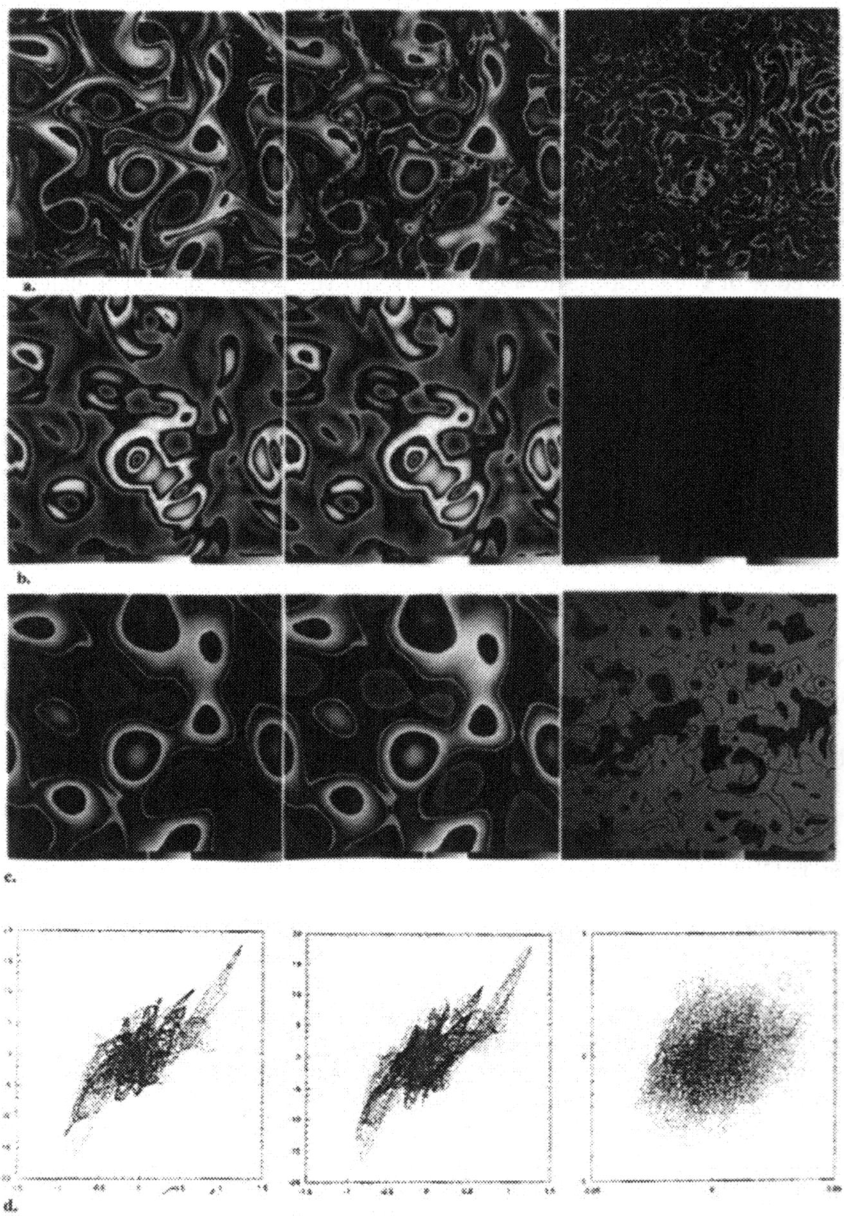
a.
b.
c.
d.

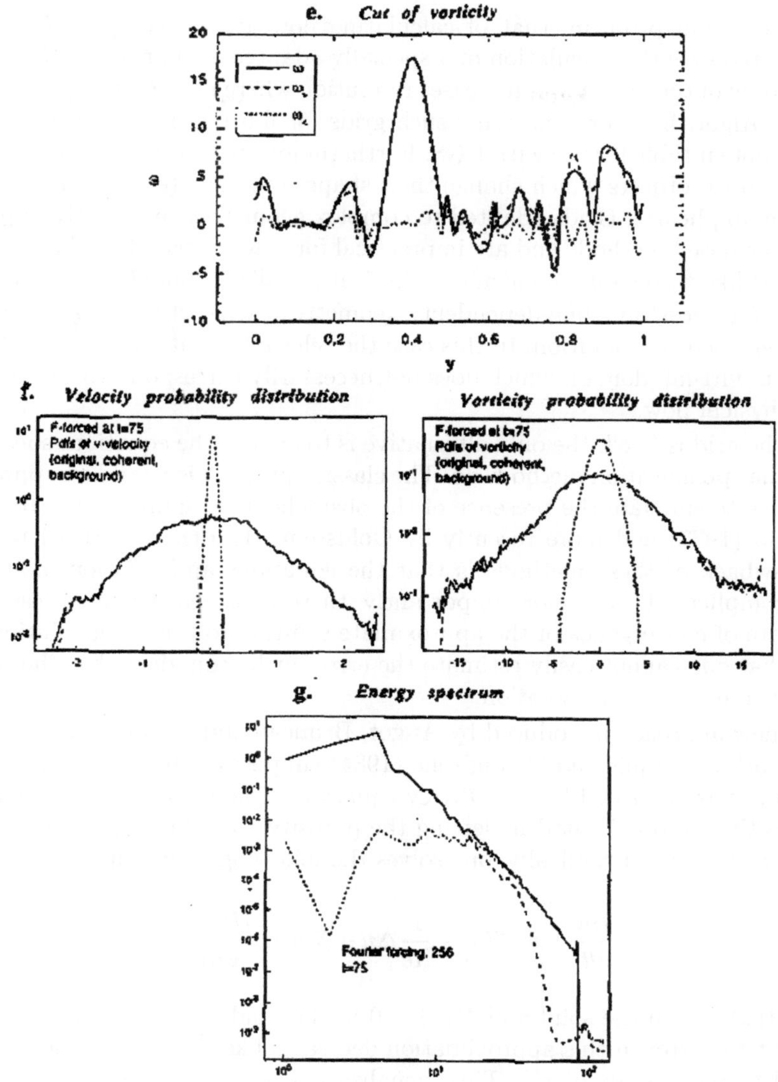
e. Cut of vorticity
ω
y
f. Velocity probability distribution
F-forced at t=75
Pdfs of v-velocity
(original, coherent, background)
Vorticity probability distribution
F-forced at t=75
Pdfs of vorticity
(original, coherent, background)
g. Energy spectrum
Fourier forcing, 256
t=75

The second problem, that of calculating flow around complicated objects, normally requires the calculation of a specially adapted grid or mesh, that follows the contours of the object and increases resolution where necessary (e.g. near sharp corners). Algorithms for generating such grids for finite element codes exist, but they are not suitable for use with DNS. Furthermore, such grid-based approach are impractical for objects which change their shape over time (e.g. a swimming fish, or a pumping heart). Grids adapted to complex geometries are clearly impossible for Fourier-based methods and are impractical for wavelet-based methods. Ideally, one would like to be able to calculate physically realistic boundary conditions for complicated (possibly time-dependent) geometries while still using a Cartesian grid to represent the solution. In this case the relevant equations are solved over a Cartesian 'virtual' domain which does not necessarily correspond to the geometry of the physical flow.

If the grid is fixed, the only alternative is to change the equations solved, and this is what penalisation methods do. The classic approach is to add an appropriate body force to simulate the presence of the obstacle. This approach has been used by Peskin (1977) and more recently by Goldstein, Handler and Sirovich (1993). The drawback of these methods is that the equations and solution method are rather complicated, and, more importantly, there is no mathematical estimation of the rate of convergence of the approximate solution to the exact solution. This means that one cannot easily estimate the error, or be confident that the solution will converge to the true solution.

A new approach introduced by Angot, Bruneau and Fabrie (1997), based on earlier work by Arquis and Caltagirone (1984) supposes that the fluid is a porous medium and is described by the D'Arcy equations. The interaction of a fluid with a solid is then approximated by letting the porosity go to infinity in the fluid and to zero in the solid. Practically, one solves the following equation

$$\frac{\partial u}{\partial t} + u \cdot \nabla u - \frac{1}{Re} \Delta u + \nabla p = \frac{H(x) u}{\epsilon Re} \tag{10}$$

where $H(x) = 1$ in the solid and $H(x) = 0$ in the fluid. It can be shown theoretically that the error in this approximation decreases like $\epsilon^{3/4}$ (in practice the error is found to be only about ϵ^1). This 'penalisation by artificial porosity' approach allows the flow around complicated objects (which may change shape over time) to be easily computed by changing the mask function $H(x)$. By combining this treatment of complicated geometries with the Coherent structure/random background simulation approach proposed above we hope to develop simulation methods for high Reynolds number flows that will be useful to engineers.

Acknowledgment

We acknowledge support from the Newton Institute for Mathematical Sciences in Cambridge (England).

References

P. Angot, C.-H. Bruneau and P. Fabrie 1997. A penalization method to take into account obstacles in incompressible flows. *Preprint no. 97017, Mathématiques Appliquées, Université de Bordeaux.*

E. Arquis and J. P. Caltagirone 1984 Sur les conditions hydrodynamiques au voisinage d'une interface milieu fluide – milieu poreux: application à la convection naturelle. *C. R. Acad. Sci. Paris* II *1–4.*

C. Bardos and J. M. Ghidaglia, 1998. Detachment of turbulent boundary layer and defect measure. *Preprint C.M.L.A., ENS-Cachan, in preparation.*

A. Chorin, 1996. Vortex methods. *Computational Fluid Mechanics, ed. Lesieur, Comte, Zinn-Justin, Elsevier.*

D. Donoho, 1993. Wavelet Shrinkage and W. V. D., *Progress in Wavelet Analysis and Applications, ed. Y. Meyer and S. Roques, 109–128.*

M. Farge, 1992. Wavelet Transforms and their Applications to Turbulence. *Ann. Rev. of Fluid Mech., 24, pp. 395–457.*

M. Farge, E. Goirand, Y. Meyer, F. Pascal and M. V. Wickerhauser, 1992. Improved Predictability of Two-dimensional Turbulent flows using Wavelet Packet Compression. *Fluid Dyn. Res., 10, pp. 229–250.*

M. Farge, M. Holschneider and J.-F. Colonna, 1990. Wavelet Analysis of Coherent Structures in Two-dimensional Turbulent Flows. *'Topological Fluid Mechanics', ed. K. Moffatt and A. Tsinober, Cambridge University Press, pp. 765–776.*

M. Farge, N. Kevlahan, V. Perrier and E. Goirand, 1996. Wavelets and Turbulence. *IEEE Proc., 'Special Issue on Wavelets', ed. I Daubechies and J. Kovasevic, vol. 84, 4, 1996, pp. 639–669.*

M. Farge, N. Kevlahan, Valérie Perrier and Kai Schneider, 1997. Turbulence Analysis, Modelling and Computing using Wavelets. *'Wavelets and Physics', ed. Hans van den Berg, Cambridge University Press, accepted.*

M. Farge, C. Meneveau, Y. Guezennec, C.-M. Ho, 1990. Continuous Wavelet Transform of Coherent Structures. *Center for Turbulence Research, Stanford University and NASA-Ames, pp. 331–348.*

M. Farge et G. Rabreau, 1988. Transformée en ondelettes pour détecter et analyser les structures cohérentes dans les écoulements turbulents bidimensionnels. *C. R. Acad.Sci. Paris, 307, série II, pp. 1479–1486.*

M. Farge and R. Sadourny, 1989. Wave-vortex Dynamics in Rotating Shallow Water. *J.of Fluid Mech., 206, pp. 433–462.*

M. Farge, K. Schneider and N. Kevlahan, 1997. Coherent Structure Eduction in Wavelet-forced Two-dimensional Turbulent Flows. *'Dynamics of Slender Vortices', ed. E. Krause, Kluwer, accepted.*

J. Fröhlich and K. Schneider, 1996. Numerical simulation of decaying turbulence in an adapteive wavelet basis. *Appl. Comput. Harm. Anal., 3, 393–397.*

D. Goldstein, R. Handler and L. Sirovich, 1993. Modeling a no-slip flow boundary with an external force field. *J. Comput. Phys.* **105**, 354–366.

N. Kevlahan and M. Farge, 1997. Vorticity Filaments in Two-dimensional Turbulence: Creation, stability and Effect. *J. of Fluid Mech., 346, pp. 49–76.*

C. Peskin, 1977. Numerical analysis of blood flow in the heart. *J. Comput. Phys.* **25**, 220–252.

Kai Schneider and Marie Farge, 1997. Wavelet Forcing for Direct Numerical Simulation of Two-dimensional Turbulence. *C. R. Acad. Sci. Paris, 325, Série IIb, pp. 263–270.*

Kai Schneider, Nicholas Kevlahan and Marie Farge, 1997. Comparison of and Adaptive Wavelet Method and Nonlinearly Filtered Pseudo-spectral Methods for the Two-dimensional Navier-Stokes Equations. *Theoret. and Comput. Fluid Dynamics, 9, pp. 191–206.*

W. Sweldens, 1997. The lifting scheme: A custom-design construction of biorthogonal wavelets. *Appl. Comput. Harmon. Anal.* **3**, 186–200.

Trends in Mathematics, © 1999 Birkhäuser Verlag Basel/Switzerland

Evolution of decaying two-dimensional turbulence and self-similarity

J. R. Herring, Y. Kimura, J. Chasnov

Abstract

We examine the consequences of self-similarity of the energy spectrum of two-dimensional decaying turbulence, and conclude that traditional closures (such as EDQNM and TFM) are consistent with this principle only if the regions of space contributing significantly to energy and enstrophy transfer comprise an ever diminishing region of space as time proceeds from the initial time of Gaussian chaos. Results of modifying the TFM according to this assumption are compared to the recent high resolution DNS of Chasnov.

1 Two-dimensional turbulence: Its scaling and self-similarity

Chasnov's (1996) 4096-resolution DNS suggests that decaying two-dimensional turbulence evolves *via* an approximate self-similar state. This surmise includes large scales, (for which the energy spectrum $E(k,t) \sim k^3$), the inertial range, ($E(k,t) \sim k^{-3}$), and a considerable portion of the dissipation range. We argue here that the constraint of high Reynolds number energy constancy and complete self-similarity implies that the energy (and enstrophy) transfer process must have an explicit time-dependence. We associate such time-dependence with the progressive spottiness of the energy transfer process: the turbulence lives on an ever-diminishing subset of the space available to it. The proposed modified transfer function, $T(k,t)$, $(T \ni (\dot{E} + 2\nu k^2 E) = T)$, implied by complete self similarity together with energy constancy, is $T_{mod} = (t_0/t)^{1/2} T(k,t)$, as $t \to \infty$. This implies that the length scale increases as $t^{1/2}$ instead of t, *a la* Batchelor (1969). If the form of molecular dissipation changes, so must the power of the (t_0/t) factor in T_{mod}.

We write the spectral equations in a representation that is an adaptation of that of Linn (1961) to two-dimensions:

$$\partial_t U(k,t) = \mathcal{T}(k,t) - 2\nu k^2 U(k,t) \text{ with } E(k,t) \equiv 2\pi k U(k,t),\ T(k) \equiv 2\pi k \mathcal{T}(k) \quad (1)$$

Now put

$$U(k,t) = u_0^2 A(t) F(k\lambda(t), t),\ \ K \equiv k\lambda(t) \quad (2)$$

We search for functions $A(t)$ and $\lambda(t)$ for which $F(K,t)$ has no explicit dependence on t. Introducing (2) into (1) gives,

$$\dot{A}F + (A\dot{\lambda}/\lambda)K\partial F/\partial K + A\partial_t F = \mathcal{T}/(u_0^2) - 2\nu(K/\lambda)^2 AF \quad (3)$$

If (3) is to be satisfied with $\partial_t F(K,t) = 0$, the remaining terms must balance. Accordingly, we put

$$\dot{A}/A = C\dot{\lambda}/\lambda \Rightarrow A = \lambda^C \tag{4}$$

Note that the dimensions of $\mathcal{T} \rightarrow [U\sqrt{k^4 U}]$, so that if $\mathcal{T}$ has the same time-dependence as the LHS of (3),

$$CA\dot{\lambda}/\lambda = \gamma A^{3/2}/\lambda^2, \ \lambda\dot{\lambda} = (\gamma/C)\lambda^{C/2} \tag{5}$$

If $\int_0^\infty dk E(k,t)$ is constant (as $t \rightarrow \infty$), (3) implies $C = 2$. Then

$$\lambda(t) = u_0 t + \lambda_0, \ A = \lambda^2, \ \dot{A} = 2\lambda u_0, \dot{A}/A = 2u_0/\lambda \tag{6}$$

If (6) is used, (3) becomes,

$$F + (1/2)KF_K + (\lambda/(2u_0)F_t = (1/(2\lambda u_0^3))\mathcal{T} - \nu K^2 F/(\dot{\lambda}\lambda) \tag{7}$$

However, without assuming any form for λ, we may use (4) to write (3) in a more convenient form for advancing $F(K,t)$ in time;

$$F_t = (\dot{\lambda}/\lambda)\{-(1/K)d(K^2F)/dK + (1/(u_0^2\lambda\dot{\lambda}))\mathcal{T}(k) - 2\nu K^2 F/(\dot{\lambda}\lambda)\} \tag{8}$$

Equation (8) cannot as yet be used to infer scaling of its various terms, because the argument of $\mathcal{T}$ is k, not $K \equiv \lambda k$. We make this conversion in the next section.

2 TFM analysis of $\mathcal{T}$ and its ingredients

For the sake of brevity, we state this conversion without any derivation:

$$\mathcal{T}(k) = \lambda u_0^3 \hat{\mathcal{T}}(K), \ \hat{\mathcal{T}}(K) \equiv \int_\triangle B_{KPQ}\hat{\Theta}_{KPQ}F(Q)(F(P) - F(K)) \tag{9}$$

$$d\{[\lambda/u_0]\hat{\Theta}_{KPQ}\}/dt = 1 - \{\hat{\eta}(K) + \hat{\eta}(P) + \hat{\eta}(Q)\}\hat{\Theta}_{KPQ} \tag{10}$$

and

$$\hat{\eta}(K) = \int_\triangle dPdQC_{KPQ}\hat{\Theta}_{KPQ}F(Q) + \nu K^2/(u_0\lambda) \tag{11}$$

So that (8) may be rewritten as,

$$F_t = (\dot{\lambda}/\lambda)\{-(1/K)d(K^2F)/dK + (u_0/\dot{\lambda})\hat{\mathcal{T}}(K) - 2\nu K^2 F/(\dot{\lambda}\lambda)\} \tag{12}$$

In order for $\partial_t F(K,t) = 0$ at small K, the viscous term may be neglected, and where we may use a $K \rightarrow 0$ expansion to $T(K)$ (Lesieur (1990)), we must have

$$(1/K)\partial(K^2F(K))/\partial K = \hat{\mathcal{T}}(K) \rightarrow \int_0^\infty dP\Theta_{0PP}F^2(P)/P^2 \tag{13}$$

This imposes an overall constraint on $F(K)$ if self-similar decay is to prevail over all K.

3 Scaling of transfer needed for self-similarity

Consider (8) which steps $F(K,t)$ forward in time, and let us take (6) for $\lambda(t)$. The first factor on the *rhs* simply changes dependent variable $[t]$ to something else $[ln(u_0 t + \lambda_0)]$, and this is of no real consequence. But within the $\{\cdot\}$, the effect of ν is progressively discounted (as compared to the first two terms) as $t \to \infty$. Hence, the impossibility, according to TFM, for a completely self-similar evolution of $E(k) = 2\pi k U(k,t)$.

On the other hand, we may ask what is necessary for $E(k,t)$ to be completely self-similar. According to (3), in order for the dissipation term (the last term on the *rhs*) to be $\sim$ the kinematic terms (the first two *lhs* terms) is:

$$A\dot{\lambda}/\lambda \sim 1, \quad \Rightarrow \lambda \sim t^{1/2} \tag{14}$$

where we use $C = 2$, which follows if at large R_λ, $E(t)$ is constant. If we also insist that $\mathcal{T}$ scales similarly with the kinematic and viscous terms, the non-dimensional factor, γ, in (5) must be some power of t, say,

$$\gamma(t) \sim (t/t_0)^p \tag{15}$$

Here, t_0 is a value of t less than which universal self-similarity is not obtained. Equation (5), with $C = 2$, $\lambda \sim t^{1/2}$ imply

$$p = -1/2 \tag{16}$$

Note that $\lambda \sim t^{1/2}$ assumes viscous losses $\equiv L \sim k^2 E(k)$. For $L \sim k^n E(k)$, we would obtain $\lambda \sim t^{1/n}$, with $-p = 1 - 1/n$.

To summarize, the requirement that the evolving energy spectrum be completely self-similar (as found by Chasnov (1996)), and that the energy transfer function, $\mathcal{T}$, have the functional form as given by TFM ((9)–(11)), requires $\mathcal{T}(k,t)$ to be re-scaled as:

$$\mathcal{T}(k,t) \to\sim (\lambda_0/\lambda(t))\mathcal{T}(k,t) \tag{17}$$

with

$$\lambda(t) = \sqrt{\lambda_0^2 + 2u_0\lambda_0 t} \tag{18}$$

With this re-scaling, each term within {} in (8) has no explicit dependence on time. Hence, we may expect that as $t \to \infty$, this term will be driven to zero. If the molecular dissipation law is changed from νk^2 to a hyper-viscosity (i.e.: viscous dissipation $\sim \nu_{hyper} k^4$), the re-scaling to give self-similarity must be changed.

There remains the question of why such re-scaling is a plausible expectation for two-dimensional turbulence. The basic point is that (17) implies a continuous decrease of the relative importance of the inertial transfer with time[1]. We suggest that such may be expected if the turbulence is progressively confined to smaller regions of available space, as time proceeds. A model of how this may happen is sketched in Sec. 5. Before that, we show some numerical results for the TFM, integrated according to the formulas of Sec. 2 and 3.

[1] McWilliams (1990) demonstrated through an analysis of his DNS that the presence of intense vortices suppresses transfer.

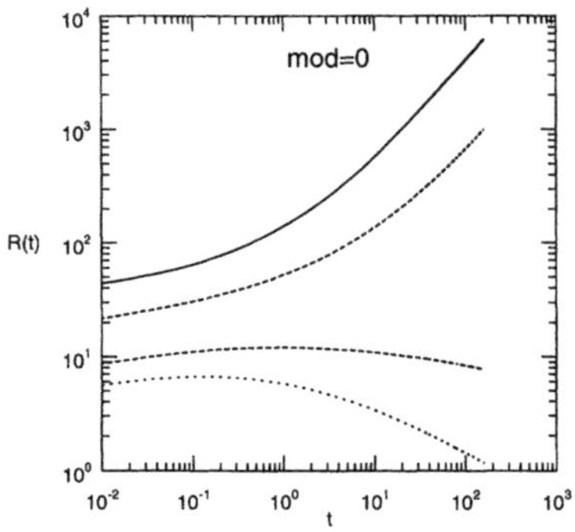

Fig. 1 $R_\lambda(t)$ for TFM (left panel) and TFM′ (right panel) and various values of ν: solid line, $\nu = .228$; short-dashed $\nu = .5000$; dot-dashed, $\nu = 1.142$; long-dashed, $\nu = 2.28$.

4 Numerical results

We present next some numerical results illustrating the issues discussed in the previous sections. First we explore how TFM and TFM′ behave for small R_λ, where

$$R_\lambda = \sqrt{E(t)}\ell/\nu$$

here

$$E(t) = \pi \int_0^\infty dKKF(K,t) \equiv \int_0^\infty dKE(K,t)$$

and

$$\ell^2 \equiv E(t)/\int_0^\infty dKK^2E(K,t)$$

We note that Chasnov (1997) found from his DNS that, below a critical value of initial R_λ, the flow reverted to a final period of decay in which $T(k,t)$ played no role. For $E(k,t=0) \sim k^3$, $k \to 0$, this condition implies $E(t) \sim t^{-2}$, and $R_\lambda(t) \sim t^{-1}$. The issue is explored in Fig. 1 (a,b) for an initial $E(K) \sim K^3/(1+K^6)$. Fig. (1a) shows results for TFM, while (1b) pertains to TFM′.

The amplitude of E is such that (13) is satisfied at small K. The value of $R_\lambda(0)$ is controlled by assigning various values of ν. We note that for TFM if $R_\lambda(0) \geq 20$, $R_\lambda(t) \sim t$ (in accord with Batchelor), whereas for initial values below this the final period of decay is recovered ($R_\lambda(t) \sim 1/t$). The TFM′ results appear to have a critical R_λ somewhat larger (~ 30), with the same time-dependence in the final period of decay. Above this value, $R_\lambda \sim t^{1/2}$, as expected. The short-time behavior of R_λ shown here differs from that of Chasnov in that our $R_\lambda(t)$ do not have an initial decrease. This is explained by the fact that Chasnov's simulations have a peak energy at much higher wave numbers than ours, so that viscous effects – which act immediately after the initial time, as contrasted to the more slowly

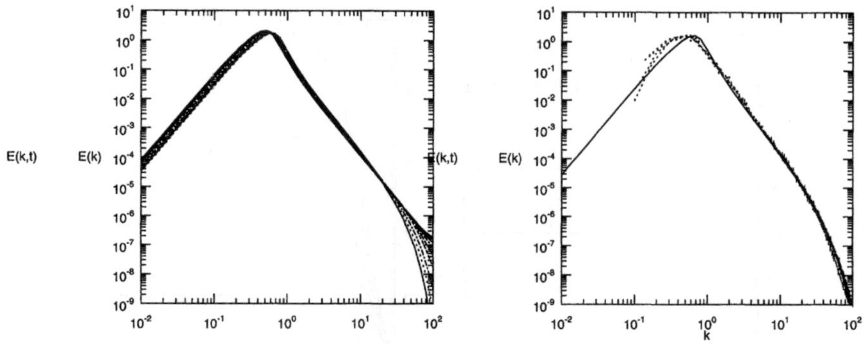

Fig. 2 $E(k,t)$ for TFM (left panel) and TFM′ (right panel) for $\nu = .010$ and for $t = .5,\ 1,\ 5,\ 10,\ 20,\ 30,\ 40,\ 50$.

developing energy transfer – are much more effective (near $t = 0$) in his case than ours.

We next discuss high R_λ spectra. For this purpose, it is convenient to have a measure of the extent of the inertial range. We do this by introducing another Reynolds number, ρ, which remains constant during the decay process. We first note that an integral measure of the extent of an inertial range that is $\sim k^{-3}$ would be indicated by:

$$\rho(t) \equiv \{\int E(K)dK \int E(K)K^4 dK\}^{1/2} / \int E(K)K^2 dK \tag{22}$$

In terms of the variable k, this is

$$\rho = u(k_s^2/k_\eta^3)/\nu \tag{23}$$

where

$$k_s^2 = \int k^4 dk E(k) / \int k^2 dk E(k), \text{ and } k_\eta = (\eta)^{1/6}/\sqrt{\nu}.$$

and η is the enstrophy dissipation rate, $2\nu \int dk E(k) k^4$. An interesting formula relating R_λ and ρ is

$$dR_\lambda/dt = -2\sqrt{\epsilon}(1 - \frac{1}{2}\rho^2) \tag{24}$$

Here, $\epsilon \equiv \int_0^\infty k^2 E(k) dk$. We note that Chasnov's highest R_λ DNS have $\rho \sim 8.5$, and we may match this value by the choice $\nu = 0.0100$, and an initial spectrum centered at $K = 1$.

Evolution of TFM and TFM' spectra, as evolved from (9) and (12), are shown in Fig. 2. The corresponding evolution of $R_\lambda(t)$ is shown in Fig. (3a), while $\rho(t)$ from Chasnov's DNS is depicted in Fig. (3b). The TFM spectrum shows a modest $E(K) \sim K^{-4}$ range just beyond the energy peak, as suggested by Lesieur

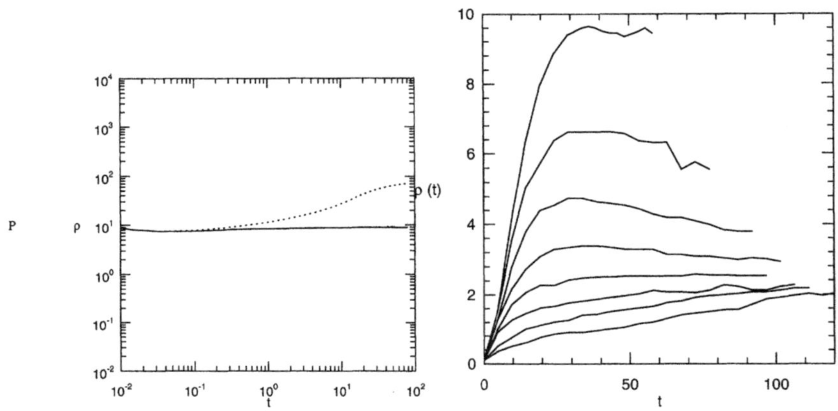

Fig. 3 Left panel: $\rho(t)$ for TFM′ (solid line) and TFM (dashed line) for the run described in Fig. 2. Right panel: $\rho(t)$ for Chasnov's simulation.

and Herring (1987). Note that the evolution of TFM is not self-similar at high K, where the viscous effects continue to weaken with time, according to the remarks of Sec. 3. The TFM' spectrum, on the other hand, quickly becomes self-similar in the variable K over the full range of K. Here, the energy transfer is much weaker. The evolution of R_λ (Fig. 3) shows late time behavior $R_\lambda \sim t$ for TFM, and $R_\lambda \sim \sqrt{t}$ for TFM'. The former is the classic Batchelor result, where the latter is that needed for complete self-similarity. The course of $E(t)$ for TFM' is $\sim \mathrm{C} - \ln(t)$, as predicted by Chasnov, and as it must be if $\epsilon(t) \sim 1/t$.

The comparison of TFM' with Chasnov's high R_λ case (Fig. (2b)) seems satisfactory for scales larger than the wave-number peak, but TFM' and the DNS differ for scales larger than the energy peak. The DNS seems a much more rounded spectrum at very large scales. Some of the difference may be attributable to a lack of large-scale resolution in the DNS.

If self-similarity is an approximate characterization of decay, it is of interest to explore the consequences of assuming $\rho(t) = \rho_0$, during the entire time of decay. In fact, this *ansatz* implies an equation for $E(t)$ through the use of $\rho(t) = \rho(0) \equiv \rho_0$, (1), (2) and the fact that $\int k^2 dk T(k,t) = 0$ for two-dimensional turbulence. Thus we have,

$$d\langle \mathbf{u}^2 \rangle / dt = -2\nu \langle \omega^2 \rangle \text{ and } d\langle \omega^2 \rangle / dt = -2\nu \langle (\nabla\omega)^2 \rangle \qquad (25\mathrm{a,b})$$

Using (22) for $\rho(t)$, and

$$\langle (\nabla\omega)^2 \rangle = 2 \int_0^\infty k^4 E(k,t) dk \qquad (26)$$

we find

$$\frac{d^2 \langle \mathbf{u}^2 \rangle}{dt^2} - \frac{\rho_0^2}{\langle \mathbf{u}^2 \rangle} \left(\frac{d\langle \mathbf{u}^2 \rangle}{dt} \right)^2 = 0 \qquad (27)$$

whose solution is:

$$\langle \mathbf{u}^2 \rangle = u_0^2 \left[1 + 2(\rho_0^2 - 1)R_0^{-1}\omega_0 t\right]^{-1/(\rho^2-1)} \tag{28}$$

An equivalent equation may be found for $\langle \omega^2 \rangle$:

$$\langle \omega^2 \rangle = \omega_0^2 \left[1 + 2(\rho_0^2 - 1)R_0^{-1}\omega_0 t\right]^{-\rho_0^2/(\rho_0^2-1)} , \tag{29}$$

with $R_0 = u_0^2/\omega_0\nu$. The asymptotic solution for enstrophy ($t \to \infty$, $R_0, \rho_0 \gg 1$) is

$$\langle \omega^2 \rangle = \frac{R_0\omega_0}{2\rho_0^2} t^{-1} \tag{30}$$

5 Dynamics of decay in terms of spectra

The underlying idea emerging from our discussion is that small scales of two-dimensional turbulence are parts of large scale structures. Hence, large and small scales decay together, with both length scales growing as $\sim \sqrt{t}$. (Such a growth is characteristic of pure viscous dissipation.) That large and small scales may be glued together is reasonable for two-dimensional flow, since the eddy turnover time for an enstrophy inertial-range eddy ($\sim \sqrt{k^3 E(k)}$) is roughly independent of scale (except for possible logarithmic corrections). Hence large and small scales may become coherent with impunity. In three dimensions, on the other hand, the inertial range eddy turn over time decreases with decreasing scale size, so that small scales may achieve a much higher degree of statistical independence from the large, energy-containing scales.

We first recall that the overall flow pattern revealed by DNS is a system of isolated vortices, surrounded by strain regions in which transfer is substantial (McWilliams, (1984), Benzi *et al.*, (1987), and Siegel and Weiss (1997)). Since the regions of strong strain are circumferential with respect to the isolated vortices, we would expect their number to be equal to the number of vortices. If we suppose the turbulence exist on independent patches, on each of which a statistical theory such as TFM holds, then the decay properties of the total system are the same as for each individual patch. If, however some of the patches – those corresponding to isolated vortices – have very little strain, and hence little transfer to small scales, the net transfer for the overall system is reduced. In this way, the net transfer, averaged across patches, decreases with time. The question of whether this simple picture is sensible may be answered by scanning a DNS to see if there are regions where some measure of enstrophy transfer is unusually small. A convenient measure would be the rate of production of mean-square vorticity gradients,

$$\frac{1}{2}\frac{d(\nabla\omega)^2}{dt} = (\nabla\omega)^T S(\nabla\omega) \tag{31}$$

where ω is the vorticity, and S the strain matrix,

$$= \begin{pmatrix} -\psi_{xy} & -\psi_{yy} \\ \psi_{xx} & \psi_{xy} \end{pmatrix} \tag{32}$$

Note that the normalized average value of the right hand side of (28) is, for isotropic turbulence, the two-dimensional skewness

$$\mathcal{S} \equiv -2\langle(\partial u/\partial x)(\partial\omega/\partial x)^2\rangle/\{\langle(\partial u/\partial x)^2\rangle^{1/2}\langle(\partial\omega/\partial x)^2\rangle^{1/2}\}$$

6 Summary and concluding comments

Classical closures such as TFM applied to high Reynolds number two-dimensional turbulence do not yield self-similar decay. The reason is traceable to (near) energy conservation at high Reynolds numbers. The two-dimensional DNS of Chasnov, on the other hand, exhibit a high degree of self-similarity, with a near constancy of total energy at the largest R_λ reported. The earlier DNS of McWilliams (1984), and the comparison of that DNS with TFM by Herring and McWilliams (1985) – if examined carefully – would have indicated the same result, although the resolution there was too limited to be conclusive. In three dimensions, self-similar decay is possible, and indeed was proposed many years ago as the universal mode of decay for high Reynolds number flows (see i.e. Batchelor, (1959)). But in that case energy decays as t^{-1}. The DNS and closure may be partially reconciled by the assumption that the energy (and enstrophy) transfer takes place in progressively smaller sub-regions of high strain, surrounding intense vortices. Such physics is indicated by a number of DNS (McWilliams (1984), Benzi *et al.* (1987)). If complete self-similarity is assumed to rule the decay process, the length scale grows as $\sqrt{t}$, and the energy transfer decreases as $1/\sqrt{t}$. The growth in length scale differs from Chasnov's (1996) numerical finding of $t^{.4}$ at the highest R_λ simulated.

Our discussion here has focused on the idea that progressive diminution of transfer is associated with a corresponding decrease of the space on which the turbulence lives. An alternate approach is to posit a Reynolds number dependence of the transfer process. This is explored elsewhere (Chasnov and Herring (1998)).

We have noted (in Sec. 3) that if the functional form of dissipation is modified, completely self-similar decay would dictate a modification in the equation that determines the length scale, $\lambda(t)$. Thus, if $\nu_{hyper}k^n$ represents dissipation, $\lambda \sim t^{1/n}$, if the statistics of second moments are self-similar. Whether the slower λ-growth rates for hyper-viscosity DNS examined by Carnavale *et al.* (1991) would be explained this way remains to be seen.

References

Batchelor, G. K. 1953: *The theory of homogeneous turbulence*, (Cambridge University Press), Cambridge, England, 195 pp.

Batchelor, G. K., 1969: *Phys. Fluids Suppl.*, **12** II, 233–239.

Benzi, R., Paternello, and P. Santangelo 1987: *Europhys. Lett.*, **3**, 379.

Carnavale, G.F., J.C. McWilliams, Y. Pomeau, J.B. Weiss, and W.R. Young 1991: Phys. Rev. Lett., **66**, 2736–2737.

Chasnov, J. 1996: *Phys. Fluids*, **9**, 171–180.

Chasnov, J. and J. R. Herring 1998: ETC7
Herring, J. R. and R. M. Kerr 1992: in *New Approaches and Concepts in Turbulence, Monte Verità* Birkhäuser Verlag Basel, 367–376.

Herring, J. R. and J. C. McWilliams 1985: *J. Fluid Mech.* **158** 229–242.

Lesieur, M. 1990: Turbulence in Fluids 2nd Edition, Kulwar Academic Publishers, Dordrecht, 412 pp.

Lesieur, M. and J. R. Herring 1985: *J. Fluid Mech.*, **161**, 77–95.

Linn, C. C. 1961: Statistical Theories of Turbulence. *Princeton University Press*, Number 10 Princeton Aeronautical Paperbacks, C. *duP* Donaldson Ed., 60 pp.

McWilliams, J.C 1984: *J. Fluid Mech.*, **146**, 21–43.

McWilliams, J. C. 1990: *Phys. Fluids*, A, **2**, 547–552.

Siegel, A. and J. B. Weiss 1996: Preprint, to appear in: *Phys. Fluids*

Trends in Mathematics, © 1999 Birkhäuser Verlag Basel/Switzerland

Linear and nonlinear models for turbulence in the presence of waves

C. Cambon

Context

Some issues in turbulence description and modelling, with possible connection with stability analysis, are discussed in this paper. In the absence of a general theory for turbulence, the different issues will be illustrated by the case of 'weak' turbulence, in which structuring effects, induced by the presence of body forces (e.g. Coriolis and buoyancy forces) or mean velocity or temperature gradients, play an essential role. Three points are addressed in the following.

The first is a discussion of the relative importance of linear and nonlinear terms in rapid and slow unsteady evolution. Linear dynamics as in 'Rapid Distortion Theory', is dominated by waves for rotating and stably stratified turbulence, whereas the 'slow' dynamics mediated by resonant triads is examined both from the point of view of statistical spectral closures and from the point of view of asymptotic analysis.

The second point considers some low-dimension manifolds, which are shown to concentrate some properties relevant for dynamics. As previously, RDT and weakly nonlinear approaches to wave turbulence are used.

The third point is advocating for a fully anisotropic description in statistical spectral approach, in order to capture relevant information on dimensionality and directional structure.

Importance of linear dynamics

Relatively recently it has become apparent that many aspects of a surprisingly wide class of turbulent flows can be successfully modelled by treating the turbulence as a small perturbation about the mean flow. The technique of linearisation about the mean flow is often referred to as 'Rapid Distortion Theory' or RDT [2], but this is perhaps a misleading nomenclature, given the growing number of applications (without 'distortion') in which linear mechanisms other than mean shear act on the turbulence with sufficient strength to render self-interaction a secondary effect.

The starting point of both the so-called RDT for homogeneous turbulence [2],[21], and of recent stability analyses [9] is as follows. For a basic flow with uniform velocity gradients

$$\frac{\partial \overline{U}_i}{\partial x_j} = \lambda_{ij}(t)x_j \text{ with } \lambda_{ii} = 0 \tag{1}$$

linearised equations admit solutions for the velocity disturbances of the form

$$u_i(\mathbf{x},t) = \widehat{u}_i(\mathbf{k}(t),t)\exp[\imath\mathbf{k}(t)\cdot\mathbf{x}] \qquad (2)$$

where the time-dependent wavevector and the amplitude satisfy the system of equations

$$\dot{k}_i = -\lambda_{ji}k_j \qquad (3.a)$$

$$\dot{\widehat{u}}_i + L_{ij}\widehat{u}_j = 0 \text{ with } L_{ij} = (\delta_{il} - 2\frac{k_i k_l}{k^2})\lambda_{lj} \qquad (3.b)$$

The system of equations (3-a-b) is referred to as Kelvin-Townsend equations, and the particular Fourier modes with a time-dependent wave-vector (2) are referred to as Lagrangian Fourier modes (for a recent review see [8]). Stability of steady, rotational and two-dimensional, basic flows, was particularly investigated. In this case, λ_{ij} takes the form

$$\lambda = \begin{pmatrix} 0 & S - W/2 & 0 \\ S + W/2 & 0 & 0 \\ 0 & 0 & 0 \end{pmatrix} \qquad (4)$$

when axes are chosen appropriately, where $S > 0$. Such steady plane flows combine vorticity W and irrotational straining S. According to the sign of $S^2 - (W/2)^2$, which also gives the pressure Laplacian, three classes are distinguished. The limiting case, $S = |W/2|$, corresponds to simple shearing (referred to as pure plane shear below), and forms the borderline between two distinct regimes, namely those in which the basic flow (1) streamlines are closed and elliptic about the stagnation point at the origin ($S < |W/2|$, elliptic case hereinafter) and those for which they are open and hyperbolic ($S > |W/2|$). The first two cases will be briefly rediscussed below. Of course, equation (1) characterises a very restricted class of unbounded flows. For practical applications, however, it can be considered to hold in a narrow domain only, assuming that the size of turbulent eddies ℓ is small compared with the overall length scales of the mean flow, $\mathcal{L}$, for instance, the size of a body encountering fine-scale free-stream turbulence (Hunt and coworkers, see [8]). In that case, one uses a local frame of reference convected with the mean velocity and approximates the mean velocity gradients as uniform, but time-varying, as given by (1). This idea has been formalised in the context of flow stability [15] using an asymptotic approach based on the classical WKB method, which is traditionally used to analyse the ray theoretic limit (i.e. short waves) in wave problems. In this approach, the system of equations (3) is recovered at leading order of the asymptotic expansion, but is related to a phase-gradient and a complex amplitude more general than $\mathbf{k}$ and $\widehat{\mathbf{u}}$. In more complicated flows, in which different zones (e.g. vortical, strained) are separated by singular points [19], such as the arrays of vortices encountered in mixing layers and wakes, short-wave linear stability analysis shows the importance of local stagnation points (elliptic and hyperbolic) of the basic flow as sources of instability.

Dynamics dominated by low-dimension manifolds

Narrow domains in spectral space are shown to be important in RDT for the three 'canonic' cases presented above (hyperbolic, rectilinear and elliptic streamlines). The case of pure plane shear has been particularly investigated [21]. The RDT solution for the initial-value problem (3) was solved in the more general way, by calculating the Green's function $G_{ij}(\mathbf{k}, t, t')$. This Green's function exhibits algebraic growth, together with an asymptotic concentration near the wave-plane $k_1 = 0$, associated with elongated structures in the streamwise direction (see [20] for an asymptotic study). This case is often referred to as plane Couette flow in the stability literature. Expressed in the language of stability theory, inviscid algebraic growth derives from the continuous spectrum and is the subject of ongoing research in the context of bypass transition in boundary layers. Only looking at flow structure, the works by Lee, Kim & Moin [14] showed a striking analogy between RDT results and DNS results for both homogeneous shear (constant high shear rate) and channel (near-wall high shear rate).

Among the three canonic cases, only the case of pure rotation, ($S = 0$, W is twice the angular velocity Ω), is stable, the disturbances consisting of inertial modes, but the anisotropic dispersion law of these modes

$$\sigma_k = \pm 2\boldsymbol{\Omega} \cdot \mathbf{k}/k = \pm 2\Omega \frac{k_{\parallel}}{k} \tag{5}$$

may allow some complex behaviours, such as resonances and confinment of energy in narrow domains. Regarding only linear effects, a weak additional ($S << W$) strain yield the elliptical flow instability, which is a parametric instability with exponential amplification of a thin angular band of oblique wavevectors ($k_{\parallel}/k = \pm 1/2$) at vanishing strain [3]). Another manifestation of the inertial waves with anisotropic dispersion law is the appearance of conical ray structures (St Andrew cross-shaped) in a rotating tank, when a harmonic stirring is applied locally with a frequency lower than Ω [12].

We will now focus on nonlinear effects in homogeneous turbulence subjected to rotation and/or stable stratification, in the absence of forcing, mean gradients or preexisting organised structures. In Fourier space, the generic form of the Navier-Stokes equations for velocity and temperature fluctuations is

$$\frac{\partial \widehat{v}_i}{\partial t} + L_{ij}\widehat{v}_j + \nu k^2 \widehat{v}_i = \int_{\mathbf{k}=\mathbf{p}+\mathbf{q}} M_{imn}\widehat{v}_m(\mathbf{p})\widehat{v}_n(\mathbf{q}) d^3\mathbf{p} \tag{6}$$

where $\widehat{v}$ is the Fourier transform of velocity $\mathbf{v} = \mathbf{u}$ (vorticity can similarly be used) in the case of pure rotation, and gathers both velocity and density fluctuations if stratification is present [10]. For convenience, the fluctuating flow is seen in the rotating frame of reference, so that the wavevectors are no longer time-dependent (since eq. (3.a) reflects convection by the basic flow in a Galilean frame of reference) and the linear matrix includes the projection of the Coriolis force. Indeed, pressure is removed from consideration using classic projection operators, and the nonlinear terms (right-hand side) appear as a convolution product. A more tractable form of (6) is found by decomposing the basic variable $\mathbf{v}$ onto the basis of the eigenvectors of the linear regime, so that

$$\widehat{v}_i = \sum_{\epsilon=0,\pm1} \xi_\epsilon N_i^\epsilon \tag{7}$$

and the equation (5) becomes

$$(\frac{\partial}{\partial t} - \imath\,\epsilon\,\sigma_k + \nu k^2)\xi_\epsilon = \sum_{\epsilon,\epsilon',\epsilon''} \int m_{\epsilon\epsilon'\epsilon''}\xi_{\epsilon'}(\mathbf{p})\xi_{\epsilon''}(\mathbf{q})d^3\mathbf{p} \tag{8}$$

in which, indeed, the linear operator is diagonalised, whereas the nonlinear term involve a new matrix ($m_{\epsilon\epsilon'\epsilon''}$ instead of M_{ijm}) but conserves a form similar to (6). Using (8), the 'rapid distortion limit' is simply

$$\xi_\epsilon(\mathbf{k},t) = \exp[\imath\epsilon\sigma_k(t-t')]\xi_\epsilon(\mathbf{k},t') \tag{9.a}$$

instead of $\widehat{v}_i(\mathbf{k},t) = G_{ij}(\mathbf{k},t,t')\widehat{v}_j(\mathbf{k},t')$ (general RDT solution, without convection, see [8]). Replacing the initial data at fixed $t'=0$ in (9.a) by a new unknow variable, say a_ϵ, so that

$$\xi_\epsilon(\mathbf{k},t) = \exp[\imath\epsilon\sigma_k(t-t')]a_\epsilon(\mathbf{k},t) \tag{9.b}$$

an equation for a_ϵ is readily derived from (8). Using the above transformation (which amounts to the 'Poincaré transformation' used by [1] in the case of pure rotation), the nonlinear dynamics of a_ϵ is easily shown to be driven by the term

$$\exp[\imath(\epsilon\sigma_k + \epsilon'\sigma_p + \epsilon''\sigma_q)(t-t')] \tag{10}$$

Indeed, the zero value of the phase of the above complex exponential characterises the resonant condition, and the simultaneous conditions

$$\epsilon\sigma_k + \epsilon'\sigma_p + \epsilon''\sigma_q = 0 \text{ with } \mathbf{k}+\mathbf{p}+\mathbf{q}=\mathbf{0}$$

gives the resonant surfaces. Different works have been carried out in the case of 'weak' wave turbulence [1], [4], [5], [23], including an asymptotic analysis in the vicinity of resonant triad in order to capture some 'slow' dynamics at leading order. These effects have been analysed in connection with statistical models in spectral space (see [7] for a review) with both numerical computation and asymptotic analysis (see [8] and J. Scott & J. Lenoble, unpublished results) of the models. Three cases will be considered in the following.

In the case of pure rotation, the previous analyses have given a new insight to the problem of two-dimensionalisation of an initially 3D turbulent flow. Two remarks help stating the issue. First, the Proudman theorem, by neglecting nonlinear, unstationary and viscous terms, only says that the linear stationary mode is the two-dimensional manifold. Looking at (5), (8) and (7), it is obvious that the linear stationary mode is given by $k_\parallel = 0$, which characterize structures without variability onto the axial direction (or $\partial/\partial x_\parallel = 0$) in physical space. Note that the two eigenmodes are wavy in (7) ($\epsilon = \pm 1$ only) for pure rotation, and correspond to the 'helical modes' extensively used to analyse triadic interactions (e.g. [22]), so that the stationary linear mode is singular. It does not correspond to the generic mode associated with $\epsilon = 0$, only occuring when stratification is present, but correspond to 'degenerated' wavy modes ($\epsilon = \pm 1$), in the limit of vanishing frequency in (5). A second, less drastic, approximation consists of dropping out nonlinear and viscous terms only, in (8), keeping the unsteady term. The result is the linear regime of inertial waves (including the singular 2D mode at vanishing frequency), as in (9.a), and this unsteady regime conserves exactly the detailed density of energy, $\widehat{u}_i^*\widehat{u}_i$, for each wavevector, so that an initial 3D equidistribution is conserved, and there is no possible transition towards 2D structure.

These two remarks show that the transition 3D-2D should be reflected in spectral space by an angular drain of spectral energy from linear unsteady modes ($k_\parallel/k \neq 0$) towards the linear steady one ($k_\parallel/k = 0$). In addition, this angular drain should involve nonlinear interactions, so that only an unsteady and nonlinear approach can explain this transition. At small Rossby number, the long-time behaviour is dominated by near-resonant interactions, and a qualitative analysis [22] has shown how resonant interactions can concentrate energy towards the 2D manifold. At significant Rossby number, both EDQNM2 (anisotropic EDQNM model [18] consistent with resonance conditions) and high resolution LES [7] exhibit preferential concentration, by computing a complete axisymmetric energy distribution $e(k, k_\parallel/k, t)$ (proportional to $< \widehat{u}_i^*\widehat{u}_i >$) created during a nonlinear evolution, which is started with pure isotropic initial conditions $e(k, k_\parallel/k, 0) = E(k)/(4\pi k^2)$.

The case of stably stratified turbulence is very different, even if the gravity waves present strong analogies with inertial waves. An additional element is the presence of the 'vortex' mode, which is the steady linear mode and correspond to $\epsilon = 0$ in (7). According to the Craya-Herring decomposition, it is present for any wavevector orientation, and contains half the total kinetic energy in the isotropic case. Accordingly, pure vortex interactions are found to be dominant (resonant conditions are obtained with $\epsilon = \epsilon' = \epsilon'' = 0$ in (10) with no need to restrict to resonant surfaces as for resonant wave interactions. EDQNM2 [10] and DNS results [11] have shown that the spectral energy concentrates towards vertical wavenumbers $k_\perp \sim 0$. These wavenumbers correspond to horizontally (because $\mathbf{k}$ and $\widehat{\mathbf{u}}$ are perpendicular) stratified turbulent structures with dominantly horizontal, low-frequency motions. As for the '2D transition' expected in pure rotation, a new dynamical insight is given to the collapse of vertical motion expected in stably stratified turbulence, but the long-time behaviour essentially differ from a 2D one. The case of combined effects of rotation and stratification presents particular interest in the geophysical context for large enough scales. In this case, the linear steady mode associated with $\epsilon = 0$ in (7) is the geostrophic mode, whereas the two wavy modes ($\epsilon = \pm 1$) are associated with inertio-gravity waves whose dispersion law is $\sigma_k = (4\Omega^2(k_\parallel/k)^2 + N^2(k_\perp/k)^2)^{1/2}$. Hence, the above analyses for pure rotation or pure stratification can be extended without problems of principle. I would like just to point out the different columnar or pancake-shape structures obtained by Kimura & Herring using DNS in the three cases. (Proc. Monte Verita II colloquium?).

Dimensionality in a fully anisotropic description

Statistical spectral models mentioned above deal with the second-order spectral tensor $\Phi_{ij}(\mathbf{k}, t)$, which is the Fourier transform of the two-point covariance matrix

$$< u_i(\mathbf{x}, t) u_j(\mathbf{x}', t) > \tag{11}$$

and its more general expression for homogeneous anisotropic turbulence is

$$\Phi_{ij} = \begin{pmatrix} 0 & 0 & 0 \\ 0 & e + Z_r & Z_i - \imath\mathcal{H}/k \\ 0 & Z_i + \imath\mathcal{H}/k & e - Z_r \end{pmatrix} \tag{12}$$

using an orthonormal frame of reference, associated with polar-spherical coordinates (k, θ, ϕ) of $\mathbf{k}$ in Fourier space, in close connection with the Craya-Herring decomposition for the velocity field. Hence Φ_{ij} can be expressed as a sum of different contributions, in term of the scalars e (energy spectrum), Z (polarisation anisotropy), with $Z = Z_r + \imath Z_i$ complex, and $\mathcal{H}$ (helicity spectrum), which all depend on $\mathbf{k}$. Isotropic turbulence is characterized by $e(k, \theta, \phi, t) = E(k)/(4\pi k^2)$, $Z = \mathcal{H} = 0$, so that the real part of Φ_{ij}, which is involved in classic one-point correlations, involve e and Z contributions as follows

$$Re\Phi_{ij} = \underbrace{\frac{E(k)}{4\pi k^2} P_{ij}}_{\text{Pure isotropic part}} + \underbrace{(e(k, \theta, \phi) - \frac{E(k)}{4\pi k^2}) P_{ij}}_{\text{Directional anisotropy}} + \underbrace{Re(ZN_iN_j)}_{\text{Polarisation anisotropy}} \tag{13}$$

(see [7] for details). It is clear from the above equation that the anisotropy is twofold. A lessening of dimensionality is only reflected by a nonzero value of the second term. Indeed, the directional anisotropy, which is expressed by a departure of e from a spherical equidistribution, is extreme in a pure two-dimensional state, where e is concentrated onto the plane $k_\parallel = 0$. From this viewpoint, the trend towards two-dimensionality (with e tending to concentrate near the plane $k_\parallel = 0$) in rotating turbulence has nothing to do with the trend towards the collapse of vertical motion (with e tending to concentrate along the axis $k_\perp = 0$) encountered in stably stratified turbulence. Starting from (12), all the classic one-point correlations can be obtained and thus analysed in terms of both directional and polarisation anisotropies, including the Reynolds stress tensor, the vorticity correlation tensor and other quantities. Among them, the integral length scales $L_{ij}^{(n)}$, related to different velocity correlations (subscripts i, j) which are separated along different directions (superscript n), give an unvaluable information. The streaklike tendency in a shear flow can be easily found in calculating both the $L_{11}^{(1)}$ component, which gives the streamwise length of the streaks, and $L_{11}^{(3)}$, which gives the spanwise separation length of the streaks (as usual, 1 and 3 reffer to streamwise and spanwise coordinates, respectively). In pure homogeneous RDT at constant shear rate, both length scales can be calculated analytically and their ratio (elongation parameter) is found to increase as $(St)^2$. Another example is provided by rotating turbulence, in which the integral length scale related to transverse velocity components and axial separation is enhanced with respect to all the other components L_{ij}^n, whereas the Reynolds stress tensor is hardly anisotropic, with only a slight dominance of its axial component. The latter results were obtained by different approaches, including high resolution LES (see [7]) $128 \times 128 \times 512$ which allow the development of axial integral length scales without interference with periodic boundary conditions. They show that the nonlinear tendency to create columnar structures is very subtle and cannot yield to well organised arrays of 2D vortices. The eventual appearance of such 'rotors' in other low resolution DNS or LES in periodic boxes is only a numerical artefact. In an actual experiment or – explicitly inhomogeneous – numerical simulation, however, the nonhomogeneous forcing and/or the presence of solid boundaries can enforce the creation of such organised vortices.

As a final comment, I would like to underline that a fully anisotropic spectral (or two-point) description carry out a very large amount of information, even if it only concerns second-order statistics. In the inhomogeneous case, the pod decomposition has renewed the interest for second-order two-point statistics [16], but this technique is never applied completely in the homogeneous case. It is only said that pod spatial modes associated with (11) are Fourier modes in the homogeneous case, but the true spectral eigenvectors corresponding to pod modes are not considered. This can be obtained in diagonalizing the tensor Φ_{ij}, and is an easy task using the above $e - Z$ decomposition (12)–(13) (the angular position of the principal axes, which are associated with the nonzero principal components $e + |Z|$ and $e - |Z|$, is fixed by the phase of Z, at each $\mathbf{k}$).

References

[1] Babin, A., Mahalov, A. & Nicolaenko, B. 1996. Global splitting, integrability and regularity of 3D Euler and Navier-Stokes equations for uniformly rotating fluids. *Eur. J. Mech. B/Fluids* **15**, 291–300

[2] Batchelor, G. K. & Proudman, I. 1954. The effect of rapid distortion in a fluid in turbulent motion. *Q. J. Mech. Appl. Maths*, **7**, 83

[3] Bayly, B. J. 1986. Three-dimensional instability of elliptical flow. *Phys. Rev. Lett.* **57**, 2160–2163

[4] Benney, D. J. & Saffman, P. G. 1966. Nonlinear interactions of random waves in a dispersive medium. *Proc. Roy. Soc. London A.* **289**, 301–320

[5] CAILLOL, P. & ZEITLIN, V. 1998. On the stationary energy spectra of the stratified fluid in a weak turbulence regime. *Proc. Monte Verita II Colloquium.*

[6] CAMBON, C., BENOIT, J. P., SHAO, L. & JACQUIN, L. 1994. Stability analysis and large eddy simulation of rotating turbulence with organized eddies. *J. Fluid Mech.*,**278**, 175–200

[7] CAMBON, C., MANSOUR, N. N. & GODEFERD, F. S. 1997. Energy transfer in rotating turbulence. *J. Fluid Mech.* **337**, 303–332

[8] CAMBON, C. & SCOTT, J. F. 1998. Linear and nonlinear models of anisotropic turbulence. *Ann. Rev. Fluid Mech.* **61**. To appear

[9] CRAIK, A. D. D. & CRIMINALE, W. O. 1986 Evolution of wavelike disturbances in shear flows: a class of exact solutions of the Navier-Stokes equations. *Proc. R. Soc. London* A 406, 13.

[10] GODEFERD, F. S. & CAMBON, C. 1994. Detailed investigation of energy transfers in homogeneous turbulence. *Phys. Fluid* **6**, 2084–2100

[11] GODEFERD, F. S., MALIK, N. A., CAMBON, C. & NICOLEAU, F. 1997. Eulerian and Lagrangian statistics in homogeneous stratified flows. *Applied Scientific Research* **57**, 319–335

[12] GREENSPAN, H. P. 1968. The theory of rotating fluids. *Cambridge University Press.*

[13] JACQUIN, L., LEUCHTER, O., CAMBON, C., & MATHIEU, J. 1990. Homogeneous turbulence in the presence of rotation. *J. Fluid Mech.*, **220**, 1–52.

[14] LEE, J. M., KIM, J. & MOIN, P. 1990. Structure of turbulence at high shear rate. *J. Fluid Mech.* **216**, 561–583

[15] LIFSCHITZ, A. & HAMEIRI, E. 1991. Local stability conditions in fluid dynamics. *Phys. Fluid A.* **3**, 2644–2651

[16] LUMLEY, J. L. 1967. The structure of inhomogeneous turbulent flows. In *Atmospheric Turbulence and Radio Wave Propagation*, ed. A. M. Yaglom & V. I. Tatarski. Moscow: NAUCA

[17] MONIN, A. S. & YAGLOM, A. M. 1975. *Statistical Fluid Mechanics I.* section 17.3, MIT Press

[18] ORSZAG, S. A. Analytical theories of turbulence. *J. Fluid Mech.* **41**, 363–386.

[19] PERRY, A. E. & CHONG, M. S. 1987 A study of eddying motions and flow patterns using critical point concepts. *Ann. Rev. Fluid Mech.* **19**, 125–155.

[20] Rogers, M. M. 1991. The structure of a passive scalar field with a uniform gradient in rapidly sheared homogeneous turbulent flow. *Phys. Fluid A.* **3**, 144–154

[21] Townsend, A. A. 1956. The structure of turbulent shear flow. *Revised version* 1976, Cambridge University Press.

[22] Waleffe, F. 1993. Inertial transfers in the helical decomposition. *Phys. Fluid* A 5, 677–685

[23] Zakharov, V. E., L'vov, V. S., Falkowich, G. 1992. *Kolmogorov spectra of turbulence. I. Wave turbulence.* Springer series in nonlinear dynamics. Springer Verlag

Trends in Mathematics, © 1999 Birkhäuser Verlag Basel/Switzerland

On the stationary energy spectra of the stratified fluid in a weak turbulence regime

P. Caillol, V. Zeitlin

A characteristic feature of stratified fluid dynamics is omnipresence of internal gravity waves. In the case of strong stratification the effective non-linearity parameter becomes small and, hence, one may look for wave-dominated turbulent regimes where turbulence is thought of as a "soup" of random weakly interacting internal gravity waves. Such an approach, called weak turbulence (for a recent review see [1]) is well-known, especially in plasma physics (cf. [2]) and consists in using the perturbation theory in wave amplitudes and Gaussian statistics for an ensemble of weakly nonlinear waves in order to get a kinetic equation for the wave amplitude density. We apply this technique to the stratified fluid in the Boussinesq approximation and obtain in this way essentially anisotropic (in vertical and horizontal components of the wave number) Kolmogorov-type energy spectra as stationary solutions of the kinetic equations.

Our starting point is stratified ideal fluid equations in the Boussinesq approximation written for velocity $\mathbf{v}$ and buoyancy $\xi = -g\frac{\rho-\rho_0}{\rho_0}$, $\rho_o(x_3)$ is a background vertical density stratification, g is acceleration due to gravity (ξ should be replaced by potential temperature in atmospheric applications)

$$\begin{aligned} \dot{\mathbf{v}} + \mathbf{v}\nabla\mathbf{v} + \nabla P - \xi\,\hat{\mathbf{x}}_3 &= 0 \\ \dot{\xi} + \mathbf{v}\nabla\xi + N^2 w &= 0 \\ \nabla\cdot\mathbf{v} &= 0. \end{aligned} \tag{1}$$

Here w denotes the vertical component of velocity, $N^2 = -g\frac{d\rho_0}{dx_3}/\rho_0$ is the square of the Brunt-Väisälä frequency which is supposed to be constant and $\hat{\mathbf{x}}_3$ is the unit vector in vertical direction. Technically, it is convenient to use the Craya-Herring decomposition for velocity:

$$\mathbf{v} = \nabla_H \times \psi\,\hat{\mathbf{x}}_3 + \nabla_H\phi + w\,\hat{\mathbf{x}}_3 \tag{2}$$

which in the linear approximations gives a separation of the flow in a vortical (ψ) and wave (ϕ, w) parts. The condition of divergenceless of the flow implies

$$\Delta_H\phi + w_{x_3} = 0\,. \tag{3}$$

Here and above the subscript H denotes the horizontal part.

As it is well-known, by introducing Fourier transforms $\psi_{\mathbf{p}}$, $\xi_{\mathbf{p}}$, $\phi_{\mathbf{p}}$ and $w_{\mathbf{p}}$ and linearizing the system (1) one gets internal gravity waves solutions with a dispersion relation

$$\omega^2 = N^2 \frac{\mathbf{p}_H^2}{\mathbf{p}^2} \tag{4}$$

where ω is the wave frequency, $\mathbf{p} = (p_1, p_2, p_3)$ is the wavenumber, $p \equiv |\mathbf{p}|$ and $\mathbf{p}_H = (p_1, p_2)$ is its projection on the horizontal plane. Note that $\psi_\mathbf{p}$ represents a zero mode of the linear system, i.e. is time-independent in the first approximation, and that mean flow/stratification corrections proportional to $\delta(p_H)$ must be added to get a general solution of the linear system. We shall discard this contributions altogether by the reasons explained below. Note also that the wave part of $\phi_\mathbf{p}$ may be always eliminated in favor of $w_\mathbf{p}$ via (3). By introducing wave amplitudes $b_\mathbf{p} = e^{i\Omega_p t}\chi^+_\mathbf{p}$ corresponding to the two branches of the dispersion equation (4) , where Ω_p is the positive-sign root of (4), we get

$$w_\mathbf{p} = b_\mathbf{p} + \bar{b}_{-\mathbf{p}}\,,\ \xi_\mathbf{p} = iN\frac{p}{p_H}(b_\mathbf{p} - \bar{b}_{-\mathbf{p}}) \tag{5}$$

The energy of the system in the absence of vortical part and mean buoyancy is

$$\begin{aligned} E &= \frac{1}{2}\int d\mathbf{x}\ (-\psi\Delta_H\psi - \phi\Delta_H\phi + w^2 + N^{-2}\xi^2) \\ &= \frac{1}{2}\int d\mathbf{p}\ (p_H^2\,\phi_\mathbf{p}\,\phi_{-\mathbf{p}} + w_\mathbf{p} w_{-\mathbf{p}} + N^{-2}\xi_\mathbf{p}\,\xi_{-\mathbf{p}}) \\ &= \frac{1}{2}\int d\mathbf{p}\ (2\,\frac{p^2}{p_H^2}\ (b_\mathbf{p}\,\bar{b}_\mathbf{p} + b_{-\mathbf{p}}\,\bar{b}_{-\mathbf{p}})) \end{aligned} \tag{6}$$

In what follows we limit ourselves by a study of purely wave regimes, thus, excluding completely the vortical mode and mean flow/stratification corrections. A physical basis for this separation is the Lagrangian (pointwise) conservation of the potential vorticity $q = (\nabla \times \mathbf{v})(\nabla\xi + N^2\hat{\mathbf{x}}_3)$. For a single harmonic wave potential vorticity is identically zero. This is not true anymore for a superposition of waves and, in fact, waves do produce vorticity and mean bouyancy already in the first order of perturbation theory. However, for a Gaussian ensemble of waves (see below) the statistically averaged potential vorticity is still zero which is not the case in the present of vortical modes or mean buoyancy corrections. We, thus, have a separation criterion and by limiting ourselves by zero mean potential vorticity configurations we discard all non-wave corrections. The first order corrections, therefore, are

$$\begin{aligned} w^{(1)}(\mathbf{p}, t) = -\int d\mathbf{k}\, d\mathbf{l}\ \delta(\mathbf{k} + \mathbf{l} - \mathbf{p}) \sum_\pm B^\pm_{kl,p} \\ \Big[\frac{b_\mathbf{k} b_\mathbf{l}}{\Omega_k + \Omega_l \mp \Omega_p - i\delta} + \frac{b_\mathbf{k}\bar{b}_{-\mathbf{l}}}{\Omega_k - \Omega_l \mp \Omega_p - i\delta} \\ - \frac{\bar{b}_{-\mathbf{k}} b_\mathbf{l}}{\Omega_k - \Omega_l \mp \Omega_p + i\delta} - \frac{\bar{b}_{-\mathbf{k}}\bar{b}_{-\mathbf{l}}}{\Omega_k + \Omega_l \mp \Omega_p + i\delta}\Big] \end{aligned} \tag{7}$$

$$\xi^{(1)}(\mathbf{p},\,t) = -iN\frac{p}{p_H}\int d\mathbf{k}\,d\mathbf{l}\;\delta(\mathbf{k}+\mathbf{l}-\mathbf{p})\sum_{\pm}(\pm)B^{\pm}_{kl,p}$$
$$\Big[\frac{b_{\mathbf{k}}b_{\mathbf{l}}}{\Omega_k+\Omega_l\mp\Omega_p-i\delta}+\frac{b_{\mathbf{k}}\overline{b}_{-\mathbf{l}}}{\Omega_k-\Omega_l\mp\Omega_p-i\delta}$$
$$+\frac{\overline{b}_{-\mathbf{k}}b_{\mathbf{l}}}{\Omega_k-\Omega_l\mp\Omega_p+i\delta}+\frac{\overline{b}_{-\mathbf{k}}\overline{b}_{-\mathbf{l}}}{\Omega_k+\Omega_l\mp\Omega_p+i\delta}\Big] \tag{8}$$

where

$$B^{\pm}_{kl,p} = (p_H^2 + p_3 l_3\frac{\mathbf{p}\cdot\mathbf{l}}{l_H^2} \pm pl\frac{p_H}{l_H})\frac{H_{kl}}{2p^2}\;,\, H_{kl} = \frac{\mathbf{k}\cdot\mathbf{l}}{k_H^2}k_3 - l_3. \tag{9}$$

We next make a standard random phase approximation by supposing that due to the presence of a large amount of weakly interacting waves the complex wave amplitudes $b_{\mathbf{k}}$ assume Gaussian statistics (cf. [3])

$$\begin{aligned} < b_{\mathbf{k}}\,\overline{b}_{\mathbf{l}} > &= N_{\mathbf{k}}\;\delta(\mathbf{k}-\mathbf{l})\,,\\ < b_{\mathbf{k}}\,b_{\mathbf{l}}\,\overline{b}_{\mathbf{m}}\,\overline{b}_{\mathbf{n}} > &= N_{\mathbf{k}}\;N_{\mathbf{l}}(\delta(\mathbf{k}-\mathbf{m})\delta(\mathbf{l}-\mathbf{n})+\delta(\mathbf{k}-\mathbf{n})\delta(\mathbf{l}-\mathbf{m})), \end{aligned} \tag{10}$$

where the wavenumber density $N_{\mathbf{k}} = N_{-\mathbf{k}}$ is a real-valued function. In what follows we denote abusively the quadratic mean N_k even though it is not a function of $|\mathbf{k}|$ only. From (1) we obtain the following (exact) equations for averages of the Fourier components of the wave field:

$$\begin{aligned} <\dot{w}_{\mathbf{p}}\,w_{\mathbf{q}}> &= \frac{p_H^2}{p^2}<\xi_{\mathbf{p}}\,w_{\mathbf{q}}> - \frac{i\,p_H^2}{p^2}\int d\mathbf{k}\,d\mathbf{l}\;\delta(\mathbf{k}+\mathbf{l}-\mathbf{p})\;I_{kl,p}<w_{\mathbf{k}}\,w_{\mathbf{l}}\,w_{\mathbf{q}}>,\\ <\dot{\xi}_{\mathbf{p}}\,\xi_{\mathbf{q}}> &= -N^2<w_{\mathbf{p}}\,\xi_{\mathbf{q}}> - i\int d\mathbf{k}\,d\mathbf{l}\;\delta(\mathbf{k}+\mathbf{l}-\mathbf{p})\;H_{kl}\,<w_{\mathbf{k}}\,\xi_{\mathbf{l}}\,\xi_{\mathbf{q}}>, \end{aligned} \tag{11}$$

where $I_{kl,p} = (1 + \frac{p_3 l_3}{l_H^2}\frac{\mathbf{p}\cdot\mathbf{l}}{p_H^2})\;H_{kl}$. By adding these two equations and their complex conjugates, we get the following kinetic equation for the energy spectral density $\epsilon_{\mathbf{p}} = \frac{p^2}{p_H^2}(b_{\mathbf{p}}\,\overline{b}_{\mathbf{p}} + b_{-\mathbf{p}}\,\overline{b}_{-\mathbf{p}})$

$$\begin{aligned} < \dot{\epsilon}_p > &= \frac{i}{2}\int d\mathbf{k}\,d\mathbf{l}\;\delta(\mathbf{k}+\mathbf{l}-\mathbf{p})\;I_{kl,p}\;[< w_{-\mathbf{k}}\,w_{-\mathbf{l}}\,w_{\mathbf{p}} > - < w_{\mathbf{k}}\,w_{\mathbf{l}}\,w_{-\mathbf{p}} >]\\ &+ \frac{i}{2N^2}\int d\mathbf{k}\,d\mathbf{l}\;\delta(\mathbf{k}+\mathbf{l}-\mathbf{p})\;H_{kl}\;[< w_{-\mathbf{k}}\,\xi_{-\mathbf{l}}\,\xi_{\mathbf{p}} > - < w_{\mathbf{k}}\,\xi_{\mathbf{l}}\,\xi_{-\mathbf{p}} >] \end{aligned} \tag{12}$$

and calculate the r.h.s. of this equation perturbatively using (10) and Sokhotsky formulas

$$\frac{1}{A\,\pm\,i\delta} = P\frac{1}{A}\;\mp\;i\pi\delta(A). \tag{13}$$

Non-resonant contributions to (12) cancel out pairwise and only the resonant terms remain. We get

$$
\begin{aligned}
<\dot{\epsilon}_{\mathbf{p}}> \quad \sim \quad & 2\pi\frac{p^2}{p_H^2}\int d\mathbf{k}\, d\mathbf{l}\ \delta(\mathbf{k}+\mathbf{l}-\mathbf{p})\{\\
& B^{+}_{\{kl\},p}\ [B^{+}_{\{kl\},p}N_kN_l - D^{-}_{lp,k}N_pN_l - D^{-}_{kp,l}N_pN_k]\ \delta(\Omega_k+\Omega_l-\Omega_p) + \\
& D^{+}_{lk,p}\ [D^{-}_{pk,l}N_pN_k + C^{+}_{pl,k}N_pN_l + D^{+}_{lk,p}N_kN_l]\ \delta(\Omega_l+\Omega_p-\Omega_k) + \\
& D^{+}_{kl,p}\ [D^{-}_{pl,k}N_pN_l + C^{+}_{pk,l}N_pN_k + D^{+}_{kl,p}N_kN_l]\ \delta(\Omega_k+\Omega_p-\Omega_l)\} \qquad (14)
\end{aligned}
$$

where $C^{\pm}_{kl,p} = B^{\pm}_{kl,p} - B^{\pm}_{lk,p}$, $D^{\pm}_{kl,p} = B^{+}_{kl,p} \pm B^{-}_{lk,p}$. This is a typical kinetic equation with two δ-functions assuring that the collision process conserves energy and momentum. However, some transformations are necessary in order to cast it to the standard form [1]. A change of variables (cf. [4]) should be made in order to have a standard *frequency*$\times$*wavenumber density* form for $\epsilon_{\mathbf{p}}$:

$$
N_p \rightarrow N\frac{p_H^3}{p^3}N_p
$$

thus giving $\epsilon_p = 2\Omega_p N_p$ in terms of new N_p which is usually interpreted as "quasi-particle" density [1]. We correspondingly modify the interaction coefficients

$$
\mathcal{B}_{kl,p} = B^{+}_{\{kl\},p}(\frac{p}{kl})^{\frac{3}{2}}, \qquad \mathcal{C}_{kl,p} = C^{+}_{kl,p}(\frac{p}{kl})^{\frac{3}{2}} \qquad \mathcal{D}^{\pm}_{kl,p} = D^{\pm}_{kl,p}(\frac{p}{kl})^{\frac{3}{2}} \qquad (15)
$$

and finally get

$$
\begin{aligned}
& <\dot{N}_{\mathbf{p}}> \sim \pi N \int d\mathbf{k}\, d\mathbf{l}\ \delta(\mathbf{k}+\mathbf{l}-\mathbf{p})\{\\
& \mathcal{B}_{\{kl\},p}\ \left[\frac{k_H^3 l_H^3}{p_H^3}\mathcal{B}_{\{kl\},p}N_kN_l - k_H^3\mathcal{D}^{-}_{lp,k}N_pN_l - l_H^3\mathcal{D}^{-}_{kp,l}N_pN_k\right]\ \delta(\Omega_k+\Omega_l-\Omega_p) + \\
& \mathcal{D}^{+}_{lk,p}\ \left[k_H^3\mathcal{D}^{-}_{pk,l}N_pN_k + l_H^3\mathcal{C}^{+}_{pl,k}N_pN_l + \frac{k_H^3 l_H^3}{p_H^3}\mathcal{D}^{+}_{lk,p}N_kN_l\right]\ \delta(\Omega_l+\Omega_p-\Omega_k) + \\
& \mathcal{D}^{+}_{kl,p}\ \left[l_H^3\mathcal{D}^{-}_{pl,k}N_pN_l + k_H^3\mathcal{C}^{+}_{pk,l}N_pN_k + \frac{k_H^3 l_H^3}{p_H^3}\mathcal{D}^{+}_{kl,p}N_kN_l\right]\ \delta(\Omega_k+\Omega_p-\Omega_l)\} \quad (16)
\end{aligned}
$$

This is a complete kinetic equation for three-dimensional random internal gravity waves and it may be used, for example, for direct numerical studies. Yet, we will proceed analytically by making additional hypothesis allowing to simplify considerably otherwise too complicated interaction coefficients in (16). Namely, we limit ourselves by systems of waves (wave packets) propagating almost vertically, i.e. $k_3 \gg k_H$ and, thus, having low frequencies. In this case we get

$$
\frac{k_H^3 l_H^3}{p_H^3}\mathcal{B}_{kl,p} = k_H^3\mathcal{D}^{-}_{kp,l} = l_H^3\mathcal{D}^{-}_{lp,k},\ \ l_H^3\mathcal{C}_{pl,k} = -k_H^3\mathcal{D}^{-}_{pk,l} = -\frac{k_H^3 l_H^3}{p_H^3}\mathcal{D}^{+}_{lk,p} \qquad (17)
$$

and the collision integral in the r.h.s. of (16) greatly simplifies

$$\begin{aligned} < \dot{N}_{\mathbf{p}} > \quad \sim \quad \pi N & \int d\mathbf{k}\, d\mathbf{l}\ \delta(\mathbf{k}+\mathbf{l}-\mathbf{p})\{ \\ & X_{kl,p}[N_k N_l - N_p N_k - N_p N_l]\ \delta(\Omega_k+\Omega_l-\Omega_p)+ \\ & Y^{+}_{kl,p}[N_k N_l + N_p N_k - N_p N_l]\ \delta(\Omega_l+\Omega_p-\Omega_k)+ \\ & Y^{+}_{lk,p}[N_k N_l + N_p N_l - N_p N_k]\ \delta(\Omega_k+\Omega_p-\Omega_l)\} \end{aligned} \tag{18}$$

where

$$X_{kl,p} = (\tilde{k}_H + \tilde{l}_H + \tilde{p}_H)^2 \frac{(p_3^2 - k_3 l_3)^2}{16\, k\, l\, p\, k_H\, l_H\, p_H} \Big(\frac{\tilde{p}_H^2 - \tilde{k}_H \tilde{l}_H}{p_3^2 - k_3 l_3} p_3 - \frac{\tilde{l}_H^2}{l_3} - \frac{\tilde{k}_H^2}{k_3} \Big)^2, \tag{19}$$

$$Y^{\pm}_{kl,p} = (\tilde{k}_H - \tilde{l}_H + \tilde{p}_H)^2 \frac{(k_3^2 + l_3 p_3)^2}{16\, k\, l\, p\, k_H\, l_H\, p_H} \Big(\frac{\tilde{k}_H^2 \pm \tilde{l}_H \tilde{p}_H}{k_3^2 + l_3 p_3} k_3 - \frac{\tilde{p}_H^2}{p_3} \pm \frac{\tilde{l}_H^2}{l_3} \Big)^2 \tag{20}$$

and $\tilde{k}_H = sign[k_3] k_H$. We will look for horizontally isotropic and vertically averaged stationary solutions of (18) by assuming that wavenumber densities N_p depend neither on orientation of the horizontal wavenumber in the horizontal plane nor on the sign of the vertical wavenumber. In this case we may perform explicitely all angular integrations in the r.h.s. of the kinetic equation (cf. [5], [2]) and, by assuming a scaling form for solutions

$$N_p \sim p_H^{\alpha}\ |p_3|^{\beta} \tag{21}$$

we get

$$\begin{aligned} < \dot{N}(p_H, p_3) > \sim & \int dk_H\, dl_H\, dk_3\, dl_3\ \delta(k_3+l_3-p_3) \frac{k_H l_H}{\triangle} \{ \\ & X_{kl,p} \Big[k_H^{\alpha} l_H^{\alpha} k_3^{\beta} l_3^{\beta} - k_H^{\alpha} p_H^{\alpha} k_3^{\beta} p_3^{\beta} - l_H^{\alpha} p_H^{\alpha} l_3^{\beta} p_3^{\beta} \Big] \\ \times \Big[1 - (\frac{p_H}{k_H})^{2\alpha+6} (\frac{p_3}{k_3})^{2\beta+2} & - (\frac{p_H}{l_H})^{2\alpha+6} (\frac{p_3}{k_3})^{2\beta+2} \Big]\ \delta\Big(\frac{k_H}{k3} + \frac{l_H}{l_3} - \frac{p_H}{p_3} \Big) \} + \\ & 2 \int dk_H\, dl_H\, dk_3\, dl_3\ \delta(k_3+l_3-p_3) \frac{k_H l_H}{\triangle} \{ \\ & Y^{+}_{kl,p} \Big[k_H^{\alpha} l_H^{\alpha} k_3^{\beta} l_3^{\beta} + k_H^{\alpha} p_H^{\alpha} k_3^{\beta} p_3^{\beta} - l_H^{\alpha} p_H^{\alpha} l_3^{\beta} p_3^{\beta} \Big] \\ \times \Big[1 - (\frac{p_H}{k_H})^{2\alpha+6} (\frac{p_3}{k_3})^{2\beta+2} & + (\frac{p_H}{l_H})^{2\alpha+6} (\frac{p_3}{l_3})^{2\beta+2} \Big]\ \delta\Big(\frac{l_H}{l_3} + \frac{p_H}{p_3} - \frac{k_H}{k_3} \Big) \} \end{aligned} \tag{22}$$

with $\triangle = \frac{1}{4}[2(p_H^2 k_H^2 + p_H^2 l_H^2 + l_H^2 k_H^2) - p_H^4 - k_H^4 - l_H^4]^{\frac{1}{2}}$ and look for the values of scaling exponents α, β annihilating the r.h.s. of this equation. A direct substitution shows that there exist two sets of scaling exponents providing solutions. The one annihilating the first bracket is a Rayleigh-Jeans type solution $\alpha = -1$, $be = 1$ leading to the equipartition of energy:

$$N_k \sim \frac{k_3}{k_H}, \qquad \epsilon_k = const \tag{23}$$

The one annihilating the second bracket is a Kolmogorov-type solution $\alpha = -\frac{7}{2}$, $\beta = -\frac{1}{2}$ giving

$$N_k \sim k_H^{-\frac{7}{2}} k_3^{-\frac{1}{2}}, \qquad \epsilon_k \sim k_H^{-\frac{5}{2}} k_3^{-\frac{3}{2}}. \tag{24}$$

A simple check of dimensions shows that it corresponds to a constant energy flux through the wave spectrum and, therefore, presumes the existence of sources and sinks in the wavenumber space.

Thus, we have shown that a system of low-frequency horizontally isotropic internal gravity waves admits stationary Rayleigh-Jeans and Kolmogorov spectra as solutions of the energy transfer equation in the random-phase approximation. The first, as usual, corresponds to the system's tendency towards the state of absolute thermodynamic equilibrium, while the second is a non-equilibrium solution corresponding to a constant energy flux. The most important characteristics of this spectrum is its anisotropy in horizontal and vertical wavenumbers - a feature which is relatively easy to check in the direct numerical simulations of the stratified turbulence in the Boussinesq approximation and *in situ* measurements. The spectrum (24) by construction satisfies the separability hypothesis often used in geophysical applications [6]. Note that, if represented in the form

$$N_k \sim k_H^{-4} \Omega_k^{\frac{1}{2}} \sim \Omega_k^{-\frac{7}{2}} k_3^{-4} \tag{25}$$

and used to integrate the energy density by frequencies with a proper measure [7] this spectrum gives the -3 power laws in horizontal or vertical wavenumber, respectively. It is interesting to note that both spectra are observed in *in situ* measurements [8], [9].

References

[1] Falkovich, G., L'vov, V. and Zakharov, V.E. 1992, *Kolmogorov spectra of turbulence*, Springer, Berlin.

[2] Zakharov, V.E. 1984, Kolmogorov spectra in weak turbulence problems, *Handbook of plasma physics*, **2**, eds. Rosenbluth M.N. and Sagdeev R.Z., p. 3.

[3] Prigogine, I. 1962, *Nonequilibrium Statistical Mechanics*, Interscience, New-York.

[4] Daubner, S. and Zeitlin, V. 1996, On the stationary energy spectra for unidirectionally propagating internal gravity waves, *Phys. Lett. A* **214**, p. 33–39.

[5] Kuznetsov, E.A., 1972, Turbulence of ion sound ain a plasma located in a magnetic field, *Zh. Eksp. Teor. Fiz.*, **62**, p. 584.

[6] Weinstock, J. 1990, Saturated and unsaturated spectra of gravity waves and scale-dependent diffusion, *J. Atmos. Sci*, **47**, p. 2221–2225.

[7] Garret, C. and Munk, W. 1975, Space-Time scales of internal waves: a progress report, *J. Geoph. Res.*, **80**, No. D3, p. 291–297.

[8] Gregg, M.C. 1977, A comparison of finestructure spectra from the main thermocline, *J. Phys. Oceanogr.*, **7**, p. 33–40.

[9] Bacmeister, J.T. *et al*, 1996, Stratospheric horizontal wavenumber spectra of winds, potential temperature, and atmospheric tracers observed by high-latitude aircraft, *J. Geoph. Res.*, **101**, No. D5, p. 9441–9470.

Trends in Mathematics, © 1999 Birkhäuser Verlag Basel/Switzerland

On a possible mechanism of anomalous diffusion in geophysical turbulence

F. Dupont, R. I. McLachlan, V. Zeitlin

The main characteristic features of the geophysical turbulence are differential (so called β-effect) rotation and stratification. Each of these phenomena is responsible for a specific type of waves: planetary or Rossby waves and internal gravity waves, respectively. It is well-known that e.g. for the β- plane turbulence [1] one may have vortex-dominated (i.e. close to the 2d turbulence), wave-dominated or crossover regimes depending on the value of the characteristic nonlinearity parameter. Hence, if the Lagrangian transport in geophysical turbulence is studied the waves will play a rôle and it is important to know what are, precisely, their transport properties. In this connection it is known that, e.g., electromagnetic waves of sufficiently large amplitude in plasma may trap and effectively transport charged particles [2]. The crucial difference with fluid dynamics, however, is that nonlinear interaction of electromagnetic waves is weak even for large wave amplitudes and, usually, may be safely neglected while it is not the case for the above-mentioned waves. Thus, it is necessary to include nonlinear effects which will be done perturbatively in what follows. Note that previously the nonlinear interactions among waves were not taken into account neither in the studies of mixing by finite number of waves [3, 4, 5] nor in the numerical studies of diffusion by an ensemble of waves [6, 7].

Below we shall limit ourselves to Rossby waves, the conclusions for internal gravity waves being similar. The dynamical system we are interested in is an equation for the streamfunction ψ of the two-dimensional velocity field $\mathbf{v} = (-\partial\psi/\partial y, \partial\psi/\partial x)$

$$\frac{\partial \Delta\psi}{\partial t} + J(\psi, \Delta\psi) + \beta\frac{\partial\psi}{\partial x} = 0 \tag{1}$$

where β is the Coriolis parameter, $J(A,B) = \frac{\partial A}{\partial x}\frac{\partial B}{\partial y} - \frac{\partial A}{\partial y}\frac{\partial B}{\partial x}$, $\Delta = \frac{\partial^2}{\partial x^2} + \frac{\partial^2}{\partial y^2}$ and $\mathbf{x} = (x,y)$ are Cartesian coordinates on the β-plane. The linear part of this equation describes propagation of Rossby waves of the form $\psi = a\cos(\omega t - \mathbf{k}\cdot\mathbf{x} + \varphi)$ with dispersion relation $\omega = \Omega(k_x, k_y) = -\beta\frac{k_x}{\mathbf{k}^2}$ As is well known [8], in the lowest order of the perturbation expansion in wave amplitude the Rossby waves' interactions are dominated by resonant triads. Indeed, if one takes a triad of waves

$$\psi_1 = \sum_{i=1}^{3} a_i \cos\Phi_i \tag{2}$$

satisfying the resonance conditions

$$\Phi_1 + \Phi_2 = \Phi_3 \quad \forall \mathbf{x},\ t, \tag{3}$$

where

$$\Phi_i = \Omega_i t - \mathbf{k}_i \cdot \mathbf{x}, \tag{4}$$

$\Omega_i = \Omega(k_{x_i}, k_{y_i})$, ϵ is the small parameter, and $a_i = a_i(\tau)$ where $\tau = \epsilon^{-1} t$ is the slow modulation time, as a first-order approximation to the perturbative expansion $\psi = \varepsilon\psi_1 + \varepsilon^2\psi_2 + \dots$ of the solution of (1), then the integrability conditions for the next approximation, ψ_2, give modulation equations for the wave amplitudes which admit solutions in terms of elliptic functions

$$a_1(\tau) = a_{1_0}\,\mathrm{dn}(\tau|\kappa), \quad a_2(\tau) = a_{2_0}\,\mathrm{sn}(\tau|\kappa), \quad a_3(\tau) = a_{3_0}\,\mathrm{cn}(\tau|\kappa). \tag{5}$$

In terms of the initial amplitudes and interaction parameters the modulus κ of the elliptic functions is given by

$$\kappa^2 = \frac{a_{3_0}^2}{a_{1_0}^2}\left|\frac{\gamma_1}{\gamma_2}\right|, \tag{6}$$

where γ_i, $i = 1, 2, 3$ are the interaction parameters

$$\gamma_1 = \frac{\mathbf{k}_2 \times \mathbf{k}_3}{\mathbf{k}_1^2}, \quad \gamma_2 = \frac{\mathbf{k}_3 \times \mathbf{k}_1}{\mathbf{k}_2^2}, \quad \gamma_3 = \frac{\mathbf{k}_1 \times \mathbf{k}_2}{\mathbf{k}_3^2}, \tag{7}$$

and the cross product of two vectors $\mathbf{a}$ and $\mathbf{b}$ is defined as $\mathbf{a} \times \mathbf{b} = a_x b_y - a_y b_x$. The modulation period, T_M, is equal to the period $K(\kappa)$ of the elliptic functions, where K is the complete elliptic integral of the first kind. Note that one may solve explicitely the resonance conditions (3) and parameterize the triad space by two angles in the wavevectors' triangle, namely by α, the angle formed by $\mathbf{k}_1$ and $\mathbf{k}_3$, and by γ, that formed by $\mathbf{k}_2$ and $\mathbf{k}_3$ – a parameterization which will be used to explore the triads' space in what follows.

Advection of the passive tracer by the divergenceless velocity field associated with a streamfunction ψ is given by

$$\dot{x}(t) = -\frac{\partial\psi(x, y; t)}{\partial y}; \ \dot{y}(t) = \frac{\partial\psi(x, y; t)}{\partial x}. \tag{8}$$

where $x(t), y(t)$ denote the particle position at time t. These equations represent a "one and a half" degree of freedom Hamiltonian system which may be studied by standard methods once ψ, the Hamiltonian, is known as a function of x, y and t. Before considering advection by resonant triads (2) we first summarize the dynamics of (8) in the presence of two harmonic waves. In this case

$$\psi = A_1 \cos(\Omega_1 t - \mathbf{k}_1 \cdot \mathbf{x}) + A_2 \cos(\Omega_2 t - \mathbf{k}_2 \cdot \mathbf{x}) \tag{9}$$

equations (8) are integrable as the time dependence may be eliminated by choosing the phases $\Phi_{1,2}$ as new dynamical variables (provided the matrix $K = \begin{pmatrix} k_{1_x} & k_{1_y} \\ k_{2_x} & k_{2_y} \end{pmatrix}$ is nonsingular). These variables are defined on the torus $\mathbf{T}^2$ but may be lifted to its universal cover $\mathbf{R}^2$. We get a system of equations

$$\begin{aligned} \dot{\Phi}_1 &= \Omega_1 + (\mathbf{k}_1 \times \mathbf{k}_2) A_2 \sin \Phi_2 \\ \dot{\Phi}_2 &= \Omega_2 - (\mathbf{k}_1 \times \mathbf{k}_2) A_1 \sin \Phi_1 \end{aligned} \tag{10}$$

which is an integrable canonical one-degree-of-freedom Hamiltonian system with Hamiltonian

$$H = \Omega_1 \Phi_2 - \Omega_2 \Phi_1 - (\mathbf{k}_1 \times \mathbf{k}_2)(A_1 \cos \Phi_1 + A_2 \cos \Phi_2). \tag{11}$$

We call $\mathbf{x}$ *physical space*, and $\Phi = (\Phi_1, \Phi_2)$ *phase space*. They are related by

$$\mathbf{x} = (K^{-1}\Omega)t - \Phi. \tag{12}$$

If a particle has $\mathbf{x}(t)$ bounded we call it *frozen*; if it is not bounded, but $\lim_{t\to\infty} \mathbf{x}(t)/t = 0$, we say it is *diffusive* (and, generally, $\mathbf{x}(t) \sim t^{1/2}$); if a non-zero $\lim_{t\to\infty} \mathbf{x}(t)/t$ exists , we say the motion is *ballistic*. Ballistic motion corresponds to particles being advected with, on average, a constant velocity; we call these particles *shooters*. In general, the situation where $\mathbf{x}(t) \sim t^\alpha$ with $\alpha \neq 1/2$ is known as anomalous diffusion. For advection by a pair of harmonic waves (9 - 11) the motion in phase space is along the isoenergetic curves $H = const$ with an average (over $\Phi \in \mathbf{T}^2$) velocity $\Omega = (\Omega_1, \Omega_2)$ which gives no average (over Φ) displacement in physical space. However, two different situations may take place according to the topology of the isoenergetic curves which is determined by the interplay of the linear and nonlinear terms in (11). In the case when the linear term is dominating the isolines of H are open and each point in the phase space is moving with a velocity close to the average velocity. In physical space any particle therefore remains close to its initial position and is frozen. In the case when linear and nonlinear terms are comparable, fixed points, each with a surrounding island of periodic orbits (elliptic islands) appear in the phase space. A fixed point in Φ-space, by virtue of (12), corresponds to a ballistically advected particle in physical space. It is clear that the whole elliptic island around a given fixed point will be also ballistically advected. The relative number of shooters is, thus, defined by the fraction of phase space occupied by the elliptic islands. All other points moving along the open orbits in phase space stay close to the origin in the physical space; but, in order to maintain zero average velocity, the cloud of "normal" points must drift in the direction opposite to the shooters' motion.

If confirmed, the ballistic advection phenomenon would provide a mechanism for the anomalous diffusion in geophysical turbulence. However, the condition for the appearance of shooters means that the wave amplitudes are relatively large and, thus, it is necessary to take into account the nonlinear effects. We therefore consider a resonant triad engendered by an initial pair of waves. The resonance condition (3) allows us to eliminate the third phase Φ_3 and, again, to get a Hamiltonian system defined in the phase space (Φ_1, Φ_2). The equations of motion and the Hamiltonian are, respectively:

$$\begin{aligned} \dot{\Phi}_1 &= \Omega_1 + (\mathbf{k}_1 \times \mathbf{k}_2)\,(a_2(t)\sin\Phi_2 + a_3(t)\sin(\Phi_1+\Phi_2)) \\ \dot{\Phi}_2 &= \Omega_2 - (\mathbf{k}_1 \times \mathbf{k}_2)\,(a_1(t)\sin\Phi_1 - a_3(t)\sin(\Phi_1+\Phi_2)) \end{aligned} \tag{13}$$

$$H = \Omega_1\Phi_2 - \Omega_2\Phi_1 - (\mathbf{k}_1 \times \mathbf{k}_2)\,(a_1(t)\cos\Phi_1 + a_2(t)\cos\Phi_2 + a_3(t)\cos(\Phi_1+\Phi_2)) \tag{14}$$

and represent a one-and-a-half degree of freedom dynamical system which is not integrable due to the explicit time-dependence of the a_i. Note that the modulation $a_i(t)$, although periodic with period T_M, is not harmonic, except for the degenerate case of the isosceles triads . A standard method of investigation for this type of dynamical system is a numerical integration and a study of the phase portrait resulting from the iterations of the Poincaré map f_P over the modulation period T_M.

In general, for the non-integrable system (13) one might expect a chaotic sea with, possibly, some islands of regular behavior as a typical phase portrait. As $\beta \to \infty$ is an integrable limit of (13), one might also expect that for any triad a more and more regular pattern of dynamical behavior emerges with increasing β. As in the integrable case of a pair of waves, the presence of elliptic islands would signify a presence of shooters. Indeed, suppose the Poincaré map $f_P : \mathbf{T}^2 \to \mathbf{T}^2$ of (13) has a period n fixed point Φ_0 with a winding number $\mathbf{m}$ about the torus and n is the discrete time. This means that in terms of the lift map $\tilde{f}_P$ on $\mathbf{R}^2$ we have

$$\tilde{f}_P^n(\Phi_0) = \Phi_0 + 2\pi\mathbf{m} \tag{15}$$

and the distance travelled in time nT_M in physical space is

$$\mathbf{x} = K^{-1}\cdot(\Omega n T_M - 2\pi\mathbf{m}) + \mathbf{x}(0). \tag{16}$$

Hence, the fixed points and the regular orbits around them belonging to the stability islands having the same average winding number correspond to shooters. Chaotic orbits correspond to slow diffusion. The presence of stability islands results in a slow drift in physical space of the diffusive cloud away from the shooters, so as to compensate for the fast escape of the shooters and to keep the average winding number zero.

We performed a direct numerical investigation of (13) for a large range of triads using the above-mentioned explicit parametrization of these latter. In order to be consistent with perturbation theory we were trying to minimize the ratio

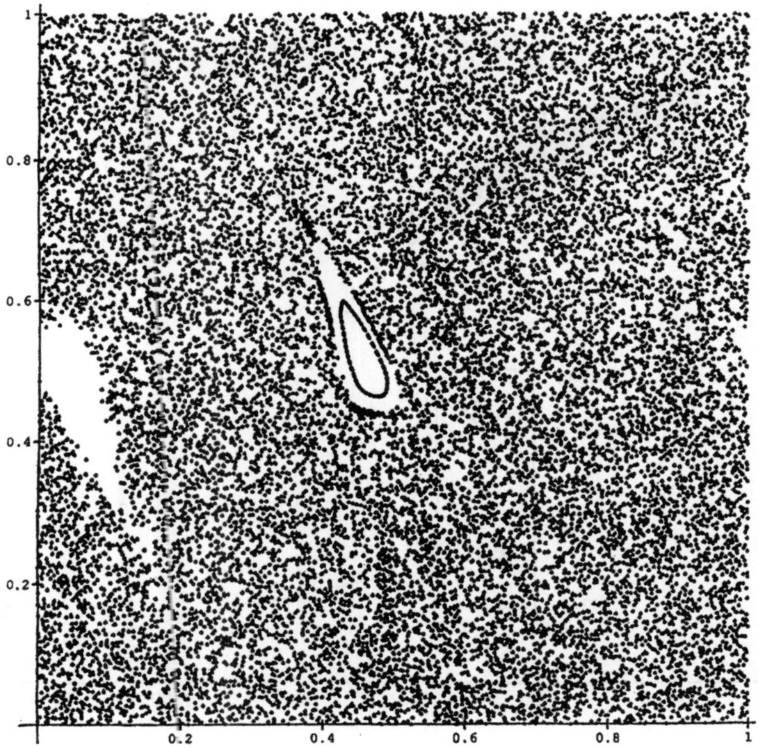

Figure 1: The phase portrait of the triad $\gamma = 0.7, \alpha = 1$ at $\beta = .5$ after 2000 T_M; $T_M = 47.14$. The wave periods T_i are 22.66, 10.98, 21.27, respectively. An elliptic trajectory inside the island is also shown.

$\frac{\max(T_i)}{T_M}$ by working in the correspondingly chosen region of the triads' parameters space. The smallness of this ratio may be considered as a rough criterion of applicability of the resonant triad description of the wave field generated by an initially active pair of waves 1 and 3. As we wish to follow many orbits of (13) for long times at many different parameter values we use a symplectic integrator [9], thus ensuring that the computed Poincaré map f_P is exactly (up to roundoff error) area-preserving and that no spurious non-Hamiltonian-like dynamics or bifurcations will be induced. We, generically, observe two types of scenario. The first one consists in a direct transition to quasi-uniform chaos from the regular behavior while decreasing β. The second one is characterized by a non-mixing ergodic behavior for large β, as expected, which gives place to chaos by more and more vigorous distortion of the phase trajectories and, then, to the birth of multiple relatively small and/or narrow islands. These primary islands disappear giving rise to chaos from which secondary "fat" islands are born (Fig. 1). It is these fat islands that give rise to relatively large numbers of shooters and a corresponding large

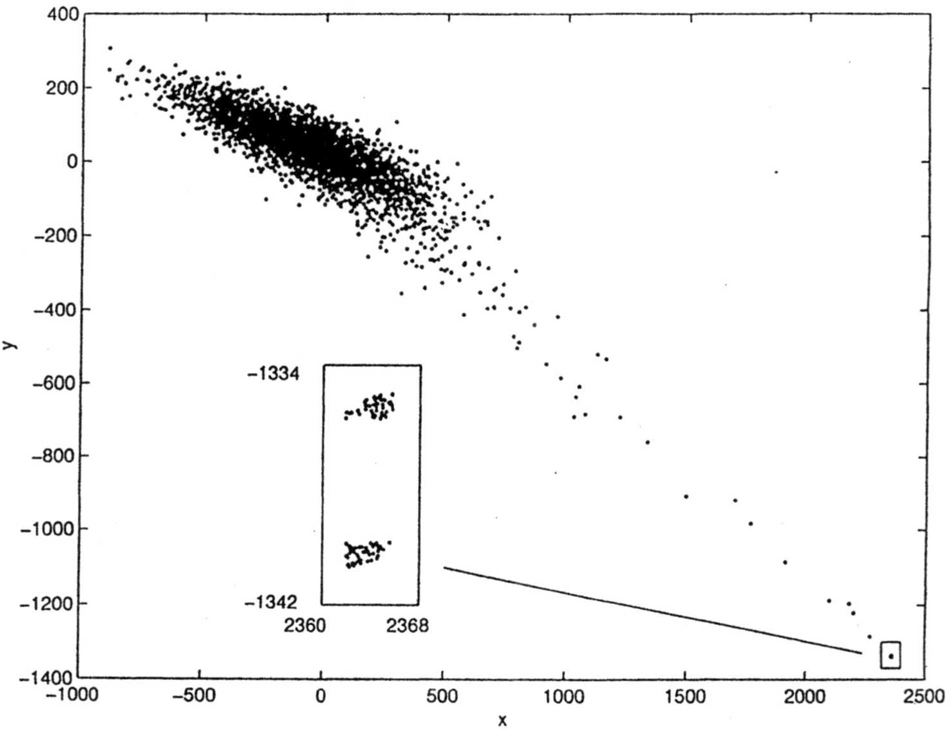

Figure 2: The dispersion of tracer particles, initially uniformly distributed in the $(2\pi)^2$ box in the phase-space after 500 modulation periods. The triad is the same as in Fig. 1.

amount of anomalous diffusion of tracer particles (Fig. 2). Although by technical reasons we did not explore the whole of parameter space and, thus, are unable to estimate a measure of islands-bearing triads we, nevertheless, see that these triads and, hence, the phenomenon of the ballistic transport are common. At the same time, a chaotic mixing related to the chaotic part of the phase portrait is universal. It should be noted that elliptic islands appear when nonlinearity is well-developed (the ratio $\max(T_i)/T_M \sim \frac{1}{3}$ or greater) while the threshold for chaotic mixing corresponds to higher values of β.

Thus, we have shown that chaotic mixing accompanied in many cases by ballistic advection of a fraction of tracer particles characterize the Lagrangian transport by resonant triads of Rossby waves. The fact that the nontrivial dynamical behavior we observed takes place in the domain of parameters corresponding to the transition from wave-dominated to vortex-dominated regimes in β-plane turbulence [1] suggests that this crossover regime may have nontrivial transport properties. This remains to be checked in direct numerical simulations.

References

[1] Rhines, P. B. (1975) "Waves and turbulence on a beta-plane" J. Fluid Mech. **69**, 417–443.

[2] Hirshman, S. P. (1980), "Two-dimensional electrostatic $E \times B$ trapping," Phys. Fluids **23**, 562–565.

[3] Pierrehumbert, R. T. (1991) "Chaotic mixing of tracer and vorticity by modulated travelling Rossby waves", Geoph. Astrophys. Fluid. Dyn. **58**, 285–319.

[4] Pierrehumbert, R. T. (1991) "Large-scale horizontal mixing in planetary atmospheres", Phys. Fluids A **3**, 1250–1260.

[5] del Castillo-Negrete, D. and Morrison, P. J. (1993) "Chaotic transport by Rossby waves in shear flow", Phys. Fluids A **5**, 948–965.

[6] Crisanti, A., Falcioni, M. Paladin, G. and Vulpiani, A. (1991) "Lagrangian chaos, transport, mixing and diffusion in fluids" Riv. Nuovo Cim. **12** #12, p. 1–80.

[7] Ottaviani, M. (1992) "Scaling laws of test particle transport in two-dimensional turbulence" Europhys. Lett. **20**, 11–116.

[8] Longuet-Higgins, M. S. and Gill, A. E. (1967) "Resonant interactions between planetary waves" Proc. Roy. Soc. Lond. A **299**(1456), 120–140.

[9] McLachlan, R. I. (1995), "On the numerical integration of ordinary differential equations by symmetric composition methods", SIAM J. Sci. Comput. **16**, 151–168.

5 Basic aspects of turbulence modelling

Trends in Mathematics, © 1999 Birkhäuser Verlag Basel/Switzerland

Basic issues of turbulence modeling

M. Germano

Abstract

Two simple problematic issues in turbulence modeling are exposed and presented to the discussion. The first is related to the different decompositions of the turbulent stress and the turbulent kinetic energy. The second to the estimation and the processing of the data, and in particular to the extraction of statistical quantities both from the Direct Numerical Simulations and the Large Eddy Simulations of turbulent flows.

1 Introduction

Turbulence modeling is located at the crossing of different points of view and attitudes. From a numerical point of view one can interpret this topic as a temporary annoying problem possibly overcome when powerful computers will permit the direct simulations of complex turbulent flows without restrictions on the geometry and the Reynolds number; from a basic point of view the problem of closure is still trying to find a reasonable rigorous physical mathematical foundation and finally from the applied point of view it is a real nightmare for people submerged by different models and variants. If we assume as a basic starting point in every theory of turbulence its *representation* in terms of spectral coefficients, statistical or physical averages, or more generally simple *objects* conditionally extracted by a weak background, turbulence modeling could be defined, reductively, as the art of writing the equations that produce directly such quantities. Let us assume that the progress in computer has finally reached the level predicted above. It could be always better to solve the Reynolds equations if we are only interested to statistical averages and the closures have experienced a parallel progress. As such turbulence modeling should expand its range of interest, and to relate more deeply its procedures with experimental and theoretical techniques based on very complex quantities or local scale decompositions, like POD or wavelets.

The multiscale structure of a turbulent field can be examined in a lot of different ways. We can classify the different techniques presently used to analyze the turbulence in terms of the nature of the decomposition that they operate on the original turbulent field. The Reynolds decomposition extracts from the chaotic signal the long time average, and for statistically steady fields we presume that this average converges to the ensemble average. The decomposition in mean ensemble values and fluctuations gives useful practical informations, but no information on coherent structures. The Fourier decomposition extracts the different spectral contributions in space or in time, but turbulence is usually intermittent, local and nonperiodic, so that also this technique can be in some cases physically ambiguous.

In order to overcome such inconveniences and to circumvent such problems new local decompositions able to detect and analyze the coherent structures present in a turbulent field have been proposed. Periodic phase averages [1], conditional averages [2], proper orthogonal decompositions [3], wavelet decompositions [4], zone decompositions [5] are particular examples of multilevel schemes recently applied to the analysis of data. The interest in multiscale approaches is not limited to the description of the turbulence, but is also present in the prediction of the dynamics of turbulent flows. New multiscale models have been recently proposed, [6], [7], and also in the Large Eddy Simulation some tentatives in this direction have been done in order to produce multiscale subgrid models more accurate in predicting some particular features of the turbulent flow like the backscatter, [8], [9].

2 Different decompositions of the turbulent stress and the turbulent energy

A simple way to classify all these decompositions is in *structured* and *unstructured* ones. We call structured decompositions the representations of a turbulent field in terms of a set of orthogonal functions φ_k

$$u_i = \sum_1^\infty u_{ik}\varphi_k \tag{1}$$

and we reserve the term *unstructured* to decompositions that do not resort to a basis. The simplest one is the *double decomposition* given by

$$u_i = U_i + u_i' \tag{2}$$

where U_i is a reference velocity and u_i' the associated fluctuation. As we know a basic issue of turbulence modelling is to represent the quantity

$$\tau_f(u_i, u_j) = \langle u_i u_j \rangle_f - \langle u_i \rangle_f \langle u_j \rangle_f \tag{3}$$

where $\langle \cdots \rangle_f$ stands for a generic filtering or averaging operator that usually we assume linear and commuting with the derivatives. If we look at the current literature we notice that the important issue of the decomposition of $\tau_f(u_i, u_j)$ connected to (2) is controversial. In particular we have

- the decomposition proposed by Leonard [10] given by

$$\tau_f(u_i, u_j) = L_f(u_i, u_j) + C_f(u_i, u_j) + R_f(u_i, u_j) \tag{4}$$

  with $U_i = \langle u_i \rangle_f$, and where

$$\begin{aligned} L_f(u_i, u_j) &= \langle U_i U_j \rangle_f - \langle u_i \rangle_f \langle u_j \rangle_f \\ C_f(u_i, u_j) &= \langle U_i u_j' \rangle_f + \langle U_j u_i' \rangle_f \\ R_f(u_i, u_j) &= \langle u_i' u_j' \rangle_f \end{aligned} \tag{5}$$

- the *consistent* decomposition proposed by Germano [11] given by

$$\tau_f(u_i,u_j) = \tau_f(U_i,U_j) + \tau_f(U_i,u'_j) + \tau_f(u'_i,U_j) + \tau_f(u'_i,u'_j) \tag{6}$$

where

$$\begin{aligned}
\tau_f(U_i,U_j) &= \langle U_iU_j\rangle_f - \langle U_i\rangle_f\langle U_j\rangle_f \\
\tau_f(U_i,u'_j) &= \langle U_iu'_j\rangle_f - \langle U_i\rangle_f\langle u'_j\rangle_f \\
\tau_f(u'_i,U_j) &= \langle u'_iU_j\rangle_f - \langle u'_i\rangle_f\langle U_j\rangle_f \\
\tau_f(u'_i,u'_j) &= \langle u'_iu'_j\rangle_f - \langle u'_i\rangle_f\langle u'_j\rangle_f
\end{aligned} \tag{7}$$

The properties of the consistent decomposition can be better appreciated if we write

$$U_i = V_i + u''_i \tag{8}$$

and if we consider the *triple* decomposition given by

$$u_i = V_i + u''_i + u'_i \tag{9}$$

where V_i is another reference velocity that we assume to remain unaltered by the filtering process. [1] We obtain

$$\begin{aligned}
\tau_f(U_i,U_j) &= \langle u''_iu''_j\rangle_f - \langle u''_i\rangle_f\langle u''_j\rangle_f \\
\tau_f(U_i,u'_j) &= \langle u''_iu'_j\rangle_f - \langle u''_i\rangle_f\langle u'_j\rangle_f \\
\tau_f(u'_i,U_j) &= \langle u'_iu''_j\rangle_f - \langle u'_i\rangle_f\langle u''_j\rangle_f \\
\tau_f(u'_i,u'_j) &= \langle u'_iu'_j\rangle_f - \langle u'_i\rangle_f\langle u'_j\rangle_f
\end{aligned} \tag{10}$$

3 The statistical estimation of a DNS database

We can assume that a Direct Numerical Simulation, or an experiment resolved in all details, produces the quantities u_i, p that satisfy the Navier Stokes equations

$$\frac{\partial u_k}{\partial y_k} = 0 \quad ; \quad \frac{\partial u_i}{\partial t} + \frac{\partial u_i u_k}{\partial y_k} = -\frac{\partial p}{\partial y_i} + \frac{\partial \sigma_{ik}}{\partial y_k}$$

where

$$\sigma_{ik} = -2\nu s_{ik} \quad ; \quad s_{ik} = \frac{1}{2}\left(\frac{\partial u_i}{\partial y_k} + \frac{\partial u_k}{\partial y_i}\right)$$

The statistical estimation is produced by applying to the DNS data the statistical operator $\mathcal{E}$ defined in some particular way. Following Aubry [12] we notice that the definition of the probability space of the flow realizations is quite flexible, and for example it consists of

[1] It means that the components V_i behave like constants when subjected to the filtering process.

- a time domain in statistically stationary flows,
- a space domain for homogeneous flows,
- a time domain conditionally determined on the basis of the phase of a background periodicity, or an intermittency function, or more generally a turbulence indicator,
- a set of initial conditions.

As regards the velocity field, the more relevant quantities that can be extracted are the following

- The statistical values $\langle u_i \rangle_e$
- The statistical, Reynolds, stress $\tau_e(u_i, u_j)$
- The turbulent kinetic energy at the statistical level

$$K_e = \frac{1}{2}\tau_e(u_i, u_i)$$

- The turbulent energy production at the statistical level

$$P(K_e) = -\tau_e(u_i, u_j)\langle s_{ij} \rangle_e$$

4 The statistical estimation of an LES database

We can assume that a Large Eddy Simulation produces the quantities $\langle u_i \rangle_f, \langle p \rangle_f$ that satisfy the filtered Navier Stokes equations

$$\frac{\partial \langle u_k \rangle_f}{\partial y_k} = 0$$

$$\frac{\partial \langle u_i \rangle_f}{\partial t} + \frac{\partial \langle u_i \rangle_f \langle u_k \rangle_f}{\partial y_k} = -\frac{\partial \langle p \rangle_f}{\partial y_i} - \frac{\partial \tau_f(u_i, u_k)}{\partial y_k} + \frac{\partial \langle \sigma_{ik} \rangle_f}{\partial y_k}$$

where

$$\langle \sigma_{ik} \rangle_f = -2\nu \langle s_{ik} \rangle_f \quad ; \quad \langle s_{ik} \rangle_f = \frac{1}{2}\left(\frac{\partial \langle u_i \rangle_f}{\partial y_k} + \frac{\partial \langle u_k \rangle_f}{\partial y_i} \right)$$

and the generalized turbulent stress

$$\tau_f(u_i, u_j) = \langle u_i u_j \rangle_f - \langle u_i \rangle_f \langle u_j \rangle_f$$

is approximated by a particular subgrid scale model

$$\tau_f(u_i, u_j) = M_f(\langle u_i \rangle_f, \langle u_j \rangle_f)$$

Formally we can assume that an LES is produced by applying to the velocity and pressure field a filtering operator $\mathcal{F}$

$$\langle u_i \rangle_f = \mathcal{F}(u_i) \qquad \langle p \rangle_f = \mathcal{F}(p)$$

but the explicit form of it is practically largely undetermined.

The statistical estimation is produced by applying to the LES data the statistical operator $\mathcal{E}$. We notice that in this case we have a hierarchy of two operators

$$\mathcal{F} \qquad \mathcal{E}$$

the LES filter and the statistical one. The statistical quantities of more interest that can be extracted *from the filtered component* of the velocity $\langle u_i \rangle_f$ are now *only* the following

- The $\mathcal{FE}$-level values $\langle u_i \rangle_{fe} = \langle \langle u_i \rangle_f \rangle_e$
- The resolved stress $\tau_e(\langle u_i \rangle_f, \langle u_j \rangle_f)$
- The resolved turbulent energy

$$R_{fe} = \frac{1}{2}\tau_e(\langle u_i \rangle_f, \langle u_i \rangle_f)$$

and we notice that only when $\mathcal{FE} = \mathcal{E}$ the $\mathcal{FE}$-level is equivalent to the statistical level.

From the basic point of view the statistical estimation of an LES database is a *multilevel* estimation, and as such it is very important to examine the energy transfer between the different levels. We notice that the turbulent kinetic energy at the $\mathcal{FE}$-level can be operatorially decomposed as

$$K_{fe} = \langle K_f \rangle_e + R_{fe}$$

where K_f, the turbulent kinetic energy at the $\mathcal{F}$-level, is given by

$$K_f = \frac{1}{2}\tau_f(u_i, u_i)$$

Also the production can be decomposed in two terms

$$P(K_{fe}) = -\langle \tau_f(u_i, u_l) \rangle_e \langle s_{il} \rangle_{fe} - \tau_e(\langle u_i \rangle_f, \langle u_l \rangle_f) \langle s_{il} \rangle_{fe}$$

and if we consider the balance of $\langle K_f \rangle_e$ and R_{fe} their productions are respectively given by [13]

$$\begin{aligned} P(\langle K_f \rangle_e) &= -\langle \tau_f(u_i, u_l) \rangle_e \langle s_{il} \rangle_{fe} - \tau_e\Big(\tau_f(u_i, u_l), \langle s_{il} \rangle_f\Big) \\ P(R_{fe}) &= -\tau_e(\langle u_i \rangle_f, \langle u_l \rangle_f) \langle s_{il} \rangle_{fe} + \tau_e\Big(\tau_f(u_i, u_l), \langle s_{il} \rangle_f\Big) \end{aligned}$$

It is clear the importance of the term

$$\tau_e\Big(\tau_f(u_i, u_l), \langle s_{il}\rangle_f\Big)$$

that represents the energy exchange between the two levels. If we recall that

$$K_{fe} = \langle K_f\rangle_e + R_{fe}$$

we conclude that there is no possibility to extract from an LES the turbulent energy at the statistical level, also if we assume $K_{fe} = K_e$. This is a very delicate point, because the assumption that the LES capture a consistent fraction of the turbulent kinetic energy

$$\langle K_f\rangle_e << R_{fe}$$

is in many cases not so accurate.

5 Conclusions

In the paper we have examined two issues that are very simple but also very important in turbulence modeling. They apply to the filtering technique of analysis, largely used in the Large Eddy Simulations of turbulent flows. This technique extends to a generic filtering operator the statistical formalism [14], and in a sense is a *local* version of it. Some controversial points related in particular to different techniques of decomposition of the turbulent stress and the turbulent energy are presented to the discussion.

References

[1] Hussain A. K. M. F. & Reynolds W. C. 1970 *The mechanics of an organized wave in turbulent shear flow* J. Fluid. Mech. 41, 241–258

[2] Antonia R. A. 1981 *Conditional sampling in turbulence measurement.* Ann. Rev. Fluid Mech. 13, 131–156

[3] Lumley J. L. 1970 *Stochastic tools in turbulence*, Academic Press, New York

[4] Farge M. & Rabreau G. 1988 *Transformée en ondelettes pour détecter et analyser les structures cohérentes dans les écoulements turbulents bidimensionnels* C. R. Acad. Sci. Paris 307, Série II, 1479–1486

[5] Dopazo C. 1977 *On conditioned averages for intermittent turbulent flows* J. Fluid. Mech. 81, 433–438

[6] Schiestel R. 1987 *Multiple-time-scale modeling of turbulent flows in one point closure* Phys. Fluids 30, 722–731

[7] Wilcox D. C. 1988 *Multiscale model for turbulent flows* AIAA Journal 26, 1311–1320

[8] Bardina J. , Ferziger J. H. , & Reynolds W. C. 1980 *Improved subgrid scale models for Large Eddy Simulation* AIAA Paper 80–1357

[9] Germano, M. , Piomelli, U. , Moin P. & Cabot W. H. 1991 *A dynamic subgrid-scale eddy viscosity model* Phys. Fluids A 3, 1760–1765

[10] Leonard A. 1974 *Energy Cascade in Large-Eddy Simulations of Turbulent Fluid Flows* Adv. in Geophysics,18A, (pp. 237–248)

[11] Germano M. 1986 *Differential filters of elliptic type* The Physics of Fluids 29, 1757–1758

[12] Aubry N. 1991 *On the hidden beauty of the proper orthogonal decomposition,* Theoret. Comput. Fluid Dynamics 2, 339–352

[13] Germano M. 1994 *Dynamic models*, ERCOFTAC Bulletin 22, 15–18

[14] Germano M. 1992 *Turbulence: the Filtering Approach*, J. Fluid Mech. 238, 325–336

Trends in Mathematics, © 1999 Birkhäuser Verlag Basel/Switzerland

Physics and modeling of small scale turbulence for large eddy simulation

C. Meneveau, J. O'Neil, F. Porte-Agel, S. Cerutti, M. B. Parlange

Abstract

An important open question for Large Eddy Simulation (LES) of turbulent flow is whether the subgrid-scales (smaller than Δ) and the associated stresses can be described by universal models. This question is examined in terms of the subgrid dissipation rate, which represents the rate at which large-scale turbulent kinetic energy or scalar fluctuations are transferred to, or from, the unresolved scales. This variable is of central importance in LES. In order to ascertain the dependence of SGS dissipation upon large scales and the degree of universality of subgrid models, various high Reynolds number flows are studied. Hot-wire data (yielding, so far, 1-D surrogates only) in the turbulent plane wake (at $Re_D = 8 \cdot 10^4$) shows that the phase-averaged surrogate SGS dissipation rate is very much affected by the presence of coherent (Kármán) vortices which occur at scales much larger than the filter scale, and that the effect persists down to very small scales. Strong effects of large-scale coherent structures are also observed in the atmospheric surface layer, as deduced from sonic anemometry data of the surrogate subgrid temperature variance dissipation. In comparing how different models reproduce the observed distribution of conditional SGS dissipation, the similarity model is found to capture the main features of the observed conditional SGS dissipation, while the Smagorinsky model gives less realistic distributions. The dynamic Smagorinsky model is able to reproduce the distribution, if the coefficient is obtained by means of conditional averaging.

Thus, the experimental results (as well as numerical results by several other authors) show that there exists strong dependence of small scales upon the large ones. While the results weaken hopes for universality, they show that 'learning' from resolved scales about the local state of turbulence appears to improve the realism of the models. However, much depends on the details of how this learning process is implemented. For instance, in the localized dynamic models, problematic features such as unphysically high levels of intermittency can occur. This is demonstrated by detailed analysis of DNS of isotropic turbulence.

1 Introduction

Prospects for quantitative predictions of high-Reynolds number turbulent flows using Large-Eddy-Simulation (LES) have improved in recent years due to various developments such as numerical methods, computer hardware, and new models. Compared to the Reynolds averaged approach (RANS), in LES mean flow variables are generally predicted with a smaller number of adjustable coefficients. Intrinsic advantages of LES exist when one is interested in higher-order quantities, such as root-mean-square and spectra of pressure fluctuations, probabilities of extreme events, instantaneous flow response to control strategies, etc.. In LES one seeks solutions to the Navier-Stokes equations filtered at some scale Δ (if possible Δ should fall in the inertial range), which requires the subgrid-scale (SGS) stress tensor to be parameterized as function of the resolved velocity field[1, 2]. The SGS tensor is defined according to $\tau_{ij} \equiv \widetilde{u_i u_j} - \tilde{u}_i\tilde{u}_j$, where the *tilde* represents spatial filtering at a scale Δ. The evolution of kinetic energy $K = \frac{1}{2}\tilde{u}_i^2$ of the resolved scales is governed by[3, 4]:

$$\frac{\partial K}{\partial t} + \tilde{u}_j \frac{\partial K}{\partial x_j} = f_i \tilde{u}_i + \tau_{ij}\tilde{S}_{ij} - 2\nu \tilde{S}_{ij}\tilde{S}_{ij} - \frac{\partial}{\partial x_i}\left(\frac{1}{\rho}\tilde{p}\,\tilde{u}_i + \tau_{ij}\tilde{u}_j - 2\nu\tilde{u}_j\tilde{S}_{ij}\right). \quad (1)$$

Here $\tilde{S}_{ij} = \frac{1}{2}(\partial\tilde{u}_i/\partial x_j + \partial\tilde{u}_j/\partial x_i)$ is the resolved strain-rate tensor, and ν,ρ are respectively the kinematic viscosity and density of the fluid. The terms in the last bracket on the right-hand side represent spatial transport (i.e. energy moving from one place to another but not across scales), while the first term represents the energy injected by the forcing. The third term is the dissipation of energy due to viscous effects, and the second term, $-\tau_{ij}\tilde{S}_{ij}$, is a local sink (or source) term, called the SGS dissipation. It represents the effect of the unresolved scales tending to change (on average, typically, to decrease) the resolved kinetic energy, without simply transporting it from one position to another. The SGS dissipation rate is one of the most important properties of the small scales, since it affects how energetic the resolved scales will be in a simulation.

When a passive scalar θ is transported in a turbulent flow, the resolved scalar fluctuations' energy $\mathcal{K} \equiv \frac{1}{2}\tilde{\theta}^2$ obeys

$$\frac{\partial \mathcal{K}}{\partial t} + \tilde{u}_j \frac{\partial \mathcal{K}}{\partial x_j} = q_j \frac{\partial\tilde{\theta}}{\partial x_j} - \gamma\left(\frac{\partial\tilde{\theta}}{\partial x_j}\right)\left(\frac{\partial\tilde{\theta}}{\partial x_j}\right) - \frac{\partial}{\partial x_j}\left(q_j\tilde{\theta} - \gamma\partial\tilde{\theta}/\partial x_j\right). \quad (2)$$

Above, $q_i \equiv \widetilde{u_i\theta} - \tilde{u}_i\tilde{\theta}$ is the subgrid-scale scalar flux. At high molecular Peclet number the dominant term on the right hand side of Eq. 2 is the SGS dissipation rate of scalar-variance, $\chi \equiv -q_i\partial\tilde{\theta}/\partial x_i$.

It is very important that models accurately predict the kinetic energy and scalar dissipation rate[5, 3]. A rigorous justification for this statement can be formulated in isotropic, high Reynolds number turbulence, where the following relationship can be derived from the filtered Navier-Stokes equations[6]:

$$\left\langle [\tilde{u}_L(\mathbf{x} + r\mathbf{e}_L) - \tilde{u}_L(\mathbf{x})]^3 \right\rangle = \frac{4}{5}\left\langle \tau_{ij}\tilde{S}_{ij}\right\rangle r - 6\left\langle \tilde{u}_L(\mathbf{x} + r\mathbf{e}_L)\tau_{LL}(\mathbf{x})\right\rangle. \quad (3)$$

We have shown[6] that for $r >> \Delta$ the last term (the two-point correlation function of the longitudinal SGS stress and the longitudinal velocity) decays quickly to zero. Hence, the above equation can be interpreted as follows: In order for LES to properly reproduce third-order statistics of the resolved velocity field at scales much larger than the grid scale, the necessary and sufficient condition is for a model to correctly predict the average rate of energy dissipation. An analogous expression can be derived for the mixed scalar-velocity structure function involving the SGS dissipation of scalar variance. Again, accurate prediction of the dissipation rate $-\left\langle q_i \partial\tilde{\theta}/\partial x_i \right\rangle$ is a necessary and sufficient condition to obtain realistic third-order moments at large scales.

Such arguments apply to the globally averaged dissipation rates. For particular SGS models such as the Smagorinsky model, one can use the condition of correctly predicting mean SGS dissipation rates for calibrating unknown model coefficient[5]. More stringent tests of models and their ability to reproduce SGS dissipation rates must be made in non-homogeneous flows, in which particular flow conditions can be isolated using conditional averaging. To evaluate the effect of a certain flow condition Q (for instance the presence of large-scale coherent structure) upon the rate of SGS dissipation, one can compute the conditionally averaged SGS dissipation rate $-\left\langle \tau_{ij}\tilde{S}_{ij}|Q \right\rangle$.

In the next few sections, we summarize experimental results that use conditional averaging to elucidate the effects of coherent structures.

2 Effects of coherent structures in a plane wake

In order to examine the effects of coherent structures on the spatial distribution of SGS dissipation of kinetic energy, we focus on the 'intermediate wake' of a circular cylinder at high Reynolds number. Such a flow displays large-scale vortical structures separated by highly strained regions.

Experiments described in more detail in [7] were performed in the Corrsin Wind Tunnel[8]. A $D = 5.08$ cm diameter smooth cylinder was used, and measurements were taken at $x/D = 25$, at a centerline speed of U_{CL}=21.5 m/s. The Taylor-scale Reynolds number at the centerline is about $R_\lambda \sim 550$. The wake profile was sampled at 15 discrete y-positions, from the centerline to the edge of the wake. Streamwise velocity time series were measured using a subminiature hot-wire probe, supplemented with x-wire measurements of two velocity components. The passage of coherent structures is clearly visible in the spectrum of transverse velocity which exhibits a distinct peak at a dimensionless Strouhal frequency of $St = 0.19$. Thus phase-averaging can be performed based on the phase Θ of the Strouhal peak mode in the transverse velocity signal. More details are given in Ref.[7].

The single hot-wire data is filtered in time, which is interpreted as a one-dimensional filtering at some scale Δ using Taylor's hypothesis. Assuming isotropy of the subgrid-scale statistics, a one-dimensional surrogate of the phase-averaged SGS dissipation is formed. The resulting expression, $-15/2\left\langle \tau_{11}\tilde{S}_{11}|\Theta \right\rangle$, is denoted the one-dimensional surrogate of the conditional SGS dissipation rate. Data at each cross-stream location y are phase discretized into 20 bins. The phase-averaged velocities are expressed in a coordinate system convecting with the rollers at a speed U_c, defined as the mean velocity at the height at which the peak vorticity is observed. Using Taylor's hypothesis, a pseudo 2-D distribution is constructed from the measurements at various phases and y. To enhance visual clarity, the data are extended using periodicity in phase, and phase-reversal about the centerline (since measurements were only performed on one side of the wake). Figure 1 displays contours of the phase-averaged (1-D surrogate) SGS dissipation $-\frac{15}{2}\left\langle \tau_{11}\tilde{S}_{11}|\Theta, y\right\rangle$, normalized with centerline molecular dissipation ϵ_{CL}. Results are for a Gaussian filter, at a scale $\Delta/\eta = 100$. It is immediately observed that the large-scale organized structures have a strong impact upon the distribution of surrogate SGS energy dissipation, even at filter scales well inside the inertial range. Similar qualitative results are obtained from scales between 200 down to 20 η. We conclude that large-scale coherent structures have important effects on the energy transfer occurring at much smaller scales.

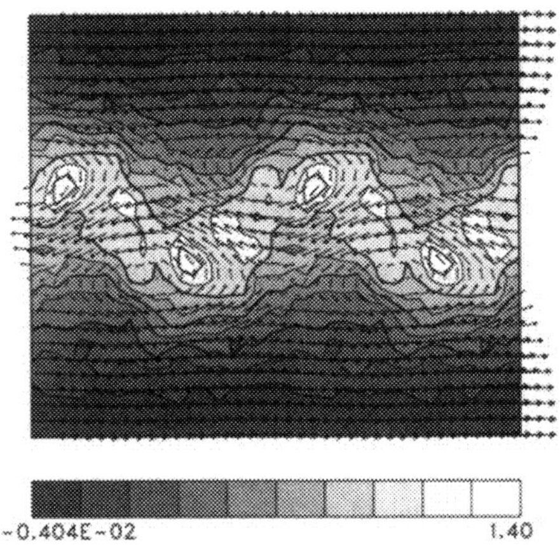

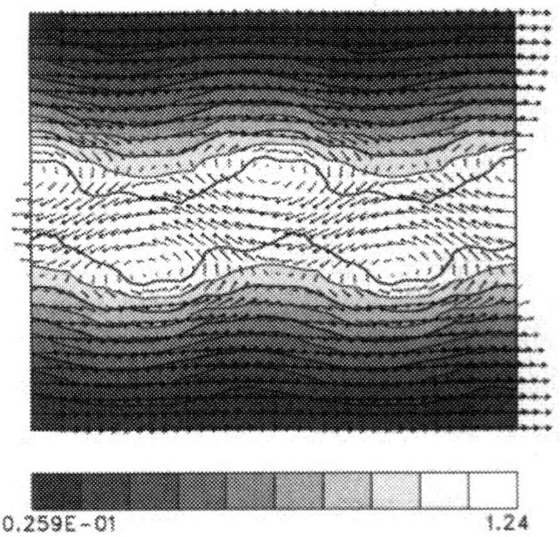

Figure 1: Left: contours of phase-averaged (1-D surrogate) SGS dissipation $-\frac{15}{2}\left\langle \tau_{11}\tilde{S}_{11}|\Theta, y\right\rangle$, normalized with centerline dissipation ϵ_{CL}. Results are for a Gaussian filter, at $\Delta/\eta = 100$. Vectors correspond to phase-averaged velocity. The flow is the intermediate wake behind a circular cylinder. Right: contours of phase-averaged surrogate Smagorinsky dissipation $15^{3/2}(0.06\Delta)^2\left\langle |\tilde{S}_{11}|^3|\Theta, y\right\rangle$

The next question is how different models are able to cope. We consider three models. The first is the traditional Smagorinsky model $-2(c_s\Delta)^2|\tilde{S}|\tilde{S}_{ij}$, where c_s is a coefficient adjusted so as to reproduce the globally averaged dissipation rate. The next is the similarity model[9, 10], in which the stress is modeled by the resolved stress $L_{ij} = \overline{\tilde{u}_i\tilde{u}_j} - \bar{\tilde{u}}_i\bar{\tilde{u}}_j$. The bar denotes test-filtering at a scale 2Δ. Then, the dynamic Smagorinsky model[11, 12] is considered. In this model, the coefficient is obtained from the resolved scales according to $C_{dyn}^2 = \langle L_{ij}M_{ij}\rangle / \langle M_{ij}M_{ij}\rangle$, with $M_{ij} = -2\Delta^2 \left(4|\bar{\tilde{S}}|\bar{\tilde{S}}_{ij} - \overline{\tilde{S}|\tilde{S}_{ij}}\right)$. For this study, the averaging of both numerators and denominators is performed based on the phase rather than globally.

Figure 1(right) displays the predictions of the (1-D surrogate) Smagorinsky model, which gives a more uniform (i.e. unrealistic) distribution. As seen in Fig. 2 the similarity model [10, 9] reproduces the main trends of the distribution. A similar result (not shown here) is obtained for the conditionally-averaged dynamic model[7].

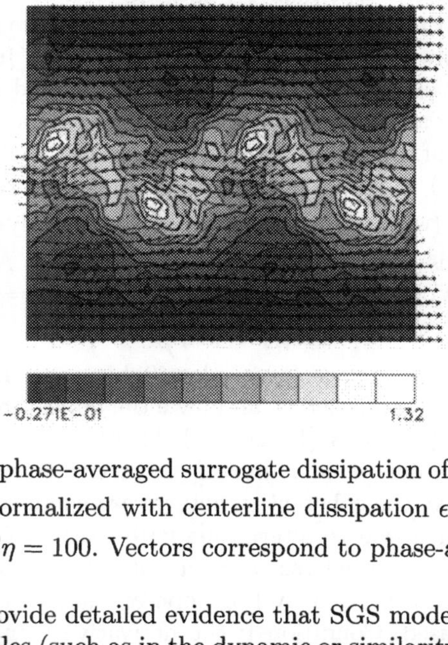

Figure 2: Contours of phase-averaged surrogate dissipation of the similarity model, $-\frac{15}{2}\left\langle L_{11}\tilde{S}_{11}|\Theta, y\right\rangle$, normalized with centerline dissipation ϵ_{CL}. Results are for a Gaussian filter, at $\Delta/\eta = 100$. Vectors correspond to phase-averaged velocity.

These results provide detailed evidence that SGS modeling based on "learning" from resolved scales (such as in the dynamic or similarity models) should lead to more realistic predictions than the simple eddy viscosity model, when turbulence is spatially complex and when coherent structures affect the SGS dynamics directly. At the level of phase-averaged dissipation, both the dynamic Smagorinsky or the similarity model appear to work well, so that any linear combination of the two should also provide a good model to reproduce phase-averaged dissipation. In recent years, such mixed models have shown promise in LES (e.g. Refs. [13, 14, 15]). The mixed model is also advantageous to model the response of SGS turbulence to rapid distortion[16].

3 Effects of coherent structures in atmospheric surface layer

To study effects of large-scale structures on SGS dissipation at much higher Reynolds number, and to study the properties of SGS dissipation of scalar variance, analysis of atmospheric data has been performed. The atmospheric data were obtained from a field experiment carried out over a bare soil surface at the Campbell Tract research field of the University of California at Davis during the summer of 1996. Further details of the site are discussed in Parlange *et al.* (1993) and Albertson *et al.* (1997). Longitudinal velocity (u_1) and temperature θ were measured using a sonic anemometer placed at a height (z) of 1.70 m. The data presented and analyzed here were obtained at 18.3 Hz over a 60 minute period. More details about the data and analysis procedures are given in [17]. The stability parameter z/L was -0.048 (L is the Obukhov length), indicating unstable atmospheric stability conditions.

The streamwise SGS heat flux, q_1, and the filtered temperature $\tilde{\theta}$ are computed by filtering in time, which again is interpreted as one-dimensional spatial filtering in the streamwise direction by invoking Taylor's frozen flow hypothesis. The SGS dissipation of scalar variance is then evaluated as $-q_1 \partial\tilde{\theta}/\partial x_1$, where streamwise derivatives are approximated by time derivatives using, again, Taylor's hypothesis. In terms of models, we consider the traditional eddy-diffusion model (its 1-D surrogate[17]) $q_1 = -(c_s^2/Pr_t)\Delta^2\sqrt{15}|\tilde{S}_{11}|\partial\tilde{\theta}/\partial x_1$, the similarity model $\mathcal{L}_1 = c_L\left(\overline{\tilde{u}_1\tilde{\theta}} - \overline{\tilde{u}}_1\overline{\tilde{\theta}}\right)$ (with $c_L = 1$), and the dynamic eddy-diffusion model[18] in which the coefficient is computed from the resolved field. A 1-D surrogate of the dynamic coefficient is obtained according to $(c_s^2/Pr_t)_{dyn} = \langle\mathcal{L}_1 N_1\rangle / \langle N_1^2\rangle$, where $N_1 = -\Delta^2\sqrt{15}\left(4|\overline{\tilde{S}}_{11}|\partial\overline{\tilde{\theta}}/\partial x_1 - \overline{|\tilde{S}_{11}|\partial\tilde{\theta}/\partial x_1}\right)$.

In the atmospheric surface layer, identifiable coherent structures occur in scalar (e.g. temperature) time series, and have the appearance of "ramps" or sawtooth shaped features (e.g. Refs. [19, 20, 21]). These are associated with the sweep-ejection motion of the flow. Under diurnal convective (buoyant) conditions, over a relatively hot surface, the coherent structures are composed of ejection motions transporting warm air upwards, alternating with sweeps of cool and dry air from aloft. The comparison between real and modeled SGS dissipation rates is here based on a simple criterion, namely the slope of the filtered temperature signal. Two conditions are used. Condition I: the derivative being larger than its root mean square value ($\partial\tilde{\theta}/\partial x_1 > \sigma_{\tilde{\theta}_x}$), which using Taylor's hypothesis corresponds to a temperature drop in the time series. This condition tends to highlight the decreasing part of the ramp structures. Condition II: $\partial\tilde{\theta}/\partial x_1 < -\sigma_{\tilde{\theta}_x}$, which corresponds to a temperature increase in the time series, tending to highlight the rising part of the ramp structures. The selection of the conditioning threshold, $\sigma_{\tilde{\theta}_x}$, is arbitrary and is large enough to isolate the local effects, and small enough to ensure the convergence of conditional averages. The resulting trends are robust with respect to changes in threshold. A conditional SGS dissipation signal is obtained by averaging a segment of the signal (of width W), whenever the condition is met at the midpoint of the segment. W is set to a value large enough so that the conditional average asymptotes to the global average as one moves away from the center of the averaging window (at $x' = 0$).

The conditional averages presented below correspond to a Gaussian filter of width $\Delta = 4.27$ m and an averaging window of $W = 40.5$ m. The solid lines in Fig-

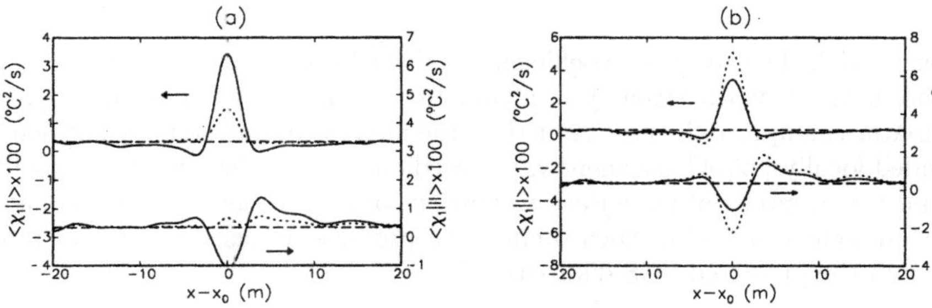

Figure 3: (a) Top curves: Conditionally averaged surrogate SGS dissipation of temperature variance, for condition I ($\partial\tilde{\theta}/\partial x_1 > \sigma_{\tilde{\theta}_x}$). Solid line: real dissipation, dashed line: eddy-diffusion model. The global mean is indicated by a long-dash line. Bottom curves: same as above, but for condition II ($\partial\tilde{\theta}/\partial x_1 < -\sigma_{\tilde{\theta}_x}$). (b) Comparison with the similarity model. Dashed lines: similarity model. Solid lines: same as in (a).

ure 3(a) show the results for the surrogate conditional dissipation under condition I (top curves) and condition II (bottom curves). A larger than average positive dissipation occurs near regions of strong negative temperature gradient, indicating that there is a higher rate of temperature variance being transferred from the large (resolved) to the small (subgrid) scales of the flow. The eddy-diffusivity model shown tends to follow the same trend, but significantly underpredicts the peak dissipation. For condition II, the real dissipation displays a negative value which indicates that there is backscatter of temperature variance, i.e. temperature variance being transferred from the subgrid scales of the flow to the resolved field. The eddy-diffusivity model fails to reproduce this effect since it is fully dissipative by construction. More details are given in [17].

Figure 3(b) shows a similar comparison for the similarity model. As is evident, the predictions are improved, showing some overpredictions in both cases. A linear combination between the similarity and eddy-diffusion model (such as in the mixed model) is likely to be a good compromise. Comparable results (not shown) are obtained when using a spectral cutoff filter. Also, we have tested the dynamic model in which the coefficient is obtained by means of conditional averaging (i.e. three distinct values, for condition I, II, and their complement). Results are qualitatively similar to those of the similarity model. For condition II, however, the dynamic eddy-diffusion coefficient becomes negative and would cause numerical instability if implemented in a simulation.

4 Effects of small-scale intermittency

Having illustrated that large-scale events can strongly impact the subgrid scales, and having shown that the procedure of learning about this impact from the resolved scales can yield realistic predictions of this impact, we now briefly point out a well-known difficulty associated with the dynamic eddy-viscosity or eddy-diffusion model. The dynamic coefficient is given by a ratio of two tensor contractions, both of which strongly oscillate. In regions of low values of M_{ij}, the denominator can approach zero faster than the numerator and so the coefficient is unbounded locally [22]. Consequently, the predicted stress displays strong oscillations and the accuracy of numerical evaluations of its divergence is questionable. From a physical standpoint, such oscillations produce unrealistic levels of intermittency in the predicted SGS dissipation rate.

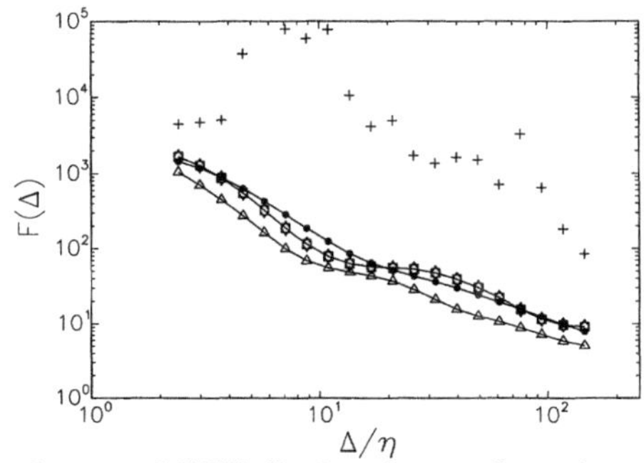

Figure 4: Flatness factor of SGS dissipation as function of filter scale Δ, in a 256^3-mode DNS of isotropic turbulence. Shown are results for the measured dissipation (full circles), Smagorinsky model (squares), similarity model (triangles), volume-averaged dynamic model (rhombi), local dynamic model (+'s).

A quantitative characterization of intermittency can be made by computing the flatness factor. We consider DNS of forced isotropic turbulence on 256^3 modes, at $R_\lambda = 150$[23]. The SGS dissipation of kinetic energy and its flatness factor is computed at various filter scales, for both the real SGS dissipation, as well as for various models. Note that now the filtering can be done in 3-D and the full tensor contractions are available from the DNS. Figure 4 shows that in the (short) inertial range, the flatness increases with decreasing scale from about 10 to 100, i.e. the spatial distribution of SGS dissipation becomes more and more intermittent at small scales. The Smagorinsky, similarity, and volume-averaged dynamic models display roughly similar trends as those of the real dissipation, while the local dynamic model displays much stronger, and thus unrealistic, levels of intermittency. A detailed analysis of anomalous scaling properties of the SGS dissipation can be found in Ref. [23].

5 Closing remarks and open questions

From the various empirical data that have been analyzed it is concluded that large-scale coherent structures can strongly affect details of the subgrid scales. This causes variations in SGS dissipation rates and/or of small-scale isotropy. Classical eddy-diffusion models have difficulties in reproducing these effects, while models based on learning from the resolved scales, such as the similarity (and mixed) and dynamic models, fare much better. For the dynamic model, some averaging must be undertaken since otherwise (with local dynamic models) the level of predicted subgrid intermittency far exceeds that of the real flow.

While the present focus has been on SGS dissipation, which is directly relevant for predicting second-order statistics of resolved scales, other statistical quantities are of considerable interest as well, such as pressure spectra, probability density functions, higher-order moments, etc.. The following question remains open: What properties of τ_{ij} are required to reproduce particular statistical features of the resolved field? At a very fundamental level, it can be shown[24] that the full multipoint conditional moment $\langle \tau_{ij}(\mathbf{x}) | \mathbf{V}(\mathbf{x}_1), \mathbf{V}(\mathbf{x}_2), \mathbf{V}(\mathbf{x}_3), \ldots \rangle$ must be predicted in order to reproduce the full multipoint distribution function of the resolved field. This condition also provides the best stochastic estimate for the rate of change of velocity in a single realization[25]. However, conditioning on the full resolved field is impractical and additional averaging over resolved degrees of freedom must be undertaken. This is the approach followed in this paper, where conditional averaging is performed based on simple conditions of the resolved field (as opposed to an entire realization). A consequence of the associated loss of information is that present results can only be interpreted as *necessary* conditions, rather than *sufficient* conditions[6]. For example, if a model does not reproduce conditional statistics (such as the eddy-diffusion models discussed above), it can be stated unambigously that the resolved conditional kinetic energies or scalar variances will be incorrectly predicted. However, the converse is not true. Even if a model reproduces the correct conditional averages (e.g. the mixed model) the simulation may still generate wrong results since other terms, such as spatial transport terms, may be incorrect. Clearly, much progress is still needed in order to relate basic understanding of turbulence small scales with turbulence modeling for LES, and to relate statistical and structural features of the small scales to those of large scales (and vice-versa).

Acknowledgements: The authors gratefully acknowledge the support of NSF, ONR, and the UC Davis Superfund. F.P-A. received partial support from the 'La Caixa' Fellowship Program (Barcelona, Spain).

References

[1] R. Rogallo and P. Moin. Numerical simulation of turbulent flows. *Ann. Rev. Fluid. Mech.*, 16:99, (1984).

[2] M. Lesieur & O. Metais. New trends in large-eddy simulations of turbulence. *Annu. Rev. Fluid Mech.*, 28:45–82, (1996).

[3] A. Leonard. Energy cascade in large-eddy simulations of turbulent fluid flows. *Adv. Geophys.*, 18:237, (1974).

[4] U. Piomelli, W.H. Cabot, P. Moin, and S. Lee. Subgrid-scale backscatter in turbulent and transitional flows. *Phys. Fluids*, A3:1766, (1991).

[5] D.K. Lilly. The representation of small-scale turbulence in numerical simulation experiments. In *Proc. IBM Scientific Computing Symposium on Environmental Sciences*, page 195, (1967).

[6] C. Meneveau. Statistics of turbulence subgrid-scale stresses: Necessary conditions and experimental tests. *Phys. Fluids*, 6:815, (1994).

[7] J. O'Neil and C. Meneveau. Subgrid-scale stresses and their modeling in a turbulent plane wake. *J. Fluids Mech.*, 349:253–293, (1997).

[8] G. Comte-Bellot and S. Corrsin. The use of a contraction to improve the isotropy of grid generated turbulence. *J. Fluid Mech.*, 25:657, (1966).

[9] J. Bardina. Improved turbulence models based on large eddy simulation of homogeneous, incompressible, turbulent flows. *Ph.D. thesis*, report TF-19:Mechanical Engineering, Stanford University, (1983).

[10] S. Liu, C. Meneveau, and J. Katz. On the properties of similarity subgrid-scale models as deduced from measurements in a turbulent jet. *J. Fluid Mech.*, 275:83, (1994).

[11] M. Germano, U. Piomelli, P. Moin, and W.H. Cabot. A dynamic subgrid-scale eddy viscosity model. *Phys. Fluids A*, A 3:1760, (1991).

[12] D.K. Lilly. A proposed modification of the Germano subgrid scale closure method. *Phys. Fluids A*, 4:633, (1992).

[13] Y. Zang, R.L. Street, and J. Koseff. A dynamic mixed SGS model and its application to turbulent recirculating flows. *Phys. Fluids* A:5, 3186 (1993).

[14] X. Wu and K. Squires. Large eddy simulation of a canonical three-dimensional boundary layer. *In: Proc. Tenth Symp. on Turbulent Shear Flows, Penn State (ed. F. Durst, B.E. Launder, F.W. Schmidt & J.W. Whitelaw)*, (1995).

[15] B. Geurts B. Vreman and H. Kuerten. On the formulation of the dynamic mixed subgrid-scale model. *Phys. Fluids*, 6:4057–4059, (1994).

[16] S. Liu, J. Katz, and C. Meneveau. Evolution and modeling of subgrid scales during rapid straining of turbulence. *J. Fluid Mech*, submitted, (1997).

[17] F. Porte-Agel, C. Meneveau, and M.B. Parlange. Some basic properties of the surrogate subgrid-scale heat flux in the atmospheric boundary layer. *B. Layer Met.*, submitted, (1997).

[18] P. Moin, K. Squires, W. Cabot, and S. Lee. A dynamic SGS model for compressible turbulence and scalar transport. *Phys. Fluids A*, 3:2746, (1991).

[19] M.R. Raupach, R.A. Antonia, and S. Rajagopalan. Rough-wall turbulent boundary layers. *Appl. Mech. Rev.*, 44:1–25, (1991).

[20] W. Gao, R.H. Shaw, and K.T. Paw U. Observation of organized structure in turbulent flow within and above a forest canopy. *B. L. Met.*, 47:349 (1989).

[21] G. Katul, G. Kuhn, J. Schieldge, and C.-I. Hsie. The ejection-sweep character of scalar fluxes in the unstable surface layer. *B. L. Met.*, 83:1–26, (1997).

[22] S. Ghosal, T.S. Lund, and P. Moin. A local dynamic model for large eddy simulation. *in Center for Turbulence Research, Annual Research Briefs, Stanford University*, 1992:3, (1992).

[23] S. Cerutti and C. Meneveau. Intermittency and relative scaling of the subgrid dissipation rate in isotropic turbulence. *Phys. Fluids*, (March 1998).

[24] S. Pope. Personal communication (1997).

[25] R.J. Adrian. Stochastic estimation of subgrid scale motions. *Appl. Mech. Rev.*, 43:S214–S218, (1990).

Trends in Mathematics, © 1999 Birkhäuser Verlag Basel/Switzerland

On dynamical systems obtained via Galerkin projections onto low-dimensional bases of eigenfunctions

D. Rempfer

Abstract

In this paper, some implications of the technique of projecting the Navier-Stokes equations onto low-dimensional bases of special eigenfunctions are explored. Such low-dimensional bases are typically obtained by truncating a particularly well-suited complete set of eigenfunctions at very low orders, arguing that a small number of such eigenmodes already captures a large part of the dynamics of the system. In addition, in the treatment of inhomogeneous spatial directions of a flow, often eigenfunctions are used that do not satisfy the boundary conditions, and in the Galerkin projection the corresponding boundary conditions are ignored. We will show that both the severe truncation as well as an improper treatment of boundary conditions can severely restrict the validity of these models. As a particular example of an eigenfunction basis, systems of Karhunen-Loève eigenfunctions will be discussed in more detail, although most of the results presented are valid for any basis.

1 Introduction

In this paper, some aspects of the approximation afforded by certain models for the dynamics of complex fluid flows will be studied. The particular type of model we will be focussing on here is constructed by Galerkin projection onto bases of eigenfunctions obtained from the method of Proper Orthogonal Decomposition [1] which are also known as Karhunen-Loève bases [2]. The appeal of such models mainly stems from the fact that using this type of techniques it is possible to derive optimally low-dimensional models for the behavior of complex transitional and turbulent flows. As the low-order models of ordinary differential equations (ODEs) that describe these models typically are much easier to analyse than the full Navier-Stokes equations, it is hoped that by studying such models some basic insights into the mechanisms of dynamically complex and notably turbulent flows can be obtained. In parantheses, we want to make it very clear at this point already that these models are by no means intended as a replacement for direct numerical simulations, for two reasons: First of all, typically this type of model normally requires detailed information on the behavior of the flow be already at hand, information that often can only be obtained through a "traditional" direct numerical simulation, and second, the severe truncations involved in these models make them unlikely candidates for an accurate prediction of a particular flow. Thus the point of models of this type is to be able to study a relatively simple system that may reveal, one hopes, some important basic mechanisms that drive the flow in question. The fundamental understanding obtained this way may then allow us to develop improved turbulence models for the accurate prediction of technically relevant flows, as well as possibly even strategies for the active control of turbulent flow, aiming at, for instance, the reduction of turbulent drag.

However, when performing studies of this type, one has to be aware of the range of validity of this kind of models which can be severely restricted for certain types of flows and/or models. Thus in this paper we will focus on the question of "In what sense do models obtained from Galerkin projection onto truncated bases of special eigenfunctions approximate the behavior of a flow?". To do this, in the next section we will first remind the reader very briefly of some of the basic properties of the eigenfunctions of Proper Orthogonal Decomposition, mainly to introduce the notation that will be used later on. In section 3 we will then go into some of the mathematical details involved in deriving this type of model, and in a section 4, some examples for the statements derived here will be discussed.

2 Properties of Karhunen-Loève eigenfunctions

The basis functions of Proper Orthogonal Decomposition can be obtained by finding the eigenfunctions $\boldsymbol{\sigma}(\boldsymbol{x})$ of the correlation tensor $\boldsymbol{R}$ of a flow,

$$\int_D \boldsymbol{R}(\boldsymbol{x}, \boldsymbol{x}') \cdot \boldsymbol{\sigma}(\boldsymbol{x}')\, d\boldsymbol{x}' = \lambda \boldsymbol{\sigma}(\boldsymbol{x}). \tag{1}$$

The kernel of this equation is the correlation function $\boldsymbol{R}$ defined by

$$R_{ij}(\boldsymbol{x}, \boldsymbol{x}') = \langle u_i(\boldsymbol{x}) u_j(\boldsymbol{x}') \rangle, \tag{2}$$

and the brackets $\langle \cdot \rangle$ denote a time average. This set of functions – that are also known as Karhunen-Loève (or KL-) eigenfunctions – has a number of remarkable properties of which the following will be important for the understanding of the material presented below:

- The eigenfunctions form an orthogonal system, which is convenient for the Galerkin procedure to be used later on.

- It is particularly important to note here that the system of KL-eigenfunctions is complete only in a special and restricted sense: These eigenfunctions are complete in the sense that using these models it is possible to exactly represent the velocity fields $\boldsymbol{u}(\boldsymbol{x}, t)$ *of the flow that were used to compute the correlation tensor forming the kernel of the eigenvalue problem the* $\boldsymbol{\sigma}_i$ *were determined from.* Note that in this paper we use the label "KL-eigenfunction" only for those eigenfunctions of (1) that do not belong to the null space of the kernel, i. e. the ones with strictly positive eigenvalues. Intuitively, this means that we are restricting this discussion to the "physical" eigenmodes of the flow, describing flow patterns that can actually be observed in the flow being studied. In contrast, in a Galerkin procedure for incompressible flow, one would like to use a set of basis functions that is complete in the sense that it can represent all solenoidal velocity fields that meet the boundary conditions. This is clearly a more general concept of completeness, and depending on the nature of the analysed flow (that is decomposed into POD eigenfunctions) POD bases may or may not possess this property. For instance, if we are looking at a transitional flow during the early stages of evolution [3], then we may find that this flow only contains a finite number of modes, maybe corresponding to one or two instability modes of linear stability theory. So in this case, we would find a finite (and actually very low-dimensional) basis, that is clearly far from being able to represent all velocity fields that meet the boundary conditions. As an aside, it is an interesting question to ask whether a fully developed turbulent flow would give rise to an infinite-dimensional POD basis that is complete in the general sense described above. While this is not immediately clear, note that, on the other hand, even if it turns out that such fully turbulent flows indeed possess finite-dimnensional attractors (and current research indicates that this is indeed the case [4]), then, because this attractor is nonlinear, one may in fact need an infinite number of POD modes to describe a flow moving along such an attractor. Paradoxically, as we will see below, POD-Galerkin models work better in a certain sense for these "infinitely complex" flows than for the ones possessing finite-dimensional bases.

- One should also keep in mind the fact that POD eigenfunctions can be shown to be linear combinations of instantaneous velocity fields of the flow. Therefore, these functions inherit all those properties of the flow that can be given as linear and homogeneous equations in the velocities. For an incompressible flow this means that each of the $\boldsymbol{\sigma}_i$ represents a solenoidal vector field. Also, if boundary conditions of a problem can be described by linear homogeneous equations, which is the case for no-slip conditions, or periodic boundary conditions, then each of the eigenfunctions will meet the boundary conditions individually.

3 Galerkin projections

We now need to derive low-dimensional models for the dynamics of a flow using Karhunen-Loève eigenfunctions. This means that we have to find a set of ordinary differential equations that describe the behavior in time of the expansion coefficients $\zeta_i(t)$ in the Karhunen-Loève series

$$\boldsymbol{u}(\boldsymbol{x},t) = \sum_j \zeta_j(t)\boldsymbol{\sigma}_j(\boldsymbol{x}). \tag{3}$$

A standard technique to find such a system of ODEs is the method of Galerkin projection [5]. This method bascially involves at its core projecting both the left-

and right-hand side of the Navier-Stokes equations which, in a highly symbolic fashion, can be written as

$$\frac{\partial \boldsymbol{u}}{\partial t} = \mathcal{F}(\boldsymbol{u}), \quad \boldsymbol{u} = \boldsymbol{u}(\boldsymbol{x}, t), \qquad t \geq 0, \boldsymbol{x} \in D, \tag{4}$$

where $\mathcal{F}$ stands for the well-known nonlinear differential operator, onto a set of basis functions. Ordinarily, these basis functions have to meet the following requirements:

- The basis functions must be complete in the general sense that they can span the entire phase space of the original equation – in our case the Navier-Stokes equations plus appropriate boundary conditions, or at least a linear subspace of it.
- For the solution to be unique, the basis functions have to be linear independent.
- For a standard Galerkin method, the eigenfunctions must meet the boundary conditions of the problem.

The projection of the Navier-Stokes equations onto such a system of basis functions then results in a system of ordinary differential equations with a nonlinear function $\boldsymbol{F}$ on its right-hand side:

$$\frac{d\zeta_i}{dt} = F_i(\zeta_1, \zeta_2, \ldots). \tag{5}$$

For the Navier-Stokes equation, the functions F_i are quadratic in the ζ_i, due to the convective terms in the equation.

We now want to look at what happens if instead of the ideal function basis described above we are using KL-eigenfunctions that – as we have seen – may not meet the above completeness requirement, and that may also not meet the boundary conditions of the problem. Suppose first that we are starting from an initial velocity field $\boldsymbol{u}(\boldsymbol{x}, t_0)$ that is part of the ensemble of velocity fields that were used to compute the correlation tensor the eigenfunctions of which give our set of basis functions. Due to the completeness property of KL bases discussed in the previous section it is then clear that this particular velocity field can be represented exactly using these eigenfunctions. In addition, because we know for the same reason that the velocity field $\boldsymbol{u}(\boldsymbol{x}, t_0 + dt)$ can be represented exactly (as this velocity field, too, can be assumed to be part of the ensemble used to determine the $\boldsymbol{\sigma}_i$), we can also conclude that $d\boldsymbol{u}(\boldsymbol{x}, t_0) = \boldsymbol{u}(\boldsymbol{x}, t_0 + dt) - \boldsymbol{u}(\boldsymbol{x}, t_0)$ lies in the span of our eigenfunction basis. Now, as $d\boldsymbol{u}$ is proportional to the right-hand side of the Navier-Stokes equations, we can thus see that the Navier-Stokes operator is captured exactly by the system of ODEs we obtained form the Galerkin projection. Obviously, this argument can be used for any velocity field $\boldsymbol{u}(\boldsymbol{x}, t)$ that is part of the ensemble of flow fields that were originally used to find the KL-eigenfunctions. What we have found, therefore, is that *if we start from a velocity field that is an element of the set that was used in determining the KL modes*, then the (exact) solution of the system of ODEs will reproduce the dynamical behavior of the flow for all times.

Note that there was no mention of boundary conditions in the argument above, and in fact, the argument given there is completely independent of whatever kind of boundary conditions apply. If at a given time the velocity field is represented exactly, then so are the boundary conditions, and if we can integrate our system of ODEs exactly in time, this will be true for all times. Interestingly, what this means in particular is that we can derive dynamical systems models along these lines for arbitrary subregions of a flow field. So, for instance, it does make sense to just look at a thin layer near the wall of a turbulent flow [6], or at consecutive cross-stream slices of a spatially evolving transitional flow [7].

Now, the question that could be asked at this point is the following: "If these POD-Galerkin models seem to work so well for the situation described above, does this mean that the ODEs for the POD-Galerkin model are equivalent to the original Navier-Stokes equation?" Of course, it is easy to see that this question will have a negative answer in general, but it is quite interesting to see exactly why such models can fail in this respect.

First of all, if we go back to the chain of arguments above, then we can see that it crucially depends on the fact that the right-hand side of the Navier-Stokes equations can be represented in our basis of eigenfunctions. This is exactly where the restricted completeness property of POD bases creates a problem. Because of this restricted completeness, the right-hand side of the Navier-Stokes operator in general can be represented only as long as the instantaneous velocity field is an element of the ensemble used to compute the POD basis (which is what we have shown above). Thus, if we are interested in the behavior of our flow on a trajectory other than the one the flow was on when the ensemble for the calculation of the POD basis was taken, then the system of ODEs from our Galerkin projection on the POD basis does not necessarily represent the Navier-Stokes operator anymore. In fact, even worse, if we add just a small disturbance to a velocity field on the original trajectory – and such a disturbance might be created by a small numerical error in the integration of the ODEs, then the right-hand side of the system of ODEs in general will not faithfully represent the right-hand side of the Navier-Stokes equations anymore. The sketch in Figure 1 serves to illustrate this situation: We started by obtaining a POD basis from a flow following the dashed trajectory in Figure 1. As this trajectory covers just a linear subspace (a plane in this case) of the complete phase space, we obtain a POD basis the eigenfunctions of which will only be able to represent fields that are restricted to this plane. Note that in a certain sense it is actually the optimality of the POD basis that causes this behaviour: POD bases are optimized in such a way that they select the smallest linear subspace within which the observed behavior can be represented [8]. It is then clear from the figure that as soon as we leave the dashed trajectory, in general we will find time derivatives that are pointing outside of the POD subspace and thus cannot be captured by our function basis.

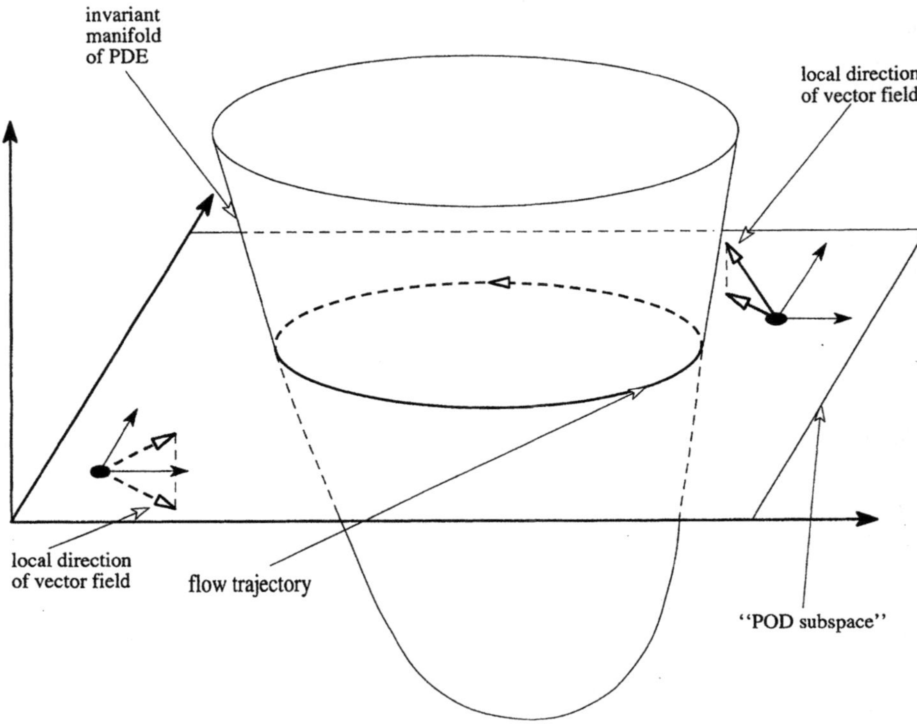

Figure 1: Projection of a flow onto a linear subspace.

Of course, Figure 1 is just an intuitive sketch, but because of the nonlinearity of the Navier-Stokes equations it is highly likely that for a given flow situation (defined by initial and boundary conditions) the attractor of these equations will be curved in phase space, so that any proper linear subspace will not be invariant, which means that we will encounter a situation that at least qualitatively resembles the one shown in Figure 1.

As the drawing in Figure 2 shows, such a situation can have grave consequences. What is shown is a two-dimensional cross-section through some higher-dimensional phase space. The horizontal line symbolizes the linear subspace described by a POD basis, and the small filled circle is a trajectory of the system that is transverse to the cross-section. The structure of the phase space is indicated by the inward spiral in the figure, which means that the trajectory that is shown will be stable in the full phase space. However, if we project the spiral down onto the POD subspace – which is exactly what a POD-Galerkin model does – then all of the derivatives are pointing away from the trajectory, which means that in POD subspace, this trajectory is in fact unstable.

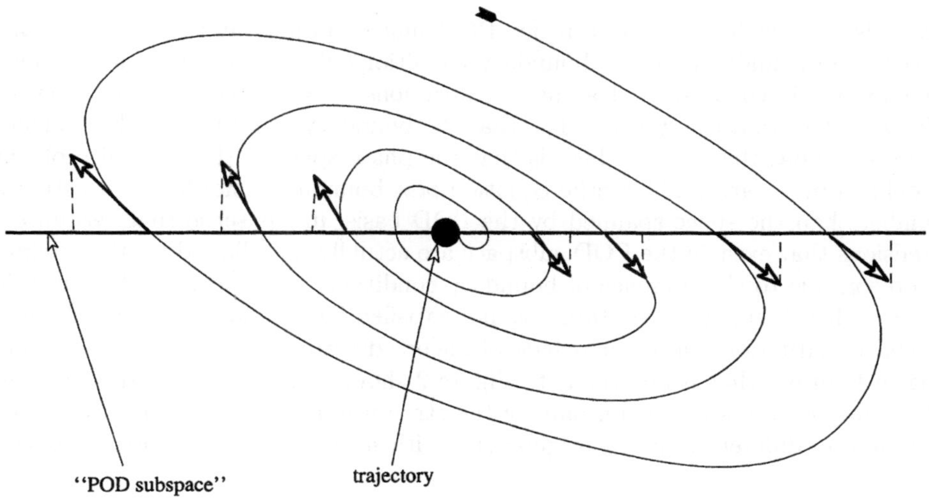

Figure 2: Two-dimensional cross-section through Navier-Stokes phase space.

In algebraic terms, what the figure shows is a dynamical system that, in its simplest form, might be described by a linear operator as in

$$\begin{pmatrix} \dot{\xi} \\ \dot{\eta} \end{pmatrix} = \begin{pmatrix} \mu & -2\omega \\ \omega/2 & \mu \end{pmatrix} \begin{pmatrix} \xi \\ \eta \end{pmatrix}. \tag{6}$$

The phase portrait of this linear oscillator for small damping $\mu < 0$ resembles a horizontal ellipse in the ξ-η-plane with an axis ratio of 4. If we turn this ellipse by 45 degrees, then the dynamical system has the form

$$\begin{pmatrix} \dot{x} \\ \dot{y} \end{pmatrix} = \begin{pmatrix} \mu + (3/4)\omega & -(5/4)\omega \\ (5/4)\omega & \mu - (3/4)\omega \end{pmatrix} \begin{pmatrix} x \\ y \end{pmatrix}, \tag{7}$$

which of course still has the same eigenvalues as before, $\mu \pm i\omega$. However, if we project this system on the line $y = 0$, then we arrive at just

$$\dot{x} = \mu + (3/4)\omega x. \tag{8}$$

So we can see that if the damping is not too large, the projected system will be unstable. In other words, a POD-Galerkin model would see an unstable trajectory when in fact there is a stable one.

This observation also brings us back to the issue of boundary conditions mentioned before. We said above that POD-Galerkin models would work equally well regardless of whether individual eigenfunctions meet the boundary conditions or not, as long as we are starting from a velocity field that is part of the ensemble that was used to determine the POD basis. While this statement is indeed true, there is a catch here: Because if our POD modes do not satisfy boundary conditions individually, then the boundary conditions of the problem translate into "compatibility constraints" leading to restrictions on what linear combinations of POD modes are actually allowed so that the boundary conditions of the problem are met. What this means, then, is that the phase space of the original problem – defined by a partial differential equation *plus* boundary conditions – is actually smaller than the space spanned by the POD basis, in the sense that certain dimensions that exist in the POD subspace are actually not allowed for the original problem, due to the presence of boundary conditions. So we can see that in this case – when boundary conditions are not satisfied by individual eigenfunctions – we have a situation that is the reverse of the one described just now, and the drawing in Figure 3 that is analogous to Figure 2 shows that this can lead to exactly the same problems as before, namely the structure of the projected phase space can be very different from the original one, with stabilities of trajectories reversed.

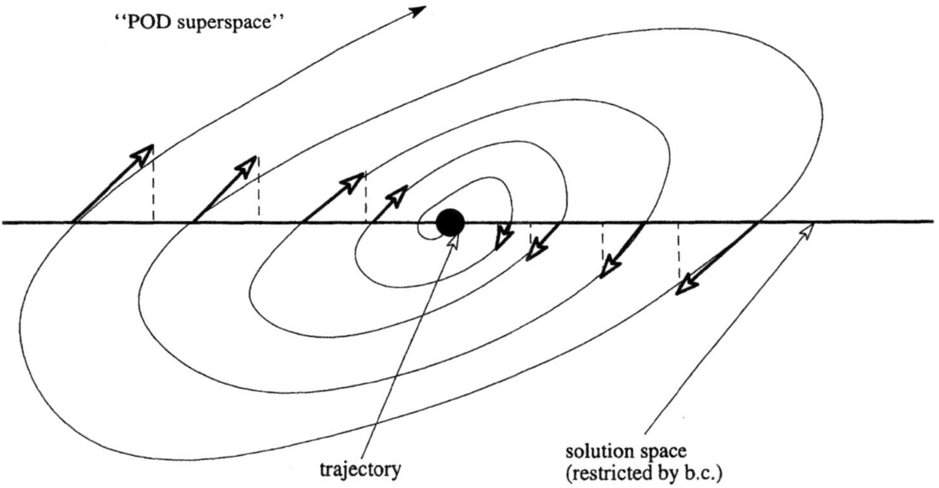

Figure 3: Two-dimensional cross-section through Navier-Stokes phase space.

In parantheses it should be added that, of course, if the boundary conditions of the problem are known, then the problem described here can be solved by explicitly adding equations that enforce the above compatibility constraints. Often, however, one would like to be able to study POD models for just a subregion of a flow field, and in that case the boundary conditions are typically unknown.

4 Behavior of a POD-Galerkin model in a transitional boundary layer

In this section an example of an actual POD-Galerkin model that exhibits just this type of behavior will be discussed. This model describes the dynamics of a transitional boundary layer within a certain cross-section of the flow (see also [7]). In this case the model included 16 POD modes, and for the trannsitional flow considered here these 16 modes essentially capture the velocity field exactly. That the model based on these modes is indeed accurate in the sense described above can be see from the plot in Figure 4 that shows a comparison of the time derivatives of the expansion coefficients as found from data of a direct numerical simualation to the ones predicted by the POD-Galerkin model using expansion coefficients that were obtained by projecting the same solution of the direct simulation onto the POD modes (thus the system of OFDEs is *not* actually integrated in time here). Obviously we can see that the model is indeed correct – as it should be if the algebra was done correctly.

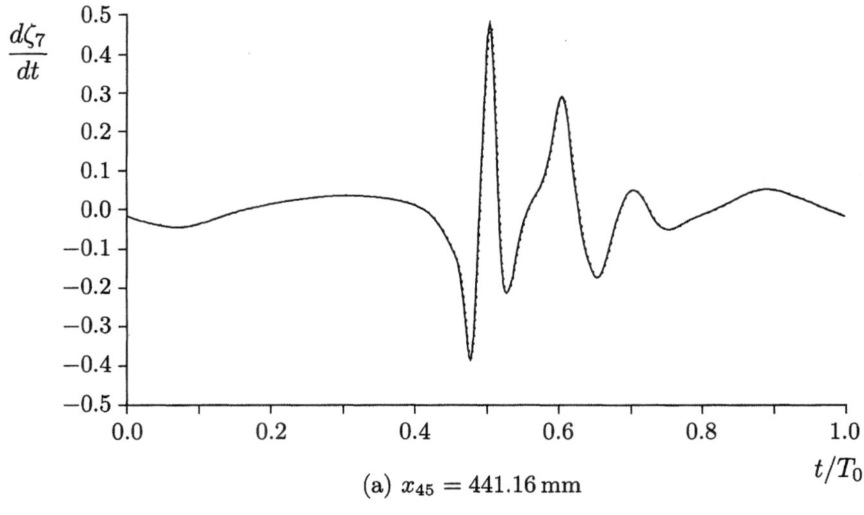

(a) $x_{45} = 441.16\,\mathrm{mm}$

Figure 4: Comparison of time derivatives for coefficient ζ_7 (Galerkin model versus direct simulation). ————, simulation,; ··············, model.

The plots in Figure 5 show how the model performs when autonomously integrated in time. We can see that initially the solution follows the correct solution, but soon an instability develops and the energy of the model increases without bound. The problem that exists here is illustrated by the plot in Figure 6. What is shown there is the error in the representation of the right-hand side of the Navier-Stokes equations as a function of the distance from the trajectory the system was on when the POD basis was determined. This error is shown for a truncated system of ODEs (10 modes) and for the full 16-mode system. We can see that right

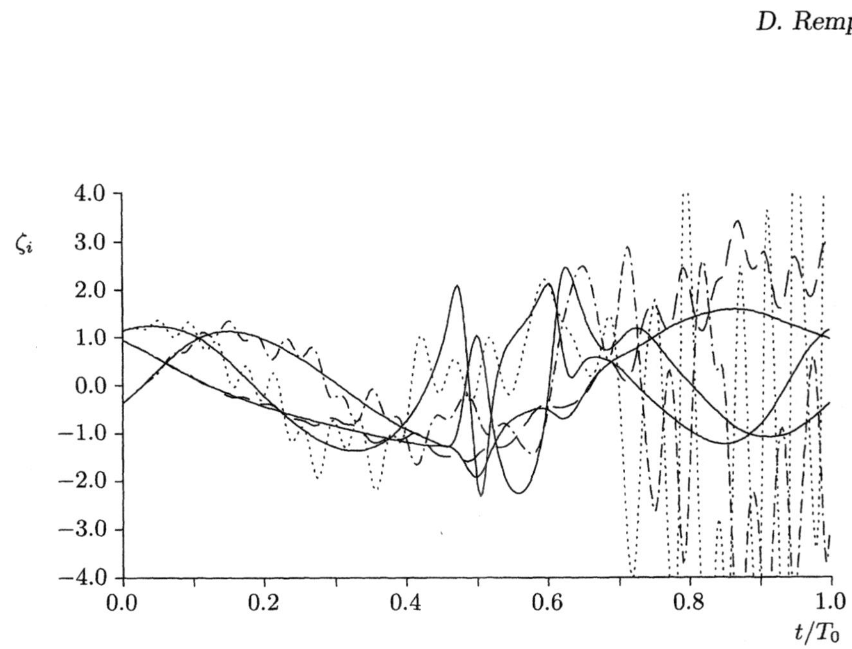

(a) Comparison of coefficients, ————, ζ_i^S; — — —, ζ_2^M; —·—·—·—, ζ_4^M; ············, ζ_6^M.

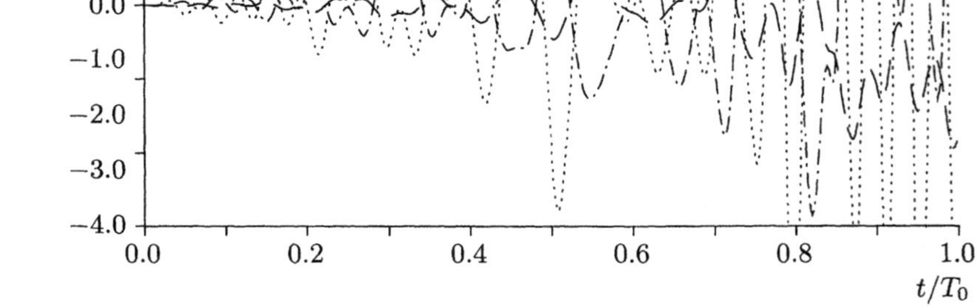

(b) Difference of coefficients, — — —, $\zeta_2^M - \zeta_2^S$; —·—·—·—, $\zeta_4^M - \zeta_4^S$; ············, $\zeta_6^M - \zeta_6^S$.

Figure 5: Comparison of dynamics generated by model to "true" dynamics from direct numerical simulation.

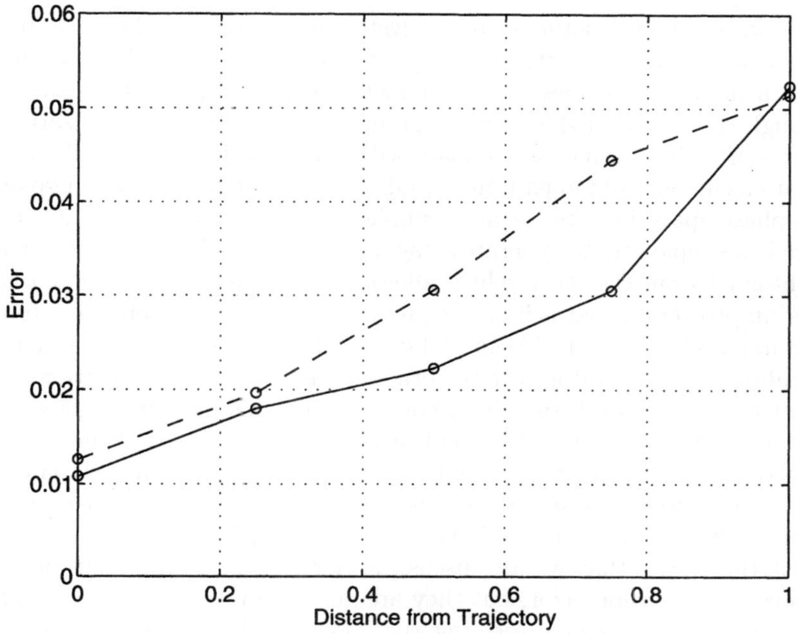

Figure 6: Error in representation of Navier-Stokes operator. — — —, truncated model, $i_{\max} = 10$; ————, full model, $i_{\max} = 16$.

on the trajectory, the error of the truncated system is larger than the one of the full system, but away from the trajectory both errors quickly rise, and beyond a certain distance the approximation of the full model is as bad as the one of the truncated system.

5 Conclusions

So, in summary what are the consequences of these observations for studies of the dynamics of POD-based models?

We have seen that this depends to a certain extent on the type of flow investigated. If we are interested in a flow that can be represented exactly by a finite-dimensional basis of POD eigenfunctions – examples of such flows are most non-turbulent and transitional flows – then the issues decribed above are definitely relevant. This means that for such flows the ODE models may exhibit unphysical stability properties, and the structure of the phase space of the model may be different from the one of the real flow.

For completeness, we note that one important way in which the phase space of the POD-Galerkin model can differ from the phase space of the PDE problem occurs when the original PDE is equivariant under the action of a certain linear group G. Generically, the attractor of such a system does not have the full symmetry G, but is only invariant with respect to some subgroup Σ, which means that there are a number of copies of this attractor in phase space that can be obtained through the action of the quotient group G/Σ. Any specific realization of the dynamics of such a system will remain within one of the copies of the attractor, so that eigenmodes obtained from a POD analysis can in general not span the complete phase space (see [9] for an example). However, as shown in [9], in this case there is a simple way to generate a "symmetrized" version of the original POD basis obtained from the data, which allows to generate a new POD basis giving a Galerkin projection that is indeed equivariant under the action of the full symmetry group G. Therefore, in this special case, it is possible to obtain a more complete eigenfunction basis using just the original data plus the information on inherent symmetries of the problem. Note, however, that this measure in general still cannot guarantee that the new POD subspace is invariant. Also, if the problem is symmetric with respect to an infinite group (as in the case of invariance with respect to continuous translations in a homogeneous coordinate that occurs for many flows), then the method described in [9] is not applicable.

Note that the effects that we are discussing here have nothing to do with numerical issues or truncation error, but they are an integral part of the method which in these cases consists of projecting the Navier-Stokes equations onto a subspace that is not invariant.

Although we have thus seen that such POD-Galerkin models – for fundamental reasons – can not be equivalent to the original Navier-Stokes equations, this does not mean that such models are useless. The main point of models of this type is to be able to study a low-dimensional description of complex flows, which can help us identify important basic mechanisms that would be very difficult to detect in the much more complicated setting of a description based directly on the Navier-Stokes equations. So these models do have their place, but one has to be aware of the fact that some aspects of the dynamics of a flow may not be mirrored correctly by models of this type. Wherever possible, a prudent way to check one's findings would be to look at the behavior of an extended model, that includes eigenfunctions of the null space of the kernel in (1).

A possible way to do this is the following: We start by computing a set of eigenfunctions that belong to the null space of the kernel in (1). These can be generated by performing a Gram-Schmidt orthonormalization on a set of basis functions formed by the KL-eigenfunctions plus a number of eigenfunctions from a complete space (e. g. trigonometric functions). Having obtained such an extended basis, one possible way to check the representation within the KL basis is to perform Floquet analyses on a number of extended systems of successively lower order (down to the original KL system), and to look at the convergence of the Floquet multipliers for this sequence of analyses. What one would like to see, then, is that the Floquet multipliers for the coordinates corresponding to the KL eigenfunctions are represented well within the pure KL basis. In addition, multipliers corresponding to eigenfunctions from the null space should be inside the unit circle (otherwise the orbit of the POD-Galerkin model would be unstable for the full PDE problem).

Finally, if we are interested in a flow that generates an infinite-dimensional basis of eigenfunctions then the problems discussed here do not apply. So in such cases it may be possible to derive valid statements on the structure of the phase space of the POD-Galerkin model, and such statements do carry over to the Navier-Stokes phase space. Often times in such a situation one will be forced to look at properties of truncated models, for which the same remarks apply as for the finite-dimensional models above. This means that one again has to make sure that the observed behavior is not an artifact of the truncation process. In other words, one must show explicitly that the behavior in question does persists at arbitrary orders of the truncated model. The Floquet analyses described above may be helpful in such a case as well.

References

[1] LUMLEY, J. L. (1970) *Stochastic Tools in Turbulence*. Academic Press, New York.

[2] LOÈVE, M. (1955) *Probability Theory*. Van Nostrand, Princeton.

[3] REMPFER, D.; FASEL, H. (1994) Evolution of Three-Dimensional Coherent Structures in a Flat-Plate Boundary Layer. *J. Fluid Mech.* **260**, pp. 351–375.

[4] KEEFE, L.; MOIN, P.; KIM, J. (1992) The Dimension of Attractors Underlying Periodic Turbulent Poiseuille Flow. *J. Fluid Mech.* **242**, pp. 1–29.

[5] FLETCHER, C. A. J. (1984) *Computational Galerkin Methods*. Springer, New York.

[6] AUBRY, N.; HOLMES, P.; LUMLEY, J. L.; STONE, E. (1988) The Dynamics of Coherent Structures in the Wall Region of a Turbulent Boundary Layer. *J. Fluid Mech.* **192**, pp. 115–173.

[7] REMPFER, D. (1996) Investigations of Boundary Layer Transition via Galerkin Projections on Empirical Eigenfunctions. *Phys. Fluids* **8** (1), pp. 175–188

[8] BERKOOZ, G.; HOLMES, P; LUMLEY J. L. (1993) The Proper Orthogonal Decomposition in the Analysis of Turbulent Flows. *Ann. Rev. Fluid Mech.* **25**, pp. 539–575.

[9] SMAOUI, N.; ARMBRUSTER, D. (1997) Symmetry and The Karhunen-Loève Analysis. *SIAM J. Sci. Comp.* **18** (5).

Trends in Mathematics, © 1999 Birkhäuser Verlag Basel/Switzerland

Symmetries of the Navier-Stokes equations and their implications for subgrid-models in large-eddy simulation of turbulence

M. Oberlack

Abstract

Necessary and sufficient conditions are derived for the subgrid-scale and filter function in large-eddy simulation of turbulence in order to preserve all symmetry transformation properties of Navier-Stokes equations.

1 Introduction

In [4] a new approach for inhomogeneous turbulent shear flows has been developed. The results therein unify many known scaling laws and several new scaling laws have been derived. The approach is based on Lie group analysis, also called symmetry analysis (see e.g. [1]), which guarantees that all scaling laws, based on classical point transformations, will be obtained. The scaling laws, or in the methodology of symmetry analysis, the invariant solutions, are solely based on Navier-Stokes equations. No ad hoc empirical assumptions have been used for its derivation which go beyond Navier-Stokes equations.

The most important result in [4] is the ordinary differential equation

$$\frac{d\bar{u}_1}{(a_1 - a_2)\bar{u}_1 + a_3} = \frac{dx_2}{a_1 x_2 + a_4}, \tag{1}$$

where $\bar{u}_1$ is the mean velocity in a parallel shear flow and x_2 is the coordinate perpendicular to it.

For various parameter combinations of a_1–a_4 the integrals of (1) are solutions, which include an algebraic law found in the center of a channel flow and in the near wall region, the common logarithmic law of the wall, the linear viscous sublayer, the linear mean velocity in the center of a Couette flow, the linear mean velocity in the center of a rotating channel flow, and an exponential mean velocity profile. The latter describes the outer part of a boundary layer flow over a flat plate, and it is an explicit form of the velocity defect law. Each of the parameters a_1–a_4 correspond to classical transformation properties of the Navier-Stokes equations, also called symmetries, to be explained below.

The mathematical background of the above mentioned scaling laws are symmetries of the Navier-Stokes equations which are transformations that do not alter the form of the equations written in the new variables. Suppose the system of partial differential equations under investigation is given by

$$\boldsymbol{F}(\tilde{\boldsymbol{x}}, \tilde{\boldsymbol{y}}, \tilde{\boldsymbol{y}}^{(1)}, \tilde{\boldsymbol{y}}^{(2)}, \ldots) = 0, \tag{2}$$

where $\tilde{\boldsymbol{x}}$ and $\tilde{\boldsymbol{y}}$ are the vectors of the independent and the dependent variables, respectively and $\tilde{\boldsymbol{y}}^{(n)}$ refers to the nth-order derivative of $\tilde{\boldsymbol{y}}$ with respect to $\tilde{\boldsymbol{x}}$. A transformation

$$\tilde{\boldsymbol{x}} = \tilde{\boldsymbol{x}}(\boldsymbol{x}, \boldsymbol{y}) \quad \text{and} \quad \tilde{\boldsymbol{y}} = \tilde{\boldsymbol{y}}(\boldsymbol{x}, \boldsymbol{y}) \tag{3}$$

is called a symmetry of the equation (2) if the following equivalence holds

$$\boldsymbol{F}(\tilde{\boldsymbol{x}}, \tilde{\boldsymbol{y}}, \tilde{\boldsymbol{y}}^{(1)}, \tilde{\boldsymbol{y}}^{(2)}, \ldots) = 0 \quad \Leftrightarrow \quad \boldsymbol{F}(\boldsymbol{x}, \boldsymbol{y}, \boldsymbol{y}^{(1)}, \boldsymbol{y}^{(2)}, \ldots) = 0, \tag{4}$$

i.e. the transformation (3) substituted into (2) does not change the form of equation (2) written in the new variables.

The Navier-Stokes equations admit several such symmetries also called symmetry transformations, each of them illuminating the axiomatic properties of classical mechanics: time invariance, rotation invariance, reflection invariance, two scaling invariances, and generalized Galilean invariance which encompasses frame invariance with respect to finite translation and classical Galilean invariance. In addition the two-dimensional Navier-Stokes equations are material indifferent under a constant rotation rate where the rotation axis is aligned with the axis of independence.

In equation (1) a_1 and a_2 correspond to scaling invariance of space and time, a_3 has its origin in the Galilean invariance and a_4 comes from translation invariance in space. In equation (1) the latter transformations are given in infinitesimal form (see e.g. [1]). To derive symmetries of equations it is very convenient to work with infinitesimal transformations. However, for the present purpose of investigating symmetry properties of subgrid-scale (SGS) models in large-eddy simulation (LES) of turbulence only global transformations are employed.

The key statement of the present contribution is that any turbulence model has to obey all the above stated invariance properties of Navier-Stokes equations. This is a necessary condition in order to adequately describe turbulence. In addition these properties are the foundations of all the above mentioned scaling laws.

A very problematic issue of the above results with respect to turbulence models is the fact, that many models do not account for the latter properties of Navier-Stokes equations. It can be shown that many common SGS models in LES violate certain symmetries of Navier-Stokes equations. All scaling laws in [4] have been solely derived from the above mentioned symmetries. It can be concluded that in computations where a SGS model has been adopted, which violates these symmetries, the above mentioned scaling laws can never be captured. The widely used Smagorinsky model violates scaling invariance. The inability of the Smagorinsky model to properly emulate the near-wall behavior can be traced back to the latter deficiency. The problem is caused by the explicit dependency on the local grid length Δ. Only a few SGS models have been developed which recognize all proper symmetries of the Navier-Stokes equations one of which is the dynamic model in [2] and several of its successors. In contrast to the Smagorinsky model the dynamic model requires explicit filtering on a test filter length scale

$\tilde{\Delta} > \Delta$. In the dynamic model $\tilde{\Delta}$ and Δ only appear as a ratio and hence as a non-dimensional number.

For SGS models in LES of turbulence which have been proposed in conjunction with explicit filtering, it is common practice to filter over adjacent grid points. In case of wall-bounded flows these are usually pencil-like volumes with high aspect ratio. However, this violates frame invariance, also called co-variance principle, which is one of the most fundamental axioms in physics. If all symmetries of the Navier-Stokes equations are employed, and this is what should be done in order to properly model the Navier-Stokes equations, only a very restrictive set of filter functions can be used in LES.

2 Invariant models in LES of turbulence

The analysis to follow is based on the filtered Navier-Stokes equations

$$\bar{\mathcal{N}}_i = \frac{\partial \bar{u}_i}{\partial t} + \bar{u}_k \frac{\partial \bar{u}_i}{\partial x_k} + \frac{\partial \bar{p}}{\partial x_i} - \nu \frac{\partial^2 \bar{u}_i}{\partial x_k^2} + \frac{\partial \tau_{ik}}{\partial x_k} = 0, \tag{5}$$

the equations for the velocity fluctuations

$$\mathcal{N}_i = \frac{\partial u_i}{\partial t} + \bar{u}_k \frac{\partial u_i}{\partial x_k} + u_k \frac{\partial \bar{u}_i}{\partial x_k} - \frac{\partial \tau_{ik}}{\partial x_k} + \frac{\partial u_i u_k}{\partial x_k} + \frac{\partial p}{\partial x_i} - \nu \frac{\partial^2 u_i}{\partial x_k^2} = 0 \tag{6}$$

and the continuity equations

$$\bar{\mathcal{C}} = \frac{\partial \bar{u}_k}{\partial x_k} = 0 \text{ and } \mathcal{C} = \frac{\partial u_k}{\partial x_k} = 0. \tag{7}$$

The SGS stress is defined as

$$\tau_{ik} = \overline{\bar{u}_i \bar{u}_k} - \bar{u}_i \bar{u}_k + \overline{u'_i \bar{u}_k} + \overline{\bar{u}_i u'_k} + \overline{u'_i u'_k}, \tag{8}$$

where the filter operation is given by

$$\bar{f} = \int_V G(\boldsymbol{x}, \boldsymbol{y}) f(\boldsymbol{y}) d^3 y \quad \text{with} \quad f' = f - \bar{f} \quad \text{and} \quad \int_V G(\boldsymbol{x}, \boldsymbol{y}) d^3 y = 1. \tag{9}$$

The scaling laws mentioned in the introduction can be traced back to the symmetries of the set of equations (5)–(7). That means, transformations of the form

$$\begin{aligned} t^* = t^*(\boldsymbol{X}, \boldsymbol{Y}),\ \boldsymbol{x}^* = \boldsymbol{x}^*(\boldsymbol{X}, \boldsymbol{Y}),\ \bar{\boldsymbol{u}}^* = \bar{\boldsymbol{u}}^*(\boldsymbol{X}, \boldsymbol{Y}),\ \boldsymbol{u}'^* = \boldsymbol{u}'^*(\boldsymbol{X}, \boldsymbol{Y}), \\ \bar{p}^* = \bar{p}^*(\boldsymbol{X}, \boldsymbol{Y}),\ p'^* = p'^*(\boldsymbol{X}, \boldsymbol{Y}),\ \nu^* = \nu^*(\boldsymbol{X}, \boldsymbol{Y}), \end{aligned} \tag{10}$$

with

$$\boldsymbol{X} = \{t, \boldsymbol{x}, \nu\},\ \boldsymbol{Y} = \{\bar{\boldsymbol{u}}, \boldsymbol{u}', \bar{p}, p'\}, \tag{11}$$

exist, which preserve the form of the equations so that the following equivalence holds

$$\{\bar{\mathcal{N}}_i = 0, \mathcal{N}_i = 0, \bar{\mathcal{C}} = 0, \mathcal{C} = 0\} \quad \Leftrightarrow \quad \{\bar{\mathcal{N}}_i^* = 0, \mathcal{N}_i^* = 0, \bar{\mathcal{C}}^* = 0, \mathcal{C}^* = 0\}. \tag{12}$$

The asterisks at the equations on the right hand side refer to the fact that the equations have been written in the new variables introduced in (10).

In the following a list of all known point symmetry transformations will be given which preserve the functional form of the latter equations written in the new variables, subsequently denoted by "*".

I. Time invariance

An arbitrary time shift of the amount a

$$t^* = t + a, \;\; \boldsymbol{x}^* = \boldsymbol{x}, \;\; \bar{\boldsymbol{u}}^* = \bar{\boldsymbol{u}}, \;\; \boldsymbol{u}'^* = \boldsymbol{u}', \;\; \bar{p}^* = \bar{p}, \;\; p'^* = p', \;\; \nu^* = \nu, \tag{13}$$

has no effect on the functional form of the equations (5)–(7).

II. Rotation invariance

Rotating the coordinate system and the velocity vectors

$$t^* = t, \;\; x_i^* = A_{ij}x_j, \;\; \bar{u}_i^* = A_{ij}\bar{u}_j, \;\; u_i'^* = A_{ij}u_j', \;\; \bar{p}^* = \bar{p}, \;\; p'^* = p', \;\; \nu^* = \nu. \tag{14}$$

by a finite but arbitrary angle in space, where $\boldsymbol{A}$ is the rotation matrix with $\boldsymbol{A}\boldsymbol{A}^T = \boldsymbol{A}^T\boldsymbol{A} = \boldsymbol{I}$ and $|\boldsymbol{A}| = 1$, preserves the form of the equations (5)–(7) written in the new variables.

III. Reflection invariance

The reflection symmetry in any direction x_α is given by

$$\begin{gathered} t^* = t, \;\; x_\alpha^* = -x_\alpha, \;\; \bar{u}_\alpha^* = -\bar{u}_\alpha, \;\; u_\alpha'^* = -u_\alpha', \\ x_\beta^* = x_\beta \;, \bar{u}_\beta^* = \bar{u}_\beta \;, u_\beta'^* = u_\beta' \text{ with } \beta \neq \alpha, \;\; p^* = p, \;\; \nu^* = \nu, \end{gathered} \tag{15}$$

where the index α can be any of 1, 2, and 3 and β refers to the remaining two.

IV. Generalized Galilean invariance

Substituting

$$\begin{gathered} t^* = t, \;\; \boldsymbol{x}^* = \boldsymbol{x} + \hat{\boldsymbol{x}}(t), \;\; \bar{\boldsymbol{u}}^* = \bar{\boldsymbol{u}} + \frac{d\hat{\boldsymbol{x}}}{dt}, \;\; \boldsymbol{u}'^* = \boldsymbol{u}', \\ \bar{p}^* = \bar{p} - \boldsymbol{x} \cdot \frac{d^2\hat{\boldsymbol{x}}}{dt^2}, \;\; p'^* = p', \;\; \nu^* = \nu. \end{gathered} \tag{16}$$

into the equations (5)–(7), where $\hat{\boldsymbol{x}}(t)$ is any twice differentiable time dependent vector-function, it does not alter their functional form. (16) covers two classical symmetries: (i) Invariance with respect to finite translation in space is obtained for $\hat{\boldsymbol{x}}(t) = \boldsymbol{a}$, where $\boldsymbol{a}$ is a constant. (ii) The classical Galilean invariance is recovered if $\hat{\boldsymbol{x}}(t)$ is a linear function in time.

V. Scaling invariance

The two parameter transformation

$$\begin{aligned} t^* &= \xi t, \;\; \boldsymbol{x}^* = \gamma \boldsymbol{x}, \;\; \bar{\boldsymbol{u}}^* = \frac{\gamma}{\xi}\bar{\boldsymbol{u}}, \;\; \boldsymbol{u}'^* = \frac{\gamma}{\xi}\boldsymbol{u}', \\ \bar{p}^* &= \left(\frac{\gamma}{\xi}\right)^2 \bar{p}, \;\; p'^* = \left(\frac{\gamma}{\xi}\right)^2 p', \;\; \nu^* = \frac{\gamma^2}{\xi}\nu. \end{aligned} \tag{17}$$

is a symmetry of the equations (5)–(7), where ξ and γ are arbitrary numbers.

VI. Pressure invariance

The pressure invariance recognizes the property of the equations (5)–(7) to be unaltered by a temporally varying background pressure $\varphi(t)$

$$t^* = t, \;\; \boldsymbol{x}^* = \boldsymbol{x}, \;\; \bar{\boldsymbol{u}}^* = \bar{\boldsymbol{u}}, \;\; \boldsymbol{u}'^* = \boldsymbol{u}', \;\; \bar{p}^* = \bar{p} + \varphi(t), \;\; p'^* = p', \;\; \nu^* = \nu. \tag{18}$$

VII. 2D material indifference under constant rotation rate

Consider the two-dimensional form of the equations (5)–(7) in a constantly rotating frame about the x_3-direction with the rotation rate Ω_3. Here two-dimensionality only refers to the fact that the velocities and the pressure only depend on x_1, x_2 and t but the velocity vectors can still have three components. Considering these restrictions the transformation which leaves (5)–(7) form invariant is given by

$$\begin{aligned} t^* = t, \;\; x_i^* &= B_{ij}(t)x_j, \;\; \bar{u}_i^* = B_{ij}(t)\bar{u}_j + \dot{B}_{ij}(t)x_j, \;\; u_i'^* = B_{ij}(t)u_j', \\ \bar{p}^* &= \bar{p} + 2\,\Omega_3 \int_Q (\bar{u}_1 dx_2 - \bar{u}_2 dx_1) + \tfrac{1}{2}\Omega_3^2(x_1^2 + x_2^2), \\ p'^* &= p' + 2\,\Omega_3 \int_Q (u_1' dx_2 - u_2' dx_1), \;\; \nu^* = \nu, \end{aligned} \tag{19}$$

where $\boldsymbol{B}(t)$ is the rotation matrix about x_3 with $\boldsymbol{B}\boldsymbol{B}^T = \boldsymbol{B}^T\boldsymbol{B} = \boldsymbol{I}$, $|\boldsymbol{B}| = 1$, $\dot{B}_{ik}B_{jk} = \varepsilon_{3ij}\Omega_3$ and Ω_3 is a constant. The integrals in (19) are the usual two-dimensional stream-function.

2.1 Implications for the subgrid-scale stress model

It is common practice in LES of turbulence to model the SGS stress as a functional of the resolved velocity $\bar{\boldsymbol{u}}$ and the spatial coordinate $\boldsymbol{x}$, i.e.

$$\tau_{ik} = \mathcal{F}_{ik}(\bar{\boldsymbol{u}}, \boldsymbol{x}). \tag{20}$$

The key to physically correct models in LES of turbulence is that the SGS model equation (20) put into equation (5) should reflect the same transformation properties as the unmodelled equation. This restricts the form of the SGS models. Since the unmodelled SGS stresses obey certain transformation rules coming from the transformations (13)–(19), one can conclude that the modelled SGS stresses have to obey the same properties. Hence, the following equivalence for the SGS model $\mathcal{F}_{ik}$ has to hold

$$\tau_{ik} = \mathcal{F}_{ik}(\bar{\boldsymbol{u}}, \boldsymbol{x}) \quad \Leftrightarrow \quad \tau_{ik}^* = \mathcal{F}_{ik}^*(\bar{\boldsymbol{u}}^*, \boldsymbol{x}^*) \tag{21}$$

for all symmetries (13)–(19).

Time invariance (13), rotation invariance (14), reflection invariance (15) and pressure invariance (18) are usually not violated by common SGS models. Galilean invariance (16) in its usual form, i.e. $\hat{\boldsymbol{x}}(t) = \boldsymbol{a}+\boldsymbol{b}t$, has already been discussed in [5]. Modern SGS model obey this transformation property. However, scaling invariance and 2D material indifference are violated by several SGS models. Hence, it will be focused only on these two symmetries.

It is important to note that all symmetries (13)–(19) are independent of the fact whether the equations (5)–(7) are written in dimensional or in non-dimensional form. This is also true for the scaling symmetry if viscosity ν is replaced by $1/Re$. It addition it should be mentioned that this fact is independent of the variables used to non-dimensionalize the equations.

Scaling invariance

From (8) and (17) one can conclude

$$\boldsymbol{\tau}^* = \left(\frac{\gamma}{\xi}\right)^2 \boldsymbol{\tau}, \tag{22}$$

which has to be valid for any SGS model in order to preserve scaling invariance. Again it should be mentioned that this property has to hold regardless whether the model is written in dimensional or non-dimensional form. As has already been mentioned above that the Smagorinsky model

$$\tau_{ik} - \frac{1}{3}\delta_{ik}\tau_{mm} = C\Delta^2 \frac{\partial}{\partial x_k}\left(|\bar{\boldsymbol{S}}|\bar{S}_{ik}\right), \quad \text{with} \quad \bar{S}_{ik} = \frac{1}{2}\left(\frac{\partial \bar{u}_i}{\partial x_k} + \frac{\partial \bar{u}_k}{\partial x_i}\right), \tag{23}$$

implemented into equation (5), violates scaling symmetry. It has been argued that the grid scale Δ has to obey the same scaling property as the spatial coordinate, i.e. $\Delta^* = \gamma\Delta$. However, this leads to a contradiction since Δ is not an intrinsic variable of the equations. In order to clearly see the problem of the Smagorinsky model it will be investigated in non-dimensionalized form where Δ and an appropriate time scale t_t has been introduced into the modelled equations. As a result, the grid scale Δ disappears from the equation. Hence, the Smagorinsky model, for convenience now written in the new variables, is given by

$$\tau_{ik}^* - \frac{1}{3}\delta_{ik}\tau_{mm}^* = -C|\bar{\boldsymbol{S}}|^* \bar{S}_{ik}^*. \tag{24}$$

On the left hand side the exact scaling transformation for the stresses (22) will be employed because this is the transformation property of the unmodelled terms in the filtered Navier-Stokes equations. On the right hand side the transformation properties for the strain will be taken from the equations (17). If multiplied by $(\gamma/\xi)^2$, the arbitrary scaling factor ξ^2 cancels from both sides and its remains

$$\tau_{ik} - \frac{1}{3}\delta_{ik}\tau_{mm} = -C|\bar{\boldsymbol{S}}|\bar{S}_{ik}\gamma^2. \tag{25}$$

It is apparent that the model equation is not scaling invariant with respect to space since γ^2 does not cancel from both sides. This has a very prejudicial influence on the overall statistical behaviour of the model. In particular in flow regions where the grid size gets close to the integral length scale, such as in wall bounded flows, the error of the model becomes dominant. This problem is well known from computations where the Smagorinsky model has been employed and is usually addressed by introducing empirical near-wall treatment such as damping functions. One of the crucial ingredients for the derivation of the law of the wall in [4] is scaling invariance. Hence, the inability of the Smagorinsky to properly emulate the logarithmic law of the wall can be traced back to the violation of scaling invariance.

2D material indifference under constant rotation rate

Similar to the SGS model constraint with respect to scaling invariance one can conclude from (19) and (8) that any SGS model should transform as

$$\tau^*_{ik} = B_{im} B_{kn} \tau_{mn}, \tag{26}$$

where $\boldsymbol{B}(t)$ obeys the definitions given below (19), in order to be material indifferent in the two-dimensional limit.

A model which violates both scaling invariance and material indifference is the structure-function model in [3]. The proposed SGS model is of the form

$$\tau_{ik} - \frac{1}{3}\delta_{ik}\tau_{mm} = C^{SF}\Delta\langle(\bar{\boldsymbol{u}}(\boldsymbol{x}+\boldsymbol{r}) - \bar{\boldsymbol{u}}(\boldsymbol{x}))^2\rangle^{1/2}\bar{S}_{ik}, \tag{27}$$

where C^{SF}, and $\langle\rangle$ are, respectively, a model constant and a spatial average. Using the condition (26) in conjunction with the transformation (19) yields

$$\tau_{ik} - \frac{1}{3}\delta_{ik}\tau_{mm} = C^{SF}\Delta\langle(\bar{\boldsymbol{u}}_{(m)}(\boldsymbol{x}+\boldsymbol{r}) - \bar{\boldsymbol{u}}_{(m)}(\boldsymbol{x}) - \varepsilon_{3l(m)}\Omega_3 r_l)^2\rangle^{\frac{1}{2}}\bar{S}_{ik}. \tag{28}$$

The latter expression is not of the form (27) since it contains an additional rotation term. Hence, the structure-function model is not material indifferent. As for the Smagorinsky model one can also show that (27) is not scale invariant.

A SGS model which captures *all* the invariance requirements derived in the latter subsection is the dynamic SGS model in [2]. They proposed a procedure which, used in conjunction with the classical Smagorinsky model, results in the following SGS model

$$\tau_{ik} - \frac{1}{3}\delta_{ik}\tau_{mm} = \frac{(\widetilde{\bar{u}_m\bar{u}_n} - \tilde{\bar{u}}_m\tilde{\bar{u}}_n)\bar{S}_{mn}}{\left(\frac{\tilde{\Delta}}{\Delta}\right)^2 |\tilde{\bar{\boldsymbol{S}}}|\tilde{\bar{S}}_{mn}\bar{S}_{mn} - \widetilde{|\bar{\boldsymbol{S}}|\bar{S}_{pq}}\bar{S}_{pq}}|\bar{\boldsymbol{S}}|\bar{S}_{ik}. \tag{29}$$

Here, all the tilded quantities refer to the "test"-filter

$$\tilde{h}(\boldsymbol{x}) = \int_V \tilde{G}(\boldsymbol{x},\boldsymbol{y})h(\boldsymbol{y})d^3y, \tag{30}$$

which corresponds to the filter length $\tilde{\Delta}$ and $\tilde{\Delta} > \Delta$. The test-filter quantities are explicitly computed from the flow field. The resolved quantities are still denoted by an overbar. The dynamic model contains the ratio of two length scales which is a dimensionless number and therefore no external length scale is imposed. Consequently, the scaling invariance (22) is recovered, as can be shown by using (17), provided the proper filter function is utilized. It is straight to prove that frame invariance, generalized Galilean invariance and material indifference are also captured by the dynamic model.

2.2 Implications for the filter kernel G

Several LES models have been proposed in conjunction with an explicit filter, such as the dynamic model. However, not all filter functions are consistent with the symmetries of the Navier-Stokes equations. Due to space limitation only rotation and scaling invariance will be discussed in some detail.

Because of computational simplicity explicit filtering is usually applied over a volume which is determined by the form of the local grid geometry. However, this common practice violates frame invariance with respect to finite rotation. As a consequence, computational results do depend on the choice of the coordinate system which contradicts some of the most important axioms in physics.

From $\bar{u}_i^* = A_{ij}\bar{u}_j$ in (14) and the definition of the filter (9) one can deduce

$$\int_{V^*} G(\boldsymbol{x}^* - \boldsymbol{y}^*)u_i^*(\boldsymbol{y}^*)d^3y^* = A_{ij}\int_V G(\boldsymbol{x} - \boldsymbol{y})u_j(\boldsymbol{y})d^3y. \tag{31}$$

Employing the transformation property for the unfiltered velocities i.e. $\boldsymbol{u}^* = \boldsymbol{A}\boldsymbol{u}$ and $\boldsymbol{x}^* = \boldsymbol{A}\boldsymbol{x}$, where the latter results in $d^3y^* = d^3y$, the two integrals are equal except for the filter kernel. Hence, in order for (14) to hold, the condition

$$G(\boldsymbol{A}(\boldsymbol{x} - \boldsymbol{y})) = G(\boldsymbol{x} - \boldsymbol{y}) \tag{32}$$

must be satisfied. For arbitrary $\boldsymbol{A}$, the latter functional equation has a unique solution which is

$$G = G(|\boldsymbol{x} - \boldsymbol{y}|). \tag{33}$$

(33) corresponds to a known result from tensor invariant theory: A scalar function depending on vectors or tensors can only depend on their scalar invariants here the magnitude of separation vector. An additional consequence of (33) is, that the averaging volume V in (9) is restricted to a sphere with its centre in $\boldsymbol{x}$.

Another constraint for G emerges from scaling invariance (17). For the present purpose the filter function G is not normalized, denoted by the superscript "u". Using (9) one can conclude from $\bar{\boldsymbol{u}} = \gamma/\xi\bar{\boldsymbol{u}}$ in (17) that

$$\frac{\int_{V^*} G^{\mathrm{u}}(|\boldsymbol{x}^* - \boldsymbol{y}^*|)\boldsymbol{u}^*(\boldsymbol{y}^*)d^3y^*}{\int_{V^*} G^{\mathrm{u}}(|\boldsymbol{x}^* - \boldsymbol{y}^*|)d^3y^*} = \frac{\gamma}{\xi}\frac{\int_V G^{\mathrm{u}}(|\boldsymbol{x} - \boldsymbol{y}|)\boldsymbol{u}(\boldsymbol{y})d^3y}{\int_V G^{\mathrm{u}}(|\boldsymbol{x} - \boldsymbol{y}|)d^3y} \tag{34}$$

needs to hold. Using (17) the spatial scaling factor γ remains in the argument of G^{u} on the left hand side of (34). As a result, γ can only cancel out for arbitrary $\boldsymbol{u}$ if G^{u} has the following form

$$G^{\mathrm{u}}(|\boldsymbol{x}-\boldsymbol{y}|) = A|\boldsymbol{x}-\boldsymbol{y}|^{\alpha} \tag{35}$$

where A and α are arbitrary constants. Using (9), the final form of the filter G is obtained

$$\bar{f} = \frac{\alpha+3}{4\pi l^{\alpha+3}} \int_{\mathcal{R}_l} |\boldsymbol{x}-\boldsymbol{y}|^{\alpha} f(\boldsymbol{y}) d^3 y, \tag{36}$$

where $\mathcal{R}_l$ refers to a sphere with centre $\boldsymbol{x}$ and radius l. (36) preserves all the symmetry requirements (13)–(19). If the integration argument is sufficiently smooth, the integral converges for all $\alpha > -3$.

The invariances (13), (15), (18), and (19), even though not explicitly considered during the derivation, are consistent with (36).

References

[1] Bluman, G. W. & Kumei S. 1989 *Symmetries and Differential Equations.* Applied mathematical sciences, vol. 81, Springer.

[2] Germano, M., Piomelli, U., Moin, P. & Cabot, W. H. 1991 A dynamic subgrid-scale eddy viscosity model. *Phys. Fluids A* **3**, 1760–1765.

[3] Métais, O. & Lesieur, M. 1992 Spectral large-eddy simulation of isotropic and stably stratified turbulence. *J. Fluid Mech.* **239**, 157–194.

[4] Oberlack, M. 1997 Unified theory for symmetries in plane parallel turbulent shear flows. *CTR manuscript 163, also submitted to J. Fluid Mech.*

[5] Speziale, C. G. 1985 Galilean invariance of subgrid-scale stress models in the large-eddy simulation of turbulence. *J. Fluid Mech.* **156**, 55–62.

Trends in Mathematics, © 1999 Birkhäuser Verlag Basel/Switzerland

Subgrid modeling for the filtered scalar transport equation

A. Leonard

Abstract

We consider passive scalar transport for the case of small diffusivity. In particular, we are concerned with deriving an evolution equation for a smoothed, filtered, or coarse-grained scalar field, i.e., we consider the subgrid modeling problem for scalar transport. It is shown that for the case of velocity fields that are smooth but that induce chaotic advection a subgrid tensor diffusivity may be derived from first principles. However, this diffusivity can be locally strongly anti-diffusive, so that the filtered transport equation is ill-conditioned. A rationale for regularizing the equation is presented. In addition it is argued that analogous terms should appear in the filtered Navier-Stokes equations.

1 Introduction

In this paper we consider the derivation of a transport equation for a filtered scalar field. The problem arises, for example, in the numerical simulation of scalar transport when the effective resolution of the numerical scheme is not sufficient to represent all the important length-scales. Strain may steepen or may smooth a scalar gradient. Such processes result in the transfer of scalar variance between the resolved and subgrid scales and need to be accounted for in the evolution equation for the resolved scales.

Many authors have investigated various forms of the Smagorinsky model [1] for the momentum equation which assumes that the deviatoric part of the subgrid-scale stress is proportional to the rate-of-strain tensor of the large eddies, i.e., a gradient diffusion model (see, e.g. ref 2). Modeling of the scalar flux due to subgrid transfer has also generally been along the lines of an eddy diffusivity approach, i.e., subgrid flux proportional to the gradient of the large-scale field with a non-linear scalar diffusivity [3]. Such a model has useful properties – resolved scalar variance dissipation and proper invariance – and has been moderately successful in application but it has little justification otherwise.

In this paper [4] we show, however, that in many cases of interest, the subgrid scalar flux is not composed of gradient diffusion but involves a *tensor diffusivity* which can be derived from first principles. Interestingly this diffusivity can be locally very strongly anti-diffusive so that the method must be regularized in a meaningful way. We propose a particle method that provides such a regularization. Then we consider the filtered momentum equation for incompressible flow and argue that analogous terms should appear in these equations.

Scalar Transport

Consider scalar transport in an infinite domain. The convection-diffusion equation for the scalar density, ψ, is given by

$$\frac{\partial \psi}{\partial t} + \mathbf{u} \cdot \nabla \psi = \kappa \nabla^2 \psi \tag{1}$$

where $\mathbf{u}$ is the incompressible velocity field and κ is the scalar diffusivity. We would like to derive a large-eddy transport equation for the filtered density, $\overline{\psi}$, defined by

$$\overline{\psi}(\mathbf{x}) = \int G(\mathbf{x} - \mathbf{x}') \; \psi(\mathbf{x}') d\mathbf{x}' \tag{2}$$

where the filter function G is taken to be gaussian with characteristic width σ:

$$G(\mathbf{z}) = \frac{exp(-|\mathbf{z}|^2/\sigma^2)}{(\sigma\sqrt{\pi})^d} \tag{3}$$

Here d is the dimension of the spatial integral.

Using incompressibility and applying the filtering operation to (1) gives the evolution equation for the filtered scalar density,

$$\frac{\partial \overline{\psi}}{\partial t} + \nabla \cdot \overline{\mathbf{u}\psi}. = \kappa \nabla^2 \overline{\psi} \tag{4}$$

which, of course, is not closed due to the appearance of the term $\overline{\mathbf{u}\psi}$. One now has to formulate or model a relation between the time-dependent vector field, $\overline{\mathbf{u}\psi}$ and the pair of fields $(\overline{\psi}, \overline{\mathbf{u}})$. Such a connection exists, at least formally. For functions $\overline{u}$ and $\overline{v}$ over a one-dimensional domain we have

$$\overline{uv} = \int_{-\infty}^{\infty} \frac{exp\left(\frac{-(x-x')^2}{\sigma^2}\right) u(x')v(x')dx'}{\sqrt{\pi}\sigma} \quad = \sum_{k=0}^{\infty} \left(\frac{\sigma^2}{2}\right)^k \frac{1}{k!} \frac{\partial^k \overline{u}}{\partial x^k} \frac{\partial^k \overline{v}}{\partial x^k} \tag{5}$$

This result may be demonstrated by use of Taylor series or Hermite polynomial expansions. Applying this result to $\overline{u_\ell \psi}$ in d dimensions we have

$$\overline{u_\ell \psi} = \overline{u}_\ell \overline{\psi} + \frac{\sigma^2}{2} \frac{\partial \overline{u}_\ell}{\partial x_j} \frac{\partial \overline{\psi}}{\partial x_j} + \frac{\sigma^4}{8} \frac{\partial^2 \, \overline{u}_\ell}{\partial x_j \partial x_m} \frac{\partial \overline{\psi}}{\partial x_j \, \partial x_m} + \cdots \tag{6}$$

where repeated indices are summed. Thus, for sufficiently smooth velocity fields, the expansion (6) suggests that we can neglect the $0(\sigma^4)$ and higher order terms with the result,

$$\frac{\partial \overline{\psi}}{\partial t} + \overline{\mathbf{u}} \cdot \nabla \overline{\psi} = -\frac{\sigma^2}{2} \overline{S}_{ij} \frac{\partial^2 \overline{\psi}}{\partial x_i \partial x_j} + \kappa \nabla^2 \overline{\psi} \tag{7}$$

where $\overline{S}_{ij}$ is the strain-rate tensor of the filtered velocity field,

$$\overline{S}_{ij} = \frac{1}{2} \left(\frac{\partial \overline{u}_i}{\partial x_j} + \frac{\partial \overline{u}_j}{\partial x_i} \right) \tag{8}$$

Thus $\overline{S}_{ij}$ plays the role of a diffusion tensor for the filtered scalar field. Apparently (7) is now a closed equation, giving the solution of $\overline{\psi}$ for the case of smooth velocity fields, and for small molecular diffusivities we can neglect the second term on the RHS of (7). This is a nontrivial result given that even smooth velocity fields yield chaotic particle motions. Another clear application of this result is in turbulent transport at high Schmidt Number. Here σ is smaller than the Kolmogorov scale (so that u_i is well-resolved at scale σ) but larger than the Batchelor scale (so that ψ has a significant contribution from scales smaller than σ).

However, there is a price to pay. The diffusion operator is ill-conditioned as can be seen as follows. Transforming to principle coordinates of $\overline{S}_{ij}, \mathbf{x}'$, we find that the first term on the RHS of (7) for three-dimensional transport becomes

$$\frac{-\sigma^2}{2}\overline{S}_{ij}\frac{\partial^2\overline{\psi}}{\partial x_i \partial x_j} = -\frac{\sigma^2}{2}\left(\overline{\alpha}\frac{\partial^2\overline{\psi}}{\partial x_1'^2} + \overline{\beta}\frac{\partial^2\overline{\psi}}{\partial x_2'^2} + \overline{\gamma}\frac{\partial^2\overline{\psi}}{\partial x_3'^2}\right) \tag{9}$$

where the eigenvalues of $\overline{S}_{ij}, (\overline{\alpha}, \overline{\beta}, \overline{\gamma})$ satisfy

$$\overline{\alpha} \geq \overline{\beta} \geq \overline{\gamma} \qquad \overline{\alpha} + \overline{\beta} + \overline{\gamma} = 0 \tag{10}$$

so that along the stretching direction(s), x_1' (and possibly x_2') we have effectively negative diffusion. Thus, for example, if $u_1' \approx \overline{\alpha}x_1'$ and $\overline{\psi}(t=0) = e^{ikx_1'}$ then

$$\overline{\psi}(t) = e^{ikexp\,(-\overline{\alpha}t)x_1'}\, exp\left[\overline{\alpha}\sigma^2 k^2(1 - exp\,(-2\overline{\alpha}t))/4\right] \tag{11}$$

i.e., as the wavenumber of the field is decreased due to stretching its filtered amplitude will increase initially and the initial rate of increase becomes arbitrarily large as $k \to \infty$. Keeping the molecular diffusion does not avert this catastrophe unless the stretching rate $\overline{\alpha}$ is small or, specifically unless $\sigma^2\overline{\alpha} < 2\kappa$.

As a demonstration of this effect we have used a fully dealiased spectral method to solve (7) for the case $\kappa = 0$ with velocity field

$$u = \cos x\ \cos y \qquad v = \sin x\ \sin y + \varepsilon \sin \omega t \tag{12}$$

For $\overline{\psi}$ fields that are an axisymmetric Gaussian distribution at $t = 0$, the initial evolution of the scalar field appears reasonable but later contains strong unphysical oscillations before blowing up.

2 Regularization – Particle method

It seems clear that we require a numerical technique to solve (7) that maintains control over the high frequency content of the solution. We propose the use of lagrangian particles, based on an anisotropic gaussian distribution,

$$\overline{\psi}(\mathbf{x}, t) = \mathop{\Sigma}_{k} \frac{a_k[det\mathbf{M}_k(t)]^{1/2}}{(\delta\sqrt{\pi})^d} \times \exp\left[-\mathbf{M}_k : (\mathbf{x} - \mathbf{x}_k)(\mathbf{x} - \mathbf{x}_k)/\delta^2\right] \tag{13}$$

in which each particle moves with the local velocity

$$\frac{d\mathbf{x}_k}{dt} = \overline{\mathbf{u}}(\mathbf{x}_k, t) \tag{14}$$

and the symmetric tensor $\mathbf{M}_k$ satisfies

$$\frac{d\mathbf{M}_k}{dt} = -(\nabla\overline{u})_k\mathbf{M}_k - \mathbf{M}_k(\nabla\overline{u})_k^T + \frac{\sigma^2}{\delta^2}\mathbf{M}_\mathbf{k}(\nabla\overline{\mathbf{u}}_k + \nabla\overline{\mathbf{u}}_k^T)\mathbf{M}_k \tag{15}$$

The representative core radius, δ, is taken to be constant, the coefficients a_k are also constants, and we have assumed $\kappa = 0$. Substituting (13–15) into (7) with $\kappa = 0$ and using Taylor series approximations we find that the truncation error for the anisotropic particle scheme is of the form

$$\varepsilon(\mathbf{x}) = [O(\delta^4) + O(\sigma^2\delta^2)]\frac{\partial^2\overline{u}}{\partial x^2}\frac{\partial^3\overline{\psi}}{\partial x^3} \tag{16}$$

In the above scheme the volume integral over each particle remains constant in time ($= a_k$ for the kth particle) as it should.

Without loss of generality we can take the initial condition for $\mathbf{M}_k$ to be the identity matrix

$$\mathbf{M}_k(t=0) = \mathbf{I} \tag{17}$$

The solution to (15) for $\mathbf{M}_k(t)$ is given rather nicely in terms of the left Cauchy Green strain tensor for particle k, $\mathbf{G}_k$. The tensor is defined as

$$\mathbf{G}_k = \mathbf{F}_k\mathbf{F}_k^T \tag{18}$$

where $\mathbf{F}_k$ is the deformation tensor satisfying

$$\frac{d\mathbf{F}_k}{dt} = (\nabla\mathbf{u})_k^T\mathbf{F}_k \qquad (\mathbf{F}_k(0) = \mathbf{I}) \tag{19}$$

Thus $\mathbf{G}_k$ satisfies

$$\frac{d\mathbf{G}_k}{dt} = \nabla\mathbf{u}_k^T\mathbf{G}_k + \mathbf{G}_k\nabla\mathbf{u}_k \qquad (\mathbf{G}_k(0) = \mathbf{I}) \tag{20}$$

One can now show that $\mathbf{M}_k$ is given in terms of $\mathbf{G}_k$ as follows:

$$\mathbf{M}_k = \delta^2/\sigma^2[\mathbf{I} + (\frac{\delta^2}{\sigma^2} - 1)\mathbf{G}_k]^{-1} \tag{21}$$

From (21) we see that because $\mathbf{G}_k$ is nonsingular and positive-definite $\mathbf{M}_k$ will also be nonsingular and positive-definite if $\delta \geq \sigma$. This is a reasonable constraint on the particle core radius anyway.

The limiting case $\delta = \sigma$ is required only when the unfiltered field ψ consists of δ functions. Then the exact ψ is given by (16) with $\delta = \sigma$ and $\mathbf{M}_k = \mathbf{I}$ for all time in agreement with (21) with $\delta = \sigma$. The only error is that the particles should be convected with the unfiltered velocity $\mathbf{u}$ to obtain the exact $\overline{\psi}$ rather than $\overline{\mathbf{u}}$ as in (14) .

The Fourier transform of $\overline{\psi}$ given by (13) is

$$\widehat{\overline{\psi}}(\mathbf{k}, t) = \Sigma a_k exp[-\delta^2 \mathbf{M}_k^{-1} : \mathbf{kk}/4] e^{i\mathbf{k}\cdot\mathbf{x}_k} \tag{22}$$

so that $\widehat{\overline{\psi}}$ appears well-behaved as $k \to \infty$. Thus, the particle representation (13) apparently gives us the desired control over the contributions to $\widehat{\overline{\psi}}$ at high wave numbers.

Initial test of this particle method using the velocity field given by (12) have been very encouraging (P. Moeleker, private communication). We note also that for the special case of velocity fields that are linear, i.e. $\nabla \underline{u}$ constant in space, the above particle method solves (7) exactly. In addition, for that special case, (7) is an exact evolution equation for the filtered field $\overline{\psi}$.

3 Navier-Stokes equations

Use of the expansion (5) on the filtered, constant density, incompressible momentum equation yields

$$\frac{\partial \overline{u}_i}{\partial t} + \frac{\partial}{\partial x_j}(\overline{u}_i \overline{u}_j) = \frac{-\sigma^2}{2} \overline{S}_{jk} \frac{\partial^2 \overline{u}_i}{\partial x_j \partial x_k} + O(\sigma^4) - \frac{1}{\rho}\frac{\partial p}{\partial x_i} + \nu \nabla^2 \overline{u}_i \tag{23}$$

Actually the subgrid contribution, given by the first term on the RHS above, was proposed some time ago [5] as an approximation to the subgrid stress due to the interaction between the resolved scales and the resolved scales plus the subgrid scales – essentially the same reasoning that has lead us once again to the same result, only this time more directly. More recently Cottet [6] showed that significant improvements to the accuracey of vortex methods could be obtained by using the diffusion tensor $\overline{S}_{ij}$ to correct for the backscatter part of the truncation error.

We believe this term to be an important component of a subgrid model. For example in homogeneous turbulence we may determine the energy of the large-eddy field by multiplying (23) by $\overline{u}_i$ and averaging over space to find

$$\frac{1}{2}\frac{d}{dt} < \overline{u}_i \overline{u}_i > = -I + \text{viscous} + \text{other subgrid dissipation} \tag{24}$$

where I is the energy loss due to the first term of the RHS of (23) and is given by

$$I = \frac{\sigma^2}{2} < \overline{u}_i \frac{\partial}{\partial x_j} \left(\frac{\partial \overline{u}_i}{\partial x_\ell} \frac{\partial \overline{u}_j}{\partial x_\ell} \right) > \tag{25}$$

Repeated integration by parts and use of continuity ($\nabla \cdot \overline{\mathbf{u}} = 0$) gives

$$I = -\frac{\sigma^2}{2} < \overline{u}_i \overline{u}_j \;\; \nabla^2 \frac{\partial \overline{u}_i}{\partial x_\ell} > \tag{26}$$

Following [5] we can relate to the above, for the case of isotropic decay, to the triple correlation function. Ultimately we find that

$$I = 0.98(-\overline{S})\varepsilon \tag{27}$$

where ε is the energy dissipation rate and $\overline{S}$ is the skewness of the velocity derivative, $\partial\overline{u}_1/\partial x_1$. Estimates of $-\overline{S}$ range from 0.40 to 0.60 so that the subgrid term under discussion accounts for 40% to 60% of the energy transfer to the subgrid scales. Exact computations on a decaying "shock" satisfying Burgers' equation show that the analogous term

$$-\frac{\sigma^2}{2}\frac{\partial\overline{u}}{\partial x}\frac{\partial^2\overline{u}}{\partial x^2} \tag{28}$$

produces 55% of the required dissipation.

Another test of a subgrid model is whether it satisfies material frame indifference. It has already been pointed out [7] that the tensor that leads to the subgrid term in (23),

$$-\frac{\sigma^2}{2}\frac{\partial\overline{u}_i}{\partial x_\ell}\frac{\partial\overline{u}_j}{\partial x_\ell} \tag{29}$$

is not frame indifferent. However its divergence, i.e., the actual vector contribution in (23) does satisfy material frame indifference.

4 Subgrid flux – Additional components

In general the subgrid flux is given by the infinity of terms or the RHS of (5) beyond the first term corresponding to the resolved flux. The complete infinite sum is formally equivalent to deconvolving (unfiltering) $\overline{u}$ and $\overline{v}$ taking the product of these results and filtering. By using only the first two terms we are, in essence, performing a partial deconvolution, and we have noted above that this should be adequate for sufficiently smooth velocity fields. We consider now the possiblitiy of significant contributions to the subgrid flux by terms beyond the first two on the RHS of (5).

In one such situation, u and v might contain high frequency components that are coherent. For example, for the inviscid Burgers' equation, $v = u$ and u might contain discontinututes so that the term (28) is not an accurate representation of the subgrid term as noted above. However, using the fact that $\overline{u}$ has explicitly known error function components, we can derive a modification to (28) that produces exact results, namely

$$\frac{-\sigma^2}{4}\frac{\partial}{\partial x}\left[R^2\left(\frac{\sigma}{\sqrt{2}\,\Delta u}\frac{\partial\overline{u}}{\partial x}\right)\frac{\partial\overline{u}}{\partial x}\frac{\partial\overline{u}}{\partial x}\right] \tag{30}$$

where

$$y\,R(y) = \sqrt{\frac{4}{\pi}\int_0^{\pi/4}\left(\frac{\sqrt{\pi}y}{2}\right)^{1/\cos^2\theta} d\theta} \tag{31}$$

and Δu is the jump in u ($\Delta u < 0$). Analogous modifications should be useful to treat shocks in the compressible Euler equations. One can generalize this result to more than one dimension. We find that the non- dimensional argument of R must be properly invariant.

In other applications the additional subgrid scale components of u and v might be significant but incoherent. In this case we might expect that an additional term of the gradient-diffusion type with an eddy diffusivity proporitonal $\sigma < \delta u >$ or $\sigma < \delta v >$ would be appropriate.

5 Summary

In summary, our investigation of filtered scalar transport equation has led us to a specific tensor diffusivity to represent the interaction of the resolved scales with the sub-grid scales. The resulting equation is ill-conditioned in the sense that high wavenumber components are subject to arbitrarily large increases in amplitude under stretching by the local strain field as the wavenumber goes to infinity. It is argued that control must be maintained over the high wavenumber components. We also argue that an analogous term should be included in subgrid modeling for the Navier-Stokes equation.

References

[1] Smagorinsky, J., 1963, "General circulation experiments with the primitive equations, in the basic experiment", *Mon. Wealth. Rev*, **91**, 99.

[2] Germano, M., Piomelli, U., Moin, P. and Cabot, W.H., 1991, "A dynamic subgrid-scale eddy viscosity model", *Phys. Fluids*, **A3**, 1760.

[3] Lesieur, M. and Métais, O., 1996, "New trends in large-eddy simulations of turbulence", *Annu. Rev. Fluid Mech.*, **28**, 45-82.

[4] Some of the material in the present paper appeared in Leonard, A., 1977, "Large-eddy Simulation of Chaotic Convection and Beyond," AIAA 97-0204.

[5] Leonard, A., 1974, "Energy cascade in large-eddy simulations of turbulent fluid flows", *Adv. Geophys.*, **18**, 237.

[6] Cottet, G.-H., 1996, "Artificial Viscosity Models for Vortex and Pariticle Methods," *J. Comput. Phys.*, **127**, 229.

[7] Liu, S., Meneveau, C. and Katz, J.,1994, "On the properties of similarity subgrid-scale models as deduced from measurements in a turbulent jet", *J. Fluid Mech.*, **275**, 83.

6 Statistical aspects

Trends in Mathematics, © 1999 Birkhäuser Verlag Basel/Switzerland

Turbulent fluctuations, predictability, and action principles

G. L. Eyink

1 Fluctuations and predictability

Strong fluctuations are a characteristic property of turbulence. Even small noise, which may arise from molecular fluctuations, ignorance of subgrid modes, etc. is magnified by nonlinearity [1, 2]. This fact has two important implications for predictability of turbulent flow.

First, predictions for mean values or for resolved modes in an LES computation are necessarily subject to large fluctuations in individual realizations. The recognition of this fact in meteorology has led to the practice of "ensemble forecasting" [3], in which a single operational forecast is replaced by an entire ensemble generated from suitable perturbations in the initial data. Likewise, models of long-term climate variability have been introduced in which slow changes of climate are explained as the integrated response to continuous random excitation by short period disturbances [4]. It is clear that some knowledge of turbulence-generated random noise is required as an input in order to achieve a statistically successful predictive capability in these areas.

A second implication is that correct prediction even of mean values requires a good description of the fluctuations. Indeed, the fluctuations in turbulent flow are of the same order as the means and make a very significant contribution to their values. The simplest manifestation of this fact is the Reynolds stress $\tau_{ij} = \langle u_i' u_j' \rangle$ in the averaged equations for the mean velocities, which gives the direct contribution of the fluctuation field u'. It is well-understood that the dissipation and transport produced by such small-scale turbulent fluctuations are the most crucial quantities to model for a successful prediction of the mean flow properties [5].

Motivated by such concerns, we have recently initiated a new approach to the theory of turbulent noise [6]–[10]. Our approach is based upon a formulation of fluctuation theory in terms of *action functionals*, pioneered by Onsager [11, 12]. Although less well known than the methods of stochastic Langevin equations or Fokker-Planck-type equations for probability distributions, the action functional provides another complete formulation of fluctuation theory. As reviewed below, there is a *Rayleigh-Ritz method* which uses the Navier-Stokes equations variationally to calculate the turbulent actions. The Rayleigh-Ritz method is very closely related to traditional *moment-closure* schemes and may be used to further exploit and evaluate them. Indeed, a key advantage of the action functional is that it provides a variational principle for the mean values, which generalizes the hydrodynamic *least-dissipation principle* of Rayleigh. It is not surprising that there should be a deep connection between dissipation and fluctuations, since both arise together from random small-scale motions. The variational principles for mean values provide new *realizability conditions*, which may be exploited in modelling as diagnostics of closure predictions. Preliminary investigations have shown [8] that a qualitatively correct description of fluctuations is necessary for an accurate prediction of mean values, exactly as suggested by our previous considerations. These points shall all be elaborated below.

2 Action principles for turbulent fluctuations

In this section we shall briefly recall the probabilistic basis of the action principles in turbulence. The large-deviations theorem of Cramér [13] (see also Frisch [14], Section 8.6.4), pertains to the fluctuations in the empirical average over N i.i.d. versions $Z_i(t)$ of a random process:

$$\overline{Z}_N(t) = \frac{1}{N}\sum_{i=1}^{N} Z_i(t).$$

The theorem, under the condition that $\langle\exp\left[\int dt\, h(t)Z(t)\right]\rangle < \infty$ (finite exponential moments), gives the probability of a fluctuation history $z(t)$ as

$$\text{Prob}\left(\overline{Z}_N(t) \approx z(t) : -\infty < t < +\infty\}\right) \sim \exp\left(-N\cdot\Gamma[z]\right).$$

This result, under modest assumptions on the tails of distributions, applies to the empirical averages that are formed from independent trials in laboratory experiments on physical systems. According to the Cramér theory, the rate function $\Gamma[z]$ is obtained as a Legendre transform

$$\Gamma[z] = \sup_h \left(< h, z > -W[h]\right)$$

of the cumulant-generating functional

$$W[h] = \log\langle\exp\left[\int dt\, h(t)Z(t)\right]\rangle.$$

In terminology borrowed from quantum field-theory, $\Gamma[z]$ is called the *effective action* on Z-histories [15]. There follow important realizability conditions: (i) (positivity): $\Gamma[z]\geq 0$, (ii) (unicity of minimizer): $\Gamma[z]=0$ iff $z(t)=\overline{z}(t)$, the ensemble-mean, and (iii) (convexity): $\lambda\Gamma[z_1]+(1-\lambda)\Gamma[z_2]\geq\Gamma[\lambda z_1+(1-\lambda)z_2]$, $0<\lambda<1$. In particular, $\Gamma[z]$ determines the mean history $\overline{z}(t)$ by a variational principle of *least effective action.* $\Gamma[z]$ also determines in distribution the entire random process $Z(t)$. For example, the effective action is a generating functional for irreducible correlation functions of $Z(t)$,

$$\Gamma_p(t_1,\ldots,t_p) = \left.\frac{\delta^p\Gamma[z]}{\delta z(t_1)\cdots\delta z(t_p)}\right|_{z=\overline{z}},$$

and these determine in turn all the connected (cumulant) functions. Cf. [15].

For statistically steady states it is natural to conjecture a similiar fluctuation principle, namely, that the empirical time-average

$$\overline{Z}_T \equiv \frac{1}{T}\int_0^T dt\, Z(t)$$

should obey

$$\text{Prob}\left(\{\overline{Z}_T \approx z\}\right) \sim \exp\left(-T \cdot V[z]\right).$$

This is a refinement of the standard ergodic hypothesis. Existence of a time-extensive limit of the effective action, with

$$V[z] \equiv \lim_{T\to\infty} \frac{1}{T}\Gamma[\{z(t) = z : 0 < t < T\}],$$

is a sufficient condition for its validity. Realizability conditions on the *effective potential* V (positivity, unicity of minimizer, convexity) follow as a consequence. It is proved [16, 17] that the principle holds for uniformly hyperbolic dynamical systems. Thus, the refined hypothesis holds within Ruelle's picture of turbulence, as described by an invariant measure on a strange attractor [18]. Actually, the hypothesis has also been proved by Donsker and Varadhan for a wide class of ergodic Markov processes [19]. Because the description of physical systems must involve random molecular noise as well as deterministic dynamics, such results are at least as relevant as those for pure dynamical systems. Chaotic behavior of the underlying dynamics is really only required to ensure validity of the theory in the zero-noise limit.

3 Rayleigh-Ritz method and moment closures

Since the work of Martin, Siggia, and Rose [20] some quarter century ago, it has been recognized that statistical dynamics may be reformulated as an equivalent (MSR) quantum field theory. It was even mentioned by those authors that variational methods could be devised within their field-theoretic scheme. Yet, it is only recently that we have formulated a *Rayleigh-Ritz method* analogous to those employed in quantum theory [7]. This method provides one approach, using the Navier-Stokes dynamics, to calculate the effective actions. The basis of the method is a theorem characterizing the effective action as a stationary point of the "nonequilibrium action":

$$\Gamma[z] = \text{st.pt} \int_{-\infty}^{+\infty} dt\, \langle \mathcal{A}(t), (\partial_t - \widehat{L})\mathcal{P}(t)\rangle$$

varying over $\mathcal{A} \in L^\infty, \mathcal{P} \in L^1$ with constraints

$$\begin{aligned} \langle \mathcal{A}(t), \mathcal{P}(t)\rangle &= 1 \\ \langle \mathcal{A}(t), \widehat{Z}\mathcal{P}(t)\rangle &= z(t). \end{aligned}$$

The idea comes from earlier work of Jackiw and Kerman [21] in quantum field theory. In this theorem the right trial function $\mathcal{P}[\mathbf{v},t]$ is a normalized state and the left trial function $\mathcal{A}[\mathbf{v},t]$ is a bounded observable. $\widehat{L}$ is the Liouville-Hopf operator generating time-evolution of PDF's and $\widehat{Z}$ is the operator of multiplication by the observable $Z[\mathbf{v}]$. The effective potential is likewise characterized variationally as the stationary point of the "non-equilibrium Hamiltonian" in MSR field theory:

$$V[z] = \text{st.pt}_{\substack{\langle \mathcal{A},\mathcal{P}\rangle = 1 \\ \langle \mathcal{A},\widehat{Z}\mathcal{P}\rangle = z}} \left[-\langle \mathcal{A},\widehat{L}\mathcal{P}\rangle\right].$$

See [7]. This is an MSR analogue of Symanzik's theorem [22] in quantum field-theory.

These theorems are the basis of a practical Rayleigh-Ritz method of approximation, which has close ties to the method of *moment closures*. The ingredients of such an approximation scheme are a guess for the probability density function in phase space, or *PDF Ansatz*

$$\mathcal{P}[\mathbf{v}] = P[\mathbf{v};\mathbf{M}]$$

as a right trial function, and a *superposition of moment functions*

$$\mathcal{A}[\mathbf{v}] = \sum_i a_i \psi_i[\mathbf{v}]$$

as a left trial function. The functions $\psi_i[\mathbf{v}]$ are the same as those which appear in the moment closure, plus the constant function $\psi_0[\mathbf{v}] \equiv 1$. The PDF *Ansatz* is conveniently parameterized by the mean values M_i which it yields for those functions. The vector variables $\mathbf{M},\mathbf{a}$ are the variational parameters of the method. Varying over the trial functions in the nonequilibrium action yields the time-dependent variational equations

$$\dot{\mathbf{a}} + \frac{\partial V_0}{\partial \mathbf{M}}(\mathbf{M},h)a_0 + \left(\frac{\partial \mathbf{V}}{\partial \mathbf{M}}\right)^{\top}(\mathbf{M},h)\cdot\mathbf{a} = V_0(\mathbf{M},h)\mathbf{a}, \tag{1}$$

$$\dot{\mathbf{M}} = \mathbf{V}(\mathbf{M},h) - V_0(\mathbf{M},h)\mathbf{M}, \tag{2}$$

$$a_0 + \mathbf{a}\cdot\mathbf{M} = 1. \tag{3}$$

$h(t)$ is a Lagrange multiplier enforcing the constraint on $\langle \Psi^L(t), \widehat{Z}\Psi^R(t)\rangle = \mathbf{a}\cdot\mathbf{Z}(\mathbf{M})$ (with $Z_i(\mathbf{M}) \equiv \langle \psi_i Z\rangle_{\mathbf{M}}$). When $h(t) \equiv 0$, these equations reduce to the standard moment closure equations:

$$\dot{\mathbf{M}} = \mathbf{V}(\mathbf{M}),$$

with $V_i(\mathbf{M}) \equiv \langle \widehat{L}^* \psi_i \rangle_{\mathbf{M}}$. Note that, in general, $\mathbf{V}(\mathbf{M},h) = \mathbf{V}(\mathbf{M}) + h\mathbf{a}\cdot\mathbf{Z}(\mathbf{M})$. From the solutions $\mathbf{a}_*(t), \mathbf{M}_*(t)$ of the variational problem the approximate effective action is obtained as

$$\Gamma[z] = \int_0^\infty dt\ \mathbf{a}_*(t)\cdot \left[\dot{\mathbf{M}}_*(t) - \mathbf{V}\left(\mathbf{M}_*(t)\right)\right], \tag{4}$$

with $h(t)$ chosen so that $\mathbf{a}_*(t)\cdot\mathbf{Z}(\mathbf{M}_*(t)) = z(t)$. A similiar approximation scheme holds also for the effective potential, setting all time-derivatives equal to zero, with

$$V_*[z] = -\mathbf{a}_*\cdot\mathbf{V}\left(\mathbf{M}_*\right).$$

For more details, see [6].

4 An example: Freely-decaying turbulence

The scheme is best illustrated by a simple example. Following [9], we consider the decay of homogeneous, isotropic turbulence under the Navier-Stokes dynamics at high Reynolds number. The theory of the mean energy decay is nicely reviewed in [14], Section 7.7. The energy spectrum is believed to rapidly attain a quasi-equilibrium form, which can be modeled by

$$E(k,t) = \begin{cases} Ak^m & k < k_L(t) \\ \alpha\varepsilon^{2/3}(t)k^{-5/3} & k_L(t) < k < k_d(t) \\ 0 & k > k_d(t) \end{cases} \tag{5}$$

The Kolomogorov $\frac{5}{3}$ spectrum with mean dissipation $\varepsilon(t)$ occurs over the inertial subrange $k_L(t) < k < k_d(t)$, while the dissipation range spectrum for $k > k_d(t)$ is set to zero as a simplification. The key assumption is that the spectrum for $k < k_L(t)$ is a power-law with a time-independent coefficient A. Frisch has called this the *principle of permance of large eddies*[14]. The parameters $\varepsilon(t), k_L(t)$ and $k_d(t)$ in this model are related at high Reynolds number by $k_L(t) = \left(\frac{\alpha}{A}\varepsilon(t)\right)^{\frac{3}{3m+5}}$ and $k_d(t) = \left(\frac{4}{3\alpha\nu}\right)^{3/4}\varepsilon^{1/4}(t)$. One independent parameter remains, which may be taken to be the mean energy $E_*(t) = \frac{1}{2}\langle v^2(t)\rangle$. Within the above model this obeys

$$\dot{E}_*(t) = -\Lambda_m \cdot E_*^p(t). \tag{6}$$

with $\Lambda_m^{-1} = \alpha^{3/2}\left(\frac{1}{m+1}+\frac{3}{2}\right)^{\frac{3m+5}{2m+2}} A^{\frac{1}{m+1}}$ and $p = \frac{3m+5}{2m+2}$. A version of this closure equation was derived already by Kolmogorov [23].

The same set of hypotheses may be further exploited in the Rayleigh-Ritz method. Within a Gaussian *Ansatz* the right trial function is taken as:

$$\mathcal{P}[\mathbf{v}] = \frac{1}{\mathcal{N}(t)} \exp\left[-\frac{1}{2}\int d^3\mathbf{k}\hat{v}_i^*(\mathbf{k})\left(E^{-1}\right)_{ij}(\mathbf{k},t)\hat{v}_j(\mathbf{k})\right]$$

with $E_{ij}(\mathbf{k},t) = \left(\delta_{ij} - \frac{k_i k_j}{k^2}\right) E(k,t)$, where $E(k,t)$ is given by the above model spectrum with a mean energy $K(t)$. Corresponding to the Kolmogorov closure equation, one takes as left trial state

$$\mathcal{A}[\mathbf{v}] = a_0(t)1 + a_1(t) \cdot \widehat{K}(\mathbf{x};\mathbf{v})$$

with $\widehat{K}(\mathbf{x};\mathbf{v}) \equiv \frac{1}{2}v^2(\mathbf{x})$. Carrying out the variation of the action under the constraints $\langle \mathcal{A}(t), \mathcal{P}(t)\rangle = 1, \langle \mathcal{A}(t), \widehat{K}\mathcal{P}(t)\rangle = E(t)$, one then obtains the Rayleigh-Ritz approximation to the effective action of the energy history:

$$\Gamma^{(Gauss)}[E] = \frac{3}{2(p-2)\Lambda_m}\int_0^\infty dt\, \frac{\left(\dot{E}(t) + \Lambda_m \cdot K^p(t)\right)\left(\dot{K}(t) + \Lambda_m \cdot K^p(t)\right)}{K^{p+1}(t)},$$

where the variational parameter $K(t)$ satisfies

$$\Lambda_m \cdot K^p(t) + \dot{E}(t) = (p-2)\Lambda_m \cdot (E(t) - K(t)) \cdot K^{p-1}(t).$$

It is not hard to prove that this equation has a unique solution for $K(t)$ whenever $E(t) > 0, \dot{E}(t) < 0$. Predictions of all p-time correlations of the energy fluctuation $\delta E(t) \equiv E(t) - E_*(t)$ can be calculated from this action.

Some insight into these results can be obtained by evaluating the action to quadratic order in δE, as:

$$\Gamma^{(Gauss)}[E] = \frac{3}{8(p-1)\Lambda_m}\int_0^\infty dt\, \frac{\left(\delta\dot{E}(t) + \Lambda_m \cdot pE_*^{p-1}(t)\delta E(t)\right)^2}{E_*^{p+1}(t)} + O(\delta E^3).$$

This is precisely the form of an *Onsager-Machlup action* [12]. The predictions of all 2-time statistics are therefore equivalent to those of the Langevin equation

$$\delta\dot{E}(t) + \Lambda_m \cdot pE_*^{p-1}(t)\delta E(t) = \sqrt{2R_*(t)}\eta(t) \tag{7}$$

obtained by linearizing the energy-decay equation around the predicted mean and by adding a white-noise random force $\eta(t)$ with coefficient given by

$$R_*(t) = \frac{2(p-1)}{3}\varepsilon_*(t)E_*(t). \tag{8}$$

In particular, the effective action is realizable to this order and all necessary positivity properties must hold. The fact that the linearized closure equation appears to describe the decay of small fluctuations of the energy from its mean is closely related to the *regression hypothesis* in [11]. The random term represents a turbulence-generated force or "eddy noise". The proportionality of the noise covariance to the mean energy dissipation is a *fluctuation-dissipation relation*, analogous to that in equilibrium.

5 Discussion and further results

It may be shown that, in *any* moment closure, the effective action for the moment variables themselves has always the Onsager-Machlup form to quadratic order [24]. Thus, a formal Langevin model holds always for small fluctuations around the mean. The "Onsager coefficient" matrix appearing there is obtained by a linearization of the closure equation and is related to transport and dissipation characteristics of the flow. Thus, the covariance of the turbulent noise in general satisfies a fluctuation-dissipation relation, as intuition would suggest. However, the predicted noise covariance need not be positive in general! The fact that $R_*(t) > 0$ in Eq.(8) above is a consistency check on the *Ansatz* employed in the calculation. The effective action in this case satisfies necessary realizability conditions. However, other superficially consistent PDF *Ansätze* may lead to negative noise covariances, violating realizability.

A case in point is the *amplitude mapping closure* for passive scalar advection [25]. The dynamical problem of interest is a chemically reacting but passively advected scalar field,

$$\partial_t \Theta + (\mathbf{v}\cdot\nabla)\Theta = Q(\Theta) + \kappa \triangle \Theta,$$

with Q the reaction rate. In the mapping closure for this problem there are infinitely-many moment functions labelled by θ,

$$\psi_\theta[\Theta] = \delta(\Theta(\mathbf{x}) - \theta)$$

whose ensemble-averages give the 1-point probability density function (PDF): $P(\theta) = \langle \delta(\Theta(\mathbf{x}) - \theta)\rangle$. The moment equation is

$$\partial_t P(\theta, t) + \frac{\partial}{\partial \theta}\left([Q(\theta) + \lambda(\theta, t)]P(\theta, t)\right) = 0,$$

where $\lambda(\theta) = \kappa\langle \triangle\theta | \theta\rangle$ is the conditional mean diffusion. This latter is the quantitity for which a closure must be provided in terms of P. The mapping *Ansatz* is

$$\Theta(\mathbf{x}, t) \sim X(\Theta_0(\mathbf{x}, t), t)$$

where Θ_0 is a reference Gaussian scalar field, with specified time-dependence, and X is the *mapping function*. The PDF is then determined from X via $P(\theta, t) = P_0(\theta_0, t)/X'(\theta_0, t)$. This provides a closure of the conditional diffusion, as

$$\lambda(\theta, t) = \kappa\xi_0^2(t)\left[-\frac{\theta}{\theta_0^2(t)}X'(\theta_0, t) + X''(\theta_0, t)\right],$$

with $\xi_0^2(t) = \langle|\nabla\Theta_0(t)|^2\rangle$. Unlike our previous example, the mapping *Ansatz* is a highly non-Gaussian model, when X is nonlinear.

According to our general remarks above, the Rayleigh-Ritz effective action of P itself in the mapping closure must have the Onsager-Machlup form to quadratic order. The equivalent Langevin equation for the PDF fluctuation δP, may be shown to be [26]:

$$\delta\dot{P}(\theta,t)+(Q(\theta)\delta P(\theta,t))'+2(\lambda_*(\theta,t)\delta F(\theta,t))''-(\chi_*(\theta,t)\delta F(\theta,t))'''=J'(\theta,t).$$

The left-hand side of the equation is just the *linearized mapping closure equation.* Here $F(\theta,t)$ is the cummulative distribution function associated to P and $\chi(\theta,t)=\kappa\langle|\nabla\theta|^2\rangle|\theta\rangle$ is the conditional mean dissipation in the closure. The term $J(\theta,t)$ on the right-hand side is a random *probability current.* It has zero mean and covariance given by

$$\langle J(\theta,t)J(\bar{\theta},\bar{t})\rangle=\mathcal{J}_*(\theta,\bar{\theta})\delta(t-\bar{t})$$

with the covariance in θ-space given as

$$\begin{aligned}\mathcal{J}_*(\theta,\bar{\theta})=&2\chi_*(\theta,t)P_*(\theta,t)\cdot\delta(\theta-\bar{\theta})+2[\lambda_*'(\theta,t)+\lambda_*'(\bar{\theta},t)][F_*(\theta\wedge\bar{\theta},t)-F_*(\theta,t)F_*(\bar{\theta},t)]\\&-[\chi_*''(\theta,t)+\chi_*''(\bar{\theta},t)][F_*(\theta\wedge\bar{\theta},t)-F_*(\theta,t)F_*(\bar{\theta},t)]-[\chi_*'(\theta\wedge\bar{\theta},t)P_*(\theta\wedge\bar{\theta},t)\\&-\chi_*'(\bar{\theta},t)P_*(\bar{\theta},t)F_*(\theta,t)-\chi_*'(\theta,t)P_*(\theta,t)F_*(\bar{\theta},t)].\end{aligned}\quad(9)$$

Note that $\theta\wedge\bar{\theta}=\min\{\theta,\bar{\theta}\}$, the minimum value. As in general, we see here, too, that noise strength is related to mean transport and dissipation characteristics.

However, we have shown in [26] that the noise covariance $\mathcal{J}_*$ above has exactly one negative eigenvalue! This indicates a "statistical instability" of the mapping *Ansatz* when it is applied to the dynamics of fluctuations. The realizability failure for amplitude mapping closure is correlated with a very poor quantitative prediction of multi-time scalar statistics. In general, the realizability conditions on the effective actions and potentials seem to be sensitive *a priori* measures of the quality of predictions. This is probably one of the most practically useful features of the method. For further examples and discussion, see [8, 10].

References

[1] R. H. Kraichnan, "Instability in fully developed turbulence," Phys. Fluids **13** 569 (1970).

[2] D. Ruelle, "Microscopic fluctuations and turbulence," Phys. Lett. A **72** 81 (1979).

[3] Z. Toth and E. Kalnay, "Ensemble forecasting at NMC: the generation of perturbations," Bull. Am. Meteor. Soc. **74** 2317 (1993).

[4] K. Hasselmann, "Stochastic climate models, I, " Tellus **XXVIII** 473 (1976).

[5] C. G. Speziale, "Analytical methods for the development of Reynolds-stress closures in turbulence," Ann. Rev. Fluid Mech. **23** 107 (1991).

[6] G. L. Eyink, "Turbulence noise," J. Stat. Phys. **83** 955 (1996).

[7] G. L. Eyink, "Action principle in nonequilibrium statistical dynamics," Phys. Rev. E **54** 3419 (1996).

[8] G. L. Eyink and F. J. Alexander, "Turbulence fluctuations and new universal realizability conditions in modelling," Phys. Rev. Lett. **78** 2563 (1997).

[9] G. L. Eyink, "Fluctuations in the irreversible decay of turbulent energy," Phys. Rev. E **56** 5413 (1997).

[10] G. L. Eyink and F. J. Alexander, "Predictive turbulence modelling by variational closure," J. Stat. Phys., to appear (1998).

[11] L. Onsager, "Reciprocal relations in irreversible processes, II," Phys. Rev. **38** 2265 (1931).

[12] L. Onsager and S. Machlup, "Fluctuations and irreversible processes," Phys. Rev. **91** 1505 (1953).

[13] H. Cramér, "Sur un noveau théorème-limite de la théorie des probabilités," Actualités Scientifiques et Industrielles **736** 23 (1938).

[14] U. Frisch, *Turbulence.* (Cambridge University Press, 1995).

[15] C. Itzykson and J.-B. Zuber, *Quantum Field Theory.* (McGraw Hill, New York, 1985).

[16] L.-S. Young, "Some large deviations results for dynamical systems," Trans. Am. Math. Soc. **318** 525 (1990).

[17] Y. Kifer, "Large deviations in dynamical systems and stochastic processes," Trans. Am. Math. Soc., **321** 505 (1990).

[18] D. Ruelle, "Measures describing a turbulent flow," Ann. N. Y. Acad. Sci. **357** 1 (1980).

[19] S. R. S. Varadhan, *Large Deviations and Applications.* (SIAM, Philadelphia, 1984).

[20] P. C. Martin, E. D. Siggia, and H. A. Rose, Phys. "Statistical dynamics of classical systems," Rev. A **8** 423 (1973).

[21] R. Jackiw and A. Kerman, "Time-dependent variational principle and effective action," Phys. Lett., **71A** 158 (1979).

[22] K. Symanzik, "Renormalizable models with simple symmetry breaking," Commun. Math. Phys. **16** 48 (1970).

[23] A. N. Kolmogorov, "On degeneration of isotropic turbulence in an incompressible viscous liquid," Dokl. Akad. Nauk. SSSR, **31** 538 (1941).

[24] G. L. Eyink, "Rayleigh-Ritz effective action and Onsager-Machlup action," in preparation.

[25] H. Chen, S. Chen, and R. H. Kraichnan, "Probability distribution of a stochastically advected scalar field," Phys. Rev. Lett. **65** 575 (1990).

[26] G. L. Eyink, "A Rayleigh-Ritz study of passive scalar decay with amplitude mapping closure: analytical results," in preparation.

Trends in Mathematics, © 1999 Birkhäuser Verlag Basel/Switzerland

Instantons in the theory of turbulence

V. Lebedev

Abstract

We describe a method for finding the non-Gaussian tails of probability distribution function (PDF) for solutions of stochastic differential equations, such as convection equation for a passive scalar, random driven Navier-Stokes equation etc. Existence of such tails is generally regarded as a manifestation of intermittency phenomenon. Our formalism is based on the WKB approximation in the functional integral for the conditional probability of a large fluctuation. Then the main contribution to the functional integral is given by a coupled field-force configuration – *instanton*. We argue that the tails of the single-point velocity probability distribution function (PDF) are generally non-Gaussian in developed turbulence. By using instanton formalism for the Navier-Stokes equation, we establish the relation between the PDF tails of the velocity and those of the external forcing. In particular, we show that a Gaussian random force having correlation scale L and correlation time τ produces velocity PDF tails $\ln \mathcal{P}(v) \propto -v^4$ at $v \gg v_{\rm rms}, L/\tau$. For a short-correlated forcing when $\tau \ll L/v_{\rm rms}$ there is an intermediate asymptotics $\ln \mathcal{P}(v) \propto -v^3$ at $L/\tau \gg v \gg v_{\rm rms}$. We consider the tails of probability density function (PDF) for the velocity that satisfies Burgers equation driven by a Gaussian large-scale force. The instantonic calculations show that for the PDFs of velocity and its derivatives $u^{(k)} = \partial_x^k u$, the general formula is found: $\ln \mathcal{P}(|u^{(k)}|) \propto -(|u^{(k)}|/\mathrm{Re}^k)^{3/(k+1)}$. We consider high-order correlation functions of the passive scalar in the Kraichnan model. Using the instanton formalism we find the exponents ζ_n of the structure functions S_n for $n \gg 1$ at the condition $d\zeta_2 \gg 1$ (where d is the dimensionality of space). At $n < n_c$ (where $n_c = d\zeta_2/[2(2-\zeta_2)]$) the exponents are $\zeta_n = (\zeta_2/4)(2n - n^2/n_c)$, while at $n > n_c$ they are n-independent: $\zeta_n = \zeta_2 n_c/4$. Besides ζ_n, we also estimate n-dependent factors in S_n and critical behavior of S_n at n close to n_c.

1 Introduction

Anomalous scaling related to intermittency phenomenon is probably the central problem of the theory of turbulence. In 1941 Kolmogorov formulated his famous theory of developed turbulence [1], where scaling behavior of different correlation functions of the turbulent velocity was predicted. Experimentally one observes deviations from the scaling exponents, proposed by Kolmogorov. It is recognized that the deviations are related to rare strong fluctuations which give the main contribution into the correlation functions [2, 3, 4]. This phenomenon which is usually called intermittency is the most striking peculiarity of developed turbulence.

The intermittency phenomenon (reflected in non-Gaussian, scaling-violating tails of PDF) is believed to be the hardest part of the yet to be built theory of turbulence. Neither the physical mechanism nor the mathematical properties of rare fluctuations responsible for intermittency are known. Now, what is the most likely force which can lead to the given rare fluctuation of the field? Our main idea is that such force is not random at all. It satisfies the well defined equation, which follows from the WKB approximation in the functional integral [5]. Asymptotically, fluctuations of the force around this most likely one are negligible. In this respect, the method is similar to the "optimal fluctuation" method used at treating properties of a solid with quenched disorder (see e.g. the book [6]). Similar approach has been used at summing high-order terms of the perturbation series in the quantum field theory [7].

The problem under consideration is quite general, it can be formulated for any field governed by a nonlinear dynamic equation and driven by a random "force". Generally, the PDF of the field depends both on the statistics of the driven force and on the form of the dynamical equation. Here, we are interested in the second dependence so that we assume the force to be Gaussian. Because of nonlinearity, the PDF of the field is non-Gaussian even for a Gaussian random force. Note that strong intermittency appears also for linear problems with "multiplicative noise", for instance, for a passive scalar advected by a random velocity field.

We start with the dynamical equation

$$\partial_t u + \mathcal{L}(u) = \phi\,, \tag{1}$$

that controls the evolution of a field $u(t, \mathbf{r})$ under the action of a random "force" $\phi(t, \mathbf{r})$. Here $\mathcal{L}(u)$ is a nonlinear term, it can be thought to be local in space. Generally, both the field u and the force ϕ have a number of components. The Gaussian statistics of the force ϕ is completely characterized by the pair correlation function

$$\langle \phi(t_1, \mathbf{r}_1)\phi(t_2, \mathbf{r}_2)\rangle = \Xi(t_1 - t_2, \mathbf{r}_1 - \mathbf{r}_2)\,. \tag{2}$$

In principle, the relations (1,2) contain all the information about the statistics of u.

The equation (1) describes e.g. thermal fluctuations in hydrodynamics where it is reduced to the well known Langevin equation [8]. Then ϕ is short-correlated in time and in space that is it can be treated as a white noise. For some systems, this thermal noise produces remarkable dynamical effects. Some examples are collected in the book [9]. Here, we will be interested in turbulence where ϕ is an external "force" correlated on large scales in space. Turbulence was first treated in terms of the equation (1) by Wyld [10] who formulated the diagram technique as a perturbation series with respect to the nonlinear term in the Navier-Stokes equation. The diagram technique cannot be applied to our problem since we are interested in non-perturbative effects. Nevertheless we can use the functional that generates the technique since it is a non-perturbative object. Such generating functional was introduced in [11, 12, 13], for the equation (1) it has the form

$$\mathcal{Z}(\lambda) \equiv \left\langle \exp\left(i \int dt\, d\mathbf{r}\, \lambda u \right) \right\rangle = \int \mathcal{D}u\mathcal{D}p \exp\left(i\mathcal{I} + i \int dt\, d\mathbf{r}\, \lambda u \right) , \quad (3)$$

where p is an auxiliary field and the effective action is

$$\mathcal{I} = \int dt\, d\mathbf{r}\, p[\partial_t u + \mathcal{L}(u)] + \frac{i}{2} \int dt\, dt'\, d\mathbf{r}\, d\mathbf{r}'\, \Xi(t-t', \mathbf{r}-\mathbf{r}')pp' . \quad (4)$$

The coefficients of the expansion of $\mathcal{Z}$ in λ are the correlation functions of u. The auxiliary field p determines response functions of the system, for instance, the linear response function (Green function) is $G = \langle up \rangle$. Note the remarkable property [14]

$$\int \mathcal{D}u\mathcal{D}p \exp(i\mathcal{I}) = 1 ,$$

related to causality. That is the reason why the normalization constant is unity in (3). This makes it possible to average directly $\mathcal{Z}$ over any additional random field if necessary.

The asymptotics of $\mathcal{Z}(\lambda)$ at large λ is determined by the saddle-point configuration (usually called classical trajectory or instanton) which should satisfy the following equations obtained by varying the argument of the exponent in (3) with respect to u and p

$$\partial_t u + \mathcal{L}(u) = -i \int dt'\, d\mathbf{r}'\, \Xi(t-t', \mathbf{r}-\mathbf{r}')p(t', \mathbf{r}') , \quad (5)$$

$$\partial_t p - \frac{\delta \mathcal{L}}{\delta u} p = \lambda . \quad (6)$$

Solutions of the equations (5,6) are generally smooth functions of t and $\mathbf{r}$. Comparing the equations (1) and (5) we conclude that the right hand side of the equation (5) just describes a special force configuration necessary to produce the instanton. If u_{inst} is a solution of (5,6) then asymptotically at large λ

$$\delta \ln \mathcal{Z}(\lambda)/\delta\lambda = iu_{\text{inst}} . \quad (7)$$

Let us discuss the boundary conditions for the saddle-point equations. The equation (5) implies that we should fix the value u_{in} for the field u at the initial time t_{in}. Due to the presence of dissipation u_{in} is forgotten for a finite time and the answer doesn't depend on it. In the following we believe that $u_{\text{in}} = 0$. Contrary, a boundary condition for the field p is implied at the remote future since, as follows from (6), it propagates backward in time. Minimization of the action generally requires $p \to 0$ at $t \to \infty$.

If one is interested in the simultaneous statistics of u then it is convenient to chose the function λ as

$$\lambda(t, \mathbf{r}) = y\delta(t)\beta(\mathbf{r}), \tag{8}$$

where y is a factor and β is an appropriate function of $\mathbf{r}$ depending on what spatial correlation functions we are going to study. In this case, we should find the solution for p satisfying the rule: $p = 0$ at $t > 0$. The system (5,6) is thus to be treated for $t < 0$ only. That corresponds to the causality principle since only processes occurring in the past could influence the value of the simultaneous correlation functions at $t = 0$. The formal ground for the rule follows from the consideration of the problem in the restricted time interval $t < t_0$. Then the minimization of $\mathcal{I} + \int dt\, d\mathbf{r}\, \lambda u$ over the final value $u(t_0)$ gives $p(t_0) = 0$ because of the boundary term originating from $\int dt\, d\mathbf{r}\, p\partial_t u$. Substituting (8) into (6) we find

$$p(t = -0, \mathbf{r}) = -y\beta(\mathbf{r}), \tag{9}$$

what can be interpreted as the boundary condition for the field p which propagates backward in time. Thus, we will treat the instantonic equations for the time interval $-\infty < t < 0$ with the boundary conditions (9) and $u \to 0$ at $t \to -\infty$.

One may be interested also in the probability distribution function (PDF) for the field u. It can be expressed via the generating functional $\mathcal{Z}(\lambda)$. It is convenient to treat the PDF of a paticular object

$$u_\beta = \int d\mathbf{r}\, \beta(\mathbf{r}) u(\mathbf{r}).$$

The PDF of the object is written as

$$\mathcal{P}(u_\beta) = \int \frac{dy}{2\pi} \exp(-iyu_\beta)\mathcal{Z}, \tag{10}$$

where the expression (8) should be substituted into $\mathcal{Z}$. We expect that the behavior of $\mathcal{P}(u)$ at large u as well as the behavior of $\mathcal{Z}(\lambda)$ for large λ is associated with some saddle-point configurations. Generally, the configurations are not always the same for both (3) and (10). Indeed, we see from (10) that the tail of $\mathcal{P}(u)$ at large u corresponds to a large value of $\delta \ln \mathcal{Z}(\lambda)/\delta\lambda$ which is related to large λ only if the tails of both PDF and the generating functional decay faster than exponent. One can also be interested directly in the high-order moments

$$\langle u_\beta^n \rangle = \int \mathcal{D}u\, \mathcal{D}p \exp\left(i\mathcal{I} + n \ln u_\beta\right).$$

The saddle-point equations for $\langle u_\beta^n \rangle$ coincide with (5,6) if to substitute

$$\lambda \to -\frac{in}{u_\beta}\delta(t)\beta(\mathbf{r}).$$

2 Single-point velocity distribution in turbulence

It is a common misconception that the single-point velocity PDF in developed turbulence is generally close to Gaussian [2, 3]. One may think that Gaussianity appears because large-scale motions (that give the main contribution into velocity statistics at a point) are connected to external forcing usually believed to be Gaussian. However, a nonlinear system often demonstrates an output statistics substantially different from the statistics of the input (be it initial conditions or external force). It is generally difficult to relate, say, the PDF of the velocity to that of the force. Our aim here is to show that it is possible, nevertheless, to relate the probabilities of rare fluctuations that is to relate the tails of the PDF-s of the velocity and the force respectively. The main results of the section are published in [15].

Let us first explain the simple physics involved and formulate the predictions following from physical arguments, then we develop the formalism which gives the predicted PDF tails. Qualitatively, one may describe the interplay between external force and nonlinearity in the following way. Force f pumps velocity $v \sim ft$ until the time $t_* \sim L/v$ when nonlinearity restricts the growth. The relation between velocity and forcing can thus be suggested as follows: $v^2 \sim fL$. Therefore, velocity's PDF can be obtained by substituting $f \sim v^2/L$ into force's PDF $\mathcal{P}_f$: $\mathcal{P}(v) \sim \mathcal{P}_f(v^2/L)$. Those arguments presume that t_* is less than the correlation time τ of forcing. If opposite is the case $t_* \gg \tau$ then the pumping is effectively Gaussian and the law of velocity growth is different $v^2 \sim f^2 t\tau$, so that $v^3 \sim f^2 L\tau$ and the velocity's PDF is $\mathcal{P}(v) \sim \exp[-(v/v_{\rm rms})^3]$.

Let us stress that the above arguments can be only applied to rare events with velocity and force being much larger than their root-mean-square values when the influence of background fluctuations can be neglected. The above predictions are thus made for PDF tails. A rigorous way to describe rare fluctuations is the instanton formalism developed for turbulence in [5] and sketched in Introduction.

We start with the forced Navier-Stokes equation

$$\partial_t v_\alpha + v_\beta \nabla_\beta v_\alpha - \nu \nabla^2 v_\alpha + \nabla_\alpha P = f_\alpha \,, \tag{11}$$

where $\mathbf{f}$ is a random force pumping the energy into the system and ν is the viscosity coefficient. Incompressibility is assumed so that $\operatorname{div} \mathbf{v} = \operatorname{div} \mathbf{f} = 0$. The field P in (11) is the pressure divided by the mass density ρ. Velocity correlation functions can be presented as path integrals which form is determined by the statistics of pumping. Let us first consider a Gaussian forcing with the correlation function $\langle f_\alpha(t,\mathbf{r}) f_\beta(t',\mathbf{r}')\rangle = \Xi_{\alpha\beta}(t-t',\mathbf{r}-\mathbf{r}')$ which is assumed to decay on the scale τ as a function of the first argument and on the scale L as a function of the second one. Then moments of the velocity can be written as path integrals:

$$\langle v^{2n}\rangle = \int \mathcal{D}p\,\mathcal{D}v\,\mathcal{D}P\,\mathcal{D}Q\, \exp\left(i\mathcal{I} + 2n \ln v\right) , \tag{12}$$

where $\mathbf{p}$ is an auxiliary field and the effective action has the following form [11, 12, 13]

$$\mathcal{I} = \int dt\, d\mathbf{r}\, \big[p_\alpha\big(\partial_t v_\alpha + v_\beta \nabla_\beta v_\alpha - \nu\nabla^2 v_\alpha + \nabla_\alpha P\big)$$
$$+ Q\nabla_\alpha v_\alpha\big] + \frac{i}{2}\int dt'dtd\mathbf{r}'d\mathbf{r}\, \Xi_{\alpha\beta}(t-t',\mathbf{r}-\mathbf{r}')p_\alpha p'_\beta\,, \tag{13}$$

related to (11). The independent fields P and Q play the role of Lagrange multipliers enforcing the incompressibility conditions $\nabla_\alpha v_\alpha = 0$ and the analogous condition $\nabla_\alpha p_\alpha = 0$ for the response field p_α.

The tails of the velocity PDF are determined by high moments with $n \gg 1$ which can be found by applying the saddle-point method to the integral (12). The configuration $\mathbf{v}(t,\mathbf{r})$, $\mathbf{p}(t,\mathbf{r})$, $P(t,\mathbf{r})$ and $Q(t,\mathbf{r})$ corresponding to a saddle point is our instanton. The extremum conditions for the argument of the exponent in (12) determining the instanton give two dynamical equations

$$\partial_t v_\alpha + v_\beta\nabla_\beta v_\alpha - \nu\nabla^2 v_\alpha + \nabla_\alpha P$$
$$= -i\int dt'\, d\mathbf{r}'\, \Xi_{\alpha\beta}(t-t',\mathbf{r}-\mathbf{r}')p_\beta(t',\mathbf{r}')\,, \tag{14}$$
$$\partial_t p_\alpha - p_\beta\nabla_\alpha v_\beta + v_\beta\nabla_\beta p_\alpha + \nu\nabla^2 p_\alpha + \nabla_\alpha Q$$
$$= -2\,i\,n\,\delta(t)\delta(\mathbf{r})v_\alpha/v^2\,, \tag{15}$$

and two incompressibility constraints

$$\nabla^2 P = -\nabla_\alpha(v_\beta\nabla_\beta v_\alpha)\,. \tag{16}$$
$$\nabla^2 Q = \nabla_\alpha(p_\beta\nabla_\alpha v_\beta - v_\beta\nabla_\beta p_\alpha)\,. \tag{17}$$

As was explained in Introduction, the system (14–17) is to be solved at negative time because it describes the prehistory that leads to a given measured value of the velocity at $t=0$. Since $\mathbf{p}=0$ at $t>0$ then the right-hand side of (15) directly produces the value of $\mathbf{p}$ at $t=-0$:

$$p_\alpha = 2in\left[\delta_{\alpha\beta}\delta(\mathbf{r}) - \frac{1}{4\pi r^3}\left(\delta_{\alpha\beta} - \frac{3r_\alpha r_\beta}{r^2}\right)\right]\frac{v_\beta}{v^2}\,, \tag{18}$$

where $\mathbf{v}$ is taken at the origin. The equations (14,15) are to be solved at $-\infty < t < 0$ with the constraints (16,17) under the boundary conditions (18) and $v\to 0$ as $t\to-\infty$. The system is quite complicated and its complete solution (which will provide an important information about the spatio-temporal domains with high velocity) is still ahead of us. The main difficulty is an effective spatial nonlocality of the constraints (16,17). What is important to state is that both fields $\mathbf{v}$ and $\mathbf{p}$ decay at moving backwards in time during some characteristic time t_* which we call the lifetime of the instanton. Actually it is the same time which we qualitatively discussed above at treating rare events with strong forcing since the instanton presents just the space-time picture of typical events contributing to $\langle v^{2n}\rangle$.

Fortunately, n-dependence of $\langle v^{2n}\rangle$ can be readily established which gives the functional form of PDF tails. The large parameter n makes the lifetime of the instanton t_* to be short as we shall see below. That makes it possible to consider the pumping correlation function Ξ as time independent. In this case, the parameter n can be excluded from (14–17) by the rescaling transformation

$$t \to X^{-1}t\,, \quad \mathbf{v} \to X\mathbf{v}\,, \quad P \to X^2 P\,, \quad Q \to X^4 Q\,,$$
$$\nu \to X\nu\,, \quad \mathbf{p} \to X^3\mathbf{p}\,, \quad X^4 = n\,. \tag{19}$$

That gives a general n-dependence of the velocity

$$\mathbf{v} = n^{1/4}\varphi(\nu^4/n)\,, \tag{20}$$

where dimensionless function φ is expected to go to some constant when it's argument goes to zero. The n-dependence of the instanton solution $v \propto n^{1/4}$ gives the following n-dependence of the moment $\langle v^{2n}\rangle \propto n^{n/2}$ which corresponds to the PDF tail

$$\ln \mathcal{P}(v^2) \propto -v^4\,. \tag{21}$$

If $L/\tau < v_{\rm rms}$ (where $v_{\rm rms}$ is the typical value of the velocity fluctuations) then the asymptotics (21) is realized at $v \gg v_{\rm rms}$ and $\ln\mathcal{P} \sim -(v/v_{\rm rms})^4$. In the opposite limit of a short-correlated pumping with $L/\tau \gg v_{\rm rms}$, an intermediate asymptotics exists where the pumping correlation function Ξ can be treated as delta-correlated in time: $\Xi(t,\mathbf{r}) = \delta(t)\chi(\mathbf{r})$. Then the only changes in (19) are $\mathbf{p} \to X^2\mathbf{p}$, $Q \to X^3 Q$ and $X^3 = n$. That leads to the law

$$\ln \mathcal{P}(v^2) \sim -(v/v_{\rm rms})^3\,, \tag{22}$$

which is valid at $L/\tau \gg v \gg v_{\rm rms}$. For larger v, the asymptotics (21) is realized. We thus see that for Gaussian pumping the velocity PDF always decreases faster than Gaussian.

Note that the above consideration can be readily extended for the consideration of velocity differences $\Delta\mathbf{v}$ at any separation between the points. There is an intermediate region $\Delta v_{\rm rms} \ll \Delta v \ll v_{\rm rms}$ where $\mathcal{P}(\Delta v)$ is not described by the naive instanton formalism. The same formulas (21,22) as well as below (29) describe the remote tails of $\mathcal{P}(\Delta v)$ at $\Delta v \gg v_{\rm rms}$, which are probably unaccessible experimentally at present. Contrary, faster-than-Gaussian decay of a single-point velocity PDF was recently observed both in experiments [16] and in numerical simulations [17, 18, 19], the qualitative reason for that (short lifetime of strong fluctuations) was discussed in [19].

The transformation (19) shows that the life time of the instanton is short for large n: $t_* \propto n^{-1/3}$ for the fast pumping $\tau \ll t_*$ and $t_* \propto n^{-1/4}$ for the slow pumping $\tau \gg t_*$. Therefore at large enough n we always deal with a slow pumping.

In other words, distant tails of the velocity PDF are determined by a simultaneous statistics of the pumping which is not necessary Gaussian.

The above procedure can be readily generalized for an arbitrary pumping statistics when instead of (12) one gets

$$\langle |\mathbf{v}|^{2n} \rangle = \int \mathcal{D}f \mathcal{D}p \mathcal{D}v \mathcal{D}P \mathcal{D}Q \mathcal{P}_f \exp\left(i\tilde{\mathcal{I}} + 2n \ln |\mathbf{v}|\right) . \tag{23}$$

Here, we substituted the simultaneous PDF of pumping $\mathcal{P}_f(f)$ since we treat the slow pumping, and the effective action $\tilde{\mathcal{I}}$ figuring in (23) gets now the following form

$$\tilde{\mathcal{I}} = \int dt\, d\mathbf{r}\, \big[p_\alpha(\partial_t v_\alpha + v_\beta \nabla_\beta v_\alpha - \nu \nabla^2 v_\alpha + \nabla_\alpha P) \\ + Q \nabla_\alpha v_\alpha - p_\alpha f_\alpha\big] . \tag{24}$$

The saddle-point equation for the velocity is now written as

$$\partial_t v_\alpha + v_\beta \nabla_\beta v_\alpha - \nu \nabla^2 v_\alpha + \nabla_\alpha P = f_\alpha , \tag{25}$$

with the relation

$$\int dt\, p_\alpha = -i\delta \ln \mathcal{P}_f(f)/\delta f_\alpha . \tag{26}$$

The equation for $\mathbf{p}$ and the incompressibility constraints are the same (15,16,17).

Let us assume the tail of $\mathcal{M} = -\ln \mathcal{P}_f$ to be scale invariant with the exponent a:

$$\mathcal{M}(X\mathbf{f}) = X^a \mathcal{M}(\mathbf{f}) , \tag{27}$$

for Gaussian statistics $a = 2$. Now, we can generalize (19)

$$t \to X^{-1} t , \quad \mathbf{v} \to X\mathbf{v} , \quad P \to X^2 P , \\ \nu \to X\nu , \quad \mathbf{f} \to X^2 \mathbf{f} , \\ \mathbf{p} \to X^{2a-1} \mathbf{p} , \quad Q \to X^{2a} Q , \quad X^{2a} = n , \tag{28}$$

keeping the form of the equations (25), (26) and removing n from the equation (15). Using the law (28) we can find how the characteristic velocity scales with n. As above the answer can be expressed in terms of the behavior of the tail of the simultaneous PDF $\mathcal{P}(v^2)$ for the velocity:

$$\ln \mathcal{P}(v^2) \propto -v^{2a} . \tag{29}$$

The lifetime t_* of our instanton scales as $t_* \propto n^{-1/(2a)}$. Of course for $a = 2$ we reproduce the Gaussian case (21). Note that velocity PDF decays faster than that of the force; in particular, Gaussian velocity would correspond to the force PDF decaying exponentially.

The expression (13) implies the homogeneous pumping. Meanwhile, the transformations (19) and (28) do not touch space. Therefore our conclusions are true also for a more physical case of an inhomogeneous pumping, in particular acting on the boundaries of the flow. One should be careful at applying our results to the flow, say, after a grid. The problem is in the presence of the mean velocity V which is usually larger than $v_{\rm rms}$. Then the statistics of the velocity fluctuations in bulk reproduce the statistics of the velocity fluctuations on the grid if $v < V$ and our reasoning are correct for $v > V$.

3 Burgers' turbulence

In the section we consider the forced Burgers' turbulence which is controlled by the following equation

$$\partial_t u + u\partial_x u - \nu\partial_x^2 u = \phi\,. \tag{30}$$

The equation (30) describes the evolution of weak one-dimensional acoustic perturbations in the reference frame moving with the sound velocity [20]. It is natural to assume the external force ϕ to be δ-correlated in time in that frame:

$$\langle\phi(t_1,x_1)\phi(t_2,x_2)\rangle = \delta(t_1-t_2)\chi(x_1-x_2)\,. \tag{31}$$

Then the statistics of ϕ is Gaussian and is completely characterized by χ. We are interested in turbulence excited by a large-scale pumping with some correlation length L so that χ does not essentially vary at $x < L$ and goes to zero where $x > L$. Besides L, the correlation function χ may be characterized by the parameter $\omega = [-(1/2)\chi''(0)]^{1/3}$ having the dimensionality of frequency. Then e.g. $\chi(0) \sim L^2\omega^3$. Developed turbulence corresponds to a large Reynolds number $\mathrm{Re} = L^2\omega/\nu \gg 1$.

The physical picture of Burgers turbulence is quite clear: arbitrary localized perturbation evolves into shock wave with the viscous width of the front, that gives k^{-2} for the energy spectrum at $\mathrm{Re} \gg kL \gg 1$ [20, 21]. The presence of shocks leads to a strong intermittency, PDF of velocity gradients is substantially non-Gaussian [22] and there is an extreme anomalous scaling for the moments of velocity differences $w = u(\rho) - u(-\rho)$: $\langle w^n\rangle \propto (\rho/L)$ for $n > 1$ [23]. Simplicity of the equation and transparency of underlying physics make it reasonable to hope that consistent formalism for the description of intermittency could be developed starting from Burgers' equation [24, 25, 26, 27]. The instanton formalism has been applied to the Burgers equation first by Gurarie and Migdal [28] who found the right tail $\ln\mathcal{P}(w) \sim -[w/(\omega\rho)]^3$ determined by inviscid behavior of smooth ramps between the shocks. That cubic right tail has been earlier predicted by Polyakov from the conjecture on the operator product expansion [24], it corresponds to the same right tail for the gradients $\ln\mathcal{P}(u') \sim -(u'/\omega)^3$ derived by Gotoh and Kraichnan [29]. The PDF's $\mathcal{P}(u')$ and $\mathcal{P}(w)$ are not symmetric, the asymmetry is due to the simple fact that positive gradients are smeared while the steepening of negative gradients could be stopped only by viscosity. Here, we describe the viscous instantons that give the left tails of the PDF's and the single-point velocity PDF. The main results of the section are published in [30].

Even though some calculations are lengthy, the simple picture appears as a result. Let us consider a rare event producing the main contribution to some tail of PDF for u. Since white forcing pumps velocity by the law $w^2 \propto \phi^2 t$ while the typical time of growth is restricted by the breaking time $t \sim L/w$ then the Gaussianity of the forcing $\ln \mathcal{P}(\phi) \propto -\phi^2/\chi(0)$ leads to $\ln \mathcal{P}(w) \sim -[|w|/(L\omega)]^3$. At a shock, $w^2 \simeq -\nu u'$ so that $\ln \mathcal{P}(u') \sim -[-u'/(\omega \mathrm{Re})]^{3/2}$. These simple estimates are confirmed by consistent calculations.

As was explained in Introduction we write the high-order moments of the velocity derivatives $u^{(k)} = \partial_x^k u$ (including velocity for $k=0$) and of the difference w as the path integrals:

$$\langle [u^{(k)}]^n \rangle = \int \mathcal{D}u\mathcal{D}p \exp\left\{ i\mathcal{I} + n \ln[u^{(k)}(0,0)] \right\}, \tag{32}$$

$$\langle w^n \rangle = \int \mathcal{D}u\mathcal{D}p \exp\left\{ i\mathcal{I} + n \ln[u(0,\rho) - u(0,-\rho)] \right\}. \tag{33}$$

Here, $\mathcal{I} = \int dt\, \mathcal{L}(u,p)$ is the effective action with the Lagrangian $\mathcal{L}$ determined by the equation (30) [11, 12, 13, 14]:

$$\mathcal{L} = \int dx\, (p\partial_t u + pu\partial_x u - \nu p \partial_x^2 u) + \frac{i}{2}\int dx_1 dx_2 p_1 \chi_{12} p_2 .$$

The main idea implemented here is that the high-order moments (for $n \gg 1$) are determined by the saddle-point configurations of the path integrals (32,33). The corresponding saddle-point (instanton) equations are

$$\partial_t u + u\partial_x u - \nu \partial_x^2 u = -i \int dx'\, \chi(x-x') p(x'), \tag{34}$$

$$\partial_t p + u\partial_x p + \nu \partial_x^2 p = y\delta(t)\beta(x)\,, \tag{35}$$

$y\beta = in[\delta(x+\rho) - \delta(x-\rho)]/w$ for differences, and $y\beta = in\delta'(x)/\partial_x u(t,0)$ for gradients. Solution of (34,35) gives the moments in the saddle-point approximation:

$$\langle [u^{(k)}]^n \rangle \sim [u^{(k)}(0,0)]^n e^{i\mathcal{I}_{\mathrm{extr}}}, \quad \langle w^n \rangle \sim [w(0)]^n e^{i\mathcal{I}_{\mathrm{extr}}}. \tag{36}$$

The boundary conditions for the instanton equations were formulated in [5] (see also the Introduction): $u \to 0$ at $t \to -\infty$ and $p = 0$ at $t > 0$. The role of the last term in (35) is then reduced to the final condition $p(x) = -y\beta$ imposed on p at $t = -0$. Viscosity smears p if to move backwards in time so that at large negative time both fields u and p are zero. We will refer to that stage as vacuum since it is realized at the absence of the source in the right-hand side of (35).

Since the Lagrangian does not explicitly depend on time, then the "energy" E is conserved by (34,35):

$$E = i\int dx (pu\partial_x u + \nu \partial_x p \partial_x u) - \frac{1}{2}\int dx_1 dx_2 p_1 \chi_{12} p_2 .$$

Since $\mathcal{L}$ does not explicitly depend on coordinates then the "momentum" J is conserved as well: $iJ = \int dx\, p\partial_x u$. Because of the conservation laws we should treat solutions of (34,35) with $E = J = 0$ since they are zero at $t \to -\infty$. That gives the following saddle-point value of the effective action in (36): $\mathcal{I}_{\rm extr} = \int dt\, dx\, p\partial_t u$.

Conservation laws help us to understand general properties of the solutions. We consider $t = 0$, substitute $p(x) = -y\beta$ and analyze the balance of different terms in E. For the gradient $a(t) = \partial_x u(t, 0)$,

$$E = -na(0) + \omega^3 n^2/a^2(0) - n\nu\partial_x^3 u(0,0)/a(0) = 0\,.$$

There is a difference between the cases of positive and negative a. For $a(0) > 0$, the viscous contribution to the "energy" is unessential and two first terms can compensate each other (see below). Contrary, the instanton that gives $a(0) < 0$ cannot exist without viscosity. For the velocity, $J = \partial_x u(0,0) = 0$ and $E = \chi(0)n^2/u^2(0) - \nu\partial_x^2 u(0,0)$. Without viscous term, "energy" cannot be zero. Note that the answer we shall obtain for the velocity PDF does not contain viscosity while it's consistent derivation requires the account of the viscous terms in the equations.

Let us first describe the effectively inviscid instantons producing the right tails of the PDFs for gradients and differences [24, 28, 29]. At $t = 0$, the field p is localized near the origin. A positive velocity slope "compresses" the field p so that one can expect that at negative time the width of p remains much smaller than L despite a viscous spreading. Then, it is possible to formulate the closed system of equations for the quantities $a(t)$ and $c(t) = -i\int dx\, x\, p(t,x)$ since for narrow p and small x we can put

$$\int dx'\chi(x-x')p(t,x') \to -i\partial_x\chi(x)c(t) \approx 2i\omega^3 xc(t)\,.$$

Then

$$\partial_t c = 2ac, \quad \partial_t a = -a^2 + 2\omega^3 c. \tag{37}$$

The instanton is a separatrix solution of (37). The initial condition $a(0)c(0) = n$ by virtue of the "energy" conservation $E = -ca^2 + \omega^3 c^2 = 0$ gives $a(0) = \omega^3 c^2(0)/n = \omega n^{1/3}$. For differences, $w = 2a(0)\rho$. One can check that $\mathcal{I}_{\rm extr} = i\int dt\, c\partial_t a \sim a(0)c(0) = n$ which is negligible in comparison with $n\ln[a(0)]$ so that $\langle(u')^n\rangle \sim [a(0)]^n \sim \omega^n n^{n/3}$ which gives the right cubic tails of the PDFs $\ln\mathcal{P}(u') \sim -(u'/\omega)^3$ [29] and $\ln\mathcal{P}(w) \sim -[w/(\rho\omega)]^3$ [24, 28]. The width of p is much less than L through the time of evolution $T \sim n^{-1/3}\omega^{-1}$ giving the main contribution into the action [28]. The right tails of $\mathcal{P}(u')$ and $\mathcal{P}(w)$ are thus universal i.e. independent of the large-scale properties of the pumping. Above consideration does not imply that the instanton is completely inviscid, it may well have viscous shock at $x \sim L$, this has no influence on the instanton answer (since p is narrow) while may influence the fluctuation contribution i.e. pre-exponential factors in the PDFs.

The main subject of this part is the analysis of the instantons that give the tails of $\mathcal{P}(u)$ and the left tails of $\mathcal{P}(u')$ and $\mathcal{P}(w)$ corresponding to negative a, w.

Even though the field p is narrow at $t = 0$, we cannot use the simple system (37) to describe those instantons. The reason is that sweeping by a negative velocity slope provides for stretching (rather than compression) of the field p at moving backwards in time. As a result, the support of $p(x)$ stretches up to L so that one has to account for the given form of the pumping correlation function $\chi(x)$ at $x \simeq L$. This leads to a nonuniversality of $\mathcal{P}(u)$ and of the left tails of $\mathcal{P}(u')$ and $\mathcal{P}(w)$ which depend on the large-scale properties of the pumping. As we shall see, the form of the tails is universal, nonuniversality is related to a single constant in PDF. Additional complication in analytical description is due to the shock forming from negative slope near the origin. The shock cannot be described in terms of the inviscid equations so that we should use the complete system (34,35) to describe what can be called viscous instantons.

Apart from a narrow front near $x = 0$, the velocity field has L as the only characteristic scale of change. The life time T of the instanton is then determined by the moment when the position of p maximum reaches L due to sweeping by the velocity u_0: $T \sim L/u_0$. Such a velocity u_0 itself has been created during the time T by the forcing so that $u_0 \sim |c|_{max} T L \omega^3$. To estimate the maximal value of $|c(t)|$, let us consider the backward evolution from $t = 0$. We first notice that the width of p (which was zero at $t = 0$) is getting larger than the width of the velocity front $\simeq u_0/a$ already after the time $\simeq a^{-1}$. After that short time, the values of c and a stay of order of their values at $t = 0$. Then, one may consider that $p(t, x)$ propagates (backwards in time) in a smooth velocity field the amplitude of which can be estimated as u_0, and

$$\partial_t c = -i \int_{-\infty}^{\infty} dx\, x u \partial_x p \sim i u_0 \int dx\, p\,,$$

where the integral $i \int dx\, p$ can be estimated by it's value at $t = 0$ which is n/u_0. Therefore, we get $c_{max} \sim nT$ so that $T \sim n^{-1/3}\omega^{-1}$ and $u_0 \sim L\omega n^{1/3}$. At the viscosity-balanced shock, the velocity u_0 and the gradient a are related by $u_0^2 \simeq \nu a$ so that $a(0) \simeq \omega \mathrm{Re}\, n^{2/3}$.

Let us briefly describe now the consistent analytic procedure of the derivation of the function $c(t)$ that confirms above estimates. We use the Cole-Hopf substitution [20] for the velocity $\partial_x \Psi = -u\Psi/2\nu$ and introduce $P = 2\nu \partial_x p/\Psi$. The saddle-point equations for Ψ and P

$$\partial_t \Psi - \nu \partial_x^2 \Psi + \nu F \Psi = 0, \tag{38}$$

$$\partial_t P + \nu \partial_x^2 P - \nu F P - 2\nu \lambda'(x) \delta(t) \Psi^{-1} = 0 \tag{39}$$

contain F determined by $\partial_x F(t, x) = -i \int dx' \chi(x - x') p(t, x')/2\nu^2$ and fixed by the condition $F(t, 0) = 0$. Calculations are straightforward if one passes to Heisenberg representation for (38) introducing the evolution operator $\hat{U}(t)$ which satisfies the equation $\partial_t \hat{U} = \hat{H}\hat{U}$ with $\hat{H}(t) = \nu(\partial_x^2 - F)$. It is remarkable that one can develop the closed description in terms of two operators $\hat{A} = \hat{U}^{-1} x \hat{U}$ and $\hat{B} = \hat{U}^{-1} \partial_x \hat{U}$:

$$\partial_t \hat{A} = -2\nu \hat{B}\,, \quad \partial_t \hat{B} = -\nu F_x(t, \hat{A})\,. \tag{40}$$

We note that all the moments of p (and therefore $F(t,x)$) can be expressed in terms of $\hat{A}$. Since we study the time interval when $p(t,x)$ is narrow, it is enough for our purpose to consider $x \ll L$ where $F(t,x) = c(t)x^2\omega^3/2\nu^2$. Further simplification can be achieved in this case and the closed equation for

$$c(t) = -(i/2)\int dx\lambda'(x)\Psi^{-1}(0,x)\hat{A}^2(t)\Psi(0,x)\,,$$

can be derived from (40):

$$(\partial_t c)^2 = 4\omega^3 c^3 + 16\xi_2^2 + 4\omega^3\xi_1^3\,.$$

Here $\xi_1 = i\int dx\lambda(x)x$ and $4\xi_2 = -i\int dx\lambda(x)\partial_x[xu(0,x)]$. Integrating we get

$$t = \frac{1}{2}\int_{c(0)}^{c}\frac{dx}{\sqrt{\omega^3x^3 + 4\xi_2^2 + \xi_1^3}}\,, \tag{41}$$

which describes $c(t)$ in an implicit form. Further analysis depends on the case considered. For the gradients, we substitute $\xi_1 = n/a_0$ and $\xi_2 = -n/2$ and see that, as time goes backwards, negative $c(t)$ initially decreases by the law $c(t) = c(0)+2nt$ until $T = \omega^{-1}(n/2)^{-1/3}$ then it grows and the approximation looses validity when $c(t)$ approaches zero and the account of the pumping form $\chi(x)$ at $x \simeq L$ is necessary. For self-consistency, we require the width of $p(x)$ at this time to be of order L and get the estimate $a(0) \simeq \omega\mathrm{Re}\, n^{2/3}$ and thus confirm the above picture. The main contribution to the saddle-point value (36) is again provided by the term $[\partial_x u(0,0)]^n$ and we find $\langle (u')^n\rangle \simeq [a(0)]^n \simeq (\omega\mathrm{Re})^n n^{2n/3}$, which corresponds to the following left tail of PDF at $u' \gg \omega\mathrm{Re}$

$$\mathcal{P}(u') \propto \exp[-C_1(-u'/\omega\mathrm{Re})^{3/2}]\,. \tag{42}$$

For higher derivatives $u^{(k)}$, by using (41) we get initial growth $c(t) = c(0)+n(k+1)t$ which gives $u^{(k)}(0,0) \sim N^{k+1}L^{1-k}\omega\mathrm{Re}^k$ leading to $\langle [u^{(k)}]^n\rangle \sim \omega\mathrm{Re}^k L^{1-k} n^{(k+1)/3}$ which can be rewritten in terms of the PDF tail:

$$\mathcal{P}\left(|u^{(k)}|\right) \propto \exp\left[-C_k\left(|u^{(k)}|L^{k-1}/\omega\mathrm{Re}^k\right)^{3/(k+1)}\right]. \tag{43}$$

Note that the degree of non-Gaussianity increases with increasing k. On the other hand, the higher k the more distant is the validity region of (43): $u^{(k)} \gg u^{(k)}_{\mathrm{rms}} \sim L^{1-k}\omega\mathrm{Re}^k$. The quantities C_k in (42,43) are constants of order unity.

For the differences, $\xi_1 = 2n\rho_0/w$ and $4\xi_2 = -n[1 + 2\rho_0 u_x(0,\rho_0)/w]$ and we get $\langle w^n\rangle \simeq (L\omega)^n n^{n/3}$ which corresponds to the cubic left tail

$$\mathcal{P}(w) \propto \exp\left\{-C_-[w/(L\omega)]^3\right\}, \tag{44}$$

valid at $w \gg L\omega$. In the intermediate region $L\omega \gg w \gg \rho\omega$, there should be a power asymptotics which is the subject of current debate [24, 29, 27]. It is natural that ρ-dependence of $\mathcal{P}(w)$ cannot be found in a saddle-point approximation; as a pre-exponential factor, it can be obtained only at the next step by calculating the contribution of fluctuations around the instanton solution. This is consistent with the known fact that the scaling exponent is n-independent for $n > 1$: $\langle w^n(\rho)\rangle \propto \rho$. For the velocity, $\lambda(x) = -in\delta(x)/u(0,0)$ is an even function so that F is a linear (rather than quadratic) function of x for narrow p: $F(x) = \chi(0)bx/2\nu^2$ with $b = -i\int dx p(x)$. Direct calculation shows that "energy" and "momentum" conservation makes b time independent: $b = n/u(0,0)$. It is easy then to get the n-dependence of $u(0,0)$: Velocity stretches the field p so that the width of p reaches L at $T \simeq L/u(0,0)$ while the velocity itself is produced by the pumping during the same time: $u(0,0) \simeq \chi(0)bT = \chi(0)nT/2u(0,0) \simeq n\chi(0)L/u(0,0)$. That gives $u(0,0) \simeq L\omega n^{1/3}$ and

$$\mathcal{P}(u) \propto \exp\{-C[u/(L\omega)]^3\}\,. \tag{45}$$

The product $L\omega$ plays the role of the root-mean square velocity $u_{\rm rms}$. The numerical factors C, C_- and C_k are constants of order unity which are determined by the evolution at $t \sim T$ i.e. by the behavior of pumping correlation function $\chi(x)$ at $x \sim L$. Therefore the factors are non-universal.

Let us summarize. At smooth almost inviscid ramps, velocity differences and gradients are positive and linearly related $w(\rho) \approx 2\rho u'$ so that the right tails of PDFs have the same cubic form [24, 28, 29]. $\mathcal{P}(\partial_x u) \sim \exp[-(\partial_x u/\omega)^3]$, $\mathcal{P}(w) \sim \exp[-(w/2\rho\omega)^3]$. Those tails are universal i.e. they are determined by a single characteristics of the pumping correlation function $\chi(r)$, namely, by it's second derivative at zero $\omega = [-(1/2)\chi''(0)]^{1/3}$. Contrary, the left tails found here contain nonuniversal constant which depends on a large-scale behavior of the pumping. The left tails come from shock fronts where $w^2 \simeq -\nu u'$ so that cubic tail for velocity differences (44) corresponds to semi-cubic tail for gradients (42). The formula (44) is valid for $w \gg u_{\rm rms} \simeq L\omega$ where $\mathcal{P}(w)$ should coincide with a single-point $\mathcal{P}(u)$ since the probability is small for both $u(\rho)$ and $u(-\rho)$ being large simultaneously. Indeed, we saw that the tails of $\ln\mathcal{P}(u)$ at $u \gg u_{\rm rms}$ are cubic as well. Note that (43) is the same as obtained for decaying turbulence with white (in space) initial conditions by a similar method employing the saddle-point approximation in the path integral with time as large parameter [31]. That, probably, means that white-in-time forcing corresponds to white-in-space initial conditions. Note also that if the pumping has a finite correlation time τ then our results, strictly speaking, are valid for $u, w \ll L/\tau$ and $u' \ll 1/\tau$.

4 Instanton for the Kraichnan model

One of the classical objects in theory of turbulence is so-called passive scalar advected by a fluid. The role of the passive scalar can be played by the temperature or the density of pollutants. Correlation functions of the scalar in a turbulent flow possess scaling properties which in the frame of the theory analogous to Kolmogorov's one were established by Obukhov [32]. Intermittency enforces deviations from the Obukhov's exponents, they appear to be even stronger than deviations from the

Kolmogorov exponents for the correlation functions of velocity [33, 34, 35, 36]. Let us also stress, that the statistical properties of the passive scalar, though valuable on their own, are directly related to that of Lagrangian trajectories in a turbulent fluid.

Unfortunately, a consistent theory of turbulence describing anomalous scaling is not constructed yet. This accounts for the difficulties associated with the strong coupling inherent to developed turbulence. This is the reason for attempts to examine the intermittency phenomenon in the framework of different simplified models. The most popular model used for this purpose is the Kraichnan model of the passive scalar advection [37], where the advecting velocity is believed to be short-correlated in time and has a Gaussian distribution. That allows one to examine the statistics of the passive scalar in more detail.

The scalar in the Kraichnan model exhibits strong intermittency even if such is absent in the advecting velocity field. The fact was proved in the works [38, 39, 40, 41, 42, 43, 44, 45], where the equation for the correlation function F_n of the order n was solved assuming that different parameters, such as ζ_2, $2-\zeta_2$, or d^{-1}, are small (remind, that ζ_2 is the exponent of the second-order structure function and d is the dimensionality of space). The correlation functions which can be examined in the framework of the methods of the noted papers are bounded from above by their order n. An attempt to solve the problem at large n was made in [46], where an n-independent asymptotics was found. Note also the work by Kraichnan [47] where a closure was assumed enabling to find ζ_n for any n. An anomalous scaling in the Kraichnan model is observed also in numerics [48, 49, 50]. Here, we are going to develop a technique formulated in [5] and described in Introduction.

Advection of the passive scalar θ by the velocity field $\mathbf{v}$ is described by the equation

$$\partial_t\theta + \mathbf{v}\nabla\theta - \kappa\nabla^2\theta = \phi\,, \tag{46}$$

where κ is the diffusion coefficient and ϕ is the source of the passive scalar (say, if θ corresponds to fluctuations of temperature, then ϕ represents the power of heaters). In a turbulent flow, $\mathbf{v}$ is a random function of time and space coordinates. The source ϕ is also assumed to be a random function. Then passive scalar correlation functions are determined by the statistics of $\mathbf{v}$ and ϕ. Usually, one is interested in simultaneous correlation functions $F_n = \langle\theta(\mathbf{r}_1)\dots\theta(\mathbf{r}_n)\rangle$, since a large-scale velocity destroys temporal correlations in the Eulerian frame, whereas simultaneous objects are not influenced by it.

It is convenient to examine the anomalous scaling in terms of the so-called structure functions

$$S_n(r) = \langle|\theta(\mathbf{r}/2) - \theta(-\mathbf{r}/2)|^n\rangle\,. \tag{47}$$

One expects that in the convective interval of scales the structure functions reveal a scaling behavior

$$S_n(r) \propto r^{\zeta_n}\,. \tag{48}$$

The exponents ζ_n are of great interest since they reflect the intermittency. In Obukhov's theory [32] $\zeta_n = (n/2)\zeta_2$ and the differences $(n/2)\zeta_2 - \zeta_n$ characterize the anomalous scaling. One can formulate an estimate [2, 3, 4]

$$S_n(r) \sim A_n (S_2)^{n/2} \left(\frac{L}{r}\right)^{(n/2)\zeta_2 - \zeta_n}, \tag{49}$$

where L is the pumping length and A_n is an n-dependent factor. The intermittency implies $(n/2)\zeta_2 - \zeta_n > 0$. Therefore, at $r \ll L$, the structure functions S_n appear to be much larger than their naive estimate following from the Obukhov's theory.

In the Kraichnan model both $\mathbf{v}$ and ϕ are assumed to be independent random functions δ-correlated in time and described by Gaussian statistics homogeneous in space. Therefore, statistical properties of the fields are entirely characterized by the pair correlation functions

$$\langle \phi(t_1, \mathbf{r}_1)\phi(t_2, \mathbf{r}_2)\rangle = \delta(t_1 - t_2)\chi(|\mathbf{r}_1 - \mathbf{r}_2|), \qquad \chi(0) = P_2, \tag{50}$$

$$\langle v_\alpha(t_1, \mathbf{r}_1) v_\beta(t_2, \mathbf{r}_2)\rangle = \delta(t_1 - t_2)\left\{\mathcal{V}_0 \delta_{\alpha\beta} - \mathcal{K}_{\alpha\beta}(\mathbf{r}_1 - \mathbf{r}_2)\right\}. \tag{51}$$

Here $\chi(x)$ is a smooth function decaying on the scale L which is the pumping length, and the constant P_2 has the physical meaning of the pumping rate of θ^2.

The velocity correlation function is assumed to possess some scaling properties. Namely, $\mathcal{K}(\mathbf{r}) \propto r^{2-\gamma}$ where the exponent γ characterizes the roughening degree of the velocity field: The field is smooth in space at $\gamma = 0$ and is extremely irregular at $\gamma = 2$. We will treat an arbitrary γ satisfying the inequality $0 < \gamma < 2$. The tensorial structure of $\mathcal{K}_{\alpha\beta}$ is determined by the incompressibility condition $\operatorname{div} \mathbf{v} = 0$ (which is implied in the Kraichnan model):

$$\mathcal{K}_{\alpha\beta}(\mathbf{r}) = \frac{2D_*}{\alpha} r^{-\gamma}\left[(r^2\delta_{\alpha\beta} - r_\alpha r_\beta) + \frac{d-1}{2-\gamma} r^2 \delta_{\alpha\beta}\right], \tag{52}$$

where d is the dimensionality of space and D_* is a constant characterizing the strength of velocity fluctuations. For convenience, we introduced into the expression (52) the factor

$$\alpha = \frac{2d(d-1)}{(2-\gamma)}, \tag{53}$$

The expression (52) is correct on scales $r \ll L_u$ where L_u is the velocity correlation length (velocity pumping length) and therefore the constant $\mathcal{V}_0$ in (52) can be estimated as $\mathcal{V}_0 \sim \mathcal{K}(L_u)$. We will believe that the velocity pumping length L_u is larger than the pumping length of the passive scalar L.

One assumes that fluctuations of the velocity are strong enough to ensure the condition

$$D_* L^{2-\gamma} \gg \kappa. \tag{54}$$

Then there exists a convective interval of scales $r_d \ll r \ll L$, where r_d is the diffusive length

$$r_d^{2-\gamma} \sim \kappa / D_*. \tag{55}$$

We will be interested in the scaling properties of correlation functions of θ only in the convective interval.

The assumption of Gaussianity and zero correlation time for the fields $\mathbf{v}$ and ϕ allows to derive a closed partial differential equation for the n-th order correlation function F_n of θ [37, 51, 40]. For the simultaneous pair correlation function $F_2(r_{12}) = \langle\theta(t,\mathbf{r}_1)\theta(t,\mathbf{r}_2)\rangle$ one can solve the equation and find the explicit expression for F_2. In the convective interval

$$S_2(r) = 2[F_2(0) - F_2(r)] \sim \frac{P_2}{D_*} r^\gamma . \tag{56}$$

Comparing (56) with (48) one concludes that the exponent γ introduced by (52) directly determines the scaling of the second order structure function: $\zeta_2 = \gamma$ [37].

However, for $n > 2$ the equations for F_n are too complicated to be integrated exactly. In [39, 40, 41, 42] the equations were analyzed at the conditions $(2-\gamma) \ll 1$ or $d\gamma \gg 1$ where the statistics of the passive scalar is close to Gaussian. The analysis led to an anomalous scaling which can be written in terms of the exponents ζ_n of the structure functions (47,48)

$$\zeta_n = \frac{n}{2}\gamma - \frac{2-\gamma}{2(d+2)} n(n-2) . \tag{57}$$

This expression covers both limiting cases $2-\gamma \ll 1$ and $d\gamma \gg 1$. The first contribution in the right-hand side of the equation (57) represents the normal scaling, whereas the second contribution is just the anomalous scaling exponent. The calculations leading to (57) are correct if the the anomalous contribution is much smaller than the normal one. That means that the inequality

$$n \ll \frac{d\gamma}{2-\gamma} , \tag{58}$$

should be satisfied.

One can develop an instanton approach to the problem which allows us to find the exponents ζ_n (48) of the structure correlation functions (47) for any order $n \gg 1$ at the same additional condition $d\gamma \gg 1$ as in [40, 41]. Below, we will present only the general scheme of the calculations and their results. Detailed explanations can be found in [54].

As previously we will be interested in simultaneous correlation functions of the object

$$\vartheta = \int d\mathbf{r}\, \beta(\mathbf{r})\theta(\mathbf{r}) , \tag{59}$$

where β is a function of coordinates. Simultaneous correlation functions of ϑ can be represented as derivatives of the generating functional $\mathcal{Z}(\lambda) = \langle \exp(iy\vartheta)\rangle$. We will believe $\kappa = 0$. Then we should be careful dealing with point objects. That means that the characteristic scale of the function β in (59) should be larger than r_d. Besides we will assume that the characteristic scale of β is much smaller than

L. This implies that it lies in the convective interval. For example $S_n \approx \langle|\vartheta|^n\rangle$ where the function (59) is

$$\beta(\mathbf{r}_1) = \delta_\Lambda\left(\mathbf{r}_1 - \frac{\mathbf{r}}{2}\right) - \delta_\Lambda\left(\mathbf{r}_1 + \frac{\mathbf{r}}{2}\right) . \tag{60}$$

Here $\delta_\Lambda(\mathbf{r})$ is the function with the width $\Lambda^{-1} \gg r_d$ and satisfying the condition

$$\int d\mathbf{r}\, \delta_\Lambda(\mathbf{r}) = 1 .$$

We will be interested in high-order correlation functions of ϑ. One could expect [5] that the tail can be calculated in the saddle-point approximation in the path integral. Unfortunately, direct application of the method to the PDF does not lead to success. To recognize the reason, let us consider the transformation

$$\mathbf{v} \to X\mathbf{v} ,\ p \to Xp ,\ t \to X^{-1}t ,\ y \to Xy ,\ \kappa \to X\kappa , \tag{61}$$

analogous to (19). One can check that under this transformation both $y\vartheta$ and the corresponding effective action acquire the factor X. That means that in the saddle-point approximation $\ln \mathcal{Z} = yf(y/\kappa)$ with some unknown function f. On the other hand, we expect that correlation functions of the scalar itself (but not of its gradient) do not depend on the diffusivity, and the results of the works [38, 39, 40, 41, 42, 44, 45, 51] confirm the expectation. Then at small κ the function f remains a κ-independent constant and we obtain

$$\ln \mathcal{Z} \propto |y| . \tag{62}$$

The expression (62) is definitely correct at large enough y. Unfortunately it cannot help to find $\mathcal{P}(\vartheta)$ since after substituting (62) into

$$\mathcal{P}(\vartheta) = \int \frac{dy}{2\pi} \exp(-iy\vartheta)\mathcal{Z}(y) , \tag{63}$$

we realize that the characteristic value of y in the integral can be estimated as $y \sim \vartheta^{-1}$. Therefore at large ϑ the main contribution to the integral is determined by the region where (62) does not work. Above arguments are based on the saddle-point approximation for the generating functional $\mathcal{Z}$. One could try to construct an instanton starting directly for structure functions. Let us explain why the approach will also fail. Obviously, a saddle-point solution is anisotropic because of the incompressibility condition $\operatorname{div}\mathbf{v} = 0$. Fluctuations related to smooth variations of the anisotropy axis in time and space are strong and destroy the saddle-point approximation. We conclude that the naive instantonic approach to the problem fails. Thus, we should transform the path integral to new variables allowing to construct an instanton with weak fluctuations on its background.

In the diffusionless case the left hand side of the equation (46) describes simply transfer of passive scalar particles along the streamlines. Then it is natural to pass into the Lagrangian frame. To do that, let us introduce Lagrangian trajectories $\mathbf{x}(t)$ which obey the equation

$$\partial_t \mathbf{x} = \mathbf{v}(t, \mathbf{x}) \,. \tag{64}$$

We will label the trajectories by the coordinates $\mathbf{r} = \mathbf{x}(t = 0)$. Remind that due to causality we should treat the instanton at negative times [5]. Therefore one can say that $\mathbf{r}$ is the terminating point of the trajectory $\mathbf{x}(t, \mathbf{r})$. In other words, we treat the evolution backward in time. The next step is to reformulate the path integral for the PDF of ϑ in terms of the Lagrangian differences

$$R_{12}(t) \equiv R(t, \mathbf{r}_1, \mathbf{r}_2) = |\mathbf{x}(t, \mathbf{r}_1) - \mathbf{x}(t, \mathbf{r}_2)| \,. \tag{65}$$

Using the conventional procedure [11, 12, 13] and the equation (64) we get the effective action

$$\begin{aligned} i\mathcal{I}_R = i \int\limits_{-\infty}^{0} dt \int d\mathbf{r}_1 \, d\mathbf{r}_2 \, m_{12} (\gamma^{-1} \partial_t R_{12}^\gamma + D_*) \\ - \frac{D_*}{d} \int\limits_{-\infty}^{0} dt \int d\mathbf{r}_1 \, d\mathbf{r}_2 \, d\mathbf{r}_3 \, d\mathbf{r}_4 \, Q_{12,34} m_{12} m_{34} \,, \end{aligned} \tag{66}$$

describing the statistics of the Lagrangian differences. Here, $m_{12} \equiv m(t, \mathbf{r}_1, \mathbf{r}_2)$ is the auxiliary field conjugated to R_{12} and $Q_{12,34}$ is a cumbersome combination of the Lagrangian differences for the points $1, 2, 3, 4$ originating from (52) [54].

The average value of $|\vartheta|^n$ can be written now as

$$\langle |\vartheta|^n \rangle = \int \frac{dy \, d\vartheta}{2\pi} \int \mathcal{D}R \, \mathcal{D}m \, \exp \left(i\mathcal{I}_R - \mathcal{F}_\lambda - iy\vartheta + n \ln |\vartheta| \right) . \tag{67}$$

$$\mathcal{F}_\lambda = \frac{y^2}{2} \int dt \, d\mathbf{r}_1 \, d\mathbf{r}_2 \, \chi(R_{12}) \beta(\mathbf{r}_1) \beta(\mathbf{r}_2) \,, \tag{68}$$

The integration in (67) is performed over functions of t, $\mathbf{r}_1$, $\mathbf{r}_2$, with some boundary conditions. The boundary condition for R_{12} follows from $\mathbf{x}(0) = \mathbf{r}$:

$$R_{12}(t = 0) = |\mathbf{r}_1 - \mathbf{r}_2| \,. \tag{69}$$

The boundary condition for the field m_{12} should be $m_{12}(-\infty) = 0$, since we deal with free integration over R_{12} at the remote past. Let us stress that due to the definition (65) the triangle inequalities

$$R_{12} + R_{23} > R_{13} \,, \tag{70}$$

should be satisfied. Actually, the inequalities are constrains which should be imposed on the field R_{12} at integrating in (67). It is very hard to take into account the constrains explicitly. Actually we ignore the constrains at treating the instanton, which is correct at two conditions. First, the inaqualities (70) should be satisfied for the instantonic solution itself. Second, fluctuations on the background of our instanton should be weak (what is the applicability condition of the instantonic formalism). We argue in [54] that the conditions are satisfied if

$$d\gamma \gg 1\,. \tag{71}$$

The inequality (71) is implied below.

We will be interested in the structure functions $S_n \approx \langle|\vartheta|^n\rangle$ with (59,60). Calculations based on the saddle-point approximation in the integral (67) are published in [54]. Below, we formulate only results.

The main peculiarity of our picture is the existence of a critical order

$$n_c = \frac{d\gamma}{2(2-\gamma)}\,. \tag{72}$$

At $n < n_c$ we obtain

$$S_n \sim \left(\frac{n}{\gamma}\frac{P_2 C_1}{D_*}L^\gamma\right)^{n/2}\left(\frac{r}{L}\right)^{\zeta_n}\,. \tag{73}$$

$$\zeta_n = n\gamma/2 - (2-\gamma)n^2/(2d)\,. \tag{74}$$

The quantity C_1 in (73) is a constant of order unity, whose value depends on the shape of χ (that is on the details of the pumping) and is consequently non-universal. Comparing the expression (74) with (57) we see that they coincide at the condition $d \gg 1$ which was implied in our derivation. Surprisingly, the n-dependence of ζ_n written in (57), is correct not only for $n \ll n_c$ (58), but up to $n = n_c$ which is the boundary value restricting the applicability region of (73,74).

At $n > n_c$ the scaling exponents ζ_n appear to be n-independent and equal to the value

$$\zeta_c = \frac{d\gamma^2}{8(2-\gamma)}\,. \tag{75}$$

For the structure functions one gets

$$S_n \sim \left(\frac{n}{\gamma}\frac{P_2 C_2}{D_*}L^\gamma\right)^{n/2}\left(\frac{r}{L}\right)^{\zeta_c}\,. \tag{76}$$

The quantity C_2 in (76) is again a nonuniversal constant of order unity. Note that two cases where $\gamma \ll 1$ or $2-\gamma \ll 1$ need a special analysis. The answer (76) should be slightly corrected in the case $\gamma \ll 1$ and keeps its form at $2-\gamma \ll 1$. One can establish the inequality

$$\ln\frac{n}{d} < \gamma \ln\frac{L}{r}\,, \tag{77}$$

which restricts the region where (76) is correct.

We can treat the structure functions S_n as continuous functions of n. Then the vicinity of the critical value $n = n_c$ needs a special treatment. The expression for the structure functions can be written as

$$S_n \sim \left(\frac{(n-n_c)^2}{\gamma n_c} \frac{P_2 C_\pm}{D_*} L^\gamma \right)^{n_c/2} \left(\frac{r}{L}\right)^{\zeta_n}, \tag{78}$$

which implies the condition $|n - n_c| \ll n_c$. The factors $C_\pm$ are nonuniversal constants of order unity which are different for the cases $n < n_c$ and $n > n_c$. The exponents ζ_n in (78) are determined by (74) if $n < n_c$ and $\zeta_n = \zeta_c$ (75) if $n > n_c$. We see that the critical behavior of the structure functions is determined by the factor

$$S_n \propto |n - n_c|^{n_c} . \tag{79}$$

At the above consideration we suggested that r/L is the smallest parameter of our theory. But if $n \to n_c$ then $|n_c - n|$ starts to compete with r/L and at small enough $n_c - n$ our consideration is broken. The criterion which determines the validity of (78,79) is

$$\gamma \ln \frac{L}{r} \gg \frac{n_c}{|n - n_c|} . \tag{80}$$

The criterion determines the narrow region near n_c where the critical behavior (79) is saturated. We conclude that the inaquality

$$\gamma \ln \frac{L}{r} \gg 1 ,$$

should be assumed for our treatment to be correct. Otherwise a special consideration is needed.

Let us summarize the results of this section. First of all, we have established the n-dependence of the scaling exponents which are determined by the expression (74) if $n < n_c$ and remain the constant (75) if $n > n_c$ where n_c is defined by (72). Note that the similar nonanalytical behavior of ζ_n is established also for Burgers' turbulence [20] where $\zeta_n = 1$ for $n > 1$. The constant (75) does not coincide with one derived in [46]. Our results contradict also to the answer obtained by Kraichnan [47] who used a closure for the structure functions. For $n \ll n_c$ our result coincides with the answer obtained perturbatively [40, 41] at large d. Surprisingly the quadratic dependence of ζ_n on n is kept up to $n = n_c$. Such n-dependence of ζ_n is well known from the so-called log-normal distribution proposed by Kolmogorov [53]. But we want to stress that the log-normal distribution is definitely inapplicable in the situation. The reason is in n-dependence of the structure functions S_n which is also established in our work. The expressions (73,76) show that the behavior of r-independent prefactors in S_n is characteristic rather of Gaussian distribution. A natural explanation of the fact can be found in terms of zero mode ideology [38, 39, 40, 41, 42, 55]. We know that for $n > 2$ the main contribution to the structure function S_n in the convective interval is related to zero modes of the equation for the n-th order correlation function of the passive scalar. The exponents of the modes are determined by the equation (and could be very sensitive to the value of n) whereas the coefficients before the modes (determining their contribution to S_n) are formed on the pumping scale where the statistics of the passive scalar is nearly Gaussian. That is the explanation of the n-dependence of the prefactors in (73,76). Probably the most striking feature of our results is the critical behavior of S_n (treated as continuous functions of n) near $n = n_c$. The behavior determined by (78,79) resembles mainly a critical behavior of, say, heat capacity near a second-order phase transition, the quantity $(n - n_c)/n_c$ playing the role of the reduced temperature. Then the inequality (80) is the analog of the size restriction (the conventional critical behavior is observed if the critical radius is less than the size of the specimen).

5 Conclusion

We finished the consideration of different cases where the instantonic formalism (sketched in Introduction) enables to estimate high moments of different turbulent quantities. In the section 2 we treated the single-point distribution of the velocity in turbulent flows. Though the corresponding instantonic equations are too complicated to be solved explicitly, they possess a remarkable rescaling symmetry which (neglecting viscosity) enables to relate the single-point PDF with the statistics of the pumping force. In the section 3 we consider the instanton for the Burgers' turbulence which cannot be constructed without viscosity. Nevertheless the answers for the tails of the single-point PDF and of the PDF for the velocity difference show that they can be obtained in the framework of the same rescaling procedure as in the section 2, neglecting viscosity. Therefore the shock presenting in the instanton can be treated for the quantities as a simple discontinuity in the velocity field. That confirms the validity of the procedure made in the section 2. Besides the PDF tails for gradients of the velocity cannot be obtained without knowing the detailed structure of the instanton, determined by interplaying different terms including the viscous one. Correspondingly the answers for the tails obtained in the section 3 essentially depend on the viscosity coefficient and cannot be established by a rescaling procedure.

In the section 4 we consider high moments of the passive scalar in the Kraichnan model. The same rescaling symmetry prevents to apply directly the instantonic technique to this case. To avoid the difficulty we performed an exact transformation of the path integral (determining the moments) to new variables (which are absolute values of the Lagrangian differences), which are suitable for the instantonic analysis. That means that we have taken into account the most dangerous part of fluctuations destroying the naive instanton. Note that a similar situation takes place for the one-dimensional (compressible) modification of the Kraichnan model [56]. Though only the case of large dimensionalities of space admits a detailed analysis of the instantonic equations some peculiarities of the picture found could be universal. For example we have encounted the n-independent asymptotics of the exponents of the structure functions S_n. Probably the most remarkable feature of the structure functions found is their critical dependence on n near a critical value n_c.

To avoid misunderstanding let us stress that all our results are valid for the forced turbulence. As to the decaying turbulence it needs a separate consideration. Our technique is definitely useless for the final stage of the evolution where only coherent structures are seen, which are preserved as a result of competition of structures with different life times. Besides we hope that our instantonic technique could be useful for describing the initial stage of the decaying turbulence where the strong nonlinearity transforms the initial distribution creating a nontrivial statistics of fluctuations.

Acknowledgements

I am grateful to E. Balkovsky and G. Falkovich for numerous valuable discussions. I acknowledge support from the ENS-Landau Institute Twinning Programme.

References

[1] A. N. Kolmogorov, C. R. Acad. Sci. URSS, **30**, 301 (1941).

[2] G. K. Batchelor, *Theory of Homogeneous Turbulence* (Cambridge University Press, New York, 1953).

[3] A. Monin and A. Yaglom, *Statistical Fluid Mechanics* (MIT Press, Cambridge 1975).

[4] U. Frisch, *Turbulence: the Legacy of A. N. Kolmogorov* (Cambridge University Press, New York, 1995).

[5] G. Falkovich, I. Kolokolov, V. Lebedev and A. Migdal, Phys. Rev. E**54**, 4896 (1996).

[6] I. Lifshits, S. Gredeskul and A. Pastur, *Introduction to the theory of disordered systems* (Wiley Interscience, New York 1988).

[7] L. N. Lipatov, Sov. Phys. JETP **45**, 216 (1977).

[8] L. D. Landau and E. M. Lifshitz, Course of Theoretical Physics VIII, E. M. Lifshitz and L. P. Pitaevskii, Statistical Physics II, Pergamon Press, Oxford, 1980.

[9] E. I. Kats and V. V. Lebedev, Fluctuational effects in the dynamics of liquid crystals, Springer-Verlag, N.-Y., 1993.

[10] H. W. Wyld, Ann. Phys. **14**, 143 (1961).

[11] P. C. Martin, E. Siggia, and H. Rose, Phys. Rev. A**8**, 423 (1973).

[12] C. de Dominicis, J. Physique (Paris) **37**, c01-247 (1976).

[13] H. Janssen, Z. Phys. B **23**, 377 (1976).

[14] C. de Dominicis and L. Peliti, Phys. Rev. B**18**, 353 (1978).

[15] G. Falkovich and V. Lebedev, Phys. Rev. Lett. **79**, 4159 (1997).

[16] A. Noullez, G. Wallace, W. Lempert, R.B. Miles and U. Frisch, J. Fluid Mech. (to be published).

[17] A. Vincent and M. Meneguzzi, J. Fluid Mech. **225**, 1 (1991).

[18] J. Jimenez, A. Wray, P. Saffman and R. Rogallo, J. Fluid Mech. **255**, 65 (1993).

[19] R. H. Kraichnan and Y. Kimura, in *Progress in Turbulence Research*, edited by H. Branover and Y. Unger (American Institute of Aeronautics and Astronautics, Washington, 1994), pp 19-27.

[20] J. M. Burgers, *The Nonlinear Diffusion Equation* (Reidel, Dordrecht 1974).

[21] R. Kraichnan, Phys. Fluids **11**, 265 (1968).

[22] T. Gotoh and R. Kraichnan, Phys. Fluids **5**, 445 (1993).

[23] T. Gotoh, Phys. Fluids **6**, 3985 (1994).

[24] A.M. Polyakov, Phys. Rev. E**52**, 6183 (1995).

[25] J. Bouchaud, M. Mézard and G. Parisi, Phys. Rev. E**52**, 3656 (1995).

[26] A. Chekhlov and V. Yakhot, Phys. Rev. E**52**, 5681 (1995).

[27] Weinan E, K. Khanin, A. Mazel and Ya. Sinai, Phys. Rev. Lett. **78**, 1904 (1997).

[28] V. Gurarie and A. Migdal, Phys. Rev. E**54**, 4908 (1996).

[29] T. Gotoh and R. Kraichnan (to be published).

[30] E. Balkovsky, G. Falkovich, I. Kolokolov, and V. Lebedev, Phys. Rev. Lett. **78**, 1452 (1997).

[31] M. Avellaneda, R. Ryan and Weinan E, Phys. Fluids **7**, 3067 (1995).

[32] A. M. Obukhov, C. R. Acad. Sci. URSS, Geogr. Geophys., **13**, 58 (1949).

[33] R. Antonia, E. Hopfinger, Y. Gagne, and F. Anselmet, Phys. Rev. A **30**, 2704 (1984).

[34] C. Meneveau and K. R. Sreenivasan, Phys. Rev. A **41**, 2246 (1990).

[35] K. R. Sreenivasan, Proc. R. Soc. Lond. A **434**, 165 (1991).

[36] I. Hosokawa and K. Yamamoto, in *Turbulence and Coherent Structures*, ed. by O. Metais and M. Lesieur (Kluwer, London 1991).

[37] R. H. Kraichnan, Phys. Fluids **11**, 945 (1968).

[38] B. Shraiman and E. Siggia, C. R. Acad. Sc. **321**, Ser. II, 279 (1995).

[39] K. Gawedzki and A. Kupiainen, Phys. Rev. Lett. **75**, 3608 (1995).

[40] M. Chertkov, G. Falkovich, I. Kolokolov and V. Lebedev, Phys. Rev. E **52**, 4924 (1995).

[41] M. Chertkov and G. Falkovich, Phys. Rev. Lett. **76**, 2706 (1995).

[42] D. Bernard, K. Gawedzki and A. Kupiainen, Phys. Rev. E **54**, 2564 (1996).

[43] B. Shraiman and E. Siggia, Phys. Rev. Lett. **77**, 2463 (1996).

[44] E. Balkovsky and D. Gutman, Phys. Rev. E **54**, 4435 (1996).

[45] E. Balkovsky, G. Falkovich, and V. Lebedev, Phys. Rev. E **55**, R4881, (1997).

[46] M. Chertkov, Phys. Rev. E **55**, 2722 (1997).

[47] R. H. Kraichnan, Phys. Rev. Lett. **52**, 1016 (1994).

[48] R. H. Kraichnan, V. Yakhot, and S. Chen, Phys. Rev. Lett. **75**, 240 (1995).

[49] A. L. Fairhall, B. Galanti, V. S. L'vov, and I. Procaccia, Phys. Rev. Lett. **79**, 4166, (1997).

[50] U. Frisch, A. Mazzino, and M. Vergassola, preprint cond-mat/9802192.

[51] B. Shraiman and E. Siggia, Phys. Rev. E **49**, 2912 (1994).

[52] L. Richardson, Proc. Roy. Soc., London A **110**, 709 (1926).

[53] A. N. Kolmogorov, J. Fluid Mech., **12**, 82 (1962).

[54] E. Balkovsky and V. Lebedev, Phys. Rev. E **58**, 5776 (1998).

[55] D. Bernard, K. Gawedzki and A. Kupiainen, preprint cond-mat/9706035.

[56] E. Balkovsky and G. Falkovich, Phys. Rev. E. **57**, 1231 (1998).

Trends in Mathematics, © 1999 Birkhäuser Verlag Basel/Switzerland

A variational approach to turbulent boundary layers

E. A. Novikov, D. G. Dommermuth

Abstract

A variational approach to turbulent boundary layers, leading to an equation for the mean velocity profile, is proposed. A solution of this equation for turbulent flow along a cylindrical surface is obtained and compared with experimental data.

For a statistical description of the Hamiltonian systems, we have the well known principle of maximum entropy. This is related to the Gibbs distribution, which is concentrated on the energy surface in phase space, see, e. g., [1]. For turbulent flows a generalization of the Gibbs ideas was suggested [2] by using the inviscid invariant manifold, which is based on the local form of Kelvin's theorem of conservation of the velocity circulations (see also [3,4]). By averaging over this manifold, the irreversible statistical tendencies in turbulent flows were described. Namely, it was shown [2] that for three-dimensional turbulent flows the energy is transferred towards small scales, while for two-dimensional flows the flux of energy is directed towards large scales and the enstrophy (vorticity) is transferred towards small scales.

In the presence of solid boundaries, the averaging over Kelvin's manifold becomes technically difficult. So, for a description of turbulent boundary layers (TBL) we need a more simple procedure. Such a procedure is suggested below.

The principal parameter of TBL is the momentum flux from the solid boundary. For the case of a flat plate, the momentum flux is defined by:

$$\nu \frac{du(z)}{dz} |_{z=0} \equiv u_*^2 \tag{1}$$

Here ν is the kinematic viscosity, $u(z)$ is the mean velocity profile, z is the distance from the boundary and u_* is the friction velocity. The thickness of the viscous sublayer is defined by $z_* = \nu/u_*$. For $z \gg z_*$, from dimensional arguments [5,6]:

$$\frac{du}{dz} = \frac{u_*}{\kappa z}, \qquad \kappa \approx 0.4, \tag{2}$$

where κ is the Karman constant. Integration of (2) gives the well known logarithmic profile

$$u = \frac{u_*}{\kappa}[\ln(z/z_*) + b], \qquad b \approx 2, \tag{3}$$

where b is another empirical constant, corresponding to a hydrodynamically smooth boundary.

For many applications we need to know the mean velocity profile in the presence of complex three-dimensional boundaries. Thus, we have to go beyond the dimensional arguments. The momentum flux, which generally is a tensor dependent on space coordinates, remains an important factor in a description of TBL. Based on averaging over Kelvin's manifold [2], it seems natural to assume, that a variational approach to TBL may be based on quadratic forms of spatial derivatives of the mean velocity. This leads us to the following equation:

$$\frac{\delta}{\delta u_i(\mathbf{x})}\int(\frac{\partial u_j}{\partial x'_k})^2 d\mathbf{x}' = 2\lambda_{ijklm}\frac{\partial u_j}{\partial x_k}\frac{\partial u_l}{\partial u_m}, \qquad (4)$$

where $\delta/\delta u$ stands for the variational (functional) derivative and summation is assumed over the repeated indexes from 1 to 3 (in the general case). Equation (4) can be written in an operator form:

$$\mathcal{D}f = 2\lambda f. \qquad (5)$$

Here f stands for the quadratic form $\partial u_j/\partial x_k \partial u_l/x_m$. Operator $\mathcal{D}$ includes the functional derivative, summation over indexes (multiplication by $\delta_{jl}\delta_{km}$, where δ_{jl} is the unit tensor) and the space integration. Thus, the tensor λ plays a role of the eigenvalue for the operator $\mathcal{D}$ (or a Liapounov exponent in functional space). The tensor λ has to be connected with the momentum flux. This can be done by an asymptotic analysis or after an appropriate solution of equation (4) is obtained (see below for simple examples). It is easy to see, with a use of the no-slip boundary condition, that the integral in (4) is equal to the enstrophy (integral of the vorticity squared). A derivation of (4), which involves a functional integration [7], will be presented elsewhere. It is assumed that flow is unlimited in any direction normal to the boundary, otherwise (4) has to be modified.

By calculating the functional derivative in (4), we get equation:

$$\Delta u_i + \lambda_{ijklm}\frac{\partial u_j}{\partial x_k}\frac{\partial u_l}{\partial x_m} = 0, \qquad (6)$$

where Δ is the Laplace operator. We can see that (6) is consistent with the logarithmic profile (2) for distances smaller than the radius of the local curvature of the boundary. At the same time, in the case of a finite three-dimensional body in an uniform stream, (6) is consistent with a potential flow at distances bigger than the size of the body.

Let us consider the TBL over the plane boundary. In this case we have only one component of the mean velocity $u \equiv u_1$, which depends only on one coordinate $z \equiv x_3$, and we have only one component of the eigenvalue $\lambda \equiv \lambda_{11313}$. Equation (6) takes the form:

$$\frac{d^2u}{dz^2} + \lambda(\frac{du}{dz})^2 = 0. \qquad (7)$$

This equation, obviously, has solution $\frac{du}{dz} = (\lambda z)^{-1}$, which coincides with (2) for $\lambda = \kappa/u_*$. We note, that (7) with constant λ, as well as (6), is not designed for the viscous sublayer ($z \leq z_*$). If we formally apply (7) to the viscous sublayer with the no-slip boundary condition, we get incorrect constant b in (3).

Now we turn to turbulent flow along a cylinder with radius a. In this case, in cylindrical coordinates, we also have only one component of the mean velocity u (in the direction of the axis of the cylinder), which depends on one coordinate r, and we have only one eigenvalue λ. Equation (6) takes the form:

$$\frac{d^2u}{dr^2} + \frac{1}{r}\frac{du}{dr} + \lambda(\frac{du}{dr})^2 = 0. \qquad (8)$$

The general solution of this equation is:

$$u\lambda = \ln[\ln(r/a) + c] + \beta, \qquad (9)$$

where c and β are nondimensional constants of integration. We require that (9) coincides with (3) in the distance diapason $z_* \ll z \equiv r - a \ll a$. This leads to: $\lambda = \kappa/u_*, c = 0, \beta = b + \ln a_+$, where $a_+ \equiv a/z_*$. Formula (9) takes the form:

$$u = \frac{u_*}{\kappa}\{\ln[a_+ \ln(z_+/a_+ + 1)] + b\}, \qquad (10)$$

where $z_+ \equiv z/z_*$.

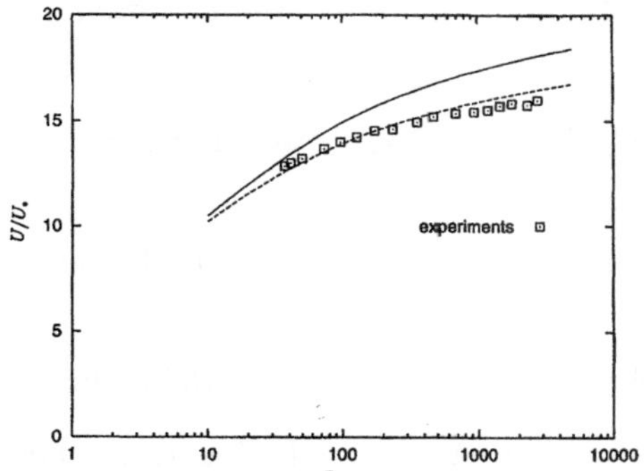

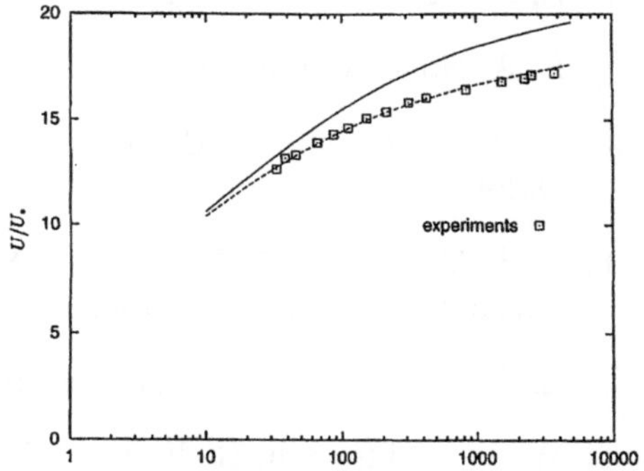

Figure 1: Comparison of the double-logarithmic velocity profile (10) with experimental data [8] for turbulent flow along a rod: (a) radius of the rod $a = 0.01$in., $a_+ \approx 46$ – estimation from data (continuous line), $a_+ = 20$ – best fit (broken line); (b) $a = 0.02$in., $a_+ \approx 84$ – estimation from data, $a_+ = 30$ – best fit.

Figures 1(a) and (b) show comparison of formula (10) with experimental data [8] for two values of a_+. We have chosen the data with the biggest deviation from the logarithmic profile. We see that theoretical curves are sufficiently close to experimental points, which are well described by the double-logarithmic profile. But, there is a systematic difference, which is interpreted as a difference in the parameter a_+. A possible explanation of this difference is that the TBL near very thin rods (with radii 0.01in. and 0.02in.) may not be well developed and the measurements of the wall shear stress, related to parameter a_+, involve some uncertainties and interpolations [8].

We plan to do analytical and numerical studies of equations (6) for various boundaries, including connecting the tensor λ with corresponding momentum fluxes. We also plan to combine equations (6) with the large-eddy simulations of turbulent flows. We hope that presented variational approach to TBL and equations (6) will open another direction in turbulence research and applications.

EAN is supported in this work by the Office of Naval Research under Grant ONR-N00014-97-0186 and by the Department of Energy under Grant DE-FG03-91ER14188. DGD is Supported by the Office of Naval research under Grant ONR-N00014-97-C-0345

References

[1] L. D. Landau and E. M. Lifshitz, *Statistical Physics, Part 1,* Pergamon Press, 1980

[2] E. A. Novikov, in *Turbulent Motions*, ed. V. Struminsky, (Nauka (Science), Moscow, 1974), p 85; Archives of Mech., **12** (7), 741 (1974)

[3] E. A. Novikov, Soviet Physics JETP, **57** (3), 566 (1983)

[4] E. A. Novikov, Phys. Rev. E, **52**, 2574 (1995)

[5] L. D. Landau and E. M. Lifshitz, *Fluid Mechanics*, Butterworth-Heinemann, Oxford, 1987

[6] A. S. Monin and A. M. Yaglom, *Statistical Hydromechanics*, v. 1, Hydrometeoizdat, St. Petersburg, 1992

[7] E. A. Novikov, Progr. Math. Scien. (Ac. Scien. USSR), **16** (2), 135 (1961)

[8] W. W. Willmarth, R. E. Winkel, L. K. Sharma and T. J. Bogar, J. Fluid Mech., **76** (1), 35 (1976)

Trends in Mathematics, © 1999 Birkhäuser Verlag Basel/Switzerland

Scaling laws of a passive scalar energy flux in fully developed turbulence

S. Ciliberto, E. Leveque, C. Baudet, G. Ruiz-Chavarria

Abstract

The cascade scaling properties of a passive scalar, here temperature, in turbulence are studied. Firstly, $\delta u_\ell \delta\theta_\ell^2/\ell \sim N_\ell$ is proposed as a statistical advection-diffusion balance at scale ℓ, between the scalar dissipation rate N_ℓ and the advective flux $\delta u_\ell \delta\theta_\ell^2/\ell$, of thermal energy. This reasoning points out a statistical similarity between kinetic and thermal energy fluxes.

The transport of passive scalars, such as a dye or temperature fluctuations, by a turbulent flow has long been a subject of fundamental [1] and pratical interests [2]. A question which remains open is how the statistical properties of a scalar field $\theta(\vec{r})$ are related with tnose of the velocity field $\vec{u}(\vec{r})$. Amog the different approaches, used to characterized the scalar statistics, a very common one is the scalar structure functions, $< \delta\theta^p >=< |\theta(\vec{r}+\vec{\ell}) - \theta(\vec{r})| >$, as a function of the scale ℓ. Here $< . >$ stands for ensemble average. At very high Reynolds number Re, $< |\delta\theta_\ell|^p > \propto \ell^{\xi_p}$ for ℓ in the inertial range, that is $L >> \ell >> \eta_\theta$ where L is the integral scale and η_θ is the scalar dissipation scale. Experimental observations have ensured the existence of such scaling laws [3] and motivated today many efforts to predict the scaling exponents ξ_p [1]. Many recent theoretical studies focused on the prediction of the exponent ξ_p in the case of a gaussian velocity field. This approach to the problem, although irrealistic because of the gausian approximation, is actually giving interesting informations on the scaling laws of a turbulent field [1].

An important difficulty in this study of the passive scalar statistics is the inexistence of an exact scaling relation for the scalar, analogous to *Kolmogorov's 4/5 law* for the velocity [4]. On the other hand, a scaling equation can be established for the scalar energy flux [5]:

$$\langle \delta u_\ell \delta\theta_\ell{}^2 \rangle = -4/3\ N\ell \tag{1}$$

where $\delta u_\ell = [\vec{u}(\vec{r}+\vec{\ell},t) - \vec{u}(\vec{r},t)].\vec{\ell}/\ell$ is the longitudinal velocity increment, $N = \chi\langle|\nabla\theta|^2\rangle$ is the (constant) mean scalar dissipation rate and χ is the diffusivity of the fluid. An other major difference between the velocity and scalar structure functions concerns the property of Extended Self Similarity [6]. Scaling laws, as mentioned previously, are strictely defined in the fully developed turbulence regime. In practice, this asymptotic regime is approached but never achieved in the sense that, one can never exactely exhibit a finite inertial-range of scales with pure power-law dependence. Deviations are observed and are related to small but non-negligeable dissipative effects. When considering structure functions against others, these deviations mainly cancel out such that one recovers power-law scalings over an extended range. In the case of the $\langle|\delta\theta_\ell|^p\rangle$'s, it has been recently demonstrated experimentally that the range of self-similarity is not as extended as for the velocity structure functions [3]. These experimental and theoretical facts motivate us to pursue a different approach by focusing on the scaling properties of the structure functions of the scalar energy flux: $\langle|\delta u_\ell \delta\theta_\ell{}^2|^{p/3}\rangle$. In this paper we present some experimental results, which provide a new insight to the scalar statistical properties. The passive scalar here considered is temperature.

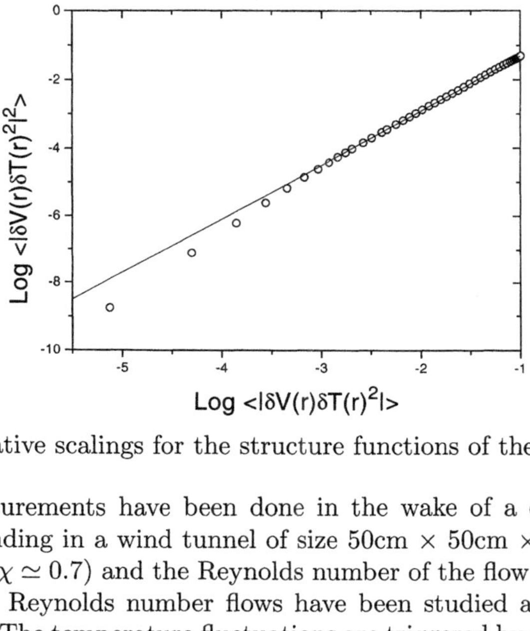

Figure 1: Relative scalings for the structure functions of the scalar energy flux.

The measurements have been done in the wake of a cylinder of diameter $D = 10$cm standing in a wind tunnel of size 50cm $\times$ 50cm $\times$ 300cm []. The fluid is air ($Pr = \nu/\chi \simeq 0.7$) and the Reynolds number of the flow is $Re = U_{rms}D/\nu \simeq$ 36000. Various Reynolds number flows have been studied and led to consistent similar results. The temperature fluctuations are triggered by the electrical heating of an array of parallel fine wires. This array, known as mandoline, permits an effective mixing of heat without mean temperature gradient. The mandoline stands vertically 40cm downstream the cylinder. In order to measure simultaneously the velocity and temperature fluctuations, two probes have been placed apart at a distance of 1mm. The temperature probe works as a cold wire, whereas the velocity probe works as a hot wire. A special attention has been paid to avoid interference between the two probes. These measurements have been made 3m downstream the cylinder. Various distances have been studied but led to similar results. Local time measurements have been transformed into spatial measurements by using Taylor's *frozen flow* hypothesis: $\ell = U_{rms}\tau$, justified for a turbulence rate $\mathcal{T} \simeq 10\%$.

As expected, the structure functions $\langle|\delta u_\ell \delta\theta_\ell{}^2|^{p/3}\rangle$, computed here as time average, do not exhibit a clear inertial-range with power-law dependence on ℓ. We rather focus on the relative scalings by studying $\langle|\delta u_\ell \delta\theta_\ell{}^2|^{p/3}\rangle$ as a function of $\langle|\delta u_\ell \delta\theta_\ell{}^2|\rangle$. In a log-log coordinate system, a clear scaling range is observed. In the dissipative range, the trivial scaling exponent $p/3$ is recovered. The property of E.S.S. is well satisfied for the structure functions of the energy scalar flux. This observation suggests that the E.S.S. property rather applies to fluxes: $\delta u_\ell{}^3/\ell$ and $\delta u_\ell \delta\theta_\ell{}^2/\ell$, and thus appears to be closely related to the process of energy transfer. It enables to access to an accurate estimation of the *asymptotic* scaling exponents Ψ_p, by taking into account that $\Psi_3 = 1$ from (1). We have actually checked that $\langle\delta u_\ell \delta\theta_\ell{}^2\rangle \sim \langle|\delta u_\ell \delta\theta_\ell{}^2|\rangle$. We have evaluated the relative scaling exponents Ψ_p/Ψ_3 by least squares fitting, with an uncertainty of the order of one percent. The Ψ_p's are reported in Table 1.

Table 1: Scaling exponents of scalar energy flux and velocity.

order p	1	2	3	4	5	6	7	8
Ψ_p	0.4	0.73	1.00	1.23	1.43	1.60	1.80	1.95
ζ_p	0.36	0.70	1.00	1.27	1.53	1.77	2.01	2.23

In the third line of Table 1, the relative scaling exponents for the velocity structure functions are very closed, as expected, to those usually measured in homogeneous and isotropic fully developed turbulence. In this experiment $\Psi_p \not\simeq \zeta_p$; the kinetical and scalar energy transfer do not follow similar scalings.

We focus now on the coarse-grained thermal energy dissipation rate, approximated by the 1D surrogate $N_\ell = 1/\ell \int_x^{x+\ell} N(x')dx'$ where $N(x) = 3\chi(\partial u/\partial x)^2$. We can test whether N_ℓ satisfies the closure condition proposed in the model of hierarchy [7]:

$$\frac{N_\ell^{(p+1)}}{N_\ell^{(\infty)}} \sim \left(\frac{N_\ell^{(p)}}{N_\ell^{(\infty)}}\right)^\beta, \tag{2}$$

where β is a constant independent of ℓ and $N_\ell^{(p)} = \langle N_\ell^{p+1}\rangle/\langle N_\ell^p\rangle$.

In order to test the relevance of (2) for the scalar energy dissipation rate, we display (Figure 2) $Y_{n+1} = N_{\ell_2}^{(n+1)}/N_{\ell_1}^{(n+1)}$ as a function of $Y_n = N_{\ell_2}^{(n)}/N_{\ell_1}^{(n)}$ for different scales ratios ℓ_2/ℓ_1 and various n [8]. The data points are aligned for fixed scale ratios ℓ_2/ℓ_1, which indicates that the closure relation (2) is well satisfied, with β given by the slope. The different lines, corresponding to different ratios ℓ_2/ℓ_1, are parallel which means that β is constant and independent of scale. A least squares fitting leads to the value $\beta = 0.4 \pm 0.02$.

A final point regarding the scalar energy dissipation moments concerns the generalisation of the Refined Kolmogorov Similarity Hypothesis (RKSH), which has been proposed for the velocity field [10, 11]. The generalisation of RKSH for the velocity structure functions simply states that

$$\langle|\delta v|^p\rangle \propto \langle\epsilon_\ell^{p/3}\rangle\langle\delta v^3\rangle^{p/3} \tag{3}$$

where ϵ_ℓ the energy dissipation on the scale ℓ. Eq. (3), which has been called Generalized ESS (G-ESS), holds from the largest to the smallest scale for all kind of

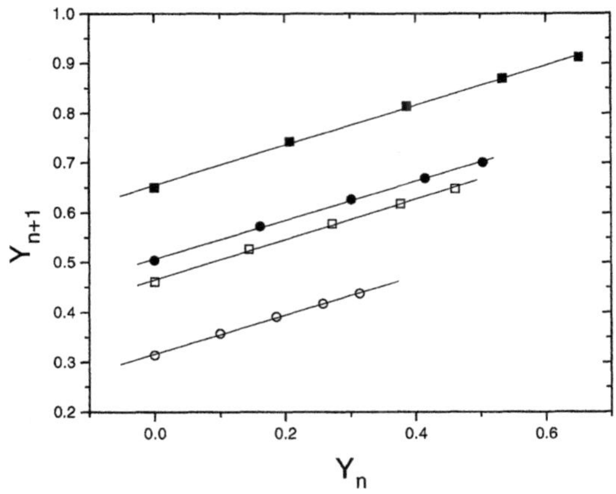

Figure 2: Y_{n+1} as defined in the text versus Y_n for various arbitrary scale ratios ℓ_2/ℓ_1. the values of n are 0, 0.25, 0.50, 0.75 and 1.0.

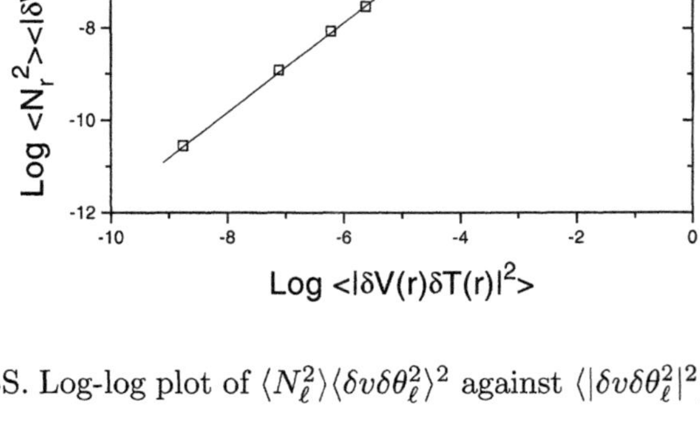

Figure 3: GESS. Log-log plot of $\langle N_\ell^2\rangle\langle\delta v\delta\theta_\ell^2\rangle^2$ against $\langle|\delta v\delta\theta_\ell^2|^2\rangle$. The straight line has slope 1.

turbulent flows, even when ESS is not verified [11]. We have checked if an equation similar to Eq. 3 could be applied to the scalar energy dissipation. Specifically we have verified the following relationship for the scalar structure functions:

$$\langle|\delta v \delta\theta_\ell^2|^p\rangle \propto \langle N_\ell^{p/3}\rangle \langle \delta v \delta\theta_\ell^2\rangle^{p/3} . \tag{4}$$

In Fig. 3 the quantity of the right hand side of Eq. (4) is plotted against the one of the left hand side in a log-log plot, for $p = 2$. The straight line of slope 1 indicates that Eq. (4) is quite well verified for all scales, indicating that GESS is also a property of the scalar energy flux. It is important to notice that a similar equation does not hold if the scalar structure functions are used instead of the scalar energy flux [3, 11].

These results provide a new insight on the scalar statistics by focusing on the cascade-related quantities $\delta u_\ell \delta\theta_\ell{}^2$ and N_ℓ. Robust properties, shared with the velocity turbulent field, are then observed: the exact scaling relation (1), the property of Extended Self-Similarity, the statistical closure relation (2) and the GESS (4). Finally, beyond these results on the scalar, this study augurs the existence of a unified framework to describe the scaling properties of a turbulent system.

References

[1] R.H. Kraichnan, *Phys. Rev. Lett.* **72**, 1016 (1994). V.S. Levov, I. Procaccia and A.L. Fairhaell, *Phys. Rev. E* **50**, 4684 (1994). G. Eyink, *Phys. Rev. E* **54**, 1497 (1996). R.H. Kraichnan, V. Yakhot and S. Chen, *Phys. Rev. Lett.* **75**, 240 (1995). S. Chen and N. Cao, *Phys. Rev. Lett.* **78**, 3459 (1997).

[2] See for review: K. Sreenivasan and R.A. Antonia, *Annu. Rev. Fluid. Mech.* **29**, 435 (1997).

[3] G. Ruiz-Chavarria, C. Baudet and S. Ciliberto,*Physica D* **99**, 369 (1996).

[4] A. N. Kolmogorov, *C. R. Acad. Sci. URSS* **30** 301 (1941).

[5] A. S. Monin and A. M. Yaglom, Statistical Fluid Mechanics, M.I.T. Press, Cambridge, Massachussetts, 1975.

[6] R. Benzi, S. Ciliberto, C. Baudet. F. Massaioli, R. Tripicione, and S. Succi, *Phys. Rev. E* **48**, 29 (1993).

[7] Z.-S. She and E. Leveque, *Phys. Rev. Lett.* **72**, 336 (1994).

[8] G. Ruiz-Chavarria, C. Baudet and S. Ciliberto, *Phys. Rev. Lett.* **74**, 1986 (1995).

[9] A. N. Kolmogorov, *J. Fluid Mech.* **13**, 82–85 (1962).

[10] R. Benzi, S. Ciliberto, C. Baudet, G. Ruiz-Chavarria, *Physica D* **80**, 385 (1995).

[11] R. Benzi, L. Biferale, S. Ciliberto, M. V. Struglia, R. Tripiccione *Physica D* **96**, 162 (1996).

Trends in Mathematics, © 1999 Birkhäuser Verlag Basel/Switzerland

Intermittency in channel-flow turbulence

F. Toschi, G. Amati, R. Piva, S. Succi, R. Benzi

In this work we present a study of intermittency in a channel, *i.e.* in a non-homogeneous, non-isotropic flow. Scaling exponents of longitudinal streamwise structure functions, ζ_p, are used as quantitative indicators of intermittency. We find that, near the core of the channel, the value of ζ_p are consistent with homogeneous/isotropic values. Near the boundaries intermittency is much higher approaching passive scalar values.

1 Introduction

Spatio-temporal intermittency is one of the most intriguing properties of fluid dynamic turbulence. Although many efforts have been devoted to the understanding of intermittency in homogeneous and isotropic flows, much less has been done in situations where the flow is no longer homogeneous and isotropic. Structure functions built with velocity differences, $S_p(r) = \langle (v(x+r) - v(x))^p \rangle$, scales homogeneously in the inertial range, *i.e.* $S_p \sim r^{\zeta_p}$. Intermittency reflects itself in anomalous values (*i.e.* different from the celebrated Kolmogorov 1941 prediction, $\zeta_p = p/3$) of the ζ_p (see for example [1]).

For homogeneous and isotropic turbulence several models have been proposed to explain the experimentally measured values of ζ_p, in particular we cite the She-Leveque model [2] as one which has proved to be quite accurate if compared with experimental data.

Here we investigate numerically a channel flow (see [4]) possibly the simplest instance of non-homogeneous and non-isotropic flow, our purpose being to start a general study on intermittency in non-homogeneous and non-isotropic flows.

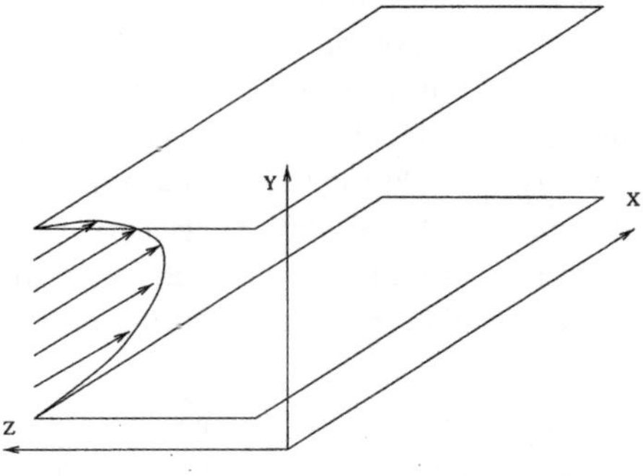

Figure 1: Sketch of the numerical setup.

The paper is organized as follows. In Section 2 we present some details of our numerical simulations and discuss some general characteristics of a channel flow. In Section 3 we present our results concerning intermittency and in Section 4 conclusions are drawn.

2 Turbulent channel flow

In order to study the intermittency properties of a channel flow [3, 4] we have performed a number of numerical simulation achieving a high statistical accuracy (of about 200 eddy turnover times of the slower scale). Numerical simulations have been performed on a APE100 massively parallel machine using a LBE (Lattice Boltzmann Equation) code. The spatial resolution of the simulation was of $L_x \cdot L_y \cdot L_z = 256 \cdot 128 \cdot 128$ grid points. Boundary condition where periodic in the x and z directions and no-slip on top ($y = 127$) and bottom ($y = 0$) planes (see Figure 1). The fluid is driven by a volume forcing (acting as a "gravitational" force) in positive x direction. The Reynolds number of our numerical simulations was $Re \sim 3000$. Further details about the numerical code can be found in [5] and references therein.

Due to the particular choice of boundary conditions and forcing, two boundary layers develop near to the top/bottom planes. Roughly speaking the flow can be divided into three regions. A laminar buffer which extends only a few grid points (~ 5) above the horizontal planes, a bulk region occupying the core of the channel and a logarithmic turbulent boundary layer region in between. In Figure 2 we report the mean value of the streamwise velocity, $\langle v_x(y) \rangle$, as a function of y. The three distinct regions are also shown in Figure 2.

We will not enter here the details about helical coherent structures which develop in the boundary layers, which represent an interesting subject on its own (e.g. see [6]). Here we just want to study the statistical properties of a channel and in particular their dependence on the distance from the walls. The following discussion has to be taken just as a simple phenomenological introduction whose aim will be to clarify our results on intermittency of Section 3.

On general grounds we expect that the core of the channel can be assumed to be (locally) homogeneous and isotropic, while this is not the case for the flow near to the boundaries. Indeed it is well known that in a channel flow, inside the boundary layers, instabilities develop giving rise to coherent structures. These structures are ultimately responsible for the development and sustainment of turbulence in the whole channel. As these structures evolve they break and mix and are eventually ejected into the core of the channel. We remark that near the walls these structures are mainly aligned streamwise, giving rise to non-trivial anisotropy effects in the $x - z$ plane.

As a result, if the flow can be locally regarded as nearly homogeneous-isotropic near the core of the channel, it cannot be considered y-homogeneous and not even $x - z$ isotropic near the walls.

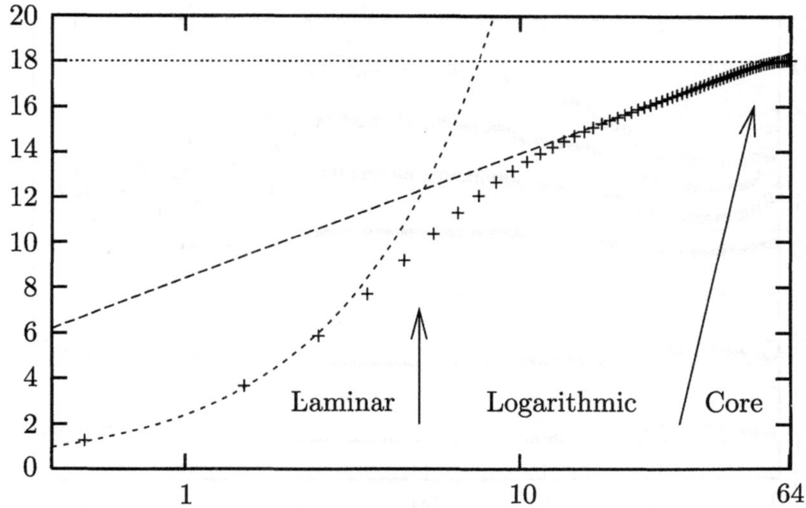

Figure 2: Log-Lin plot of mean value of streamwise velocity $\langle v_x(y) \rangle$ (in a.u.) as a functions of y (in lattice units). Only half of the channel is shown. Laminar, logarithmic and core regions are indicated.

Furthermore in the boundary layers we will have, at the same time, presence of almost homogeneous/isotropic spots of turbulent fluid and coherent structures not yet mixed. A crude approximation could be to think at "a two fluid" picture. If this picture is correct we then expect to find higher intermittency in this region than in the nearly homogeneous/isotropic one. As we want to study intermittency in the whole channel we introduce the following y-dependent longitudinal streamwise structure functions:

$$S_p(r, y) = \langle (v_x (x + r, y, z) - v_x (x, y, z))^p \rangle \tag{1}$$

The averaging procedure is of course done on slice of given y value. These $S_p(y)$ structure functions are the central objects of the following analysis.

3 Results on intermittency

Our quantitative intermittency indicators, the exponents $\zeta_p(y)$, are, in principle, y-dependent, as they are extracted from structure functions $S_p(y)$ which depend on y. In order to perform the analysis we have measured longitudinal structure functions for every y, then we have extracted intermittency exponent using the, by now standard, ESS technique (see for details [7, 8]). The need for using ESS is evident since for *Re* numbers as low as 3000 no inertial range is present if looking at structure functions in real space. Of course using this procedure we are unable to assess the y-dependence of ζ_3. The ESS exponents have been extracted performing

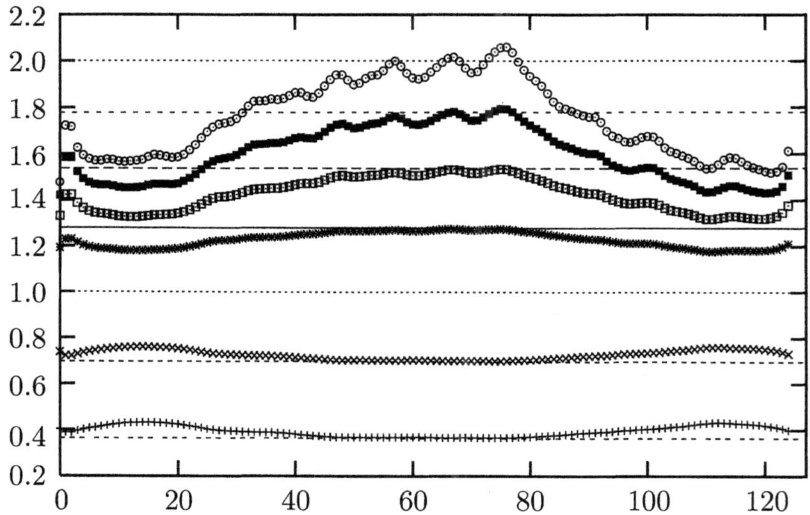

Figure 3: Behavior of $\zeta_p(y)$ exponents as a function of y, for (from bottom to top) $p = 1, 2, 4, 5, 6, 7$. Horizontal lines correspond to She-Leveque ζ_p values [2] for homogeneous/isotropic turbulence.

a fit of S_p versus S_3 for $20 \leq r < 40$ (in lattice units). This r-window has been chosen in order to be inside an ESS local-slope plateau both in the boundary layer ($y \sim 15$) and in the core of the channel ($y \sim 64$). Keeping the r-window fixed we have extracted relative intermittency exponents, $\zeta_p(y)$, for every y. In Figure 3 we plot the values of $\zeta_p(y)$ obtained for $p = 1, 2, 4, 5, 6, 7$. We want to stress that the way we extracted ζ_p/ζ_3 is not dictated by any prejudice of expecting a pure scaling law of structure functions (inside the chosen r-window) for each y value. The procedure chosen is intended to be the most objective one if one wants "effective" exponents for each y without introducing bias due an arbitrary choice of a different r-window for each y value.

A first observation is that the flow is much more intermittent near the walls than in the center of the channel. Even if $\zeta_p(y)$ don't show huge variations with y (but something around a 20%) their dependence with y is evident and definitely outside error-bars. A second observation is that homogeneous/isotropic commonly accepted values for ζ_p (drawn as straight line in Figure 3) are indeed very close to the $\zeta_p(y)$ values measured in the core of the channel. This confirms our expectation (see Section 2) that statistical properties of the core of the channel should resemble that of homogeneous/isotropic turbulence.

More interesting is the behavior of $\zeta_p(y)$, both as function of p and y, outside the core of the channel. From a detailed data analysis we found that it can be

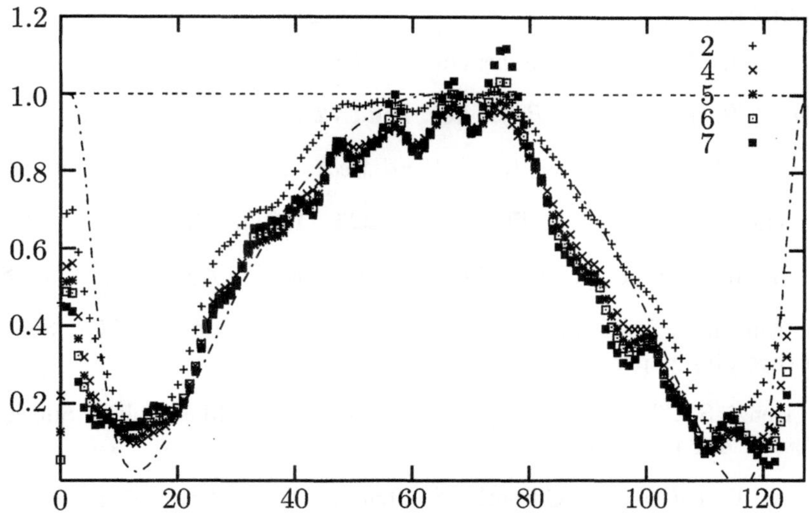

Figure 4: Data represents $\frac{\zeta_p(y)-\zeta_p^{PS}}{\zeta_p^{SL}-\zeta_p^{PS}}$ as a function of y, ζ_p^{PS} being the passive scalar [9] and ζ_p^{SL} She-Leveque values [2]. Data for $p = 2, 4, 5, 6, 7$ are plotted. Dot-dashed line is proportional to the function $\langle (v_x(y) - \overline{v}_x(y)) (v_y(y) - \overline{v}_y(y)) \rangle^2$.

linked to a particular component of the Kolmogorov stress tensor (see Figure 4):

$$\langle (v_x(y) - \overline{v}_x(y)) \cdot (v_y(y) - \overline{v}_y(y)) \rangle^2 \tag{2}$$

We remark that also the square of the vorticity, $\langle \omega^2(y) \rangle$, has a similar behavior [6]. In Figure 4 we show how all exponents, $\zeta_p(y)$, collapse on a unique curve and how they have the tendency to approach passive scalar values [9] near to the boundaries (but not so near to be inside the viscous boundary layer).

4 Conclusions

In this work we have presented clear evidences of increased intermittency (with respect to homogeneous/isotropic turbulence) for a non-homogeneous, non-isotropic flow. The indicators we have used were intermittency scaling exponents of longitudinal streamwise structure functions. Other indicators as, for example, velocity p.d.f. could as well have been used, confirming the same trend (see for example [6]). In particular we confirmed the almost homogeneous/isotropic character of turbulence around the core of the channel. We obtained inside the turbulent boundary layers exponents strongly intermittent and remarkably similar to the passive scalar values. Finally we remark that we have further evidences (also coming from different kind of flows, see [10, 11]) that this behavior is quite general in non-homogeneous and non-isotropic flow.

References

[1] U. Frisch, *Turbulence*, Cambridge U.P., Cambridge (1995).

[2] She, Z. S., Leveque, E., *Phys. Rev. Lett.*, **72-3**, (1994), 336–339.

[3] Kim, J., Moin, P., Moser, R., *J. Fluid Mech.*, **177**, (1987), 133–166.

[4] Jimenez, J., Moin, P., *J. Fluid Mech.*, **225**, (1991), 213–240.

[5] G. Amati, S. Succi, R. Piva, *Int. Jour. of Mod. Phys. C*, **8-4**, (1997), 869.

[6] G. Amati, F. Toschi, S. Succi, R. Piva, Contribution to ETC7. See also a forthcoming paper.

[7] R. Benzi, S. Ciliberto, R. Tripiccione, C. Baudet, F. Massaioli, S. Succi, *Phys. Rev. E*, **48**, **1**, (1993), R29.

[8] Benzi, R. Biferale, L. Ciliberto, S. Struglia, M. V., Tripiccione, R., *Physica D*, **96**, (1996), 162–181.

[9] Passive scalar values from S. Ciliberto private communication.

[10] Chillá, F., Pinton, J. F., R. Labbé, *Europhys. Lett.*, **35-4**, (1996), 271–276.

[11] Toschi, F., Chillá, F., Pinton, In preparation.

Trends in Mathematics, © 1999 Birkhäuser Verlag Basel/Switzerland

Structure functions in homogeneous and non-homogeneous turbulent flows

P. Orlandi, R. A. Antonia, P. G. Esposito

Abstract

The extended self-similarity or ESS method was applied to DNS data obtained in different flows, decaying isotropic turbulence and a fully developed turbulent channel flow, and shown to yield satisfactory estimates of the scaling exponents. In our view, ESS should be treated as a useful method though one which does not improve our understanding of the well documented departure from the Kolmogorov (1941a) scaling. The present results tend to confirm the importance of coherent structures in the context of this departure.

1 Introduction

One of the principal characteristics of turbulent flows is its intermittent nature, reflecting the presence of relatively rare and intense events. The intermittency can be quantified by the departure of scaling exponents of velocity structure functions from the Kolmogorov (1941a), hereafter K41, prediction. Earlier estimates of these exponents were made using $\langle \delta u^n(r) \rangle \sim r^{\zeta_p}$ ($\delta u(r)$ is the longitudinal velocity increment $u(x+r) - u(x)$; for convenience, the explicit dependence on r will not be indicated hereafter) when the separation r lies within the inertial range (IR). For K41, $\zeta_p = p/3$. Values of ζ_n obtained in this manner were reported for both experiments (e.g. Antonia *et al.*, 1982; Anselmet *et al.*, 1984) and simulations (e.g. Vincent and Meneguzzi, 1991) over a wide range of R_λ, the Taylor microscale Reynolds number.

Like Vasilenko *et al.* (1975) and Antonia *et al.* (1982), Anselmet et al. (1984) observed that by plotting $\langle (\delta u)^p \rangle$ vs $\langle (\delta u)^{p+2} \rangle$, with p even, the scatter was reduced while the range over which a power-law occurred was extended. This idea was adopted, with some modification (namely $\langle |\delta u|^p \rangle$ was plotted against $\langle |\delta u|^3 \rangle$), by Benzi *et al.* (1993). They called the technique "extended self-similarity" (ESS) emphasizing that the scaling now extended down through the dissipative range thus enabling ζ_p to be estimated even when R_λ is quite small (e.g. Camussi *et al.*, 1996). This claim was queried by Stolovitzky and Sreenivasan (1993) who noted that, as p increased, the inertial and dissipative ranges exhibited clearly different scalings. In spite of this observation, there is much evidence to suggest that ESS can potentially yield reliable estimates of ζ_p provided the limitations of the technique are understood. Recently, significant attention has been devoted to the different scaling behaviors of longitudinal and transverse velocity increments, e.g. the direct numerical simulations of isotropic turbulence of Boratav and Pelz, 1997 and Chen *et al.*, 1997. In each of these two papers, ESS was one of the methods used for estimating ζ_p.

The interpretation of results obtained by applying ESS to low Reynolds number flows is somewhat delicate. If there is any physics implicit in ESS, it would seem to be that intermittency is present even when the small scales are comparable in size to the energy containing eddies. This contradicts the usual belief that the intermittency is related to structures with strong vorticity localized in tubular regions of radius much smaller than the integral scale of turbulence (e.g. Jimenez *et al.*, 1993). The validity of these conjectures can be tested by doing numerical simulations of isotropic turbulence at small Reynolds numbers. More importantly, one could enquire as to whether the same results can be obtained in anisotropic or inhomogeneous turbulence. Only if a universal behavior of intermittency is found or if the intermittency can be linked in a satisfactory way to the external conditions of the flows, the extended ESS scaling could then allow improvements of subgrid models used in Large Eddy Simulations. Although we are sceptical about this possibility, we think that it is worth analysing several data sets from direct simulations of isotropic, homogeneous (but anisotropic) and non-homogeneous turbulence to determine whether ESS gives at least scaling exponents accounting for the different conditions. Benzi *et al.* (1995) proposed a variation of ESS, which they called "generalised extended self-similarity" or GESS. Instead of plotting $\log F_p^*$, with $F_p^* \equiv \langle|\delta u|^p\rangle$ against $\log F_3^*$ (ESS), $\log G_p$, with $G_p^* = \langle|\delta u|^p\rangle / \langle|\delta u|^3|\rangle^{p/3}$, was plotted against $\log G_q^*$. The exponents ζ_p were inferred from $\rho(p,q)$ in the relation $G_p^* = G_q^{*\rho(p,q)}$. They noted that, especially at large r, oscillations were further reduced relative to ESS. Since non-uniformities in the structure functions are expected to increase as the inhomogeneity of the flow increases, GESS should be more effective in evaluating ζ_p. In the present case, GESS was tested and the results obtained were not as reliable as those from ESS.

2 Discussion and results

The rationale (Benzi *et al.*, 1993) in using F_3^* as the reference in ESS was that, according to Kolmogorov (1941b), $\langle(\delta u)^3\rangle = 4/5\langle\epsilon\rangle r$ in the IR. Benzi *et al.* (1993) assumed that F_3^* is equivalent to $|F_3| \equiv |\langle(\delta u)^3\rangle|$. This equivalence needs to be carefully assessed. While Boratav & Pelz (1997) found that F_3 is proportional to F_3^*, Pearson and Antonia (1997) noted that $|F_3| \sim F_3^{*\alpha}$ with the constant of proportionality less than 1 and α, of order 0.9, showing a weak dependence on R_λ. In our case, because of the reduced number of samples, we found that for isotropic turbulence at three different Reynolds numbers, $|F_3|$ and F_3^* behave as in Fig. 1a. The proportionality observed at small r disappears at large r. For the present (160^3) finite-difference direct simulation of decaying isotropic turbulence, the results at small r should be trustworthy since for $r = 2\pi/160$, $|F_3|/F_2 \approx 0.5$. This absolute value of the velocity derivative skewness agrees with that usually

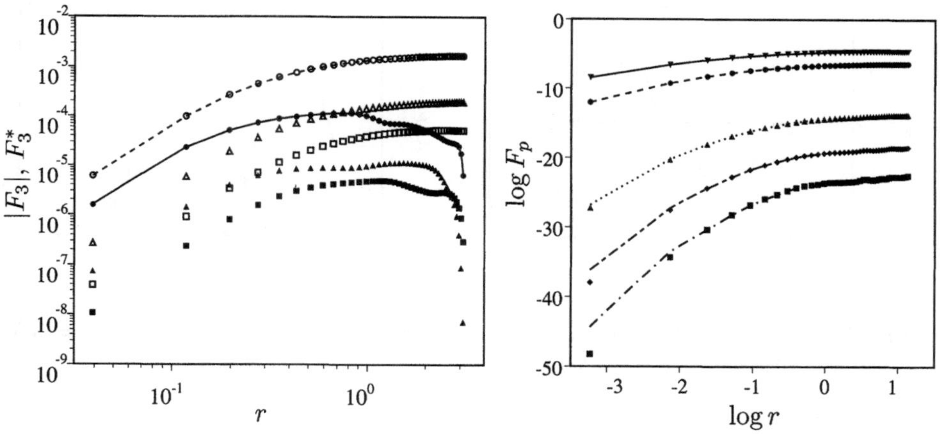

Figure 1: a) Distributions of $|F_3|$ (—— and closed symbols) and F_3^* (---- and open symbols) vs r at ○ $R_\lambda = 59.8$, △ at 40.3 and □ at 36.1; b) profiles of F_p vs r for ——, ∇ $p = 2$, ----, ○ $p = 3$, ········, △ $p = 6$, — — —, ◇ $p = 10$, —·—, □ $p = 14$. The lines are for directly evaluated moments, the symbols refer to Eq. (1).

obtained, from either experiment or numerical simulation, at comparable values of R_λ (e.g. Sreenivasan & Antonia 1997). At large values of r, the disagreement could be ascribed to the reduction in the number of independent samples as r increases, thus causing noisy pdf tails. These oscillations have a large effect on the calculation of the odd F_p but not on F_p^*. In view of this uncertainty, we decided to relate all even-order moments of δu to F_2. We are aware that this replaces one ambiguity by another, since a value for ζ_2 must be imposed; accordingly, we assumed the validity of the log-normal model, i.e. $\zeta_p = p/3 - \mu p(p-3)/18$, $\mu = 0.2$ *(or* $\zeta_2 \simeq 0.69$*)* providing satisfactory agreement with existing data for $p = 2$. We have noted however that the estimates for ζ_p, obtained in this manner, are equal to those obtained using F_3^*.

In experimental work, Taylor's hypothesis is almost invariably invoked in generating δu and thus a large number of data are needed to achieve a satisfactory closure of the integrand in

$$F_p \equiv \langle(\delta u)^p\rangle = \int_{-\infty}^{\infty} (\delta u)^p p(\delta u) d(\delta u) \quad , \tag{1}$$

where $p(\delta u)$ is the probability density function of δu. In numerical simulations of stationary turbulence, more than one field can be used, thus enabling accurate values of $p(\delta u)$ to be obtained at large r. In decaying isotropic turbulence, if a single simulation is performed, a large number of grid points, e.g. 300^3 for Boratav & Pelz (1997), is required to achieve accurate results. For the present simulation

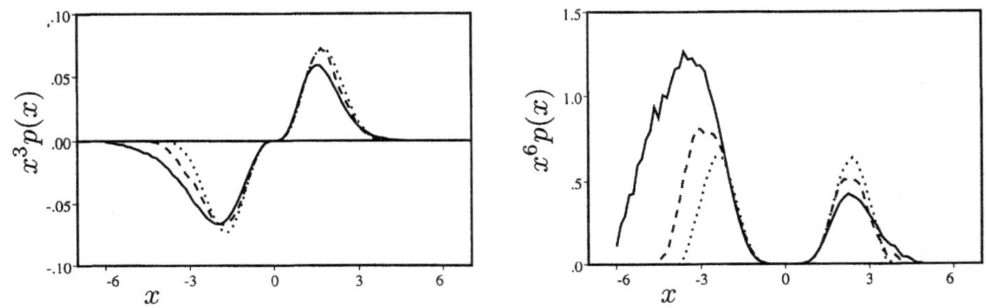

Figure 2: Integrand of F_p vs $|\delta u|$ for a) $p = 3$ and b) $p = 6$. —— $r = 3\Delta x$ ---- $r = 40\Delta x$ ········ $r = 80\Delta x$.

(160^3), we have evaluated $\langle(\delta u)^p\rangle$ directly and compared these results with those obtained from Eq. (1). Fig. 1b shows that the agreement is good for r greater than $4\Delta x$ ($\Delta x = 2\pi/160$) for all values of p (up to 20). The differences observed at small r for $p > 4$ relate to the small number of samples causing large oscillations in the tails of $p(\delta u)$. The data for $r < 4\Delta x$ were discarded when estimating ζ_p. The only way to check that the number of samples is sufficient to achieve satisfactory high-order moments is to plot the integrand of Eq. (1) at different values of p and verify that it approaches zero at large values of $|\delta u|$. Fig. 2 indeed shows that for $p = 3$ and 6 (isotropic turbulence, $R_\lambda = 40.3$) the convergence is rather good. When p increases, the number of samples is insufficient. The convergence decreases for small Δx, but since, in the evaluation of ζ_p the data for $r < 4\Delta x$ are discarded, we believe is that the results should be valid up to $p = 10$ (results are shown in subsequent figures). up to $p = 20$.

A random field with the same spectrum as that, at a given time, for decaying isotropic turbulence was generated in order to verify whether ζ_p was equal to $p/3$, i.e. check that K41 is satisfied. This is a necessary though not sufficient condition for an ESS-type technique to work satisfactorily for isotropic turbulence. It is also a demonstration that the departure from K41 reflects, inter alia, the presence of vortical structures, whose presence is a consequence of the dynamics of the Navier-Stokes equations; these structures would be absent in a random field with only kinematic constraints. The absence of structures in a random field was demonstrated by She *et al.* (1990) by flow visualization but, to our knowledge, ζ_p was not calculated.

We have calculated ζ_p for three different fields ($R_\lambda = 36.1$, 40.3, 59.8) for which the ratio of the grid size to the Kolmogorov length scale was 1.35, 1.86 and 3.02. We verified that the moments of longitudinal velocity increment were the same in each of the three directions (the results presented are actually for δu_3, with r in the x_3 direction). As a further approximation of isotropic turbulence, we have estimated ζ_p at the center of a turbulent channel flow at $R_\tau = 180$, where R_τ is the Reynolds number based on the friction velocity *and the channel half-width.*

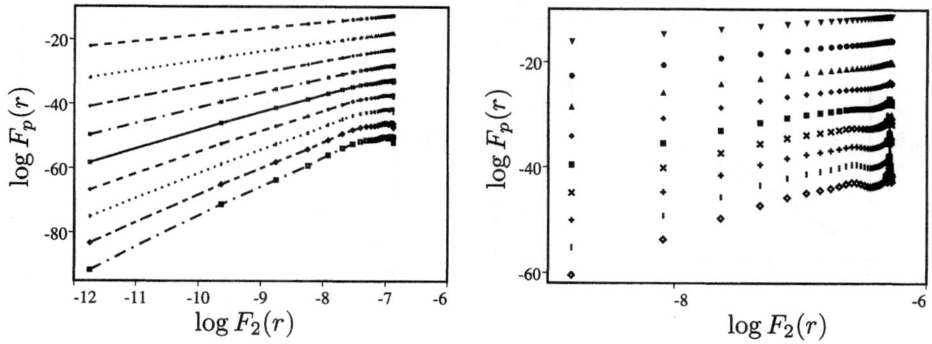

Figure 3: Structure functions of even order $p+2$ versus those of order 2, for a) isotropic turbulence, b) at the center of a plane channel flow.

In this case, the results have been compared with those measured in an experiment. The measurement technique and procedure used to evaluate the scaling exponents are described in Antonia *et al.* (1998).

The accuracy of ζ_p largely depends on the accuracy with which the slope of F_p vs F_q is estimated. Fig. 3a shows log − log plots of F_p vs F_2 $[ESS(F_p)]$ for isotropic turbulence ($R_\lambda = 40.3$) and at the center of the channel Fig. 3b. In each case, there is a range of values for which the slope can be accurately evaluated. For the channel flow, at large Δx, there are oscillations at large values of p. These are due to the reduced number of samples by comparison to isotropic turbulence. The plane channel DNS was performed with a streamwise 256 × wall-normal 150 × spanwise 64 grid. With 50 fields saved, the streamwise structure functions were evaluated using 50 × 128 × 64 samples.

Fig. 4 shows a comparison between the present results for isotropic turbulence compared with the experimental results of Anselmet *et al.* (1984) and the DNS results of Vincent & Meneguzzi (1991) in forced turbulence. In the same plots, the lines are those related to the different models. The present results show a reduced intermittency with respect to the previous one. The comparison with experiments at a much higher R_λ indicate, as expected, a bigger inertial range (at larger R_λ though with a larger departure from K41. The difference between the present results and those for forced turbulence should depend on the type of forcing that is imposed in the simulation. To confirm that the intermittency can increase when a particular type of forcing is imposed, note that the flow at the center of the channel (Fig. 4b) is indeed more intermittent than decaying isotropic turbulence. The very interesting outcome of Fig. 4b is that the experimental and numerical results show a very impressive agreement given that the scaling exponents were evaluated from completely different methods. This is a further, not previously attempted, validation of the numerical results for a plane turbulent channel flow. In Fig. 4b, the synthetic field with the same energy spectrum as that for $R_\lambda = 40.3$, but with

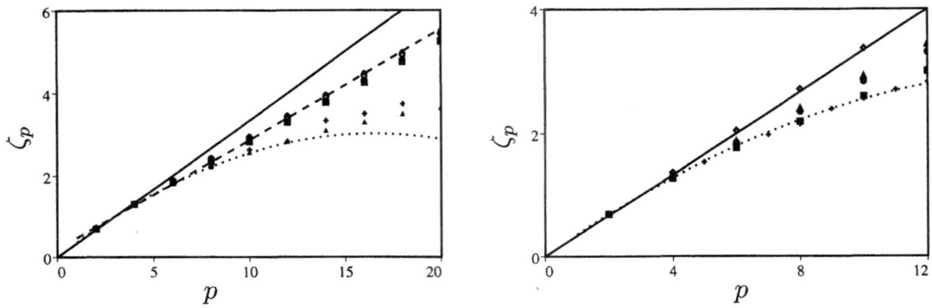

Figure 4: Scaling exponents a) isotropic turbulence □ $R_\lambda = 40.3$ ○ $R_\lambda = 36.1$ (open symbols F_p vs F_2, closed symbols F_p vs F_3^*), + Anselmet △ Vincent & Meneguzzi, ---- β model, ········ log model , —— $p/3$; b) ◇ , random field, □ $R_\lambda = 40.3$ and ○ $R_\lambda = 36.1$ isotropic decaying turbulence, □ plane channel DNS, + Antonia *et al.* experiment at $R_\tau = 180$.

a random distribution of the phases, gives scaling laws that coincides with $p/3$; this is a demonstration that the method used to evaluate the exponents is satisfactory. In a physical context, we have earlier noted that this result indicates that the vortical structures play a major part in causing the intermittency. Since isocontour plots of fluctuating velocity components scaled with the local *rms* value exhibited slightly different shapes, thus suggesting that the flow is not really isotropic, we evaluated the structure functions for $u_1(x_1, x_2, x_3) - u_1(x_1 - \Delta r_1, x_2, x_3)$ and $u_3(x_1, x_2, x_3) - u_3(x_1, x_2, x_3 - \Delta r_3)$. Identical results were obtained in isotropic turbulence. The spanwise direction increments yielded the same scaling laws as for isotropic turbulence. However, the more intermittent character of the streamwise velocity was indeed linked to bigger structures with more dominant quiescent periods.

For the plane channel flow, the previous results underline that the intermittent character of turbulence reflects the presence of the vortical structures. At small Reynolds numbers, there is only a limited range of scales so that the intermittency is not that in the strict Kolmogorov sense. It could be of interest if the scaling exponents can be used to characterize the structures. However, since the scaling exponents are defined in a statistical sense there is unlikely to be a one to one correspondence between exponents and structures, i.e. the same exponents could result from completely different flow structures. Nonetheless, the exponents could usefullly complement the characterization of turbulence when combined with other, for example, flow visualization. These ideas encouraged us to evaluate the structure functions at different distances from the wall. Antonia *et al.* (1998) considered the dependence of ζ_p on y^+, the subscript + indicating, as usual, in wall variable normalization. For $p = 4$, 6 and 8, ζ_p was minimum at $y^+ \simeq 20$, i.e. just above the location where the maximum turbulent energy production occurs. Fig. 5 shows ζ_p vs p at $y^+ = 20$ for both the streamwise and spanwise components. Near

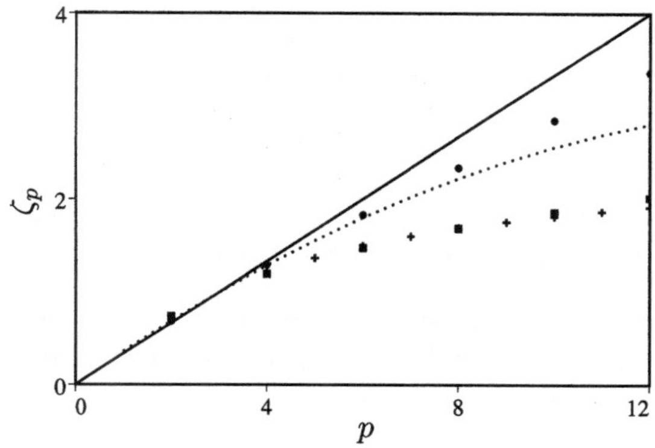

Figure 5: Scaling exponents for the longitudinal structure functions in the plane channel at $y^+ = 20$ ∘ spanwise, ▪ streamwise component, + experiments Antonia *et al.* (1998), —— $p/3$, ········ lognormal model

the wall, the numerical results compare very well with experiment. The streamwise velocity fluctuation shows a more intermittent character than the spanwise component. Flow visualizations were useful to understand that this increased intermittency is linked to the association of the streamwise velocity fluctuation with low and high speed streaks, thus producing large values of the increments very intermittently. On the other hand, isocontour plots of the spanwise velocity fluctuation are similar to those at the center of the channel; consequently, the value of ζ_p is similar to that in Fig. 4.

3 Conclusion

From the different flows investigated, the ESS procedure appears to be a very efficient method for calculating the scaling exponents. By plotting moments of order p vs moments of order q, the oscillations, present when the moments are plotted vs r, disappear, thus enabling the exponent to be more easily calculated. This behavior was previously observed by Vasilenko *et al.* (1975) but was not given any physical importance. We subscribe to this point of view and, on the basis of the present results, do not think the ESS procedure could be useful in terms of improving for example the physical content of subgrid models used in LES. With regard to the scaling laws, our view is that when they are used in combination with other observational tools, they could help in characterizing vortical structures. On their own, they are unlikely to be effective, given the non-universality of the structures. We have restricted ourselves here to longitudinal structure functions. Using the same data, we have recently evaluated transverse structure functions; the results will be presented elsewhere.

References

[1] Antonia, R. A., Satyaprakash, B. R. and Chambers, A. J.: 1982. Reynolds number dependence of velocity structure functions in turbulent shear flows, *Phys. Fluids*, **25**, 29–37.

[2] Antonia, R. A. and Pearson, B. R.: 1997. Scaling exponents for turbulent velocity and temperature increments, *Europhy. Lett.* (submitted)

[3] Antonia, R. A., Orlandi, P. and Romano, G. P.: 1998. Scaling of longitudinal velocity increments in a fully developed turbulent channel flow *Physics of Fluids A* (submitted)

[4] Anselmet, F., Gagne, Y., Hopfinger, E. J. and Antonia, R. A.: 1984. High-order velocity structure functions in turbulent shear flows, *J. Fluid Mech.*, **140**, 63–89.

[5] Benzi, R., Ciliberto, S., Tripiccione, R., Baudet, C., Massaioli, F. and Succi, S.: 1993. Extended self-similarity in turbulent flows, *Phys. Rev. E*, **48**, R29–R32.

[6] Benzi, R., Biferale, L., Ciliberto, S., Struglia, M. V. and Tripiccione, R.: 1995. On the intermittent energy transfer at viscous scales in turbulent flows, *Europhy. Lett.*, **32**, 709–713.

[7] Boratav, O. N. and Pelz, R. B.: 1997. Structures and structure functions in the inertial range of turbulence, *Phys. Fluids*, **9**, 1400–1415.

[8] Camussi, R., Barbagallo, D., Guj, G. and Stella, F.: 1996. Transverse and longitudinal scaling laws in non-homogeneous low *Re* turbulence, *Phys. Fluids*, **8**, 1181–1191.

[9] Chen, S., Sreenivasan, K. R., Nelkin, M. and Cao, N.: 1997. A refined similarity hypothesis for transverse structure functions, *Phys. Rev. Lett.*, **79**, 1253–1256.

[10] Jimenez, J., Wray, A. A., Saffman, P. G. and Rogallo, R. S.: 1993. The structure of intense vorticity in homogeneous isotropic turbulence, *J. Fluid Mech.*, **255**, 65–90.

[11] Kolmogorov, A. N.: 1941a. Local structure of turbulence in an incompressible fluid for very large Reynolds numbers, *Dokl. Akad. Nauk. SSSR*, **30**, 299–303.

[12] Kolmogorov, A. N.: 1941b. Energy dissipation in locally isotropic turbulence, *Dokl. Akad. Nauk. SSSR*, **32**, 19–21.

[13] Pearson, B. R. and Antonia, R. A.: 1997. Velocity structure functions in a turbulent plane jet, *Eleventh Symposium on Turbulent Shear Flows*, Grenoble.

[14] She, Z. S., Jackson, E. and Orszag, S. A.: 1990. Intermittent vortex structures in homogeneous isotropic turbulence, *Nature*, **344**, 226–228.

[15] Sreenivasan, K. R. and Antonia, R. A.: 1997. The phenomenology of small-scale turbulence, *Ann. Rev. Fluid Mech.*, **29**, 435–472.

[16] Stolovitzky, G. and Sreenivasan, K. R.: 1993. Scaling of structure functions, *Phys. Rev. E*, **48**, R33–R36.

[17] Vasilenko, V. M., Lyubimtsev, M. M. and Ozmidov, R. V.: 1975. Fluctuations of the turbulent energy dissipation rate and of the higher-order structure functions of the velocity field in the ocean, *Iz. Atmos. Ocean. Phys.*, **11**, 926–932.

[18] Vincent, A. and Meneguzzi, M.: 1991. The spatial structure and statistical properties of homogeneous turbulence, *J. Fluid Mech.*, **225**, 1–20.

Trends in Mathematics, © 1999 Birkhäuser Verlag Basel/Switzerland

Pressure fluctuations in a turbulent channel

O. Métais, E. Lamballais, M. Lesieur

1 Introduction

In a turbulent flow, coherent vortices are generally characterized by a high vorticity modulus and a low pressure. It was shown by Métais and Lesieur [15] that the pressure probability distribution function (pdf) was skewed in isotropic turbulence, with an exponential tail in the lows and a Gaussian one in the highs. On the other hand, the pdf of any vorticity component is symmetric, with exponential-like tails. Analogous results were found in Direct Numerical Simulation (DNS) of a mixing layer [2], and, for the pressure, in experiments of turbulence between two counter-rotating disks [6] and in a jet.

It is well established since the experiments of Kline et al.[9] in the turbulent boundary layer, that streamwise streaks of respectively low and high longitudinal velocity (with respect to the local mean velocity profile) exist close to the wall, between approximately (in wall units ν/v_*) $y^+ = 5$ and $y^+ = 40 \sim 50$. The same low and high-speed streaks were found in the turbulent channel, since the large-eddy simulations of Moin and Kim [16] (see also the DNS of Kim et al. [8]). The low-speed streaks correspond to fluid rising away from the wall with, as a result, production of a locally inflexional longitudinal velocity profile which might explain the vorticity ejections through local Kelvin-Helmholtz instabilities. It is by this ejection mechanism that vorticity at the wall is transported into the boundary layer, making it turbulent. High-speed streaks are, on the contrary, regions of descending flow sweeping to the wall. Here, the local longitudinal velocity profiles are not inflexional, and sweeps produce instantaneous local spanwise vorticities at the wall two or three times higher (in modulus) than the average value at the wall v_*^2/ν (see [4]).

The goal of the present numerical study is to examine via a Direct Numerical Simulation (DNS) how the various pdf's of pressure and velocity react to the existence of streaks and ejected hairpins in a turbulent channel flow. The details of the present study can be found in [12] and [13].

2 Numerical experiment

We study here a fluid of uniform density between two infinite parallel flat plates. Let x, y, z be respectively the longitudinal, transverse and spanwise directions. We solve numerically the unstationary Navier-Stokes equations, with periodic boundary conditions in the streamwise and spanwise directions, and adherence at the walls. We use a mixed numerical method: desaliased pseudo-spectral along

(x, z), finite differences of order six (compact schemes) along y. These schemes reach a quasi-spectral precision, while allowing more flexibility in the choice of the computational grid. Temporal integration uses Crank-Nicholson and third-order Runge-Kutta schemes, respectively for the viscous and nonlinear terms. The incompressibility condition is ensured up to the zero machine in all the computation domain and at each time step by a fractional steps method. We define the macroscopic Reynolds number $Re = U_m \; 2h/\nu$ based upon the bulk velocity U_m and the width of the channel $2h$. The width of the calculation domain is $(L_x, L_y, L_z) = (4\pi h, 2h, 2\pi h)$, which requires 128×128 Fourier (dealiased) modes in (x, z) directions, and 129 grid points along y. The initial condition corresponds to a field computed from a large-eddy simulation using a subgrid-model based upon a spectral eddy-viscosity corrected to account for the deviation of the kinetic-energy spectra from a Kolmogorov one [14]. Calculations have been done at constant U_m, and pursued up to complete statistical stationarity. The Reynolds number is $Re = 5000$, corresponding for the developed turbulence regime to a friction Reynolds number $h^+ = v_* h/\nu = 162$. This Reynolds number is subcritical from the point of view of linear-stability theory, as in the DNS of Kim et al. [8] done at $Re_m = 5600$, where the main characteristics of a turbulent channel flow were observed. The grid is refined close to the wall, with a first point at $y^+ = 1$.

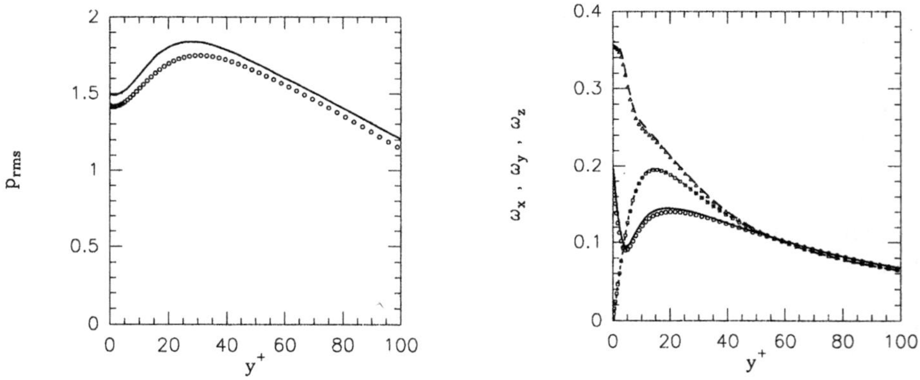

Figure 1: Statistical data obtained in DNS of a turbulent channel flow; present simulations (straight line), Kuroda [10] (symbols); r.m.s. pressure fluctuation and r.m.s. vorticity (from top to bottom, spanwise, vertical, longitudinal)

Figure 1 shows that our DNS agrees well with a previous DNS carried out by Kuroda [10] which was based on spectral methods at $h^+ = 150$ and turned out to be in good agreement with the experiments [5]. Figure 1 shows that the profile of the pressure fluctuations peaks at $y^+ \approx 30$ corresponding to the beginning of the logarithmic range. It is certainly due to low pressure associated to high vorticity at the tip of the ejected hairpin. Fig. 1 also shows the rms vorticity fluctuations, a quantity very difficult to measure precisely experimentally. It indicates that the

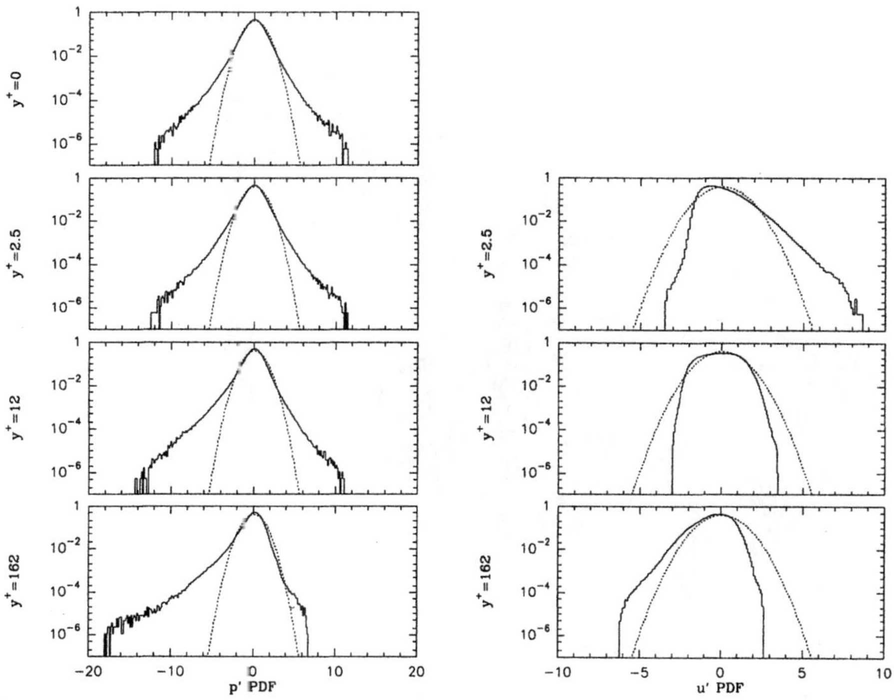

Figure 2: Probability density distibution of the pressure fluctuations (left) and of the longitudinal velocity fluctuations (right) for different y-locations

maximum vorticity produced is spanwise and at the wall, in fact under the high-speed streaks, as discussed above. The rms vorticity perpendicular to the wall is about 30% higher than the longitudinal vorticity in the region $5 < y^+ < 30$. This is another indicator of the streaks.

3 PDF's of pressure and velocity

All the pdf's presented now concern the fluctuations with respect to the mean, and the argument is normalized by the rms value. They are plotted in semi-log coordinates, and the dashed parabola indicates a Gaussian distribution of variance 1.

Figure 2 shows respectively the pdf's of pressure at the wall, very close to it in the viscous region ($y^+ = 2.5$), where the streaks have maximum amplitude ($y^+ = 12$), and in the core of the flow at the channel centre ($y^+ = 162$). Figure 2 also presents the longitudinal (u') velocity fluctuations at $y^+ = 2.5, 12$ and 162. If one accepts that asymmetric distributions are associated with vortical organization, the pdf distribution therefore does not show any trace of coherent vortices at

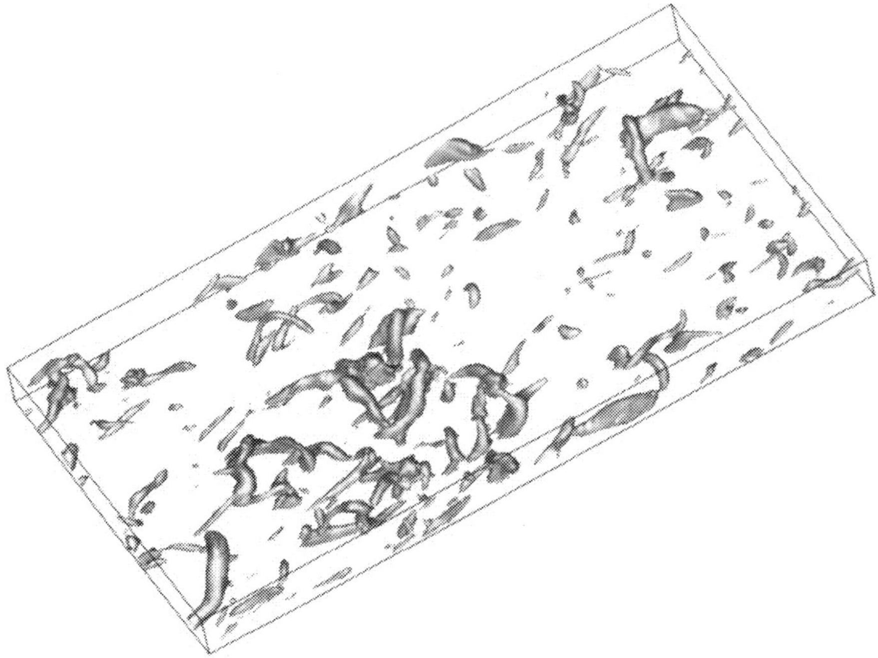

Figure 3: Low-pressure isosurfaces (only half of the channel is represented; from [12])

the wall. The pressure pdf at the wall ($y^+ = 0$) is symmetric, with exponential wings. Notice that such a pressure pdf had already been determined in analogous simulations by Kim [7], but not in logarithmic coordinates, and with a smaller number of statistical samples. Note that the wall measurements in the laboratory experiment of the flow between counter rotating disks performed by Cadot et al. [1] show asymmetric distributions. The apparent imcompatibility with the present observation comes from the very distinct generation mechanism of the flow vortices. As stressed above, the hairpin vortices of the channel flow are not created in the immediate vicinity of the wall, which implies that no vortices are present at the wall. In the experiment by Cadot et al. [1], the two counter rotating disks generate a high-shear region in the middle of the tank, which leads to intense vortices resulting from Kelvin-Helmholtz instability. The latter in their further evolution eventually impact the wall yielding very intermittent low pressure peaks and asymmetric pressure signals. Note that recent pressure measurements by Labbé *et al.* [11] in the same flow configuration as Cadot et al. have shown that, at variance with the mid-plane region, the pdf becomes symmetric in the vicinity of the rotating disk. It is probable that this flow region, far from the central high-shear region, does not include well organized vortices.

Turning back to our channel DNS, at $y^+ = 2.5$, the pressure is very close to the distribution at the wall. The longitudinal velocity is highly intermittent in the high speeds, and "sub-Gaussian" in the low speeds. Thus, high-speed streaks are much more intense and intermittent than the low-speed ones close to the wall. Such a distribution for u' at the wall is responsible for the positive skewness measured in the experiments of Comte-Bellot [3], and recovered in the DNS of Kim [7]. Another consequence of these abrupt excursions of longitudinal velocity in the high-speed streaks close to the wall is the creation of intense fluctuations of spanwise vorticity (and hence of drag) at the wall just underneath, and of same sign as the basic vorticity. This point was noticed by Ducros et al. [4] in large-eddy simulations of a weakly-compressible boundary layer spatially-developing above a flat plate. Still at $y^+ = 2.5$, we have cheked that the pdf of v' is very intermittent, with faster descents than rises. This is in agreement with the fact that the flow upwells in the low-speed regions, and sinks in the high-speed regions. As for the spanwise velocity w', it also found to be extremely intermittent (see [13]).

In the core of the streaks ($y^+ = 12$), u' has no intermittency at all, since both sides of the pdf are sub-Gaussian. It seems to indicate that the streaks are, at this level, weakly-random component of the flow. The pressure becomes asymmetric with more intermittency in the troughs than in the highs, which are still exponential. In the central region of the channel ($y^+ = 162$), the pressure pdf resembles skewed distributions encountered in isotropic turbulence or free-shear flows. Visualizations of the pressure field do show in fact that large Λ-shaped vortices are carried by the flow (Figure 3), and their existence might explain the pressure distribution. v' and w' are found very close to Gaussian (see [13]) like in three-dimensional isotropic turbulence. As for u', it is now preferably negative, with excursions of low speeds. It seems then that the intermittent sweeps of high speed at the wall are balanced by low speeds in the channel centre.

4 Concluding remarks

On the basis of a highly-resolved direct-numerical simulation of a turbulent plane channel at $h^+ = 162$, we have determined the pdf's of the pressure at the wall, and of the pressure and velocity fluctuations at $y^+ = 2.5, 12$ and 162. The pressure at the wall is symmetric with exponential tails, and seems not to be affected by the existence of coherent vortices, nor by the high-and low-speed streaks system. The pressure at $y^+ = 2.5$ is very similar to the pressure at the wall, while the velocity fluctuations show very intermittent rapidly-descending high-speed streaks, and smoother rising low-speed streaks. At $y^+ = 12$, the streaks system is weakly random with sub-Gaussian pdf's for u'. The pressure pdf has become asymmetric. At $y^+ = 162$, the pressure has an exponential tail in the lows and is close to Gaussian in the highs. This looks like very much the pressure pdf's observed in isotropic turbulence (cf Figure 4) or free-shear flows. It is generally associated with the presence of coherent vortices. In the channel, the coherent vortices are hairpin and Λ-vortices of various sizes ejected from the wall. However, the association of

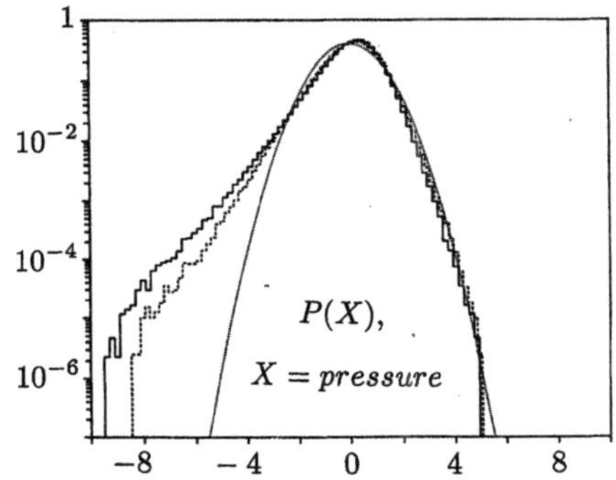

Figure 4: Pressure density function in direct numerical simulation of three-diemnsional isotropic turbulence. Continuous line, direct numerical simulation; dashed line, Gaussian kinematic field; dotted line; Gaussian distribution of same variance (from [15]).

skewed pressure pdf's with coherent vortices has to be taken cautiously. Figure 4 taken from Métais and Lesieur [15] indeed compares the pressure pdf obtained from a DNS of three-dimensional isotropic turbulence and the pdf of a fictitious pressure field calculated from a Gaussian velocity field of same kinetic energy spectrum (obtained by randomizing the phases). The highly skewed shape is also shared by the kinematic Gaussian evaluation, although the intermittency is decreased as compared with the original field. Therefore, the understanding of the signature of the coherent vortices on the low pressure statistics is still an open issue.

Acknowledgements: Computations were carried out at the IDRIS (Institut du Développement et des Ressources en Informatique Scientifique, Paris).

References

[1] O. Cadot, D. Douady and Y. Couder, Characterization of the low-pressure filaments in a three-dimensional turbulent shear flow. *Phys. Fluids* **7** (3), (1995) pp. 630–646.

[2] P. Comte, M. Lesieur and E. Lamballais, Large and small-scale stirring of vorticity and a passive scalar in a 3D temporal mixing layer. *Phys. Fluids A* 4, (1992) pp. 2761–2778.

[3] G. Comte-Bellot, *Ecoulement turbulent entre deux parois parallèles*. Publications scientifiques et techniques du ministère de l'air 419, Documentation scientifique et technique de l'armement (1965)

[4] F. Ducros, P. Comte and M. Lesieur, Large-eddy simulation of transition to turbulence in a boundary-layer developing spatially over a flat plate. *J. Fluid Mech.* **326**, (1996) pp. 1–36.

[5] F. Durst and H. Kikura, Low Reynolds number effects on a fully developed turbulent channel flow. In *Proc. 10th Symp. on Turbulent Shear Flows* (1995) P2–25.

[6] S. Fauve, C. Laroche and B. Castaing, Pressure fluctuations in swirling turbulent flows. *J. Phys II* **3**, (1993) pp. 271–278.

[7] J. Kim, On the structure of pressure fluctuations in simulated turbulent channel flow. *J. Fluid Mech.* **205**, (1989) pp. 421–451

[8] J. Kim, P. Moin and R. Moser, Turbulence statistics in fully developed channel flow at low Reynolds number. *J. Fluid Mech.* **177**, (1987) pp. 133–166

[9] S.J. Kline, W.C. Reynolds, F.A. Schraub and P.W. Runstadler, The structure of turbulent boundary layers. *J. Fluid Mech.* **95**, (1967) pp. 741–773.

[10] A. Kuroda, *Direct-numerical simulation of Couette-Poiseuille flows.* PhD thesis, University of Tokyo, (1990).

[11] R. Labbé, J.F. Pinton and S. Fauve, Power fluctuations in turbulent swirling flows. *J. Phys II France* **6**, (1996), pp. 1099–1110.

[12] E. Lamballais, *Simulations numériques de la turbulence dans un canal plan tournant.* PhD thesis, Institut National Polytechnique de Grenoble, (1996).

[13] E. Lamballais, M. Lesieur and O. Métais, Probability distribution functions and coherent structures in a turbulent channel" *Physical Review E*, **56** (6), pp. 6761–6766.

[14] E. Lamballais, O. Métais and M. Lesieur, Influence of a spanwise rotation upon the coherent-structure dynamics in a turbulent channel. In *Direct and Large-Eddy Simulation II*, J.P. Chollet, L. Kleiser and P.R. Voke eds., Kluwer Academic Publishers, (1997) pp. 225–236.

[15] O. Métais and M. Lesieur, Spectral large-eddy simulations of isotropic and stably-stratified turbulence. *J. Fluid Mech.* **239**, (1992) pp. 157–194.

[16] P. Moin and J. Kim, Numerical investigation of turbulent channel flow. *J. Fluid Mech.* **118**, (1982) pp. 341–377

[17] P. Moin and J. Kim, The structure of the vorticity field in turbulent channel flow. Part 1. Analysis of instantaneous fields and statistical correlations. *J. Fluid Mech.* **155**, (1985) pp. 441–464

Trends in Mathematics, © 1999 Birkhäuser Verlag Basel/Switzerland

Statistics of longitudinal and transverse velocity increments

W. van de Water

Abstract

So far, too little experimental information about the turbulent velocity field has been used to address the scaling question. We discuss the measurement of transverse structure functions as a next possible step. The associated scaling exponents seem to point to a stronger effect of intermittency on transverse than on longitudinal velocity increments. We finally discuss ways to estimate the error in high-order structure functions, to estimate the maximum useful order for a given number of samples and ways to improve the accuracy.

Introduction

The possible existence of universal scaling properties of velocity fluctuations is an exciting aspect of strong turbulence. I believe that this will remain so, despite the signs of structures that were discussed at this Colloquium, and despite the profound questions asked about the meaning of scaling exponents. The prime instrument to quantify scaling is the velocity structure structure function that is defined as $G_p(r) = \langle (\Delta u(r))^p \rangle$, where the velocity difference $\Delta u(r)$ is measured over a distance r, $\Delta u(r) = u(x + r) - u(x)$ and the average is done over x. There are two distinct ways of measuring Δu. In the *longitudinal* case, the measured component of $\Delta \boldsymbol{u}$ points in the same direction as $\boldsymbol{r}$, whereas these vectors are *perpendicular* in the transverse case. We will call $G_p^L(r)$ and $G_p^T(r)$ the structure functions for the longitudinal (L) and transverse (T) case, respectively.

If indeed the extreme events of turbulence are linked with structures of a well-defined kind, other statistical quantities than the structure function may become thinkable. One example was presented at this Colloquium by Patrick Tabeling who sought to single out the contribution of "worms" to a point signal of velocity readings in a strongly turbulent cryogenic flow.

So far, the discussion about scaling of turbulence was based on simple point measurements of a single component of the turbulent velocity field. It surprises that so much can be said about so little information. We will argue that more can be learned from more extended information, which will be illustrated by a comparision of longitudinal and transverse structure functions. We were the first [1] to give experimental evidence that the anomalous behavior of the exponents was stronger for the transverse velocity increments than for the longitudinal ones. If this results stands up to close scrutiny, it suggests the development of new experimental techniques that provide more extended and precise information about the small-scale velocity field and that allow the collection of adequate statistics.

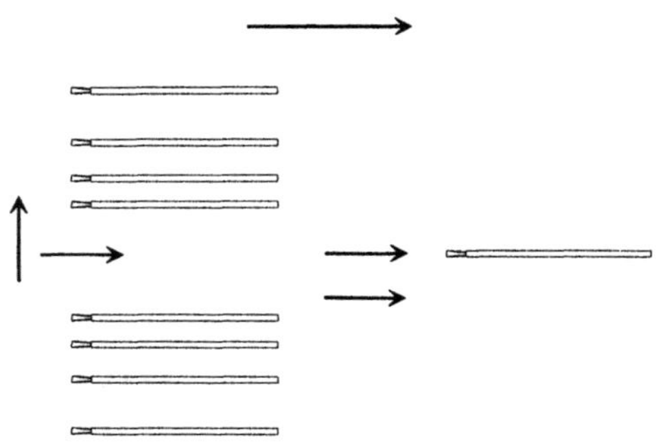

Figure 1: (a) The geometry that is used for measuring transverse structure functions. The distance vector $\boldsymbol{r}$ over which velocity differences $\Delta \boldsymbol{u} = \boldsymbol{u}(\boldsymbol{x}+\boldsymbol{r}) - \boldsymbol{u}(\boldsymbol{x})$ are measured is oriented perpendicularly to the measured velocity fluctuation $\boldsymbol{u}$. In this setup, no recourse to Taylor's hypothesis is needed. The array of probes used is approximately drawn to scale. (b) In the longitudinal case $\boldsymbol{r}$ and $\boldsymbol{u}$ point in the same direction. Distances follow from elapsed time t through invocation to Taylor's hypothesis, $\boldsymbol{r} = \boldsymbol{U}t$.

Experiments

Longitudinal velocity increments were measured in jet- and grid turbulence using a single probe. In this case time delays t are interpreted as spatial separations r using Taylor's hypothesis $r = Ut$ with U the mean velocity. Transverse velocity increments were measured in grid-generated turbulence with an array of probes oriented perpendicularly to the mean flow direction. In this case no recourse to Taylor's hypothesis is needed. The transverse measurements are more difficult as the smallest separation r is given by the smallest physical distance of two diferent probes (3 mm). Another experimental problem is the crucial importance of the probe calibration. Velocity differences are now measured with two different probes and the relative sensitivity of the probes severely matters.

Figure 2 shows longitudinal structure functions for jet turbulence and transverse structure functions for grid turbulence. In the transverse case, the odd-order moments are identically equal to zero and we have computed odd moments of the absolute $|\Delta u|$ value of the velocity difference. As Fig. 2 demonstrates, it is possible to obtain clear scaling behavior [2], without recourse to "extended self-similarity" [3], in which log log plots are made from one moment as a function of another one. For extended self- similarity to work, it is necessary to compute longitudinal moments of $|\Delta u|$ [4]. Such practise, however, ignores the skewness of the longitudinal velocity fluctuations that is intimately tied to the turbulent energy cascade. We have found that important information is contained in the proper longitudinal moments. As in the transverse case separations r correspond to physical distances between different probes, the dynamic range of distances is limited. We are currently working on arrays of microprobes that allow smaller separations. In the longitudinal case instead, those small separations come easily, as they correspond to small time delays between successive time-samples.

The anomalous character of scaling is illustrated by plotting $|G_p|^{1/p}$. In the strictly self-similar case, the slope of $|G_p(r)|^{1/p}$ in a log log plot would not depend on p.

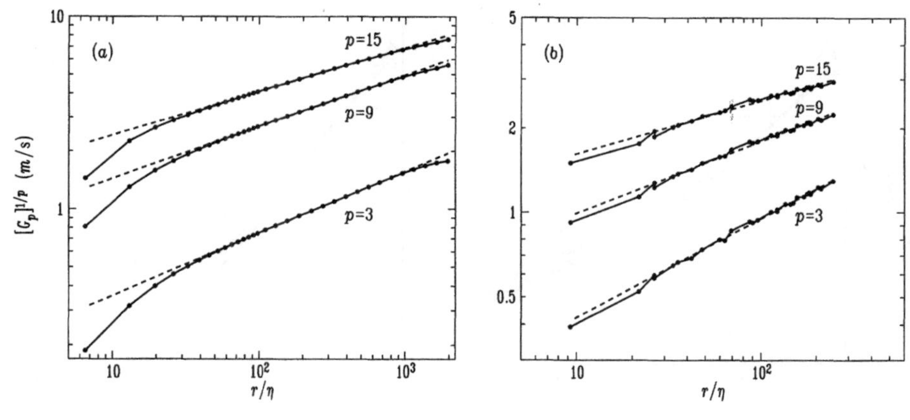

Figure 2: (a) Dots: longitudinal structure functions of order 3, 9, and 15 measured in jet turbulence. The Taylor microscale Reynolds number is $R_\lambda \approx 10^3$. The total number of velocity samples N is 3×10^8 (in integral scales L the integration time is $T = 2.5 \times 10^6 L/U$). Dashed lines: fits of ideal scaling behavior $G_p^L(r) \sim r^{\zeta^L(p)/p}$ with $\zeta^L(p)/p = 0.32, 0.27$, and 0.23, for $p = 3, 9$ and 15, respectively. (b) Dots: transverse structure functions of order 3, 9, and 15 measured in grid turbulence. The Taylor microscale Reynolds number is $R_\lambda \approx 5 \times 10^2$. The total number of velocity samples N is 1×10^8. Dashed lines: fits of ideal scaling behavior $G_p^T(r) \sim r^{\zeta^T(p)/p}$ with $\zeta^T(p)/p = 0.35, 0.26$, and 0.19, for $p = 3, 9$. and 15, respectively. Because transverse probability distribution functions are symmetric, these odd-order structure functions are defined as $G_p^T(r) = \langle |\Delta u|^p \rangle$.

Scaling exponents

The structure function has scaling behavior $G_p(r) \sim r^{\zeta(p)}$. According to Kolmogorov (1941) theory, turbulent velocity fluctuations are self-similar with $\zeta(p) = p/3$. However, measured scaling exponents differ from this prediction. Whereas Fig. 2(a) shows the proper longitudinal moments, we have also determined moments of the *absolute value* of the longitudinal velocity increments $\widetilde{G}_p^L(r) = \langle |\Delta u|^p \rangle$. In spite of the ignorance of skewness, this has become common practice. There is now concensus about the experimental value of the associated exponents $\widetilde{\zeta}^L(p)$ [5].

The scaling exponents have been plotted In Fig. 3 in a manner that most clearly shows the effect of intermittency. The longitudinal exponents $\widetilde{\zeta}^L(p)$ are very nicely parametrized by the model of She and Leveque [6]. As absolute values $|\Delta u|$ are used, also moments with $-1 < p \leq 1$ are meaningful, and Fig. 3 shows that also the small moments $p < 3$ exhibit intermittency. The exponents were normalized to $\zeta(p = 3) = 1$, which holds exactly in the longitudinal case. However, in the experiments small but significant deviations were found $\zeta^L(3) = 0.96 \cdots 1.03$ and $\zeta^T(3) = 1.08$, which we ascribe to experimental artifacts.

It is a striking observation that the transverse exponents $\zeta^T(p)$ show a larger deviation from self-similarity than the longitudinal ones, and for $p > 3$ the model of

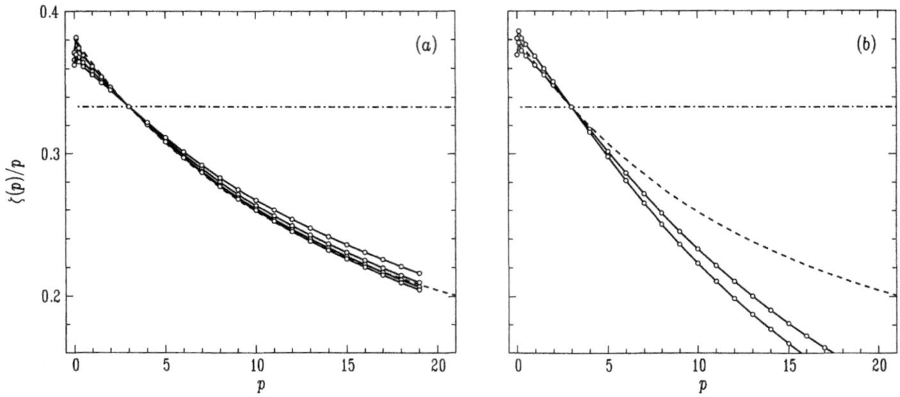

Figure 3: (a) Scaling exponents $\widetilde{\zeta}^L(p)/p$ of *longitudinal* structure functions that were measured in jet- and grid turbulence. Dash-dotted line: Kolmogorov's prediction $\zeta(p) = p/3$; dashed line: prediction of the She-Leveque model [6]. (b) Scaling exponents $\zeta^T(p)/p$ of *transverse* structure functions that were measured in grid turbulence.

Ref. [6] strongly disagrees with the transverse exponents. Several explanations are possible. For example, it has been suggested that the strongest events in turbulence are line vortices. In our experimental arrangement a longitudinal measurement cuts a line through the turbulent velocity field whereas the transverse measurement cuts a plane. In the latter case, the chances to detect line vortices are bigger than in the longitudinal arrangement.

Probability distribution functions

The moments G_p follow from the distribution functions $P(\Delta u)$ of velocity differences $\Delta \boldsymbol{u}(\boldsymbol{r}) = \boldsymbol{u}(\boldsymbol{x}+\boldsymbol{r}) - \boldsymbol{u}(\boldsymbol{x})$, as $G_p(r) = \int x^p P(x)\mathrm{d}x$. In the longitudinal case, the possibility to convey information from $\boldsymbol{x}$ to $\boldsymbol{x}+\boldsymbol{r}$ depends on the sign of $\Delta\boldsymbol{u}$, hence the longitudinal distribution functions are asymmetric. No such asymmetry exists for the transverse case. Figure 4 shows a comparision between longitudinal and transverse PDF's at various separations r. A measure for the quality of transverse experiments is the symmetry of the measured PDF under reflection of the velocity difference axis. Therefore, we show $P^T(\Delta u)$ and $P^T(-\Delta u)$ superposed on a single graph. The normalized PDF's for the longitudinal and transverse case are very similar. It appears that the negative Δu tail of the longtudinal PDF approximately coincides with the tails of the transverse PDF.

In all the PDF's shown, we have normalized the velocity scale at a particular separation r by the corresponding root- mean square velocity difference. The velocity scales $du^{T,L} = (G_2^{T,L})^{1/2}$ of P^T and P^L, respectively, are related through the isotropy relation $G_2^T = G_2^L + \frac{r}{2}\frac{\mathrm{d}G_2^L}{\mathrm{d}r}$. This relation predicts that for nominal

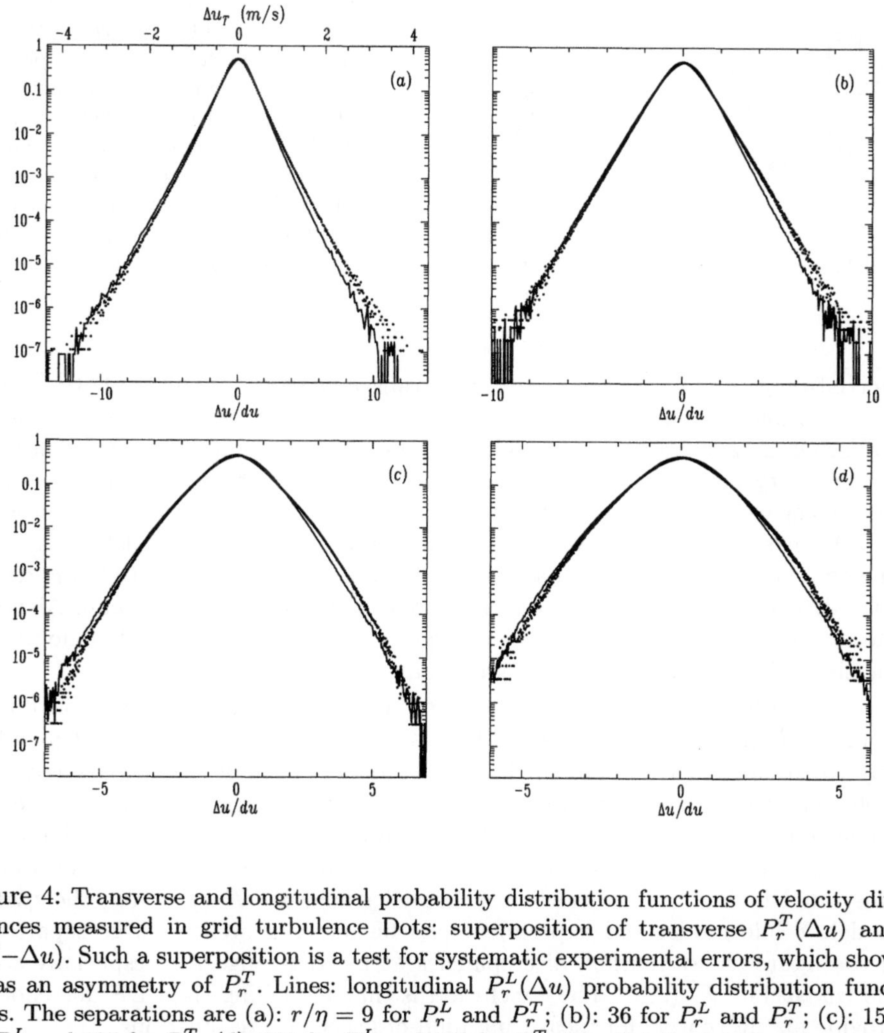

Figure 4: Transverse and longitudinal probability distribution functions of velocity differences measured in grid turbulence Dots: superposition of transverse $P_r^T(\Delta u)$ and $P_r^T(-\Delta u)$. Such a superposition is a test for systematic experimental errors, which show up as an asymmetry of P_r^T. Lines: longitudinal $P_r^L(\Delta u)$ probability distribution functions. The separations are (a): $r/\eta = 9$ for P_r^L and P_r^T; (b): 36 for P_r^L and P_r^T; (c): 150 for P_r^L and 135 for P_r^T; (d): 260 for P_r^L and 250 for P_r^T. The distribution functions have been normalized such that $\int P(x)\mathrm{d}x = 1$, $\int xP(x)\mathrm{d}x = 0$ and $\int x^2 P(x)\mathrm{d}x = 1$. The rms velocity difference du is defined as $du \equiv \langle (u(x+r) - u(x))^2 \rangle_x^{1/2}$. An absolute scale for the transverse velocity difference has been given for frame (a).

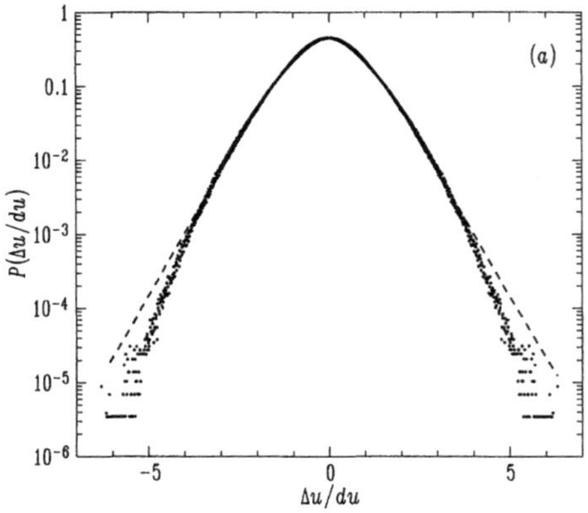

Figure 5: Comparision of probability distribution functions $P^T(\Delta u(r))$ of velocity differences with a scaling model. The velocity fluctuations are *transverse* with the separation vector $\boldsymbol{r}$ oriented perpendicularly to the measured component of $\boldsymbol{u}$. The fit between model and experiments is good when the separation r/η is large. When the separation r is in the inertial range, $r/\eta = 250$ (as shown), there are marked discrepancies.

inertial-range scaling, $G_2(r) \simeq r^{2/3}$, the transverse du^T is a factor $(4/3)^{1/2}$ larger than the longitudinal du^L, but the ratio grows to $2^{1/2}$ as r reaches the viscous scales. On an absolute velocity scale, therefore, transverse velocity differences can become very large. In Fig. 4(a) we have provided an absolute velocity scale for the transverse $P_r^T(\Delta u)$ at $r/\eta \approx 9$. For this experiment the mean velocity was $U = 13$ m/s, with the rms velocity fluctuation a mere 3.5% ($u = 0.46$ m/s). Still, during an experimental run that effectively lasted approximately 100 minutes, we have detected velocity *differences* as large as 4 m/s accross $r = 3$ mm. Although the magnitude of these differences is large, it is not anomalous as it follows from the similarity of longitudinal and transverse PDF's and isotropy.

There have been several suggestions to deduce scaling information *directly* from the distribution functions [9]. This is a tantalizing suggestion as it may be a way to circumvent the problem that the statistical accuracy of increasingly large moments is increasingly problematic. The idea is to express $P(\Delta u)$ at inertial-range separations r (which is affected by intermittency) into that at large separations which is close to Gaussian. A typical outcome of such a procedure is illustrated in Fig. 5 that shows a measured transverse $P^T(\Delta u)$ together with a model. While model and experiment can be matched well at large separations r, the figure shows marked discrepancies at inertial-range r/η; especially so for the *transverse* $P^T(\Delta u)$. The conclusion is that modelling the PDF does not yet provide an alternative for measuring high-order structure functions.

Statistical accuracy

As high-order moments are an average over increasingly large and increasingly rare velocity increments, their statistical accuracy is a problem. In all of our experiments, we employ real-time processing of velocity differences. For the longitudinal experiments we have implemented a "structurator" on a Digital Signal Processor that can collect probability distribution functions of $\Delta u(r)$ at 32 exponentially

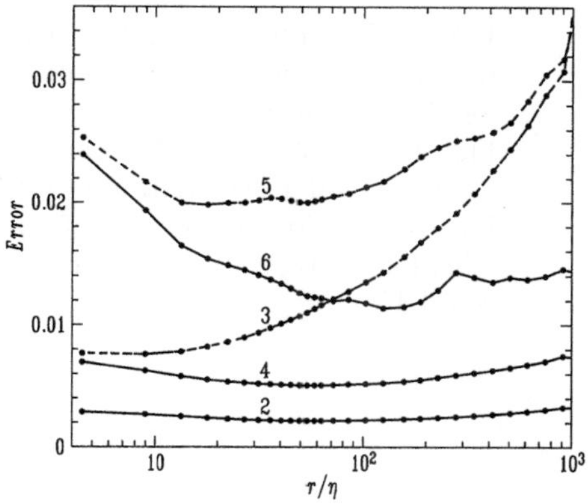

Figure 6: Error of structure function $\left\langle(\langle\Delta u^p\rangle_T - \langle\Delta u^p\rangle)^2\right\rangle^{1/2}$ as a function of the order p. It was computed from a stored time series of velocities in jet turbulence. The length of the time series is $T = 1.4 \times 10^5 L/U$. For increasing moments p, the error is largest at the smallest separations.

spaced separations r in real time at 100 kHz. Not only has this approach significantly eased the statistics problem, but the absence of stored time series and the immediate availability of the structure function has allowed systematic experimentation with the flow conditions ([10]). The maximum number of velocity samples that we have taken is 1.5×10^9, but moments with $p > 12$ remain uncertain.

It is possible to estimate the error in low-order moments due to finite integration times T [11]. Let us call $\langle\Delta u^p\rangle_T$ the estimate of $\langle\Delta u^p\rangle$ through time averaging over T. Its rms error $\mathcal{E}$ then is $\mathcal{E} \approx 2(I_p/T)^{1/2}\langle(\Delta u^p - \langle\Delta u^p\rangle)^2\rangle^{1/2}$. Clearly, what matters is the correlation time I_p which is defined as the integral over the correlation function $\rho_p(t)$ of moments of order p, $I_p = \int_0^\infty \rho_p(\tau)\mathrm{d}\tau$. The normalized function $\rho_p(t)$ involves the correlations $\langle\Delta u^p(t')\Delta u^p(t'-t)\rangle_{t'}$.

Because $\rho_p(t)$ contains moments of order $2p$, an estimate of the error $\mathcal{E}$ can only be done for low-order moments. An interesting question is how the error depends on the separation r. Intermittency implies that the tails of the PDF's become wider with decreasing separation. Therefore, the error in high-order moments should increase as the separation r decreases and the effect of intermittency grows stronger. This trend is opposite to that for low-order odd moments that gauge the deviation form Gaussianity. At large separations, the PDF is close to Gaussian, and the error should increase with *increasing* separation. The cross-over between these opposing trends is clearly seen in Fig. 6.

The tails of the distribution function evolve from exponential at small separations r to Gaussian at large separations where differences of essentially uncorrelated velocities are taken. This observation suggests a simple form for the tails of the PDF in terms of stretched exponentials $P(\Delta u) = a\exp(-\alpha|\Delta u|^\beta)$, where the parameters a, α and β depend on the separation r and on the sign of Δu (they can be different for the left and right tail of $P^L(\Delta u)$). At small separations r, $\beta \approx 1$ and $P(\Delta u)$ is close to exponential; at large separations $\beta \approx 2$ and $P(\Delta u)$ becomes Gaussian.

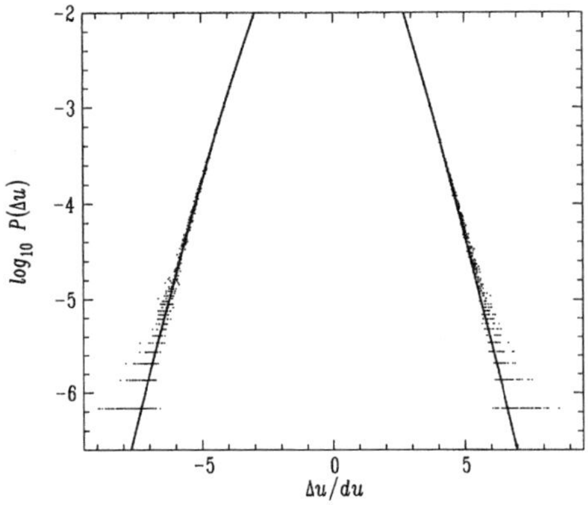

Figure 7: Dots: probability distribution function of longitudinal velocity differences with $r/\eta = 190$ Full line: fit of stretched exponential $a \exp[-\alpha|\Delta u|^{\beta}]$ over $|\Delta u/du| > 1.5$. For the left side $\Delta u/du < -1.5$, $\beta = 1.32$, for the right side, $\Delta u/du > 1.5$, $\beta = 1.13$.

The stretched exponential character of the PDF tails is illustrated in Fig. 7. Using this representation, it is possible to estimate the largest meaningful moment $p_{\max}$ for a given number N of velocity samples as $p_{\max} = \frac{1}{2}\beta \log_2 N$. This simple form again illustrates that estimates of high-order moments (that are determined by the PDF tails) are most problematic at small separations where β is smallest.

A stretched exponential representation can also be used to reach better control of statistical convergence. The contribution of the tails of the PDF to moments can be determined from stretched exponentials. Specifically, we have separated measured PDF's into three parts. One central part $[-\Delta u_0, \Delta u_0]$ whose contribution to the value of moments was computed by directly integrating and a left ($\Delta u \leq -\Delta u_0$) and right tail ($\Delta u \geq \Delta u_0$), respectively, whose contribution to the moments was computed from the parameters of a stretched exponential fit $P(x) = a \exp\left(-\alpha|x|^{\beta}\right)$. The parameters a, α and β were determined in a least squares procedure for left- and right tails separately. The statistical properties of the noise in the tails of PDF's is well understood, and it is possible to precisely assess the goodness of the stretched exponential fit. With our integration times, deviations from stretched expontial just become significant.

Figure 8 shows the odd-order $G_{19}(r) = |\langle(\Delta u(r))^{19}\rangle|^{1/19}$. The directly measured structure functon shows oscillations which make it difficult to determine the scaling exponent. These oscillations are caused by insufficient statistics; they change and often disappear when averaging over longer integration times. As Fig. 8 illustrates, they also disappear when extrapolating the tails of the PDF's using stretched exponentials. These oscillations have been interpreted in terms of an intrinsic property of fractal sets in [12]. We, however, believe that they are simply due to statistical fluctuations.

A stretched exponential representation of the tails of PDF's is a good way to reduce statistical fluctuations of high-order structure functions, but it can never be a substitute for long integration times. On the other hand, in this way true large moments ($p > 20$) will always remain elusive. From a conceptual point of

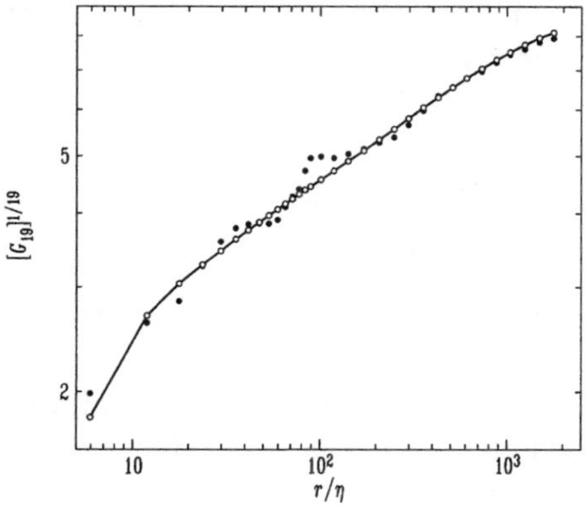

Figure 8: Dots: longitudinal structure function $|G_{19}(r)|^{1/19}$ as computed directly from measured PDFs; open circles $|G_{19}(r)|^{1/19}$ as computed from a stretched- exponential extrapolation from measured PDFs. After extrapolation the structure function shows clear scaling behavior.

view, the question about the limiting behavior $\zeta(p \to \infty)$ is extremely interesting as it points to the nature of the most intermittent structures in turbulence. We thus may never be able to answer it.

Conclusion

The interpretation of turbulence in terms of point measurements of a single component of the velocity field can now hardly be stretched further. New experimental diagnostics are needed to address the question of scaling in alternative kinematic arrangements, such as the transverse structure function. An exciting development is the use of non-intrusive laser techniques to map out the small-scale velocity field [13]. These techniques are now further developed in our laboratory, with the hope of reaching a non-intrusive measurement of the statistics of the small-scale vorticity field.

Acknowledgments

The results discussed in this contribution were obtained by Janine Herweijer as part of her PhD thesis work. It is also a pleasure to acknowledge technical assistance by Gerard Trines and Rob van der Berg.

References

[1] W. van de Water and J.A. Herweijer, Phys. Scripta **T67**, 136–140 (1996).

[2] Such scaling behavior was observed in grid turbulence in a closed recirculation windtunnel and jet turbulence. But was not seen in a turbulent cylinder wake.

[3] R. Benzi, S. Ciliberto, R. Tripiccione, C. Baudet, F. Massaioli, S. Succi, Phys. Rev. E. **48**, R29 (1993).

[4] W. van de Water and J.A. Herweijer, Phys. Rev. E. **51** 2669 (1995).

[5] A. Arneodo, C. Baudet, F. Bellin, R. Benzi, B. Castaing, B. Chabaud, R. Chavarria, S. Ciliberto, R. Camussi, F. Chillà, B. Dubrulle, Y. Gagne, B. Hebral, J. Herweijer, M. Marchand, J. Maurer, J.F. Muzy, A. Naert, A. Noullez, J. Peinke, F. Roux, P. Tabeling, W. van de Water, H. Willaime, Europhys. Lett. **34**, 411–415 (1996).

[6] Z-S She and E. Leveque, Phys. Rev. Lett. **72**, 336 (1994).

[7] O.N. Boratav and R.B. Pelz, Phys. Fluids **9**, 1400 (1997).

[8] S. Chen, K.R. Sreenivasan, M. Nelkin and N. Cao, Phys. Rev. Lett. **79**, 2253 (1997).

[9] B. Chabaud, A. Naert, J. Peinke, F. Chilla, B. Castaing and B. Hébral, Phys. Rev. Lett. **73**, 3227 (1994); R. Friedrich and J. Peinke, Phys. Rev. Lett. **78**, 863 (1997).

[10] J.A. Herweijer, F.C. van Nijmweegen, K. Kopinga, J.H. Voskamp and W. van de Water, Rev. Sci. Instrum, **65**, 1786 (1994).

[11] H. Tennekes and J.L. Lumley, *A first course in turbulence.* MIT press, 1972.

[12] L.A. Smith, J.D. Fournier and E.A. Spiegel, Phys. Lett. A **114**, 465 (1986).

[13] R. Miles, W. Lempert, and B. Zhang, Fluid Dynamics Research **8**, 9 (1991); B. Zhang, W.R. Lempert, R.B. Miles, Optics Lett. **18**, 1132 (1993);

Trends in Mathematics, © 1999 Birkhäuser Verlag Basel/Switzerland

Turbulent transport in a liquid layer heated on its free surface

A. Pumir, L. Blumenfeld

Abstract

We propose a theoretical estimate of the temperature elevation in a liquid pool, locally heated on its free surface, assumed flat, when the fluid motion is turbulent. Our treatment follows the theoretical work of Shraiman and Siggia (Phys. Rev. **A 42** , 3650 (1990)) in Rayleigh-Bénard convection. The temperature elevation is proportional to the incident power to the 2/3 power.

Heat transport by convection is an ubiquitous phenomenon [1,2,3]. Much research has been devoted to the study of Rayleigh-Bénard convection (a fluid heated from below in a closed container). Carefully controlled experiments have permitted to show that for a large range of parameters, the Nusselt number, the dimensionless heat flux, scales as $Ra^{2/7}$, where Ra is the Rayleigh number [4,5]. This scaling law can be understood theoretically by properly taking into account the specific features of the flow, namely, the presence of thin boundary layers running along the walls, superposed with a large scale coherent motion along the entire cell [6,7].

In welding or metal evaporation processes [2,3], an intense localized heat source is applied at the free surface of a metal, thereby melting the metal and setting up convective motion in the liquid. Convection in this problem is mostly driven by surface forces (Marangoni effect) [2]. It is both practically and theoretically interesting to understand the relation between the temperature elevation and the injected power (the equivalent of the Nusselt-Rayleigh relation). Here, we propose a theoretical estimate of this relation. Our approach rests on the theoretical work reviewed in [7] (see also [8]).

We consider a fluid described by the Boussinesq equations:

$$\rho\big(\partial_t\vec{u} + (\vec{u}\cdot\nabla)\vec{u}\big) = -\nabla p + \rho\alpha\vec{g}T + \rho\nu\nabla^2\vec{u} \ , \tag{1}$$

$$\nabla\cdot\vec{u} = 0 \ , \tag{2}$$

and

$$\partial_t T + (\vec{u}\cdot\nabla)T = \kappa\nabla^2 T \tag{3}$$

where $\vec{u}$ and T are the velocity and the temperature fields, ρ the mean density, ν the viscosity, κ the thermal diffusivity and α the coefficient of expansion of the fluid, which are all assumed to be constant. The strongest assumption in this work is that the upper surface remains flat, even right under the incident heating beam. It is in our view the most serious limitation of the approach presented here. The melt is assumed to fill a rectangular (2D) or cylindrical (3D) box, of depth is H

and of radius R (in 3D). The incident heating flux is confined to a limited region, resulting in the boundary condition on the upper interface ($z = 0$):

$$\kappa \partial_z T = Q_0 q(r) \tag{4}$$

where $q(r)$ is a dimensionless function of order 1 (independent of the heating power) which vanishes for $r \geq r_0$, the beam radius, and Q_0 is the injected power in the system, divided by the heat capacity in the system, assumed constant. We consider a case where r_0, R and H of the same order of magnitude. The Marangoni effect induces a shear stress due to the horizontal temperature gradient on the fluid interface:

$$\rho\nu\partial_z \vec{u}_{\|} - (\frac{d\sigma}{dT})\nabla_{\|} T = 0 \tag{5}$$

where $(d\sigma/dT)$ is the derivative of the surface tension with respect to temperature. In the applications considered here, $(d\sigma/dT)$ is negative, so the Marangoni effect tends to generate a flow from the hot to the cold regions. As a consequence, flow is pushed away from the center along the free surface.

It is convenient to use dimensionless variables defined by: $\overline{x} \equiv x/H$ (space), $\overline{t} \equiv \kappa t/H^2$ (time), $\overline{u} \equiv uH/\kappa$ (velocity), $\overline{p} \equiv pH^2/\rho\kappa^2$ (pressure) and $\overline{\theta} \equiv (T - T_M)\kappa/(HQ_0)$ (temperature). The equations of motion read:

$$\frac{1}{Pr}(\partial_{\overline{t}}\overline{u} + (\overline{u} \cdot \nabla)\overline{u} + \nabla\overline{p}) = \nabla^2\overline{u} + Ra\theta\hat{z} \tag{6}$$

$$\nabla \cdot \overline{u} = 0 \tag{7}$$

$$\partial_{\overline{t}}\overline{\theta} + (\overline{u} \cdot \nabla)\overline{\theta} = \nabla^2\overline{\theta} \tag{8}$$

where $Pr \equiv \nu/\kappa$ is the Prandtl number and $Ra \equiv g\alpha Q_0 H^4/\nu\kappa^2$ is the Rayleigh number. In addition, the boundary condition Eq. (5) becomes:

$$\partial_{\overline{z}}\overline{u}_{\|} + Ma\nabla_{\|}\overline{\theta} = 0 \tag{9}$$

where the Marangoni number, defined by $Ma \equiv -d\sigma/dT(Q_0H^2/\kappa^2\mu)$ (here, $Ma > 0$). The temperature in dimensional units, θ is equal to the dimensionless temperature $\overline{\theta}$, multiplied by Q_0 (the heat power injected in the system). In the following, we work with dimensionless variables only and for simplicity, we drop the overbars.

In a statistically steady state, the following equalities result from the equations of motion and the boundary conditions in the problem:

$$\int_{Vol} (\nabla\theta)^2 d^3x = \int_{\text{Surface}} \theta q d^2x \tag{10}$$

and:

$$\sum_{i,j} \int_{Vol} (\partial_i u_j)^2 d^3x + Ma \int_{\text{Surface}} (u_{\|} \cdot \nabla_{\|})\theta d^2x = -Ra \left(\int_{\text{Surface}} z\nabla\theta \cdot dS - \int \theta dS \right) \tag{11}$$

Eq. (10) expresses the balance between thermal dissipation $(\nabla\theta)^2$ and production, whereas Eq. (11) relates the dissipation of kinetic energy $(\partial_i u_j)^2$ to the source of motion, through the Marangoni effect (Ma term) and buoyancy (Ra term). As pointed out already the effect of buoyancy happens to be negligible compared to the Marangoni effect, so we set in the following $Ra = 0$.

In addition, we will restrict ourselves to the case of low Prandtl number, appropriate for liquid metals ($Pr \sim 10^{-2}$).

Precise experimental or numerical results on the structure of the flow in the turbulent regime are lacking. Numerical results in the laminar case show the existence of narrow boundary layers, located near the surface where the flow and the temperature gradients are located. These boundary layers are reminiscent of the situation observed in the turbulent Rayleigh- Bénard problem. We are therefore lead to the following assumptions about the flow. First, we assume that the large scale flow consists of rolls extending over the entire system. We also assume that the Marangoni effect generates jets of fluid carrying fluid away from the center along the free surface. These jets have a finite width, since turbulent jets tend to open up with a finite angle as they propagate downstream. As such, their geometry is assumed to be independent of the Marangoni number. The flow is then assumed to run along the walls, and then to concentrate under the heating beam, so as to make a vertical, upwards moving stream. In the strongly heated region, the jet splits under the heating beam, leaving a turbulent stagnation point region. This confines the temperature gradient in a narrow region of width Δ. Boundary layer theory then suggests that this region becomes narrower upon increasing the heating (Marangoni number). We denote by Θ the (dimensionless) temperature elevation in the fluid, and by U the order of magnitude of the fluid velocity inside the jets.

The temperature elevation, Θ, as well as the velocity in the jets, U, and the width of the heated region, Δ, are now predicted with the help of these assumptions, and with the method of [6,7]. We begin by recalling the estimation of the kinetic energy dissipation per unit mass: $\epsilon \sim \tilde{U}^3/\tilde{L}$, where $\tilde{U}$ and $\tilde{L}$ are the (large) velocity and length scales. In our problem, the dissipation of kinetic energy is located in the jets, and it can be estimated as:

$$Pr \sum_{i,j} \int_{Vol} (\partial_i u_j)^2 d^3x \approx U^3 \tag{12}$$

Balancing the energy dissipation, Eq. (12), with the production term, Eq. (11) leads to:

$$U^2 \sim Pr Ma \Theta \tag{13}$$

To estimate the temperature elevation as a function of the heating power, we use the boundary layer assumption for the temperature field, so Eq. (10) leads to:

$$\int_{Vol} (\nabla\theta)^2 d^3x \approx \frac{\Theta^2}{\Delta^2} \times \Delta \times (\text{Surface}) \approx \frac{\Theta^2}{\Delta} \tag{14}$$

so:

$$\Theta \sim \Delta \tag{15}$$

Finally, a last relation can be obtained by equating the advection term in the thermal boundary layer: $u \cdot \nabla\theta \sim U\Theta/\Delta$ and the diffusion term: Θ/Δ^2, leading to:

$$\Delta U \sim 1 \tag{16}$$

Combining Eq. (13, 15, 15) together, one obtains:

$$\Theta \sim \Delta \sim (PrMa)^{-1/3} \tag{17}$$

$$U \sim (PrMa)^{1/3} \tag{18}$$

In dimensional units, the temperature elevation is therefore proportional to the 2/3 power of the heating power, or equivalently, the Nusselt number in this case behaves as:

$$Nu \sim Ma^{1/3} Pr^{1/3} \tag{19}$$

Estimating the order of magnitude of the numerical prefactor in Eq. (19) would require a more precise knowledge of the geometry of the flow.

We have estimated the temperature elevation in a weld pool heated by a strong, localized heat flux. In the turbulent case, and with the simple assumptions regarding the structure of the flow spelled out above, we have found that the highest value of the temperature in the fluid scales as the intensity of the heating flux to the power 2/3. The latter prediction rests on a number of assumptions, which call for more experimental or numerical work. We note that some preliminary numerical results in 2-dimensions [9] tend to support certain approximations made in this work. Also, the preliminary exerimental results of [10] show that the 2/3 power law is in qualitative agreement with the limited available data.

In conclusion, the significant progress made in our understanding of turbulent Rayleigh-Bénard convection allows us to tackle realistic and theoretically interesting problems.

Acknowledgements

It is a pleasure to thank E. Siggia, Soubbaramayer, F. Daviaud, A. Thess, C. Karcher, T. Boeck and R. Schaller for discussions related to this work. We are grateful to the Dresden group, who communicated their results to us prior to publications.

References

[1] C. Normand, Y. Pomeau and M. Velarde, Rev. Mod. Phys. **49**, 581 (1977).

[2] T. DebRoy et S.A. David, Rev. Mod. Phys. **67**, 85 (1995).

[3] S. Schiller, U. Heisig, S. Panzer, *Electron beam technology*, John Wiley & Sons, New York (1982).

[4] B. Castaing, G. Gunaratne, F.Heslot, L. Kadanoff, A. Libchaber, S. Thomae, X-Z Wu, S. Zaleski and G. Zanetti, J. Fluid Mech. **204**, 1 (1989).

[5] G. Zocchi, E. Moses and A. Libchaber, Physica **A 166**, 387 (1990).

[6] B. I Shraiman and E. D. Siggia, Phys. Rev. **A 42**, 3650 (1990).

[7] E. D. Siggia, Ann. Rev. Fluid Mech. **26**, 137 (1994).

[8] A. Pumir and L. Blumenfeld, Phys. Rev. **E 54**, R4529 (1996).

[9] T. Boeck and A. Thess, Phys. Rev. Lett **80**, 1216 (1998).

[10] Ch. Karcher et al. private communication (1997).

Trends in Mathematics, © 1999 Birkhäuser Verlag Basel/Switzerland

On the statistics of small-scale turbulence and its universality

Ch. Renner, B. Reisner, St. Lück, J. Peinke, R. Friedrich

Abstract

We present a method of how to estimate from experimental data of a turbulent velocity field the drift and the diffusion coefficient of a Fokker-Planck equation. It is shown that solutions of this Fokker-Planck equation reproduce with high accuracy the statistics of velocity increments in the inertial range. Using solutions with different initial conditions at large scales we show that they converge. This can be interpreted as a signature of the universality of small scale turbulence in the limit of large inertial ranges.

1 Introduction

The common picture of fully developed local isotropic turbulence is that the velocity field is driven by external fields on large scales. By this driving energy is fed into the system at scales larger than the integral length L_0. A cascading process will transport this energy to smaller and smaller scales until at the viscous length scale η the injected energy is finally dissipated by viscous effects [1, 2]. It is commonly believed that this picture of a cascade leads to the universality of the statistical laws of small scale turbulence.

The standard quantity to study the statistics of turbulent fields is the longitudinal velocity increment of the length scale L_i defined as $v_i = u(x+L_i) - u(x)$, where u denotes the velocity component in the direction of the separation L_i. x is a selected reference point.

Two common ways have been established to analyze the statistical content of the velocity increments. On one side the structure functions have been evaluated by supposing a scaling behavior

$$< (v_i)^q >= \int dv_i \, (v_i)^q \, P(v_i, L_i) \propto L_i^{\zeta_q}, \tag{1}$$

where $P(v_i, L_i)$ denotes the probability density function (pdf) for v_i at the length scale L_i. Using the recently proposed so-called extended selfsimilarity [3] it has become possible to evaluate the characterizing scaling exponents ζ_q quite accurately c.f. [4, 5]. On the other side, it is interesting to find out how to parameterize directly the evolution of the probability density functions (pdf) $P(v_i, L_i)$ c.f. [6, 7].

One major challenge of the research on turbulence is to understand small scale intermittency, which is manifested in a changing form of $P(v_i, L_i)$ or equivalently in a nonlinear q-dependence of the scaling exponents ζ_q, provided that the scaling assumption is valid.

For a long time, the main effort has been put into the understanding of the q-dependence of ζ_q (for actual reviews see [2, 5]), although it is well known that due to the statistic of a finite number of data points (let say 10^7 data) it is not possible to determine accurately scaling exponents ζ_q for $q > 6$ [8], and that there are different experimental indications that no good scaling behavior is present [9, 10].

That there have been less attempts to analyze directly the pdfs may be based on the fact that up to now the scaling exponents are regarded as the simplest reduction of the statistical content and that this analysis does not depend on model assumptions. In contrast to this, proposed parameterizations of the form of the pdfs [6, 7], although they are quite accurate, are still based on some additional assumptions on the underlying statistics. Based on the recent finding that the turbulent cascade obeys a Markov process in the variable L_i and that intermittency is due to multiplicative noise [10, 11, 12], we show in Section III that it is possible to estimate from the experimental data a Fokker-Planck equation, which describes the evolution of the pdfs with L_i. We show that this Fokker-Planck equation reproduces accurately the experimental probability densities $P(v_i, L_i)$ within the inertial range. Thus an analysis of experimental data is possible which quantifies the statistical process of the turbulent cascade and which neither depends on scaling hypotheses nor on some fitting functions [13]. Having determined the correct Fokker-Planck equation for an experimental situation, we show in section VI how solutions of this equation with different large scale pdfs will converge to universal small scale statistics. This finding gives evidence for the universality of small scale turbulence.

2 Experimental data

The results presented here are based on 10^7 velocity data points. Local velocity fluctuations were measured with a hot wire anemometer (Dantec Streamline 90N10) and a hot wire probe (55P01) with a spatial resolution of about $1mm$. The sampling frequency was 8 kHz. The stability of the jet was verified by measurements of the self-similar profiles of the mean velocity according to [14]. The turbulence measurements were performed by placing the probe on the axis of a free jet of dry air developing downwards in a closed chamber of the size of 2m x 1m x1m. To prevent a disturbing counterflow of the outflowing air, an outlet was placed at the bottom of the chamber. The distance to the nozzle was 125 nozzle diameters D. As nozzle we used a convex inner profile [15] with an opening section of $D = 8mm$ and an area contraction ratio of 40. Together with a laminarizing prechamber we achieved a highly laminar flow coming out of the nozzle. At a distance of 0.25 D from the nozzle, no deviation from a rectangular velocity profile was found within the detector resolution. Based on a 12bit A/D converter resolution, no fluctuations of the velocity could be found. The velocity at the nozzle was $45.5m/s$ corresponding to a Reynolds number of 2.7 10^4. At the distance of 125 D we measured a mean velocity of $2.25m/s$, a degree of turbulence of 0.17, an integral length $L_0 = 67mm$, a Taylor length $\theta = 6,6mm$ (determined according to [16]), a Kolmogorov length $\eta = 0.25mm$, and a Taylor Reynolds number $R_\theta = 190$.

The space dependence of the velocity increments v_i was obtained by the Taylor hypotheses of frozen turbulence. For the structure functions we found a tendency to scaling behavior for $L_0 \geq L_i \geq \theta$. Intermittency clearly emerged as $L_i \to \theta$, as shown in Fig.1 by the different form of the pdfs for different scales L_i.

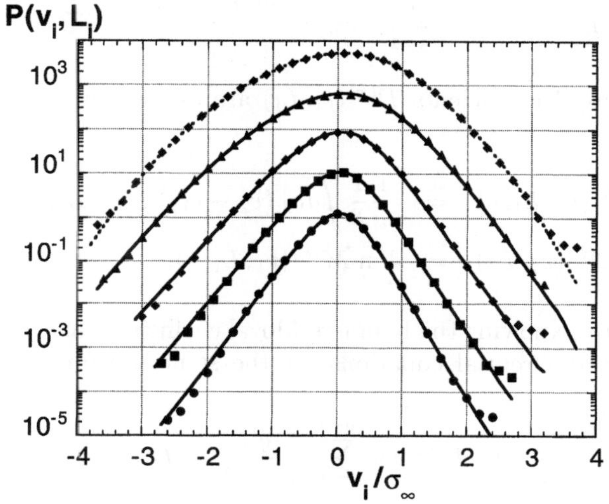

Figure 1: Probability density functions $P(v, L)$ determined from the experimental data (bold symbols) and calculated pdfs (lines) by the numerical iteration with the experimentally determined Fokker-Planck equation (5) and (6). The length scales $\lambda = ln(L_0/L)$ were: 0, 0.52, 1.04, 1.56, and 2.24 going from up to down. For the numerical iteration as initial condition, an empirical fitting function for the large scale pdf was used which is shown by a broken line. For clarity of presentation the pdfs were shifted in y-direction.

3 Measurement of Kramers-Moyal coefficients

Next, we show how to determine from experimental data appropriate statistical equations to characterize the turbulent cascade. The basic quantity for this procedure is to evaluate the cascade by the statistical dependence of velocity increments of different length scales at the same location x [17]. Either two-increment probabilities $p(v_2, L_2; v_1, L_1)$ or corresponding conditional probabilities $p(v_2, L_2|v_1, L_1)$ are evaluated from the whole data set. (Here we use the convention that $L_{i+1} < L_i$.) Investigating the corresponding three-increment statistics, we could provide evidence that the evolution of these statistics with different L_i fulfills the Chapman-Kolmogorov equation [18, 10, 11]. Furthermore evidences of the validity of the

Markov process are shown as long as the step size between different L_i is larger than a Markov length L_{mar}, which is in the order of some η [12]. Thus we know [18] that the evolution of the increments v_i is given by a master equation without the involvement of memory functions and that the evolution of the pdfs $P(v_i, L_i)$ is described by a Kramers-Moyal expansion:

$$-\frac{d}{dL_i} P(v_i, L_i) = \sum_{k=1}^{\infty} [-\frac{\partial}{\partial v_i}]^k D^{(k)}(v_i, L_i)\ P(v_i, L_i). \tag{2}$$

The Kramers-Moyal coefficients $D^{(k)}(v_i, L_i)$ are defined by the following conditional moments:

$$\begin{aligned} M^{(k)}(v_1, L_1, \delta) &= \frac{1}{n!}\frac{1}{\delta} \int dv_2\ (v_2 - v_1)^k\ p(v_2, L_2 | v_1, L_1) \\ D^{(k)}(v_1, L_1) &= \lim_{\delta \to 0} M^{(k)}(v_1, L_1, \delta), \end{aligned} \tag{3}$$

where $\delta = L_2 - L_1$. Knowing the Kramers-Moyal coefficients and thus the evolution of $P(v_i, L_i)$, the differential equations for the structure functions are obtained easily [10]

$$-\frac{d}{dL} < (v(L))^q > = \sum_{n=1}^{q-1} \frac{q!}{(q-n)!} < D^{(n)} v^{q-n} > . \tag{4}$$

If the Kramers-Moyal coefficients have the form $D^{(q)} = d_q v^q / L$ (where d_q are constants), scaling behavior of (1) is guaranteed with $\zeta_q = -\sum_{n=1}^{q-1} \frac{q!}{(q-n)!} d_n$. For a discussion of this finding in terms of multifractality and multiaffinity see [19]. The $1/L$ dependence of $D^{(q)}$ indicates that the natural space variable is lnL. Thus in the following we take $\lambda_i = ln(L_0/L_i)$ as the space variable. Furthermore we normalize the velocity increments to the saturation value of $\sigma_\infty = \sqrt{< v_\infty^2 >}$ for length scales larger than the integral length.

An important simplification is achieved if the fourth Kramers-Moyal coefficient is zero, then the infinite sum of the Kramers-Moyal expansion (2) reduces to the Fokker-Planck equation of only two terms [18]:

$$\frac{d}{d\lambda} P(v, \lambda) = \left[-\frac{\partial}{\partial v} \tilde{D}^{(1)}(v, \lambda) + \frac{\partial^2}{\partial v^2} \tilde{D}^{(2)}(v, \lambda) \right] P(v, \lambda). \tag{5}$$

Now only the drift term $\tilde{D}^{(1)}$ and the diffusion term $\tilde{D}^{(2)}$ are determining the statistics completely. From the corresponding Langevin equation characterizing the evolution of a spatially fixed increment, we know that $\tilde{D}^{(1)}$ describes the deterministic evolution, whereas $\tilde{D}^{(2)}$ describes the noise acting on the cascade. If $\tilde{D}^{(2)}$ shows a v-dependency one speaks of multiplicative noise.

In Fig. 2 we show the evaluated conditional moments $\tilde{M}^{(1)}$ and $\tilde{M}^{(2)}$ as they evolve for $\delta\lambda \to 0$. For the higher Kramers-Moyal coefficients we found that

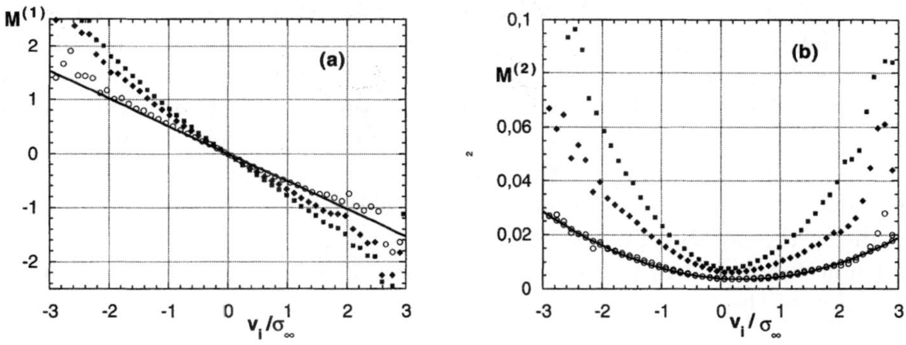

Figure 2: The conditional moments $\tilde{M}^{(1)}(v,\lambda,\delta\lambda)$, (a), and $\tilde{M}^{(2)}(v,\lambda,\delta\lambda)$, (b), for $\lambda = 1.48$ and $\Delta\lambda$= 0.04 (○), 0.06 and 0.20. The fits to the experimental data by polynomial functions (see text) are shown by solid lines. These fits were done only with respect to the interval $-2 < v < 2$).

$\tilde{M}^{(4)} < 10^{-2}\tilde{M}^{(2)}$. Thus we take the Fokker-Planck equation as the adequate description.

To estimate the $\tilde{D}^{(1)}$ and $\tilde{D}^{(2)}$ coefficient properly, we performed the following three procedures based on the assumption that we can approximate the Kramers-Moyal coefficient by the following polynomials: $\tilde{D}^{(1)} = \gamma(\lambda)$ and $\tilde{D}^{(2)} = \alpha(\lambda) + \delta(\lambda) * v_i + \beta(\lambda) * v_i^2$. First, we used the same polynomial forms for the $\tilde{M}$-coefficients and estimated the coefficients of the polynomials for $\delta\lambda \to 0$. Here problems arose due to the finite value of the Markov-length and due to the finite resolution of our detector. Secondly, inserting the polynomials of $\tilde{D}^{(1)}$ and $\tilde{D}^{(2)}$ into (4) we could determine directly the coefficients α to δ from the measured structure functions (the highest order of the structure function we used was 6). Problems arose due to the noise of the derivative of the structure functions, thus only the magnitude of the coefficient could be estimated. Thirdly, we evaluated analogously the corresponding structure functions of $P_{even}(v,L) = 1/2(P(v,L) + P(-v,L))$ and $P_{odd}(v,L) = 1/2(P(v,L) - P(-v,L))$. [20]

From all these estimations we got a good guess of the values of $\alpha(\lambda)$ to $\delta(\lambda)$. The best results we obtained for γ. The worst result was obtained for β, for which we finally used the value $\mu/18$ expected from the Kolmogorov picture of intermittency [21, 10]. As the final values we got:

$$\begin{aligned} \gamma(\lambda) &= 0.36exp\left\{(\frac{\lambda - 0.68}{2.1})^3\right\}, & \alpha(\lambda) &= 0.02exp\left\{-\frac{\lambda}{1.3}\right\} \\ \delta(\lambda) &= 0.013exp(-1.2\lambda), & \beta &= 0.017 \quad . \end{aligned} \tag{6}$$

Next, let us briefly comment this result. For γ we find that it has a constant value of 1/3 in the inertial range ($0 < \lambda < 2$). The functional dependence of γ

is given to extend the λ dependence into the dissipation range. The values δ and α are violating scaling behavior. Both, α and δ, are decaying exponentially with λ. For scales larger than the integral length these two coefficients are large and are responsible for the building up of the skewness, because these terms allow the change of the sign of the velocity increments within the stochastic process.

As further test of the validity of our approach, we calculated the evolution of the pdfs by a numerical iteration with the Fokker-Planck equation. As an initial condition the pdf $P(v_0, L_0)$ at the integral scale was approximated by an empirical function and then inserted into the numerical iteration. We found that the evolution of the pdfs depends sensitively on the chosen coefficients. (Finally, we changed the numbers in (6) in the range of some percent to obtain a best result.) The result of the numerical iteration of the Fokker-Planck equation with the above mentioned coefficients are shown in Fig. 1 by solid lines.

4 Universality

Let us now discuss how the turbulent cascade will be affected, if the statistics on large scales is changed as it may be the case for different driving forces. As an extreme case we chose a box like form for a large scale pdf (the standard deviation and the skewness was adapted to the one of our experimental pdf for L_0 (see Fig. 3)). Next, using the above mentioned coefficients (6), we iterated simultaneously the experimental approximation of $P(v_0, L_0)$, see Fig. 1, and the box-like pdf. As shown in Fig.3, we see that under the iteration the synthetic pdf becomes more and more similar to the experimental ones. This convergence was quantified by the χ^2 measure ($\chi^2 \propto \sum (P_1 - P_2)^2/P_1$). We found that χ^2 decays about exponentially with λ. Thus, we see that for a sufficient large cascade any large scale pdf will converge against universal small scale pdfs.

5 Conclusion

We have presented further evidence that the statistics of fully developed turbulence is based on a Markov process in the space variable L. It has been shown how the drift coefficient and the diffusion coefficient of a describing Fokker-Planck equation can be determined from experimental data. With these coefficients it is possible to calculate accurately the pdfs of a cascade. As a further result we have shown that situations, where the large scale statistics differ, for example due to different driving forces, will show the same small scale statistics as long as the same Fokker-Planck equations apply. Therefore, universality of small scale turbulence can best be characterized by comparing the drift and diffusion coefficients.

The next challenging questions are, how do these results depend on the Reynolds number and what happens in anisotropic turbulence. For the latter point, evidence has been found recently that for sufficiently small scales again the results of local isotropic turbulence hold [22]. We believe that these findings do not only play an important role for the complete statistical characterization of turbulence

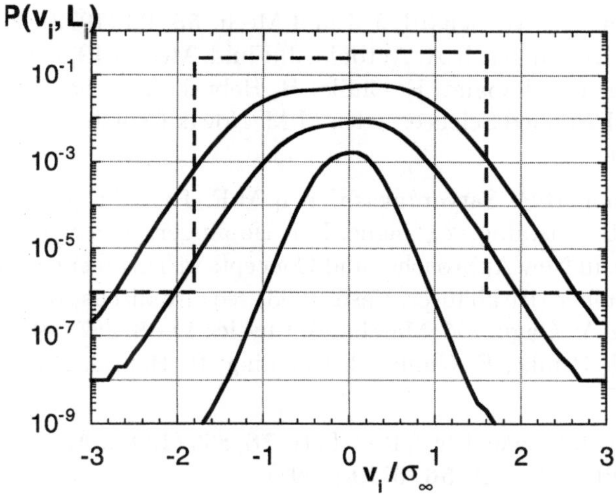

Figure 3: Evolution of a synthetic box-shaped pdf (broken line) under the numerical integration with the Fokker-Planck equation using the coefficients of (6). From top to bottom $\lambda = 0$, 1.4, 2.2 and 3.0. Presentation as in Fig.1.

but also should have a quite practical importance for numerical simulations of flow situations with high Reynolds numbers.

6 Acknowledgments

We want to acknowledge discussions with A. Naert, P. Talkner, K. Zeile, H. Brand. J.P. acknowledges the financial support of the Deutsche Forschungsgemeinschaft.

References

[1] A.S. Monin, A.M. Yaglom, Statistical Fluid Mechanics, (MIT Press, Cambridge 1975).

[2] U. Frisch, Turbulence (Cambridge 1995).

[3] R. Benzi, S. Ciliberto, C. Baudet, G.R. Chavarria, Physica **D 80**, 385 (1995).

[4] A. Arneodo, et. al. Europhys. Lett. **34**, 411 (1996).

[5] K.R. Sreenivasan, R.A. Antonia, Annu. Rev. Fluid Mech. **29**, 435 (1997).

[6] R. Benzi, L. Biferale, G. Paladin, A. Vulpiani, M. Vergassola, Phys. Rev. **67**, 2299 (1991); P. Kailasnath, K.R. Sreenivasan, G. Stolovitzky, Phys. Rev. **68**, 2767 (1992).

[7] B. Castaing, Y. Gagne, E. Hopfinger, Physica **D 46**, 177 (1990).

[8] H. Tennekes, J.C. Wyngaard, J. Fluid Mech. **55**, 93 (1972); F. Anselmat, Y. Gagne, E.J. Hopfinger, R.A. Antonia, J. Fluid. Mech **149**,63 (1984); J. Peinke, B. Castaing, B. Chabaud, F. Chilla, B. Hebral, A. Naert, in Fractals in the Natural and Applied Sciences, edt.: M.M. Novak (North Holland, Amsterdam 1994) p.295.

[9] R.A. Antonia, B.R. Satyaprakash, A.K.M.F. Hussain, J. Fluid Mech. **119**, 55 (1982); B. Castaing, Y. Gagne, E.J. Hopfinger, A new View of Developed Turbulence, in New Approaches and Concepts in Turbulence, edts. Th. Dracos and A. Tsinober (Birkhäuser, Basel 1993) see also discussion p. 47–60 therein; B. Castaing, Y. Gagne, M. Marchand, Physica **D 68**, 387 (1993); B. Chabaud, A. Naert, J. Peinke, F. Chilla, B. Castaing, B. Hebral, Phys. Rev. Lett. **73** (1994) 3227.

[10] R. Friedrich, J. Peinke, Phys. Rev. Lett. **78**, 863 (1997); A. Naert, R. Friedrich, J. Peinke, Phys. Rev. E **56**, 6719 (1997).

[11] R. Friedrich, J. Peinke, Physica **D 102**, 147 (1997).

[12] R. Friedrich, J. Zeller, J. Peinke, Europhys. Lett. **41**, 143 (1998).

[13] It has recently been shown that in an analogous way it is possible to analyse dynamical systems, and to extract the Langevin equation directly from a given data set; S. Siegert, R. Friedrich, J. Peinke, Phys. Lett. **A 243**, 275 (1998).

[14] N. Rajartnan, Turbulent Jets (Elsevier, Amsterdam 1976); Ch. Renner, Diplomarbeit (Bayreuth 1997).

[15] B. Bümel and H.E. Fiedler (Berlin) priv. communication.

[16] D. Aronson, L. Lofdahl, Phys. Fluids **A5**, 1433 (1993).

[17] J. Peinke, R. Friedrich, F. Chilla, B. Chabaud, and A. Naert, Z. Phys. **B 101**, 157 (1996).

[18] H. Risken, The Fokker-Planck Equation, (Springer-Verlag Berlin, 1984); P. Hänggi and H. Thomas, Physics Reports **88**, 207 (1982).

[19] J. Peinke, R. Friedrich, A. Naert, Z. Naturforsch. **52a**, 588 (1997).

[20] The Fokker-Planck equations for P_{even} and P_{odd} have the same form as (5), only if the coefficient $\delta = 0$, this may clear up the question of the sense of evaluating moments of the absolute values of v_i, see for example footnote on page 446 in [5].

[21] A.M. Oboukhov, J. Fluid Mech. **13**, 77 (1962); A.N. Kolmogorov, J. Fluid Mech. **13**, 82 (1962).

[22] F. Chilla et. al. to be published

Trends in Mathematics, © 1999 Birkhäuser Verlag Basel/Switzerland

A new method to characterize inhomogeneous turbulence

B. Reisner, Ch. Renner, St. Lück, J. Peinke, F. Chillá, R. Friedrich

Abstract

We present a new approach to characterize the statistics of velocity increments for two different inhomogeneous turbulent air flows. Conditional probability density functions of velocity increments for different length scales are evaluated and compared to those of a local isotropic flow. Especially on large scales we can distinct clearly the different flows.

In the recent years, much effort has been put into the research of fully developed local isotropic turbulence. One main challenge is to understand the statistics of velocity increments $v_i(x) = u(x + L_i) - u(x)$ on different length scales L_i [1]. Recently experimental evidence was given that for local isotropic flows the statistics of the increments obey a Markov process in the variable L_i [2]. There it has become evident that conditional probability density functions (cpdf) $p(v_2, L_2|v_1, L_1)$ with $L_2 < L_1$ are the central quantities to be investigated.

Inspired by the above mentioned results we have extended our analysis to inhomogeneous flow situations. Namely we analyzed experimental data from the outer regions of a free jet and of a turbulent wake. The longitudinal velocity components were measured with hot wires in different distances from the center. In the free jet ($R_\lambda = 212$) we placed the sensor in a distance of 60 nozzle diameters. The turbulent wake ($R_\lambda = 505$) was measured with an x-wire detector in the distance of 32 diameters behind a cylinder (5 cm diameter) in the windtunnel of Erlangen. In this paper we present only some exemplary results we obtained in the outer regions of the turbulent flows. For comparison we show results for one local isotropic case (see Fig. 1).

To characterize the statistics of different flow situations, we proceed as follows. Based on the Taylor hypotheses of frozen turbulence, we evaluate the cpdf $p(v_2, L_2|v_1, L_1)$ with $L_2 < L_1$. For the local isotropic case we found that there are not many structural changes in these cpdfs, thus we show it only for one pair of L_i. In Fig. 2 and 3 two pairs of cpdfs are shown, one constructed for large scales (close to an integral length) and the other constructed for small scales. There are two characteristic features. First, the cpdfs are orientated. Second, the variance changes with v_1. A significant measure of these features is the first and the second conditional moment:

$$M^{(n)}(v_1, L_1, \Delta L) = \frac{1}{n!}\frac{1}{\Delta L}\int dv_2\ (v_2 - v_1)^n\ p(v_2, L_2|v_1, L_1) \qquad (1)$$

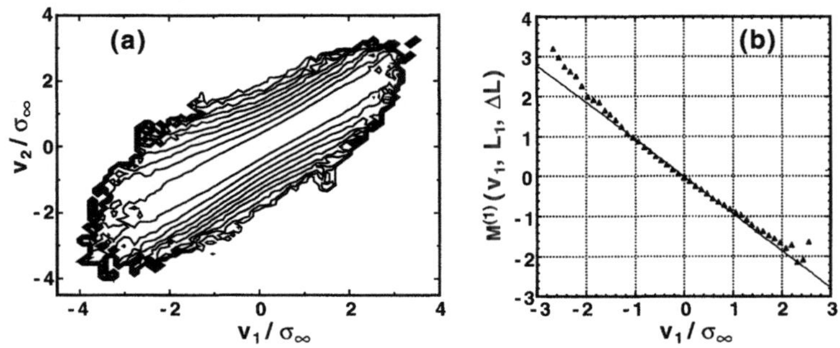

Figure 1: (a) Conditional probability density functions $p(v_2, L_2|v_1, L_1)$ and (b) conditional moments $M^{(1)}(v_1, L, \Delta L)$ determined from the experimental data of a local isotropic jet flow ($L_1 = 14.2\,mm$, $L_2 = 10.8\,mm$, $\Delta L = 1.3\,mm$).

where $\Delta L = L_2 - L_1$. If the evolution of the velocity increments with L_i obeys a Markov process, these conditional moments give for lim $\Delta L \to 0$ the Kramers-Moyal coefficients, which determine the stochastic process [3, 4]. For the isotropic turbulence we find that $M^{(1)}$ is a linear and $M^{(2)}$ a quadratic function in v_1. For the inhomogeneous case we clearly find deviations for large scales, thus $M^{(1)}$ is very asymmetric for the jet data (see Fig. 2b) and oscillates for the wake data (see Fig. 3b). Coming to small scales we note that for the jet as well as for the wake the cpdf become similar to the isotropic case, thus we see that $M^{(1)}$ becomes linear (see Fig. 2d and 3d). Only the notch at $v_1 = 0$ (see Fig. 2a,c and 3a,c) remains as one remarkable feature in the cpdf of the inhomogeneous turbulence, which is independent of the selected scale. The meaning of the notch is that small fluctuations are most likely linked to small fluctuations on smaller length scales, which is naturally expected from a laminar signal.

At last, let us comment on the asymmetric width of the cpdf of the inhomogeneous jet flow. In Fig. 2a we find that positive v_1 are linked with less noise to their corresponding v_2-values as it is the case for negative v_1. This effect must be interpreted as an asymmetry in the rising and falling part of the velocity signal. Most remarkably this feature is not present in the velocity data of the inhomogeneous wake. We interpret this effect as a consequence of the different mean flow profiles. Looking at visualizations, we found that in the outer region of a jet the turbulent spots are bent backwards leading to a triangle-like running front. If such a front passes over the detector, we find a rapid and less noisy rise in velocity which is subsequently decreasing with more noise. As we have seen above, this is only true for the large scale structure. Looking at the signal with a higher magnification, that is for smaller L_i-values, the signals become more symmetric. We note that this asymmetry is not found in the wake flow.

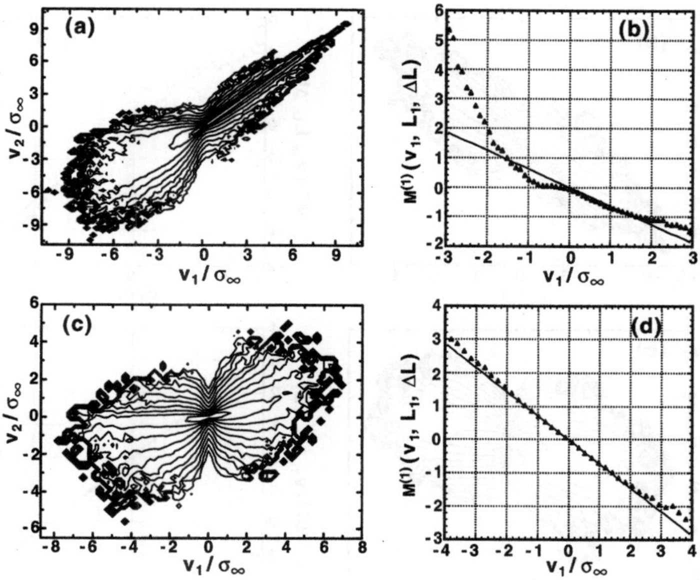

Figure 2: Cpdfs determined from the experimental data of the inhomogeneous flow of a free jet; (a) $L_1 = 23.1\,mm$, $L_2 = 20.5\,mm$ near the integral length scale and (c) $L_1 = 3.7\,mm$, $L_2 = 0.7\,mm$ for small scale. Conditional moments $M^{(1)}$, (b) $\Delta L = 1.1\,mm$ and (d) $\Delta L = 1.1\,mm$. Straight lines in (b) and (d) are presented to show the departure from linearity, as it is expected in the isotropic case.

To summarize, we have presented results on the statistics of velocity increments for inhomogeneous turbulent flows of a free jet and a wake. With the help of conditional probability densities we were able to characterize some remarkable features of inhomogeneous flow situations. On the one hand for large scales we found different asymmetries for the free jet and the wake, on the other hand for small scales we found that both systems have the tendency to show the same behavior as the local isotropic turbulence. This result fits nicely into the generally accepted picture of turbulence, namely that systems driven differently on large scales, i.e. with different large scale structures, have the tendency to generate universal small scale structures. To quantify these statements in more detail, the challenging question is, whether these findings can be embedded into the concepts of Markov-processes as it was possible for the local isotropic turbulence.

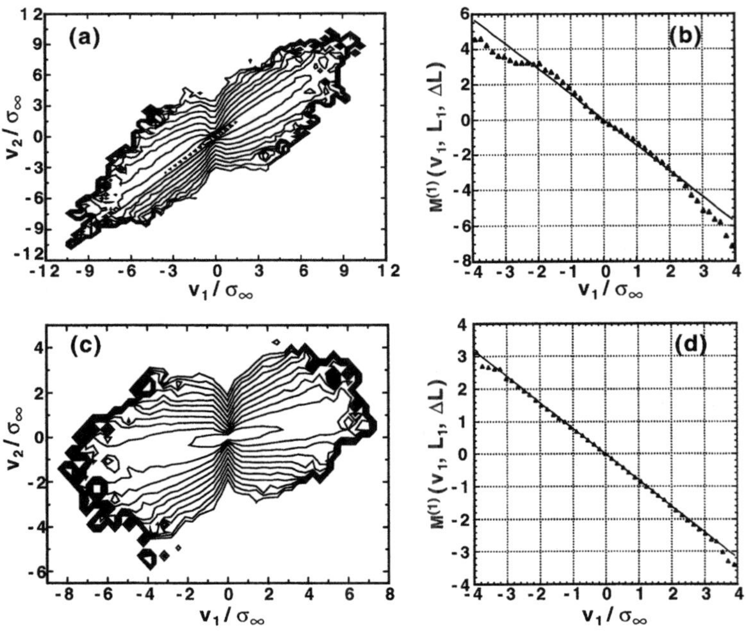

Figure 3: Cpdfs and conditional moments for the inhomogeneous flow of a wake. Analogous presentation as in Fig. 2 with (a) $L_1 = 56.8\,mm$, $L_2 = 49.7\,mm$ and (c) $L_1 = 8.9\,mm$, $L_2 = 1.8\,mm$; (b) $\Delta L = 1.8\,mm$ and (d) $\Delta L = 1.8\,mm$.

References

[1] A.S. Monin, A.M. Yaglom, Statistical Fluid Mechanics, (MIT Press, Cambridge 1975); U. Frisch, Turbulence (Cambridge 1995).

[2] R. Friedrich, J. Peinke, Phys. Rev. Lett. **78**, 863 (1997); R. Friedrich, J. Zeller, J. Peinke, Europhys. Lett. **41**, 143 (1998).

[3] Ch. Renner et.al. within this issue.

[4] H. Risken, The Fokker-Planck equation, (Springer-Verlag Berlin, 1984).

Trends in Mathematics, © 1999 Birkhäuser Verlag Basel/Switzerland

Multiscale similarity of isotropic homogeneous turbulence

M. J. Huang

Abstract

A similarity form of energy spectrum involving two lengthscales, the Kolmogorov dissipation lengthscale (η) and the empirical energy-containing lengthscale ($\mathcal{L}$), is proposed for isotropic homogeneous incompressible turbulence. The similarity form is approached by considering the energy spectrum as a summation of a series of ordered functions of variable $k\mathcal{L}$ dominant at small wave numbers and a series of ordered functions of variable $k\eta$ dominant at large wave numbers. With a corresponding energy transfer spectrum, dynamical equations for these associated expansion functions can be derived from the spectral energy equation. The results of analysis show that the expansion functions may be ordered in powers of ${Re_\lambda}^{-1/2}$, where Re_λ is the Taylor's Reynolds number and that a power-law energy decay is a necessary condition for similarity to exist. A reasonable Reynolds-number dependence of the skewness of velocity longitudinal derivatives is predicted. Constraints on the expansion functions due to definitions are discussed. An example energy spectrum is constructed and the resulting energy transfer spectrum is investigated.

1 Introduction

To reduce the extent of difficulty in analyzing turbulence, a discovery of regularities in the motion of turbulence can be very helpful. In 1938, Kármán and Howarth [1] derived an equation of evolution of the double and triple velocity correlation coefficients for isotropic homogeneous turbulence. At the same time, they proposed a self-similar solution which evolves with one lengthscale, the Taylor's microscale. A power-law energy decay with a decay exponent of one and a constant skewness of longitudinal velocity derivative come after this similarity. Barenblatt and Gavrilov [2] and George [3] modified Kármán and Howarth's similarity solution slightly. The new similarity evolves again with Taylor's microscale only but allows an arbitrary decay exponent and predicts a skewness inversely proportional to the Reynolds number. A constant skewness implies a saturation of nonlinear stretching in the turbulence. Thus, it is generally believed that Kármán and Howarth's similarity is true for turbulence at extremely large Reynolds number. A decreasing skewness with increasing Reynolds number however is not physically reasonable. Huang and Leonard [4] introduced a one parameter family of single-scale solutions, in which the decay exponent is arbitrary and an increasing skewness with increasing Reynolds number is obtained.

A similarity solution using only one lengthscale is not applicable at large Reynolds number however. The multiscale characteristics of turbulence at large

Reynolds number causes an upper bound on the validity of the single-scale similarity solutions. According to the 1941 Kolmogorov hypotheses [5], [6], [7], and their support from many experimental data [8], when Re_λ is sufficiently large, namely ${Re_\lambda}^{3/4} \gg 1$ (where $Re_\lambda \equiv q\lambda/\nu$, λ is the Taylor's microscale, $\frac{3}{2}q^2$ is the turbulent energy, and ν is the kinematic viscosity), at least three subranges of wave numbers may be distinguished — an energy-containing subrange, a universal equilibrium inertial subrange, and a universal equilibrium dissipation subrange. Thus at least three lengthscales are required to characterize motions of different scales. To be consistent, a similarity solution, if exists, of multiscales is desired.

On the other hand, a power-law energy decay may exist without similarity. Actually a power-law decay has been well observed in the lab as well as in numerical simulations [9], [10], [11], [12], [13], but a perfect similarity has not been observed yet. In the previous investigations of sinlge-scale similarities, a power-law decay usually appears as a necessary condition.

Based on the observation that energy-containing eddies can be characterized by an empirical lengthscale $\mathcal{L} \sim \lambda Re_\lambda$ [14] and the dissipation eddies can be characterized by Kolmogorov dissipation lengthscale η, an attempt in the present work thus is to explore a similarity form evolving these two lengthscales for turbulence at large Reynolds number. Once the similarity form is obtained, dynamical equations of associated similarity functions may be derived in a way similar to that employed by Huang and Leonard [4]. The power-law energy decay will be examined then. This paper is arranged as follows. In Sec.2 the single-scale similarity solutions will be reviewed. A possible multiscale similarity form will be proposed in Sec.3. Important dynamical equations of the associated similarity functions will be derived in Sec.4. Comparisons with single-scale similarities will be discussed in Sec.5. And finally given in Sec.6 are the conclusions.

2 Single-scale similarity

The Kármán and Howarth's similarity [1] is concerned with isotropic homogeneous and incompressible turbulence and can be written as follows

$$
\begin{aligned}
f(r,t) &= \widetilde{f}(r/\lambda) \\
k^{tri}(r,t) &= \widetilde{k}^{tri}(r/\lambda)
\end{aligned} \tag{1}
$$

where $f(r,t)$ and $k^{tri}(r,t)$ are the double and triple velocity correlation coefficient, defined as

$$
\begin{aligned}
f(r,t) &\equiv \overline{u(\mathbf{x},t)u(\mathbf{x}+\mathbf{r},t)}/q(t)^2 \\
k^{tri}(r,t) &\equiv \overline{u(\mathbf{x},t)^2u(\mathbf{x}+\mathbf{r},t)}/q(t)^3.
\end{aligned} \tag{2}
$$

where r is the separation of the two points, t is the time, u is the velocity component in the $\mathbf{r}$-direction. Barenblatt and Gavrilov's [2] and George's [3] similarities have the same form of

$$\begin{aligned} f(r,t) &= \tilde{f}(r/\lambda) \\ k^{tri}(r,t) &= Re_\lambda{}^{-1} \cdot \tilde{k}^{tri}(r/\lambda). \end{aligned} \tag{3}$$

Huang and Leonard [4] based on their direct numerical data generalized the similarity form to be

$$\begin{aligned} f(r,t) &= f_1(r/\lambda) + Re_\lambda{}^{\beta} \cdot f_2(r/\lambda) \\ k^{tri}(r,t) &= Re_\lambda{}^{\beta-1} \cdot k_2^{tri}(r/\lambda). \end{aligned} \tag{4}$$

When $\beta = 0$, Barenblatt and Gavrilov's or George's similarity is restored.

The dynamical equations governing the similarity functions (between $\tilde{f}$ and $\tilde{k}^{tri}$ or among f_1, f_2, and k_2^{tri}) can be derived by substituting (3) or (4) into the Kármán-Howarth equation[1] and evoking consistency. For example, the function f_1 had been found to be a regular confluent hypergeometric function [4].

3 Multiscale similarity

The analysis herein will be presented in the wave space because it is easier there to understand the multiscale property. That is a proper multiscale similarity form for the energy spectrum $E(k,t)$ and for the energy transfer spectrum $T(k,t)$, where k is the wave number, will be searched instead of the double and triple velocity correlation coefficients. The latters actually are related to the formers by

$$E(k,t) = \frac{q^2}{\pi} \int_0^\infty kr(\sin kr - kr\cos kr) f(r,t)\, dr \tag{5}$$

and

$$T(k,t) = \frac{q^3}{\pi} \int_0^\infty k^2 r \left(\frac{3\sin kr}{kr} - 3\cos kr - kr\sin kr \right) k^{tri}(r,t)\, dr \tag{6}$$

[15]. To explore a multiscale similarity, the idea origins from expanding the energy spectrum as a function of $k\mathcal{L}$ at small wave numbers (large eddies) as an inner expansion and of $k\eta$ at large wave numbers (small eddies) as an outer expansion. Matching is performed in the inertial subrange. The uniform expansion thus provides the required, proper similarity form.

3.1 Expansions of energy spectrum

Beginning with the fact that eddies of small wave numbers contain most of the turbulent energy and are characterized by lengthscale $\mathcal{L} \equiv \lambda Re_\lambda$, the expansion of the energy spectrum at small wave numbers may be written as follows

$$E(k) \sim E_i(k) = q^2 \mathcal{L} \sum_{j=0} \mu_j(Re_\lambda)\, F_j(k\mathcal{L}). \tag{7}$$

Note quantities q and λ (and thus Re_λ) are in general time-dependent. On the other hand, eddies at large wave numbers have a characteristic kinetic energy $(\epsilon\nu^5)^{1/4}$ (where ϵ is the energy dissipation rate) and a characteristic lengthscale η. Therefore, a form of

$$E(k) \sim E_o(k) \; = \; (\epsilon\nu^5)^{1/4} \sum_{j=0} \nu_j(Re_\lambda)\; G_j(k\eta) \tag{8}$$

is suggested. Again $\epsilon = -\frac{d}{dt}\left(\frac{3}{2}q^2\right)$ and η are both time-dependent. The order functions $\mu_j(Re_\lambda)$ and $\nu_j(Re_\lambda)$ are functions of Re_λ and are assumed to be ordered by

$$\mu_0 \gg \mu_1 \gg \mu_2 \gg \cdots$$
$$\nu_0 \gg \nu_1 \gg \nu_2 \gg \cdots$$

as $Re_\lambda \gg 1$.

To make the matching, it is assumed that both spectra possess Kolmogorov's $(k^{-5/3})$ spectrum in their asymptotic behavior and, according to Kolmogorov's hypotheses, the geometric mean of $\mathcal{L}$ and η is taken to be the matching lengthscale as can be identified by writing

$$\begin{aligned} \xi \; &\equiv \; k\mathcal{L} = \left(k\lambda {Re_\lambda}^{1/4}\right) \cdot {Re_\lambda}^{3/4} \\ \zeta \; &\equiv \; k\eta \sim \left(k\lambda {Re_\lambda}^{1/4}\right) \cdot {Re_\lambda}^{-3/4}, \end{aligned} \tag{9}$$

where the definitions $\lambda = \sqrt{15\nu q^2/\epsilon}$ and $\eta = \left(\nu^3/\epsilon\right)^{1/4}$ have been employed. In other words, an intermediate variable is chosen to be $\chi \equiv k\lambda {Re_\lambda}^{1/4}$ such that $E_i(k)$ and $E_o(k)$ approach Kolmogorov's spectrum as $\xi \to \infty$ or $\zeta \to 0$ when $Re_\lambda \to \infty$ with χ fixed. Consequently the lengthscale characterizing eddies in the inertial subrange is assumed to be $\lambda {Re_\lambda}^{1/4}$. If we define the upper and lower limits of the inertial subrange as the geometric means of adjacent characteristic lengthscales, namely $\lambda {Re_\lambda}^{5/8}$ and $\lambda {Re_\lambda}^{-1/8}$, the width of the inertial range is then $\mathcal{O}({Re_\lambda}^{3/4})$, slightly wider than the prediction of $\mathcal{O}({Re_\lambda}^{1/2})$ by Dimotakis for a jet turbulence [16].

To match the Kolmogorov spectrum, it is further assumed that the asymptotic behaviors of functions F_j and G_j are well ordered and are at most $\mathcal{O}(k^{-5/3})$ when $\xi \to \infty$ or $\zeta \to 0$, that is

$$\begin{aligned} F_j(\xi) \; &\sim \; f_j\, \xi^{-5/3} \qquad \text{at most as } \xi \to \infty, \\ G_j(\zeta) \; &\sim \; g_j\, \zeta^{-5/3} \qquad \text{at most as } \zeta \to 0, \end{aligned} \tag{10}$$

where f_j (or g_j) is nonzero if $F_j = \mathcal{O}(\xi^{-5/3})$ (or $G_j = \mathcal{O}(\zeta^{-5/3})$) and zero otherwise. Therefore, as $\xi \to \infty$ and $\zeta \to 0$, spectra (7) and (8) become

$$\begin{aligned} E_i(k) &\sim q^2\mathcal{L}\sum_{j=0} f_j\mu_j(Re_\lambda)\,(k\mathcal{L})^{-5/3} \\ &= 15^{-2/3}\,(\epsilon\nu^5)^{1/4}\sum_{j=0} f_j\mu_j(Re_\lambda)(k\eta)^{-5/3} \\ &\sim C_K(\epsilon\nu^5)^{1/4}(k\eta)^{-5/3} \end{aligned} \tag{11}$$

and

$$\begin{aligned} E_o(k) &\sim (\epsilon\nu^5)^{1/4}\sum_{j=0} g_j\nu_j(Re_\lambda)(k\eta)^{-5/3} \\ &\sim C_K(\epsilon\nu^5)^{1/4}(k\eta)^{-5/3} \end{aligned} \tag{12}$$

where C_K is Kolmogorov constant. Therefore, it is required that

$$\begin{aligned} \mu_0(Re_\lambda) &= \nu_0(Re_\lambda) = 1 \\ 15^{-2/3}\,f_0 &= g_0 = C_K \end{aligned} \tag{13}$$

and

$$\mu_j(Re_\lambda) = \nu_j(Re_\lambda), \quad 15^{-2/3}\,f_j = g_j, \quad \text{for j } \geq 1. \tag{14}$$

A uniform expansion for the energy spectrum can now be obtained as either

$$\begin{aligned} E(k) = q^2\lambda Re_\lambda &\sum_{j=0} \mu_j(Re_\lambda)\,F_j(k\lambda Re_\lambda) \\ &+ (\epsilon\nu^5)^{1/4}\sum_{j=0} \mu_j(Re_\lambda)\left(G_j(k\eta) - g_j(k\eta)^{-5/3}\right) \end{aligned} \tag{15}$$

or

$$\begin{aligned} E(k) = q^2\lambda Re_\lambda &\sum_{j=0} \mu_j(Re_\lambda)\left(F_j(k\lambda Re_\lambda) - f_j(k\lambda Re_\lambda)^{-5/3}\right) \\ &+ (\epsilon\nu^5)^{1/4}\sum_{j=0} \mu_j(Re_\lambda)\,G_j(k\eta). \end{aligned} \tag{16}$$

3.2 Constraints on F_j and G_j

There are some constraints which must be satisfied by the expansion functions F_j and G_j arising from the definitions of q^2 and ϵ.

i) The turbulent energy mainly contributed by large eddies is computed as

$$\frac{3}{2}q^2 = \int_0^\infty E(k)dk$$
$$= q^2 \sum_{j=0} \mu_j(Re_\lambda)\left\{\int_0^\infty F_j(\xi)d\xi + \sqrt{15}\ Re_\lambda{}^{-1}\int_0^\infty (G_j(\zeta) - g_j\ \zeta^{-5/3})d\zeta\right\}. \tag{17}$$

Because $\mu_0(Re_\lambda) = 1$, F_j and G_j must satisfy

$$\int_0^\infty F_0(\xi)d\xi = \frac{3}{2}, \tag{18}$$

and

$$\sum_{j=0} \left\{\mu_{j+1}(Re_\lambda)\int_0^\infty F_{j+1}(\xi)d\xi + \sqrt{15}\ Re_\lambda{}^{-1}\mu_j(Re_\lambda)\int_0^\infty \left(G_j(\zeta) - g_j\ \zeta^{-5/3}\right)d\zeta\right\} = 0. \tag{19}$$

ii) The energy dissipation rate ϵ contributed mostly from small eddies is computed as

$$\epsilon = 2\nu\int_0^\infty k^2E(k)dk$$
$$= 2\epsilon\sum_{j=0} \mu_j(Re_\lambda)\left\{\int_0^\infty \zeta^2G_j(\zeta)d\zeta + \frac{1}{15}\ Re_\lambda{}^{-2}\int_0^\infty \xi^2\left(F(\xi) - f_j\ \xi^{-5/3}\right)d\xi\right\}. \tag{20}$$

Again by recalling $\mu_0(Re_\lambda) = 1$, F_j and G_j are restricted to

$$\int_0^\infty \zeta^2G_0(\zeta)d\zeta = \frac{1}{2}, \tag{21}$$

and

$$\sum_{j=0} \left\{\mu_{j+1}(Re_\lambda)\int_0^\infty \zeta^2G_{j+1}(\zeta)d\zeta + \frac{1}{15}Re_\lambda{}^{-2}\mu_j(Re_\lambda)\int_0^\infty \xi^2\left(F_j(\xi) - f_j\ \xi^{-5/3}\right)d\xi\right\} = 0. \tag{22}$$

4 Spectral energy equation

Instead of considering the uniform expansion of (15) or (16), proposed is a similarity form for energy spectrum as follows

$$E(k) = q^2\lambda Re_\lambda \sum_{j=0} \mu_j(Re_\lambda)\, \widehat{F}_j(k\lambda Re_\lambda) + \ (\epsilon\nu^5)^{1/4} \sum_{j=0} \mu_j(Re_\lambda)\, \widehat{G}_j(k\eta), \quad (23)$$

since $\mu_j(Re_\lambda) = \nu_j(Re_\lambda)$. Corresponding to it, the energy transfer spectrum in its similarity form must be

$$T(k) = q^3 \sum_{j=0} \mu_j(Re_\lambda)\, \widehat{H}_j(k\lambda Re_\lambda) \ + \ (\epsilon\nu)^{3/4} \sum_{j=0} \mu_j(Re_\lambda)\, \widehat{K}_j(k\eta). \quad (24)$$

Recall for isotropic homogeneous turbulence, the equation of evolution of energy spectrum is given by

$$\frac{\partial E(k)}{\partial t} = T(k) - 2\nu k^2 E(k). \quad (25)$$

Substituting (23) and (24) into (25), one obtains

$$\sum_{j=0} \Big((2\sigma - 15)\ \mu_j\widehat{F}_j(\xi) \ + \ (2\sigma - 5)\ \mu_j\xi\widehat{F}_j'(\xi)\Big) + Re_\lambda \sum_{j=0} (\sigma - 5)\ \mu_j'\widehat{F}_j(\xi)$$

$$+ 15^{\frac{1}{4}} {Re_\lambda}^{-3/2} \left\{ \sum_{j=0}(\sigma - 5)\mu_j'\widehat{G}_j(\zeta) + {Re_\lambda}^{-1} \sum_{j=0} \left(\frac{\sigma+5}{2}\right) \mu_j \left(\zeta\widehat{G}_j'(\zeta) - \widehat{G}_j(\zeta)\right) \right\}$$

$$= \sum_{j=0} \mu_j\widehat{H}_j(\xi) - {Re_\lambda}^{-2} \sum_{j=0} 2\mu_j\xi^2\widehat{F}_j(\xi)$$

$$+ 15^{\frac{3}{4}}\, {Re_\lambda}^{-3/2} \left\{ \sum_{j=0} \mu_j\widehat{K}_j(\zeta) - \sum_{j=0} 2\mu_j\zeta^2\widehat{G}_j(\zeta) \right\}, \quad (26)$$

where $\sigma \equiv \frac{\lambda}{\nu}\frac{d\lambda}{dt}$. Recall that $1 = \mu_0 > \mu_1 > \mu_2 > \cdots$ so that the leading order terms in (26) are $\mathcal{O}(1)$. Moreover, a choice of $\mu_j = Re^{-j}$ seemingly is good enough to generate a simiarility. However, to generalize the results, because of the appearance of the factor ${Re_\lambda}^{-3/2}$, a choice of

$$\mu_j(Re_\lambda) = {Re_\lambda}^{-j/2}, \quad j = 0, 1, 2, \ldots \quad (27)$$

is selected. Furthermore, for consistency, the value of σ must be constant. That implies, by the definition of λ, turbulent energy must decay according to a power law with a decay exponent $n = 5/\sigma$.

With this choice of $\{\mu_j\}$ and assuming a constant value of σ, terms matched up to and including the order of $\mathcal{O}({Re_\lambda}^{-5/2})$ are summarized as follows:

$$\begin{aligned}
&(a)\quad \widehat{H}_0(\xi) = \mathcal{L}(\widehat{F}_0(\xi))\\
&(b)\quad \widehat{H}_1(\xi) = \mathcal{L}(\widehat{F}_1(\xi)) - \frac{1}{2}(\sigma-5)\widehat{F}_1(\xi)\\
&(c)\quad \widehat{H}_2(\xi) = \mathcal{L}(\widehat{F}_2(\xi)) - (\sigma-5)\widehat{F}_2(\xi)\\
&(d)\quad \widehat{H}_3(\xi) = \mathcal{L}(\widehat{F}_3(\xi)) - \frac{3}{2}(\sigma-5)\widehat{F}_3(\xi)\\
&(e)\quad \widehat{H}_4(\xi) = \mathcal{L}(\widehat{F}_4(\xi)) - 2(\sigma-5)\widehat{F}_4(\xi) + 2\xi^2\widehat{F}_0(\xi)\\
&(f)\quad \widehat{H}_5(\xi) = \mathcal{L}(\widehat{F}_5(\xi)) - \frac{5}{2}(\sigma-5)\widehat{F}_5(\xi) + 2\xi^2\widehat{F}_1(\xi)\\
&(g)\quad \widehat{K}_0(\zeta) = 2\zeta^2\widehat{G}_0(\zeta)\\
&(h)\quad \widehat{K}_1(\zeta) = 2\zeta^2\widehat{G}_1(\zeta)\\
&(i)\quad \widehat{K}_2(\zeta) = 2\zeta^2\widehat{G}_2(\zeta) + \left(\frac{\sigma+5}{2\sqrt{15}}\right)\left(\zeta\widehat{G}_0'(\zeta) - \widehat{G}_0(\zeta)\right)
\end{aligned} \tag{28}$$

where

$$\mathcal{L}(\widehat{F}_j(\xi)) \equiv (2\sigma-15)\widehat{F}_j(\xi) + (2\sigma-5)\xi\widehat{F}_j'(\xi). \tag{29}$$

Shown in Table 1 are the coupling relations among the expansion functions. It is seen that up to $\mathcal{O}({Re_\lambda}^{-3/2})$ the decay of the energy-containing eddies is purely through the draining of energy from large eddies to smaller eddies by inertial forces, and not until an order of $\mathcal{O}({Re_\lambda}^{-3/2})$ do the small eddies dissipate energy. This is consistent with the Kolmogorov's hypothesis that ${Re_\lambda}^{3/2} \gg 1$ is required to completely separate energy-containing eddies from dissipation eddies. In conclusion an energy spectrum of the form

$$E(k) = (q^2\lambda Re_\lambda)\sum_{j=0} {Re_\lambda}^{-j/2}\ \widehat{F}_j(k\lambda Re_\lambda) + (\epsilon\nu^5)^{1/4}\sum_{j=0} {Re_\lambda}^{-j/2}\ \widehat{G}_j(k\eta) \tag{30}$$

with a corresponding energy transfer spectrum

$$T(k) = q^3\sum_{j=0} {Re_\lambda}^{-j/2}\ \widehat{H}_j(k\lambda Re_\lambda)\ +\ (\epsilon\nu)^{3/4}\sum_{j=0} {Re_\lambda}^{-j/2}\ \widehat{K}_j(k\eta) \tag{31}$$

is proposed.

Table 1: Terms of order ${Re_\lambda}^0$ ${Re_\lambda}^{-1/2}$, ${Re_\lambda}^{-1}$, ${Re_\lambda}^{-3/2}$, ${Re_\lambda}^{-2}$ and ${Re_\lambda}^{-5/2}$ in the spectral energy equation.

$\partial E(k)/\partial t$						$T(k)$						$2\nu k^2 E(k)$					
$\widehat{F}_0$	$\widehat{F}_1$	$\widehat{F}_2$	$\widehat{F}_3$	$\widehat{F}_4$	$\widehat{F}_5$	$\widehat{H}_0$	$\widehat{H}_1$	$\widehat{H}_2$	$\widehat{H}_3$	$\widehat{H}_4$	$\widehat{H}_5$	0	0	0	0	$\widehat{F}_0$	$\widehat{F}_1$
0	0	0	0	0	$\widehat{G}_0$	0	0	0	$\widehat{K}_0$	$\widehat{K}_1$	$\widehat{K}_2$	0	0	0	$\widehat{G}_0$	$\widehat{G}_1$	$\widehat{G}_2$

The constraints (18)(19)(21) (22) now become

$$\begin{aligned} \int_0^\infty F_0(\xi)d\xi &= \frac{3}{2} \\ \int_0^\infty F_1(\xi)d\xi &= 0 \\ \int_0^\infty F_{j+2}(\xi)d\xi + \sqrt{15}\int_0^\infty \left(G_j(\zeta) - g_j\zeta^{-5/3}\right)d\zeta &= 0 \ \text{ for all } j \geq 0 \end{aligned} \tag{32}$$

and

$$\begin{aligned} \int_0^\infty \zeta^2 G_0(\zeta)d\zeta &= \frac{1}{2} \\ \int_0^\infty \zeta^2 G_j(\zeta)d\zeta &= 0 \ \text{ for } j = 1,2,3 \\ 15\int_0^\infty \zeta^2 G_{j+4}(\zeta)d\zeta + \int_0^\infty \xi^2\left(F_j(\xi) - f_j\,\xi^{-5/3}\right)d\xi &= 0 \ \text{ for all } j \geq 0. \end{aligned} \tag{33}$$

An example energy spectrum is constructed and shown in Figure 1, in which a choice of

$$F_0(\xi) = 15^{2/3}C_K \frac{b\,\xi^2}{1 + b\,\xi^{11/3}} \tag{34}$$

and

$$G_0(\zeta) = C_K\zeta^{-5/3}\exp(-a\zeta^2), \tag{35}$$

is made, where $b = \left(\frac{2}{11}\cdot 15^{2/3}\, C_K\Gamma(\frac{9}{11})\Gamma(\frac{2}{11})\right)^{-11/2}$ computed from (32) and $a = \left(C_K\Gamma(\frac{2}{3})\right)^{3/2}$ from (33). This energy spectrum has a k^2−behavior near the origin of wave numbers, a Kolmogorov spectrum ($k^{-5/3}$) in the inertial subrange and an Gaussian tail at very large wave numbers. It was ever observed that the energy spectrum when scaled by Kolmogorov scales and multiplied by the factor of $(k\eta)^{5/3}$ has a positive "bump" at the intersection of inertial subrange and dissipation range of wave numbers (Saddoughi for example [8]). The present example spectrum does not possess such a bump. The energy transfer spectrum contributed from (34) and (35) (involving $\widehat{H}_0$, $\widehat{H}_4$, $\widehat{K}_0$, and $\widehat{K}_2$) is computed through the use of (28) and is shown in Figure 2. The integral of this spectrum over the wave space however does not vanish as may be expected. Indeed,

$$\int_0^\infty T(k,t)dk \sim \mathcal{O}(\epsilon {Re_\lambda}^{-1}), \tag{36}$$

which must be cancelled by higher order contributions.

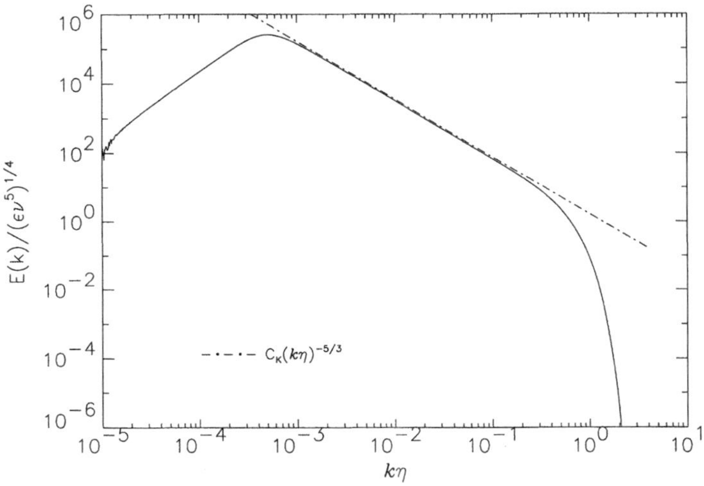

Figure 1: The example of the energy spectrum constructed by using (34) and (35) and taking $Re_\lambda = 1000$ and $C_K = 1.5$.

5 Comparisons

In equivalence, the multiscale similarity when illustrated by the double and triple velocity correlation coefficients through (5) and (6), is of the form

$$\begin{aligned} f(r) &= \sum_{j=0} {Re_\lambda}^{-j/2}\left\{ f_{1j}\left(\frac{r}{\lambda Re_\lambda}\right) + {Re_\lambda}^{-1}\cdot f_{2j}\left(\frac{r}{\eta}\right)\right\} \\ k^{tri}(r) &= \sum_{j=0} {Re_\lambda}^{-j/2}\left\{ k^{tri}_{1j}\left(\frac{r}{\lambda Re_\lambda}\right) + {Re_\lambda}^{-3/2}\cdot k^{tri}_{2j}\left(\frac{r}{\eta}\right)\right\}. \end{aligned} \tag{37}$$

The skewness is computed as

$$\begin{aligned} S &= -\lambda^3 \frac{\partial^3 k^{tri}}{\partial r^3}\bigg|_{r=0} = \frac{2}{35}\frac{\lambda^3}{q^3}\int_0^\infty k^2 T(k,t)dk \\ &= \sum_{j=0} {Re_\lambda}^{-j/2}\left(15^{3/4}\cdot\frac{d^3 k^{tri}_{2j}}{d\zeta^3}(0) + {Re_\lambda}^{-3}\frac{d^3 k^{tri}_{1j}}{d\xi^3}(0)\right) \\ &= S_0 + S_{-1/2}\,{Re_\lambda}^{-1/2} + S_{-1}\,{Re_\lambda}^{-1} + S_{-3/2}\,{Re_\lambda}^{-3/2} + \cdots, \end{aligned} \tag{38}$$

where $S_{-j/2}$'s are all constants. Notice that large eddies do not contribute until an order of ${Re_\lambda}^{-3}$. And obviously $S \to S_0$ as $Re_\lambda \to \infty$. The observation in the

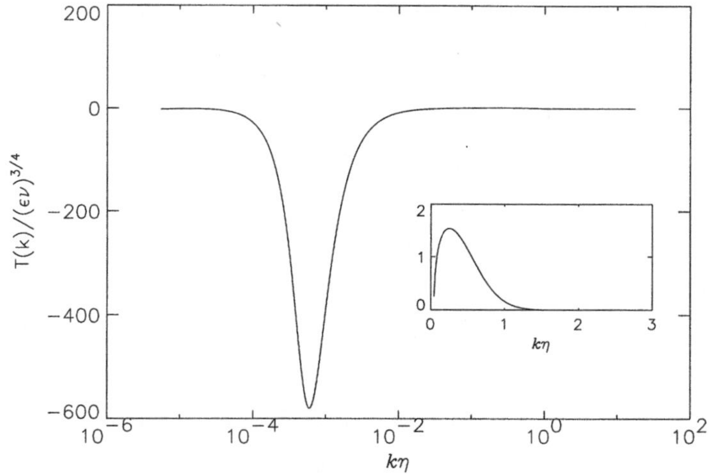

Figure 2: The energy transfer spectrum corresponding to the energy spectrum of Figure 1 with a decay exponent $n = 1.5$.

lab that $S \approx 0.5$ suggests the value of $S_0 \approx 0.5$ or at least must be positive. And if most of $S_{-j/2}$'s are negative for $j \geq 1$, the skewness approaches the limit from below. An overshooting is also possible if some $S_{-j/2}$'s are positive and some are negative. Shown in Figure 3 is a computed skewness S up to an order of ${Re_\lambda}^{-1}$ based on the similarity solution by considering a generalization of spectrum (35) to be

$$G_0(\zeta) = C_K \zeta^{-5/3} \exp -a\zeta^m \tag{39}$$

where $a = \left(2C_K\Gamma(\frac{4}{3m})/m\right)^{3m/4}$ such that (33) is satisfied. A value of $m \geq 4$ is seemingly suggested if the experimental value of skewness (≈ 0.5) is desired at large Reynolds numbers.

Another quantity that may be of interest is

$$\begin{aligned} G &\equiv \lambda^4 \left.\frac{\partial^4 f}{\partial r^4}\right|_{r=0} \\ &= \sum_{j=0} {Re_\lambda}^{-j/2} \left(15 Re_\lambda \cdot \frac{d^4 f_{2j}}{d\zeta^4}(0) + {Re_\lambda}^{-4} \cdot \frac{d^4 f_{1j}}{d\xi^4}(0) \right) \qquad (40) \\ &= C_1\, Re_\lambda + C_{1/2}\, {Re_\lambda}^{1/2} + C_0 + C_{-1/2}\, {Re_\lambda}^{-1/2} + C_{-1}\, {Re_\lambda}^{-1} + \cdots . \end{aligned}$$

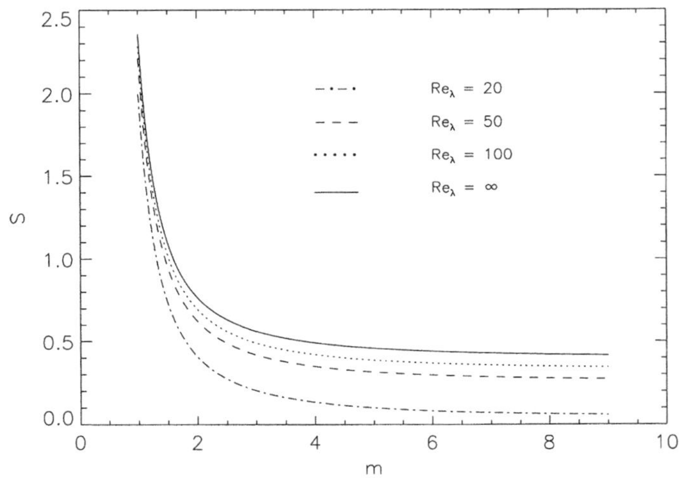

Figure 3: The skewness S contributed by small eddies with an energy spectrum of (39) with $C_K = 1.5$ and $n = 1.5$.

A nearly linear dependence on Re_λ of the quantity G was observed at low Reynolds number (see Fig.7 in the work of Huang and Leonard [4]). This leading behavior was also predicted by Huang and Leonard's similarity with a value of β close to one (slightly greater than one is required for a positive correlation between the skewness S and Re_λ). This may suggest a vanishing $C_{1/2}$ and thus $S_{-1/2}$ (from (28h)) for small to moderate Reynolds number, and furthermore may suggest a choice of the order functions μ_j to be the less general one, ${Re_\lambda}^{-j}$.

In general, this multiscale similarity is different from the single-scaling similarities. An exceptional case is when the decay exponent happens to be equal to one. In that case, the Reynolds number becomes time-independent or the length-scales are proportional to each other with proportional constants fixed in time. Therefore, all the similarities (Kármán-Howarth's, Barenblatt and Gavrilov's, George's, Huang and Leonard's, and the proposed similarity herein) coincide.

6 Conclusions

A solution in a multiscale similarity form to the spectral energy equation for isotropic homogeneous incompressible turbulence at large Reynolds number is proposed. This similarity accompanies with a power-law energy decay as single-scale similarities do. The decay exponent is arbitrary. With the use of many more degrees of freedom (many more expansion functions), some physical phenomena can be represented by the solution, such as a skewness increasing with increasing

Reynolds number with or without an overshooting and a nearly linear dependence of the quantity G on Re_λ. The future work is directed to a measurement of energy spectrum from the labs or numerical simulations and its proper decomposition into the proposed similarity form. A comparison between the simultaneously measured energy transfer spectrum and the transfer spectrum derived from the similarity analysis can be very worthwhile. Possible applications of the results to subgrid turbulent stresses are under investigation.

Acknowledgement The support of this work by the National Science Council, Taiwan, R.O.C. under contract NSC86-2621-E-002-024-T is gratefully acknowledged.

References

[1] Th. von Karman and L. Howarth. On the statistical theory of isotropic turbulence. *Proc. Roy. Soc. London*, 164A:192, 1938.

[2] G.J. Barenblatt and A.A. Gavrilov. On the theory of self-similar degeneracy of homogeneous isotropic turbulence. *Sov. Phys. JETP*, 38:399, 1974.

[3] W.K. George. The decay of homogeneous isotropic turbulence. *Phys. Fluids A*, 4:1492, 1992.

[4] M.J. Huang and A. Leonard. Power-law decay of homogeneous turbulence at low Reynolds numbers. *Phys. Fluids*, 6:3765, 1994.

[5] A.N. Kolmogorov. Dissipation of energy in locally isotropic turbulence. *C.R. Acad. Sci. U.R.S.S.*, 32:16, 1941.

[6] A.N. Kolmogorov. The local structure of turbulence in incompressible viscous fluid for very large Reynolds numbers. *C.R. Acad. Sci. U.R.S.S.*, 30:301, 1941.

[7] A.N. Kolmogorov. On degeneration of isotropic turbulence in an incompressible viscous liquid. *C.R. Acad. Sci. U.R.S.S.*, 31:538, 1941.

[8] S.G. Saddoughi and S.V. Veeravalli. Local isotropy in turbulent boundary layers at high Reynolds numbers. *CTR Manuscript 142, Stanford*, 1993.

[9] G.K. Batchelor and A.A. Townsend. Decay of turbulence in the final period. *Proc. Roy. Soc. London*, 194A:527, 1948.

[10] G.K. Batchelor and A.A. Townsend. Decay of isotropic turbulence in the initial period. *Proc. Roy. Soc. London*, 193A:539, 1948.

[11] G.K. Batchelor and A.A. Townsend. Decay of vorticity in isotropic turbulence. *Proc. Roy. Soc. A*, 190:534, 1948.

[12] Y. Sato and K. Yamamoto. Empirical equations for the structure of isotropic turbulence. *J. Chem. Engng. Japan*, 16:273, 1983.

[13] M.R. Smith, R.J. Donnelly, N. Goldenfeld, and W.F. Vinen. Decay of vorticity in homogeneous turbulence. 1993.

[14] G.K. Batchelor. Energy decay and self-preserving correlation functions in isotropic turbulence. *Quarterly of Applied Mathematics*, 6:97, 1948.

[15] G.K. Batchelor. *The Theory of Homogeneous Turbulence.* Cambridge, 1953.

[16] P.E. Dimotakis. Some issues on turbulent mixing and turbulence. *GALCIT Report*, FM93-1a, 1993.

7 Passive objects

Trends in Mathematics, © 1999 Birkhäuser Verlag Basel/Switzerland

The role of Lagrangian chaos in the creation of multifractal measures

E. Ott

Abstract

In this paper we review and discuss the role of Lagrangian chaos in creating multifractal measures associated with spatial intermittency. In particular, we consider (1) passive scalars, (2) the kinematic magnetic dynamo problem, and (3) the stability of high Reynolds number fluid flows. In all cases we take the underlying flow to be smooth in the sense that it has no fractal or power law properties of its own (its wavenumber power spectrum is peaked at some low value and decays with increasing wavenumber much faster than a power law, e.g., exponentially). Nevertheless, chaos in the Lagrangian dynamics of such flows can lead to fractals and power law wavenumber spectra for relevant physical quantities, such as the distribution of the passive scalar (problem (1)), the magnetic field (problem (2)), and the vorticity field (problem (3)). A key concept used in the quantitative treatment of these problems is that of finite time Lyapunov exponents. This paper will illustrate how consideration of finite time Lyapunov exponents can be used to quantitatively analyze these situations. Thus we provide a connection between quantities of physical interest and ergodic dynamical characterization of the underlying chaos of the flow. In the case of problem (3), we also comment on implications for high Reynolds number fluid turbulence.

1 Introduction

The purpose of this paper is to review and discuss the application of chaotic dynamics theory to the problems of obtaining dimensions of multifractal measures and power law wavenumber spectra that may result in fluid flows that are Lagrangian chaotic. We begin with some background material.

1.1 Lagrangian chaos

We consider a fluid whose Eulerian velocity is $\mathbf{v}(\mathbf{x}, t)$. We also envision that the spatial dependence of $\mathbf{v}$ is smooth in the sense that it is dominated by a single spatial scale. That is, its wavenumber spectrum is peaked about some small wavenumber and decays much faster than algebraically away from that value. This is in contrast with the situation of a fully developed turbulent flow where the wavenumber spectrum is a power law, and significant components occur on all scales between the integral scale and the dissipation scale. In particular, the flows $\mathbf{v}(\mathbf{x}, t)$ we consider have no spatial fractal properties of their own. Trajectories of infinitesimal fluid elements are given by solutions of the equation

$$d\mathbf{x}/dt = \mathbf{v}(\mathbf{x}, t) . \tag{1}$$

We shall be concerned with the situation where the solution of (1) is chaotic in the sense that two infinitesimally separated initial conditions, $\mathbf{x}_0$ and $\mathbf{x}_0 + \delta\mathbf{x}_0$, evolved by (1) typically separate exponentially with time, $\delta\mathbf{x}_0 \rightarrow \delta\mathbf{x}(t)$, where $|\delta\mathbf{x}(t)|/|\delta\mathbf{x}_0| \sim \exp(ht)$ with $h > 0$. We call this Lagrangian chaos. The evolution of the differential $\delta\mathbf{x}(t)$ can be found by taking a differential variation of (1),

$$d\delta\mathbf{x}/dt = \delta\mathbf{x} \cdot \nabla\mathbf{v}(\mathbf{x}, t) . \tag{2}$$

1.2 Maps

The traditional approaches in nonlinear dynamics for treating a system of nonlinear differential equations of the form (1) are well developed for the case where $\mathbf{v}$ is steady or time periodic. In these cases it is often useful to employ a surface of section map (for the steady case) or a stroboscopic map (for the time periodic case). In this paper we also note that a variant of the traditional map technique can be employed in the case where the flow is temporally irregular[1–7]. In particular, say $\mathbf{x}(t)$ denotes a solution of (1) for the given temporally irregular flow $\mathbf{v}$, and consider the resulting position sampled periodically at some convenient time interval T,

$$\mathbf{x}_n = \mathbf{x}(nT) .$$

The choice of T is somewhat arbitrary, but we envision that it might be of the order of the typical time it takes a particle to advect a distance of the order of the flow scale length L (e.g., $T \sim L/v_0$ where v_0 is a typical flow velocity). Since the sequence $\mathbf{x}_0$, $\mathbf{x}_1$, ... follows from solution of the deterministic equation, Eq. (1), $\mathbf{x}_{n+1}$ is uniquely determined by $\mathbf{x}_n$ via integration of Eq. (1) from time $t = nT$ to time $t = (n+1)T$. This specifies a mapping from $\mathbf{x}_n$ to $\mathbf{x}_{n+1}$,

$$\mathbf{x}_{n+1} = \mathbf{M}_n(\mathbf{x}_n) . \tag{3}$$

In the case where the flow is time periodic with period T the map does not depend on the time step, and we may replace $\mathbf{M}_n$ by $\mathbf{M}$.

1.3 Fractal dimension spectrum

We will be considering a *measure* μ which assigns a positive number to any chosen volume in the region of interest, and the measure of the entire region is one. The choice of the measure will be different for different physical problems. The D_q dimension spectrum of the measure μ is defined as follows[8,9]. Divide the volume of interest (area, if the problem is two dimensional) into a grid of cubes (squares) of edge length ϵ. Let μ_i be the measure of cube i. Then

$$D_q = (1-q)^{-1} \lim_{\epsilon \to 0} \frac{\ell n I_q(\epsilon)}{\ell n(1/\epsilon)} \tag{4}$$

where

$$I_q(\epsilon) = \sum_i \mu_i^q \ .$$

For $q \to 1$, we obtain, by L'Hospital's rule, $(1-q)^{-1}\ \ell n I_q(\epsilon) \to \sum \mu_i \ell n(1/\mu_i)$, and D_1 is called the information dimension[8–10]. If D_q varies with q, then we say that the measure μ is a multifractal measure.

In physical problems the limit $\epsilon \to 0$ in (4) can never actually be realized. Thus we say that we have a "physical fractal" if there exists a scaling range, $\epsilon_0 > \epsilon > \epsilon_*$ where $\epsilon_0/\epsilon_* \gg 1$, such that a plot of $(1-q)^{-1}\ \ell n I_q(\epsilon)$ versus $\ell n(1/\epsilon)$ is well-fitted by a straight line, and we estimate D_q by the slope of this line.

1.4 Finite time Lyapunov exponents

Say we start at a given initial position $\mathbf{x}_0$ and follow the trajectory computed from (1). Also say that we follow the position of an infinitesimally displaced point $\mathbf{x}_0 + \delta\mathbf{x}_0$. Next at some fixed time $n = m$ we examine the resulting displacement $\delta\mathbf{x}_m$ between the two orbits and maximize the quantity $|\delta\mathbf{x}_m|/|\delta\mathbf{x}_0|$ over all possible orientations of the initial displacement $\delta\mathbf{x}_0$. The finite time Lyapunov exponent for $\mathbf{x}_0$ is

$$h(\mathbf{x}_0, m) = \frac{1}{m} \max_{\delta\mathbf{x}_0} \ell n[|\delta\mathbf{x}_m|/|\delta\mathbf{x}_0|] \ . \tag{5}$$

If $\mathbf{x}_0$ is chosen at random with uniform probability in the volume of the ergodic chaotic region of phase space, then $h(\mathbf{x}_0, m)$ is a random variable, and we may define its probability distribution function $P(h, m)$. For large m, large deviation theory[11,12] yields a scaling form for $P(h, m)$

$$\ell n P(h, m) \cong -mG(h) + o(m) \ , \tag{6a}$$

where $G(h) \geq 0$ is a convex function with minimum value zero. We denote the value of h at which this minimum occurs $\bar{h}$ so that $G(\bar{h}) = 0$. Notationally, we also write (6a) as

$$P(h, m) \sim \exp -[mG(h)] \tag{6b}$$

for large m. The utility of the scaling form (6) is that it expresses the function of two variables $P(h, m)$ in terms of a (as yet unknown) function of a single variable, $G(h)$. If we expand $G(h)$ about its minimum, then (6b) yields the Gaussian, $P(h, m) \sim \exp[-m(h - \bar{h})^2 G''(\bar{h})/2]$. For large m the standard deviation of the Gaussian from $h = \bar{h}$ [namely, $(mG''(\bar{h}))^{-1/2}$] becomes small, and the expansion of G about its minimum becomes better and better, provided we restrict consideration to within a few standard deviations from $\bar{h}$. This is the central limit theorem. We shall, however, be interested in averages of quantities that place strong emphasis on the tails of the distribution function. Thus we must employ the more general form (6) rather than the Gaussian. In particular, a quantity of interest that appears frequently in the theory is

$$H(s) \equiv \lim_{m \to \infty} \frac{1}{m} \ell n \langle e^{smh} \rangle^{1/s} \ ,$$

where the angle brackets denote an average of the finite time Lyapunov exponents over initial conditions $\mathbf{x}_0$ in the ergodic chaotic region. Using (6) we have

$$\langle e^{smh} \rangle \sim \int \exp[-m(G(h) - sh)]dh ,$$

which yields

$$\langle e^{smh} \rangle \sim e^{-m[G(h_s) - sh_s]} ,$$

where h_s is the value of h at the minimum of the function $G(h) - sh$. Thus

$$H(s) = h_s - s^{-1}G(h_s) . \tag{7}$$

Use of the Gaussian form would, in general, not be correct (unless it is found that $G(h)$ is indeed parabolic over a sufficiently wide region). Note from (6b) that $P(h, m)$ approaches a delta function at $h = \bar{h}$ for $m \to \infty$. Thus, for a randomly chosen initial condition, we obtain $h(\mathbf{x}_0, m \to \infty) \to \bar{h}$ with probability one, so that $\bar{h}$ is the usual (infinite time Lyapunov exponent).

In numerical studies $G(h)$ can be found numerically by using (2) and (5) to find many values of $h(\mathbf{x}_0, m)$ for randomly chosen $\mathbf{x}_0$ values and fixed m. One can then construct an approximation to $P(h, m)$ using a histogram approach. Application of (6a) then gives an approximation to $G(h)$. To test the scaling prediction (6), the above procedure can be repeated for different large fixed m values and should yield approximately the same function $G(h)$ for each m value. See, for example, Refs. 1 and 4 for numerical results along these lines.

1.5 Outline

In the rest of this paper we discuss results for three basic situations:

(1) Passive scalar advection (Sec. 2),
(2) The fast kinematic dynamo problem (Sec. 3),
(3) The stability of smooth high Reynolds number fluid flows (Sec. 4).

In all three of these situations Lagrangian chaos of the underlying fluid flow leads to multifractal attributes for the spatial distribution of physical quantities, the passive scalar for problem (1), the magnetic field for problem (2), and the vorticity field for problem (3). It should be noted that multifractality also implies spatial intermittency. We also discuss power law wavenumber spectra that arise in some of the situations considered. In addition, in the case of problem (3), we comment on possible implications for fluid turbulence (Sec. 4).

2 Passive scalar Problems

The equation of motion describing the evolution of a passive scalar field $\phi(\mathbf{x}, t)$ is

$$\partial\phi/\partial t + \nabla \cdot (\mathbf{v}\phi) = Q . \tag{8}$$

The term Q represents effects such as sources and sinks of the scalar, diffusion, etc. In this section examples of three different situations are discussed: (i) the fractal distribution of passive scalar gradients, (ii) the fractal pattern formed by a scum floating on the surface of a moving fluid, and (iii) the pattern of advected particles entrained in an open flow past an obstacle.

For (i) and (iii) we consider the velocity $\mathbf{v}$ to be incompressible ($\nabla \cdot \mathbf{v} = 0$), in which case (8) becomes

$$\partial\phi/\partial t + \mathbf{v} \cdot \nabla\phi = Q \ .$$

That is, the time rate of change of ϕ following a fluid element trajectory is Q ($\partial/\partial t + \mathbf{v} \cdot \nabla \equiv d/dt$, and $d\mathbf{x}/dt = \mathbf{v}$).

2.1 Fractal distribution of scalar gradients of area preserving confined flows

In this section we initially restrict our attention to the simple case $Q = 0$ in (8). For $Q = 0$ the solution of (8) for an incompressible flow ($\nabla \cdot \mathbf{v} = 0$) is

$$\phi(\mathbf{x}, t) = \phi(\mathbf{x}_0(\mathbf{x}, t), 0) \ , \tag{9}$$

where $\mathbf{x}_0(\mathbf{x}, t)$ is the position of an advected particle at time $t = 0$, given that at time t it is located at the position $\mathbf{x}$. That is, ϕ does not change following the trajectory of a fluid element.

Consider the evolution of a small ϵ-radius circle initially centered at a location $\mathbf{x}$ in an incompressible two-dimensional fluid flow which exhibits Lagrangian chaos ($\bar{h} > 0$). As time progresses the circle evolves into a long thin ellipse whose major axis is of the order of $\epsilon \exp[mh(\mathbf{x}, m)]$. By area conservation the minor axis decreases and is of the order of $\epsilon \exp[-mh(\mathbf{x}, m)]$. Thus, thinking of the ellipse's minor axis, we see that two points, call them $\mathbf{a}$ and $\mathbf{b}$, initially on the circle and separated by an amount ϵ are advected to two new points $\mathbf{a}'$ and $\mathbf{b}'$ that become exponentially closer with time. According to (9) we then have for the evolution from time $t = 0$ to time $t = mT$

$$\frac{\phi(\mathbf{a}, 0) - \phi(\mathbf{b}, 0)}{|\mathbf{a} - \mathbf{b}|} \rightarrow \frac{\phi(\mathbf{a}', mT) - \phi(\mathbf{b}', mT)}{|\mathbf{a}' - \mathbf{b}'|} = \frac{\phi(\mathbf{a}, 0) - \phi(\mathbf{b}, 0)}{|\mathbf{a}' - \mathbf{b}'|} \ .$$

Since $|\mathbf{a} - \mathbf{b}|/|\mathbf{a}' - \mathbf{b}'| \simeq \exp(mh(\mathbf{x}, m))$, we conclude that $|\nabla\phi|$ increases exponentially with time. Moreover this increase is extremely spatially inhomogeneous due to the spread in the distribution of finite time exponents $h(\mathbf{x}, m)$, Eq. (6). The strongest gradients occur at positions that have experienced the largest h. As time progresses the regions experiencing the largest h values shrink in area, as reflected by the stronger peaking of P at $h = \bar{h}$ as m increases. As shown as Ref. 1, the regions of large gradient (equivalently, large h) tend to a fractal set in the following sense. Consider a smooth (non-fractal) initial distribution $\phi(\mathbf{x}, 0)$ and evolve it forward to time $t = mT$. Define a gradient measure μ of any subregion

S of the area of interest A by

$$\mu(S,m) = \left[\int_S |\nabla\phi(\mathbf{x},mT)|dA\right] / \left[\int_A |\nabla\phi(\mathbf{x},mT)|dA\right] . \tag{10}$$

The result of Ref. 1 is that, if we consider Q to be zero, then, as m increases, μ tends to a multifractal measure, and that the dimension spectrum D_q can be related to the function $G(h)$ characterizing the finite time Lyapunov exponents. In particular, D_q is

$$D_q = 2 - \frac{q-\sigma}{q-1} , \tag{11a}$$

where σ is given by

$$\min_h[G(h) - \sigma h] = q \min_h[G(h) - h] . \tag{11b}$$

Experimental confirmation for the multifractal nature of passive scalar gradients in such flows has been obtained in Ref. 13. Equations (11) have been numerically tested using a random map (3) which is area preserving and represents a confined flow, where the region of the confined flow was the surface of a two-torus (i.e., periodic boundary conditions in rectangular coordinates x and y were employed). In particular, Ref. 4 considers a simple model two-dimensional velocity field consisting of two components, $\mathbf{v} = \mathbf{v}_1 + \mathbf{v}_2$. The flow component $\mathbf{v}_1$ is a steady x-directed shear flow, $\mathbf{v}_1 = \beta y \mathbf{x}_0$, while $\mathbf{v}_2$ is a temporally varying impulsive y-directed shear flow, $\mathbf{v}_2 = [k\sum_n \delta(t-nT)]\sin(2\pi x + \theta(t))\mathbf{y}_0$, where $\delta(t)$ is a delta function and $\theta(t)$ varies irregularly with time. The equation $d\mathbf{x}/dt = \mathbf{v}$, where $\mathbf{x} = (x, y)$, can be integrated from time $t = nT + 0^+$ to time $t = (n+1)T + 0^+$ to obtain (after suitable normalization) the map,

$$x_{n+1} = (x_n + y_n) \text{ modulo } 1, \tag{12a}$$

$$y_{n+1} = \{y_n + k\sin[2\pi(x_n + y_n) + \theta_n]\} \text{ modulo } 1, \tag{12b}$$

where $\theta_n = \theta(nT)$. This map relates the position (x_{n+1}, y_{n+1}) of a tracer particle in the fluid at time $t = (n+1)T + 0^+$ to its position (x_n, y_n) at time $t = nT + 0^+$. To model the temporal irregularity we take θ_n to be randomly chosen in $(0, 2\pi)$ at each iterate n. Hence (12) is a random map.

Starting with an initial smooth density distribution $\phi(\mathbf{x}, 0)$ (proportional to $\sin(2\pi x)$ in our numerical calculations) we use our model flow to evolve $\phi(\mathbf{x}, t)$ forward in time. Figures 1(a) and 1(b) show intensity plots of a measure of $|\nabla\phi|$, the magnitude of the gradient of the tracer density, where the lightest shade corresponds to regions where $|\nabla\phi|$ is large. Figure 1(a) is taken at time $n = 10$, while Fig. 1(b) is taken at time $n = 20$. These figures show the predicted tendency of $|\nabla\phi|$ to concentrate on a fractal set as time increases.

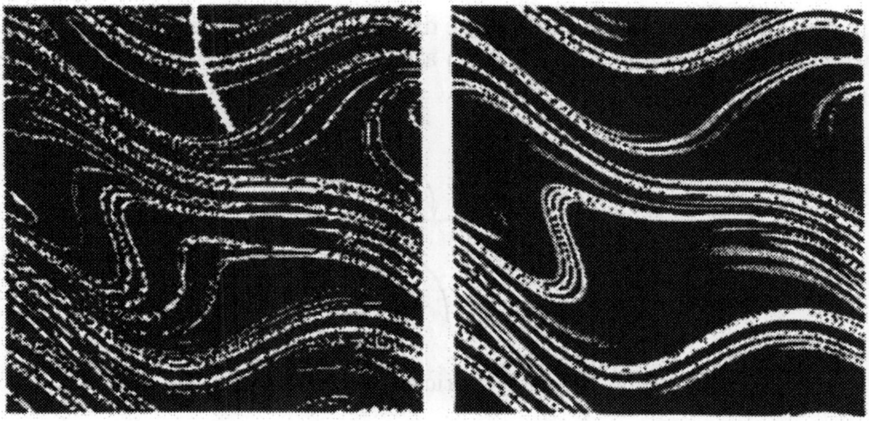

Figure 1: Grey scale intensity plot of $|\nabla\phi|$ for $k = (4\pi)^{-1}$ at (a) $n = 10$ and (b) $n = 20$.

In a real fluid, diffusion of the passive scalar is always present; $Q = D\nabla^2\phi$ in Eq. (8). Our neglect of diffusion in Eq. (9) can be justified if D is small. In particular, the observation time for the pattern should not be so long that the gradients become sharp enough that $D\nabla^2\phi$ becomes non-negligible. Thus our result (11) is valid if D is very small, and if $t = nT$ is large in the sense that the measure μ is essentially fractal over an appreciable range of scales, but nT is still small enough that the neglect of diffusion is justified.

For longer times the spatially averaged scalar variance, $C(t) = \langle[\phi(\mathbf{x}, t) - \langle\phi(\mathbf{x}, t)\rangle]^2\rangle$, decays exponentially with time, $C(t) \sim \exp(-\nu t)$. Although this decay is due to diffusion, the exponential decay rate ν is independent of D and is determined solely by the stretching properties of the flow,[4]

$$\nu = \min_h[h + G(h)].$$

Again, using random maps to numerically simulate a temporally irregular, confined, area preserving flow, excellent numerical agreement with the prediction above for the exponent ν from $G(h)$ has been obtained.

2.2 Fractal distribution of scum on the surface of a moving fluid

When particles float on the surface of a fluid, Eq. (8) must be replaced by

$$\partial\phi/\partial t + \overline{\nabla} \cdot (\bar{\mathbf{v}}\phi) = 0 \ , \tag{13}$$

where ϕ is the two-dimensional particle density in the fluid surface, $\bar{\mathbf{v}}$ is the fluid velocity evaluated at the surface, $\overline{\nabla}\cdot$ denotes the divergence restricted to the fluid surface, and any terms that would appear on the right side have been neglected. The important point is that $\bar{\mathbf{v}}$ is not incompressible even if $\mathbf{v}$ is. This is illustrated in Fig. 2 in which the arrows indicate the direction of a steady incompressible flow. Floating particles initially distributed in a cloud are compressed and tend toward the point A as time increases.

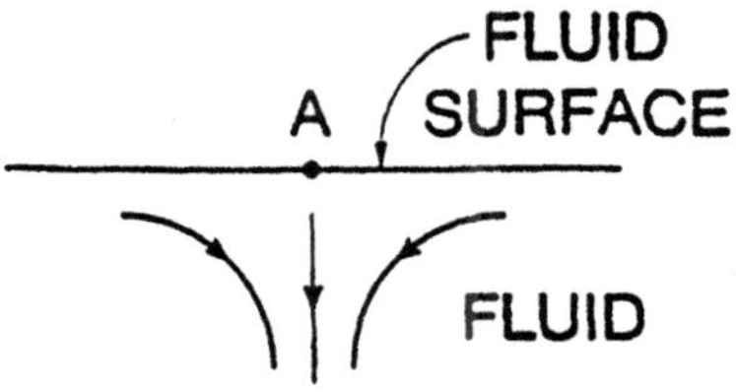

Figure 2: Floating particles are attracted to A.

In dynamical systems language, A is an attractor. The point is that, with slightly more complicated time dependent flows, the cloud can eventually coalesce onto a fractal. In fact, this is the same phenomenon as the coalescence of initial conditions of a chaotic dynamical system onto a strange attractor. In dynamical systems studies, however, the fractal strange attractor typically exists in an abstract phase space. Here, on the other hand, the relevant space is physical (namely, the fluid surface), and the fractal is consequently accessible to direct visual inspection. This was demonstrated in a real fluid experiment with an irregular pulsitile flow by Sommerer[3] and in numerical experiments using random maps in Ref. 2. In particular, if we modify our flow model of Sec. 2.1 by addition of a third component, $\mathbf{v} = \mathbf{v}_1 + \mathbf{v}_2 + \mathbf{v}_3$, where $\mathbf{v}_3$ is the surface velocity resulting from the flow shown in Fig. 2 ($\mathbf{v}_3 = -\alpha y \mathbf{y}_0$) then we obtain in place of Eqs. (12) an *area contracting* random map for tracers in the fluid surface,

$$x_{n+1} = [x_n + \alpha^{-1}(1 - e^{-\alpha})y_n] \text{modulo } 1, \tag{14}$$

$$y_{n+1} = k \sin[2\pi x_{n+1} + \theta_n] + e^{-\alpha} y_n \ . \tag{15}$$

Figure 3 shows a typical fractal pattern resulting from several applications of the map (14) and (15). The fractal nature of ϕ in this case is implied by the rigorous proof in Ref. 14 of the Kaplan-Yorke conjecture for the case of random maps. In addition, it has been confirmed[5] theoretically, numerically and experimentally, that, at long time, the Fourier wavenumber power spectrum of the density function ϕ tends to a power law, $k^{-\rho}$, where

$$\rho = D_2 - 1 \ , \tag{16}$$

and D_2 is commonly called the correlation dimension of the fractal pattern ($q = 2$ in Eq. (4)). [In this case the μ whose dimension (4) we consider is such that the measure of a set S contained within the area of interest A is $\mu = (\int_S \phi dA)/(\int_A \phi dA)$.]

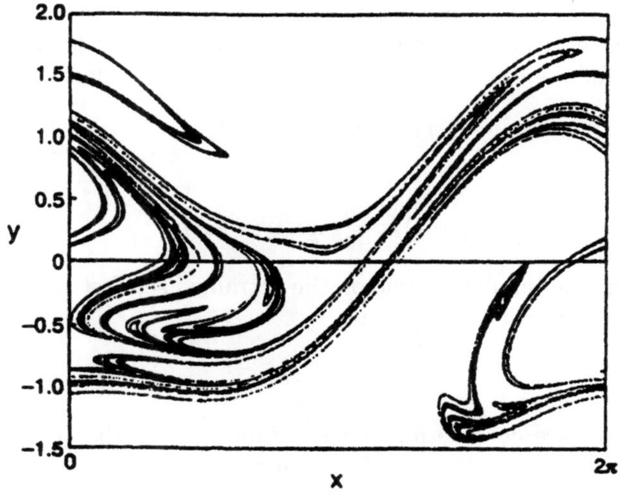

Figure 3: Fractal pattern formed by many applications of area contracting random maps to an initially smooth cloud of points. The numerical box-counting information dimension and the result from the Kaplan-Yorke formula are in excellent agreement both yielding the value $D_1 = 1.4$ (Ref. 5).

2.3 Fractal pattern of entrained scalars in flow past an obstacle

Fractal patterns also can occur in open flows as demonstrated numerically in Refs. 15 and 16, and experimentally in Ref. 17. In these references[15,17], flow past a cylinder is considered. It is well known that, as the flow velocity (equivalently, the Reynolds number) is increased, the flow bifurcates from a steady flow around the cylinder to a time periodic flow in which vortices are alternately shed from each side of the cylinder and convected downstream. References 15 and 17 consider this time periodic flow. After a cloud of dye, initially upstream from the cylinder, encounters the cylinder, some of it is entrained in the region near the cylinder. The amount of entrained dye decreases with time. Furthermore, the pattern formed by the entrained dye approaches a fractal as time increases.

This can be understood using nonlinear dynamics by considering the problem as one in chaotic scattering.[18] In particular, strobing at the period of the flow produces a two-dimensional area preserving map with a nonattracting chaotic invariant set. The entrained dye sticks close to the unstable manifold of this set, and its amount decays exponentially at the decay rate of a chaotic transient on the invariant set.[18] [This statement must be modified in the presence of other nonchaotic invariant sets (e.g., KAM surfaces).]

Now let us consider the case of a temporally irregular flow. We have a random sequence of maps $\{\mathbf{M}_n\}$ rather than a single map applied at each time. Thus there is no invariant set and consequently no unstable manifold. Hence some of the key considerations in the time periodic case no longer apply. Nevertheless, the fractal properties of the entrained dye still persist[7,16] and the result for D_q of this pattern is the same as for the case of chaotic scattering. In particular, D_q is given by[16]

$$D_q = 2 - (1-q)^{-1}\eta , \tag{17}$$

where η is given by

$$(q-1)/\tau = \max_h[\eta h - G(h)] , \tag{18}$$

and τ is the exponential decay time on the entrained set. Letting $q \to 1$, Eqs. (17) and (18) reduce the well-known result,[18]

$$D_1 = 2 - (\bar{h}\tau)^{-1} . \tag{19}$$

[In the above we are considering the D_q dimension of the same measure as in Sec. 2.2.] Equations (17)–(19) have been confirmed in Ref. 16 by comparison with box-counting results for a numerically obtained entrainment pattern generated by an area preserving random map model of an open flow.

To get an idea of the type of fractal pattern produced in such a situation, we consider the open flow, $\mathbf{v} = v_0\mathbf{x}_0 + \mathbf{v}_\omega$, discussed in Ref. 16, where the first term is simply a steady uniform flow in the x-direction. The second term is of the form $\mathbf{v}_\omega = v_\theta(r,t)\theta_0$, where (r,θ) are polar coordinates centered at $(x,y) = (0,0)$, and represents a temporally irregular soft-core vortex. For convenience we take the time dependence of v_θ to be pulsed, $v_\theta(r,t) = R(r)\sum_n \Gamma_n\delta(t-nT)$, where $R(0) = 0$ and, for small r, increases with r, subsequently decreasing as $R(r) \sim r^{-1}$ for large r. The vortex strength at time $t = nT$ is Γ_n, and this quantity is chosen randomly at each n in some range $\Gamma_+ \geq \Gamma_n \geq \Gamma_-$. Integration of this flow from $t = nT + 0^+$ to $t = (n+1)T + 0^+$ readily yields an area preserving random map. Furthermore, appropriate to our open flow situation, almost every point in the upstream region, $x < 0$, is eventually convected toward $x = +\infty$. Figure 4 from Ref. 16 shows what happens to an initial upstream tracer distribution after several iterates of the map.

3 The fast kinematic dynamo problem

We now consider the fast kinematic dynamo problem. In many respects this problem bears similarities to the passive scalar problem. The main difference is that, whereas the physical quantity for passive scalar problem is a scalar field, $\phi(\mathbf{x},t)$, now the physical quantity is a vector field, namely the magnetic flux density $\mathbf{B}(\mathbf{x},t)$. We motivate the fast kinematic dynamo problem as follows.

One of the most basic observed facts of nature is the presence of magnetic fields wherever there is flowing electrically conducting matter. In particular, magnetic fields are observed to be present in planets with liquid cores, in the Sun and stars, and in the Galaxy. A natural question is why this is so. The most common approach to this question is to consider the kinematic dynamo problem: *Will a*

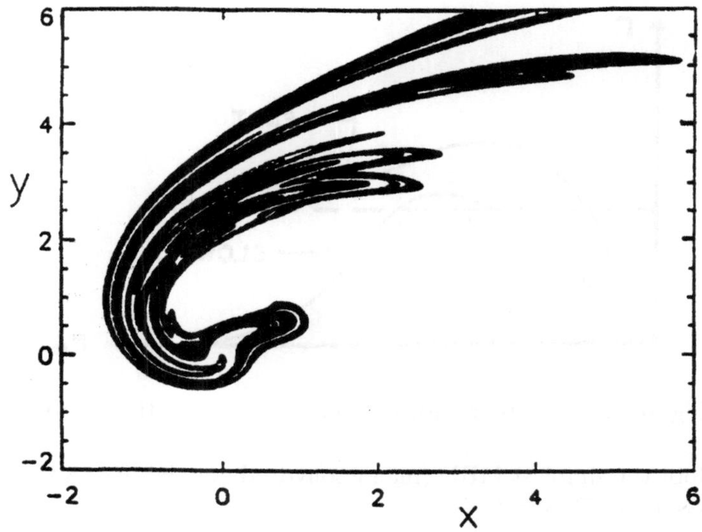

Figure 4: Distribution of tracers evolved from an initial upstream blob of tracer density after several iterations of the random map open flow model.

small seed magnetic field in an initially unmagnetized flowing electrically conducting fluid amplify exponentially in time? If the answer is yes, then it is unnatural for magnetic fields not to be present. Note that the kinematic dynamo problem is essentially a problem of linear stability. Thus the structure of magnetic fields as they are currently observed is not directly addressed, since current fields presumably have evolved to a nonlinear saturated state.

The answer to the stability question posed by the kinematic dynamo problem depends on the flow field of the fluid and on the electrical conductivity of the fluid. For a given flow field one can, in principle, ask for the conductivity dependence of the exponential growth rate Γ of a magnetic field perturbation. Vainshtein and Zeldovich[19] suggest a classification of kinematic dynamos based on the electrical conductivity dependence of Γ. In particular, if Γ approaches a positive constant as the conductivity approaches infinity, then they call the dynamo a *fast* dynamo. Otherwise they call it a *slow* dynamo. This important distinction is illustrated schematically in Fig. 5. The horizontal axis in Fig. 5 is the magnetic Reynolds number, R_m, which can be regarded as the dimensionless electrical conductivity; $R_m = \mu_0 v_0 L_0 \sigma$, where μ_0 is the (mks) magnetic permittivity of vacuum, v_0 is a typical magnitude of the flow velocity, L_0 is a typical length scale for spatial variation of the flow, and σ is the electrical conductivity of the fluid. In the Sun, for example, $R_m > 10^8$. Thus only fast kinematic dynamos are of interest in such cases. In this section we shall be concerned with fast dynamos.

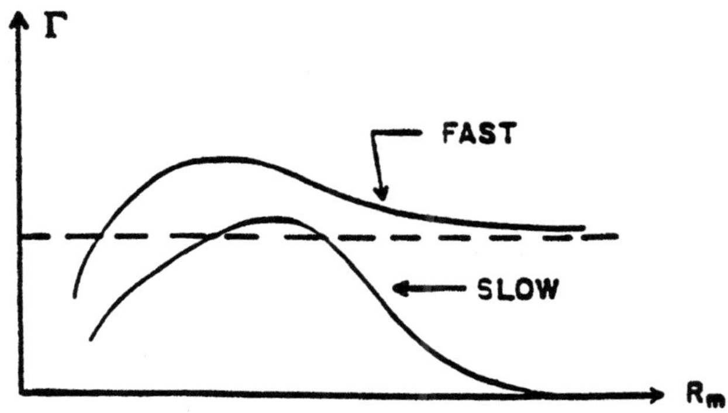

Figure 5: Γ versus R_m for fast and slow kinematic dynamos.

We adopt the simplest MHD (magnetohydrodynamic) description. The basic equation (assuming $\nabla \cdot \mathbf{v} = 0$) is

$$\partial \mathbf{B}/\partial t + \mathbf{v} \cdot \nabla \mathbf{B} = \mathbf{B} \cdot \nabla \mathbf{v} + R_m^{-1} \nabla^2 \mathbf{B} \ , \tag{20}$$

where t has been normalized to v_0/L_0, spatial scales have been normalized to L_0, and $\mathbf{v}$ has been normalized to v_0. Note that, for the kinematic dynamo problem, Eq. (20) is a linear equation in $\mathbf{B}$, because there is no linear response of the velocity, since the Lorentz force, $\mu_0^{-1}(\nabla \times \mathbf{B}) \times \mathbf{B}$, is quadratic in $\mathbf{B}$. Thus we may regard $\mathbf{v}$ as an "equilibrium" field determined by factors (e.g., convection, stirring, rotation) not appearing in Eq. (20).

Now consider Eq. (20) in the "ideal limit" which corresponds to omitting the term $R_m^{-1} \nabla^2 \mathbf{B}$,

$$d\tilde{\mathbf{B}}/dt \equiv \partial \tilde{\mathbf{B}}/\partial t + \mathbf{v} \cdot \nabla \tilde{\mathbf{B}} = \tilde{\mathbf{B}} \cdot \nabla \mathbf{v} \ , \tag{21}$$

where we use the symbol $\tilde{\mathbf{B}}$ for magnetic fields in the ideal limit. Comparing Eq. (2) for $\delta \mathbf{x}$ and Eq. (21) for $\tilde{\mathbf{B}}$ we see that the equations are the same. This is a consequence of the frozen in nature of the magnetic field at infinite conductivity, and means that the magnetic field grows in proportion to the stretching of magnetic field lines by the flow. The connection between fast dynamos and chaos is now clear: chaos implies exponential growth of $\delta \mathbf{x}$ in Eq. (2) and hence exponential field line stretching, and for a dynamo we need exponential growth of $\mathbf{B}$. There is a catch, however. In particular, the ideal equation (21) can never be fully justified even for very large R_m. What typically happens for chaotic flows is that, as R_m becomes large, $\mathbf{B}$ develops more fine scale structure, so that $R_m^{-1} \nabla^2 \mathbf{B}$ in Eq. (20) remains of the same order as the other terms in (20). This implies that $\mathbf{B}$ varies on small spatial scales of order

$$\epsilon_* \sim R_m^{-1/2} \ . \tag{22}$$

(Recall that we use the normalizations introduced in (20) so that $\mathbf{v} \sim 0(1)$ and the typical scale for spatial variation of $\mathbf{v}$ is also $0(1)$.)

In spite of this the ideal treatment is still a powerful (and correct) indication that Lagrangian chaos is the key to fast dynamo action. This point was first explicitly made in the paper of Arnold, Zeldovich, Ruzmaikin and Sokoloff[20] who considered a chaotic flow in a three dimensional abstract space of constant negative geodesic curvature (not the usual Euclidian space of classical physics), and the point was subsequently made more physically relevant by considerations for flows in ordinary Euclidian space[21,22]. By now this consideration is well-developed. An incomplete list of some representative papers is Refs. 23–43.

We now report results on how one can obtain the fractal dimension of the resulting kinematic dynamo magnetic field as well as the exponential growth rate of the instability from ergodic dynamical characteristics of the Lagrangian chaos of whatever specific velocity field $\mathbf{v}(\mathbf{x}, t)$ occurs in the dynamo problem under consideration. Note that use of the Lagrangian chaos for this purpose implies that all calculations are done from the ordinary differential equations $d\mathbf{x}/dt = \mathbf{v}$ and $d\delta\mathbf{x}/dt = \delta\mathbf{x} \cdot \nabla\mathbf{v}$ [rather than the partial differential equation, Eq. (20)]. In the remainder of this section we shall only be reporting and illustrating results, not deriving them. For derivations see Refs. 26, 33, 34 and 42.

3.1 Cancellation exponent

Before proceeding to discussions of the growth rate and fractal dimension, it is necessary to consider the extreme cancellation properties of magnetic fields in the large R_m limit. In particular, from models[27,32–34] and numerical solutions[42] it is evident that at large time the magnetic fields can undergo rapid spatial variations in which their directions flip by 180°. This results in local alternating layers with opposing fields of thickness of order $\epsilon_* \sim R_m^{-1/2}$. Thus, as R_m increases this alternation becomes more rapid.

To quantitatively characterize situations with this kind of extreme tendency for cancellation, we use the cancellation exponent introduced in Ref. 32. We consider a magnetic field distribution $\mathbf{B}$ in three dimensional space, and we choose some planar two dimensional surface S with unit normal $\mathbf{n}$. We divide S into a grid of ϵ by ϵ squares. Let ϕ_i be the magnetic flux through square i with respect to the unit normal $\mathbf{n}$ normalized to the total flux through S. If the quantity $\chi(\epsilon) = \sum_i |\phi_i|$ increases with decreasing ϵ as a power law in ϵ then we call that power the cancellation exponent and denote it κ,

$$\chi(\epsilon) = \sum_i |\phi_i| \sim 1/\epsilon^{\kappa}. \tag{23}$$

If there were no cancellation, $\phi_i \geq 0$ for all i, then, by our normalization, $\sum_i |\phi_i| = \sum_i \phi_i \equiv 1$ and $\kappa = 0$. (It is assumed that κ is generically independent of the choice of S.) If the magnetic field $B_n(\mathbf{x})$ normal to S (where $\mathbf{x}$ is on S) is a smooth bounded function, then, for small enough ϵ,

$$\sum_i |\phi_i| \cong \left[\int_S |B_n| dA\right] / \left|\int_S B_n dA\right| = (const.) < \infty,$$

independent of ϵ, and κ is again zero. In practice, the magnetic field is always smooth on small enough scale; i.e., once we consider scales as small or smaller than $\epsilon_* \sim R_m^{-1}$, Eq. (6). Thus the practical meaning of (23) is that a plot of $\ell n\chi(\epsilon)$ versus $\ell n(1/\epsilon)$ shows a linear scaling range with slope κ for $\epsilon > \epsilon_*$.

As an example illustrating the cancellation index Fig. 6 shows a plot of $\ell n\chi(\epsilon)$ for a numerical model with finite large R_m ($R_m = 10^{10}$) versus $\ell n(1/\epsilon)$. [We do not describe the model here. The interested reader should refer to Refs. 27, 32–34.] We see that the plot is well-fit by a straight line for $\epsilon > \epsilon_*$. The slope of this line is κ. For $\epsilon < \epsilon_*$ the curve flattens as expected.

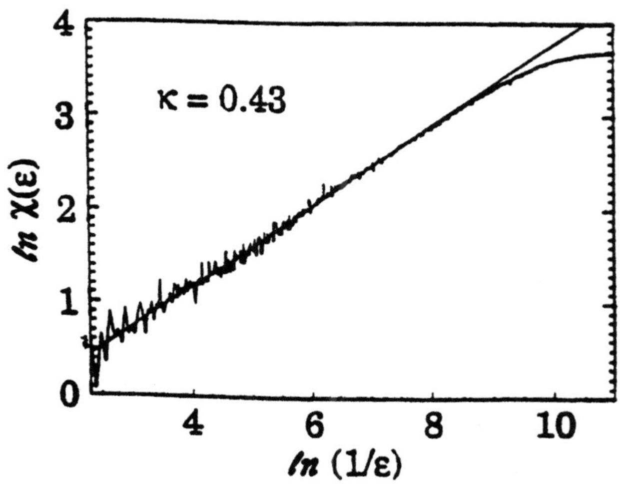

Figure 6: $\ln\chi(\epsilon)$ versus $\ln(1/\epsilon)$ for the four strip model of Refs. 22, 27, 32–34 with cancellation at $R_m = 10^{10}$.

3.2 Growth rate formula

Say we randomly sprinkle many initial conditions $\mathbf{x}_{0j}(j = 1, 2, \ldots)$ in the chaotic volume of a three dimensional dynamo flow. For each initial condition $\mathbf{x}_{0j}$ we consider a cube J of edge length δ (where δ is a differential) centered on the initial condition. We then use Eqs. (1) and (2) to evolve the cube forward in time by an amount t. This is illustrated in Fig. 7 which shows that the cube is deformed by the flow into a parallelipiped. We denote the three dimensions of the parallelipiped (the length, width and thickness) $L_{1j}\delta, L_{2j}, L_{3j}\delta$, where $L_{1j} \geq L_{2j} \geq L_{3j}$. For large t, we typically have that $L_{1j} \gg 1$ since the flow is chaotic. By volume conservation $L_{1j}L_{2j}L_{3j} = 1$, and so $L_{3j} \ll 1$. In what follows we assume that L_{2j} is typically greater than or equal to one. (In this case we expect and numerically observe that the magnetic field concentrates on a fractal set of sheet-like structures[33,34].) The quantities

$$h_i(\mathbf{x}_{0j}, t) = \frac{1}{t} \ln L_{ij} , \tag{24}$$

for $i = 1, 2, 3$, are the three finite time Lyapunov exponents from point $\mathbf{x}_{0j}$ where $h_1 \geq h_2 \geq h_3$. (The scaling result in Eq. (6) will be used subsequently for the largest exponent, $h = h_1$.)

A formula using the L_{ij} and giving the dynamo growth rate in the large R_m limit (denoted Γ_*) has been obtain in Ref. 34,

$$\Gamma_* = \lim_{t\to\infty} \frac{1}{t} < L_1 L_3^\kappa > , \tag{25}$$

where the angle brackets denote an average over the initial conditions $\mathbf{x}_{0j}$ and the number of these initial conditions is taken to infinity. In application[34,42], one uses a large number of initial conditions and plots the quantity $\ln < L_1 L_3^\kappa >$ versus t. Fitting a straight line to such a plot and obtaining its slope then gives Γ_*. The cancellation exponent κ can be obtained from the ideal equation, Eq. (21), by calculating the magnetic field on a grid of points on some conveniently chosen surface S and then obtaining the slope of $\ln \sum |\phi_i|$ versus $\ln(1/\epsilon)$. As discussed, for example, in Refs. 33 and 34, calculation of $\mathbf{B}$ on a grid on S can be done by integration along fluid element orbits and involves computations on ordinary differential equations only, namely, Eqs. (1) and (2).

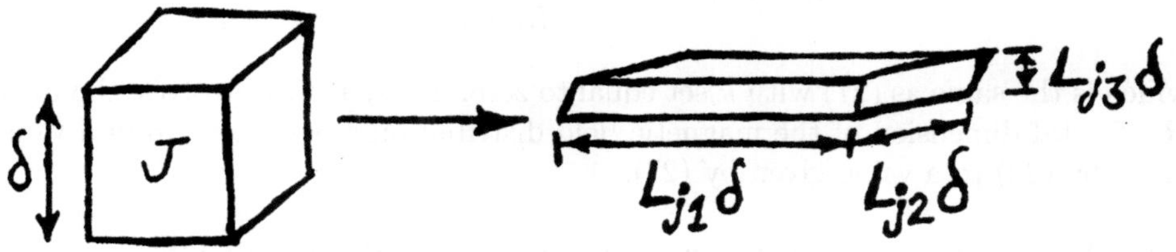

Figure 7: Deformation of the differential cube J by the flow.

Since $L_3 \ll 1$, Eq. (25) implies that cancellation always reduces the instability growth rate. Setting $\kappa = 0$, we have that if κ was positive, then

$$\Gamma_* < h_T \equiv \lim_{t\to\infty} \frac{1}{t} \ln < L_1 > . \tag{26}$$

Thus, h_T is an upper bound for the $R_m \to \infty$ growth rate Γ_*. The quantity h_T defined in (26) may, under certain circumstances, be identified with one of the fundamental quantities characterizing chaos in dynamical systems, namely the *topological entropy*[44]. Thus, we have that the topological entropy is an upper bound on the growth rate Γ_* (see Refs. 22 and 41).

3.3 Fractal dimension

Let V be any three dimensional subregion of V_0, where V_0 denotes a finite volume in which the flow is confined. Then, we define a magnetic field based measure,

$$\mu(V) = \frac{\int_V |\mathbf{B}(\mathbf{x},t)|d^3\mathbf{x}}{\int_{V_0} |\mathbf{B}(\mathbf{x},t)|d^3\mathbf{x}}.$$

For times t large enough so that the magnetic field has settled into a distribution with small scale variations at ϵ_* Du and Ott[33] obtain a formula giving the D_q dimension in terms of the L_i and the cancellation index. The result for $q = 1$ is

$$D_1 = 3 - \lim_{t\to\infty} \frac{< L_1 L_3^\kappa \ln(L_1 L_3^\kappa) > - < L_1 L_3^\kappa > \ln < L_1 L_3^\kappa >}{< L_1 L_3^\kappa \ln L_3^{-1} >}. \tag{27}$$

The case $q = 1$ is of most direct physical interest since it gives the dimension of the set on which the magnetic field concentrates. Thus, the prediction is that, if R_m is large and t is large enough so that the magnetic field varies on the limiting scale $\epsilon_* \sim R_m^{1/2}$ determined by finite resistivity, then a plot of $\sum \mu_i \ln(1/\mu_i)$ versus $\ln(1/\epsilon)$ has the slope D_1 given by (27) for ϵ small but larger than ϵ_*.

In the case where t is large but not large enough that the smallest spatial scale of magnetic field variation ϵ_* has been reached, Ott and Antonsen[26] show that there is a scaling range in ϵ for which

$$D_1 = 3 - \lim_{t\to\infty} \frac{< L_1 \ln L_1 > - < L_1 \ln L_1 >}{< L_1 \ln L_3^{-1} >} \tag{28}$$

which is the same as (27) with κ set equal to zero. Thus, as time evolves, the large R_m fractal dimension of the magnetic field distribution crosses over from a value given by (28) to a value given by (27).

3.4 Numerical computations of the kinematic dynamo PDE at large R_m

The general results discussed in the previous sections pertain to very large R_m. Although large R_m is relevant in nature, numerical computations at large R_m are difficult due to the necessity of resolving small scales, $\epsilon_* \sim R_m^{-1/2}$. Recently, Reyl et al.[6,42] have performed computations for a spatially smooth three dimensional flow at large enough $R_m (R_m = 10^5)$ that the realizations of the general properties we have discussed in this section become feasible.

The flow considered in Ref. 42 is specifically chosen so as to facilitate large R_m computation. This flow is as follows,

$$\begin{aligned} \mathbf{v}(x,y,t) &= \mathbf{x}_0 \tilde{v}_x(y) f(t) + \mathbf{y}_0 \tilde{v}_y(x) f(t - \frac{1}{3}T) \\ &+ \mathbf{z}_0 \tilde{v}_z(x) f(t - \frac{2}{3}T), \end{aligned} \tag{29}$$

where $f(t)$ is a periodic function with period $T, f(t) = 0$ for $T/3 + nT < t < (n+1)T$ with n integer, so that the flows in the x, y, and z directions are turned on sequentially. The Lagrangian chaotic dynamics generated by this flow can be analyzed by integrating $d\mathbf{x}/dt = \mathbf{v}$ over one period T. This gives a three dimensional volume preserving map relating $\mathbf{x}$ at time $t = nT$ to $\mathbf{x}$ at time $t = (n+1)T$,

$$\begin{aligned} x_{n+1} &= x_n + \tilde{v}_x(y_n), \\ y_{n+1} &= y_n + \tilde{v}_y(x_{n+1}), \\ z_{n+1} &= z_n + \tilde{v}_z(x_{n+1}), \end{aligned}$$

where $f(t)$ is normalized so that $\int_0^T f(t)dt = \int_0^{T/3} f(t)dt = 1$. Because the partial differential equation, Eq. (20), is solved using a Fourier spectral representation, it is desired that $\tilde{v}_x$, $\tilde{v}_y$, and $\tilde{v}_z$ and their convolutions with the magnetic field have simple Fourier transformations. Thus, the $\tilde{v}$'s are chosen to be sinusoidal, $\tilde{v}_x = U_x \sin(K_y y + \theta_x), \tilde{v}_y = U_y \sin(K_x x + \theta_y), \tilde{v}_z = U_z \sin(K_x x + \theta_z)$. For discussion of the numerical techniques and other details see Ref. 42.

For this flow there is a simplification of the formulae of Secs. 3.2 and 3.3. In particular, note from the map of (x_n, y_n, z_n) to $(x_{n+1}, y_{n+1}, z_{n+1})$ that the differential z-displacement remains constant, $\delta z_{n+1} = \delta z_n$. Thus $L_2 \equiv 1$, and, by volume conservation ($L_1 L_2 L_3 = 1$) we have $L_3 = L_1^{-1}$. Hence $\langle L_1 L_3^\kappa \rangle = \langle L_1^{1-\kappa} \rangle$ in Eq. (25) for the growth rate, and similarly for Eq. (27) for D_1. We can then use $L_1 = \exp(nh_1)$ and the scaling form for the distribution function [Eqs. (6) and (7)] for finite time exponents h_1 to obtain a result for the growth rate[42]

$$\Gamma_* = (1-\kappa)H(1-\kappa)/T , \tag{30}$$

where $H(s)$ is given by (7), and for D_1,

$$D_1 = 3 - (1-\kappa)\frac{H(1-\kappa) - \bar{h}}{H(1-\kappa)} . \tag{31}$$

[Note from (30) that the topological entropy of the flow ($\kappa = 0$) is $H(1)/T$ and that this is in general larger than the Lyapunov exponent, $H(1) \geq \bar{h}$.]

Using these computations of partial differential kinematic dynamo equation (20) for the flow (29), Ref. 42 addressed various large R_m issues. These include the following: (i) the predicted equality of κ obtained from the ideal equation at large time and κ obtained from solution of (20) at times after ϵ_* has been reached; (ii) the predicted equality of the theoretical growth rate result given by (25), and the growth rate from numerical solution of (20) at large R_m; and (iii) the predicted D_q from Ref. 33 for different q values (of which the $q = 1$ result is shown in Eq. (27)). In all cases, the predicted and computed results were consistent to within the limits of the numerical accuracy attained.

4 High Reynolds number instability of Lagrangian chaotic fluid flows

Taking the curl of the Navier-Stokes momentum equation for an incompressible fluid one obtains the following equation for the fluid vorticity, $\omega = \nabla \times \mathbf{v}$,

$$\frac{\partial \omega}{\partial t} + \mathbf{v} \cdot \nabla \omega = \omega \cdot \nabla \mathbf{v} + R^{-1} \nabla^2 \omega + \nabla \times \mathbf{g} , \tag{32}$$

where R is the Reynolds number, $\mathbf{g}(\mathbf{x}, t)$ is an external force, and normalizations as in Eq. (20) have been employed. Identifying $\mathbf{B}$ with ω and R_m with R, the obvious similarity of Eqs. (20) and (32) might lead one to suspect that results from the kinematic dynamo problem may be of use for gaining insight in the fluid problem. We note, however, that in the kinematic dynamo problem, $\mathbf{v}$ is regarded as given so that (20) is linear in $\mathbf{B}$, while, in contrast, in the general fluid problem $\mathbf{v}$ and ω are connected via $\omega = \nabla \times \mathbf{v}$, and Eq. (32) is inherently nonlinear. Indeed at large R, Eq. (32) with $\omega = \nabla \times \mathbf{v}$ yields high Reynolds number fluid turbulence. Nevertheless, as discussed in Sec. 4.2, it may prove worthwhile to attempt to apply the techniques of the previous section to the fluid case. For this purpose we consider the fluid problem linearized about some smooth Lagrangian chaotic flow $\mathbf{v}_0$. Equation (32) then yields

$$\partial \omega_1 / \partial t + \mathbf{v}_0 \cdot \nabla \omega_1 = \omega_1 \cdot \nabla \mathbf{v}_0 + [\omega_0 \cdot \nabla \mathbf{v}_1 - \mathbf{v}_1 \cdot \nabla \omega_0] + R^{-1} \nabla^2 \omega_1 , \tag{33}$$

where $\omega = \omega_0 + \omega_1$, $\mathbf{v} = \mathbf{v}_0 + \mathbf{v}_1$, $\omega_{0,1} = \nabla \times \mathbf{v}_{0,1}$, and ω_1 and $\mathbf{v}_1$ are linear perturbations to the unperturbed vorticity and velocity ω_0 and $\mathbf{v}_0$. (The unperturbed velocity and vorticity satisfy (32) with the replacements $\mathbf{v} \to \mathbf{v}_0$ and $\omega \to \omega_0$.) This linear problem is similar to the kinematic dynamo problem with the important difference of the appearance of additional terms on the right hand side of Eq. (33) (namely, $[\omega_0 \cdot \nabla \mathbf{v}_1 - \mathbf{v}_1 \cdot \nabla \omega_0]$). It turns out, however, that (while considerably more ingenuity is required) the appearance of these additional terms does not prevent an analysis in roughly the same spirit as that done for the kinematic dynamo. This analysis and its results are reported in Ref. 45 which also compares these results with numerical solutions for the flow, Eq. (29), for Reynolds numbers as high as $R = 10^6$. The analytical results (confirmed by the numerics) show that the ergodic properties of the Lagrangian chaos can be exploited to show that the vorticity measure,

$$\mu(V) = \frac{\int_V |\omega_1| d^3\mathbf{x}}{\int_{V_0} |\omega_1| d^3\mathbf{x}} ,$$

is multifractal for scales down to $\sim R^{-1/2}$. (As in Sec. 3.3, V_0 is the volume in which the flow is confined.) Furthermore, it is also shown[45] that the wavenumber energy density spectrum obeys a power law $k^{-\eta}$.

4.1 Analysis

The theoretical analysis[45] utilizes eikonal type equations (which are ordinary differential equations) for the evolution of the wavenumber spectrum at sufficiently high wavenumber $k \equiv |\mathbf{k}|$ ($kL_0 \gg 1$). As a result of chaos in the flow the eikonal equations cause wavenumbers to increase exponentially in time. At low k ($kL_0 \sim 1$), we do not have a simple description of the dynamics of the spectrum, but we do know that, whatever the dynamics of the spectrum is, there is an overall exponential growth of the spectrum at the linear instability growth rate Γ. We thus take Γ as an input parameter of the theory, and we attempt to find the wavenumber spectrum of the exponentially growing perturbations in the wavenumber range $kL_0 \gg 1$. Accordingly we model the dynamics at low k as an exponentially increasing source continually launching waves at some intermediate wavenumber k_0 (which we regard as not too high, but high enough that the eikonal description is valid). The launched waves then evolve by the eikonal equations propagating to higher k, until the viscosity dominated dissipation range, $k \gtrsim R^{1/2}$, is reached.

The set of equations describing the small wavelength, inviscid evolution of a "vorticity wavepacket" attached to a fluid element $\mathbf{x}(t)$ whose trajectory is given by the unperturbed flow $\mathbf{v}_0$ is obtained by a suitable expansion procedure[45] applied to (33)

$$d\mathbf{x}/dt = \mathbf{v}_0(\mathbf{x}, t) \tag{34}$$

$$d\mathbf{k}(\mathbf{x}, t)/dt = -(\nabla \mathbf{v}_0) \cdot \mathbf{k} \tag{35}$$

$$d\tilde{\omega}(\mathbf{x}, t)/dt = \tilde{\omega} \cdot \nabla \mathbf{v}_0 - \boldsymbol{\Omega} \times \tilde{\omega}, \tag{36}$$

where $\boldsymbol{\Omega}(\mathbf{x}, t) = k^{-2}(\omega_0 \cdot \mathbf{k})\mathbf{k}$. The resulting vorticity field is regarded as the superposition of wavepackets continuously launched by the source at $k \sim k_0$ and evolved by (34)–(35). An important feature is the overall exponential growth of the power spectrum, and this is taken into account by having the wavepacket emitted at time t have amplitude proportional to the exponential factor $\exp(\Gamma t)$. This picture applies for $k \lesssim R^{1/2}$. As the wavepacket wavenumber becomes larger than $R^{1/2}$, the energy in the wavepacket rapidly dissipates and is removed from consideration (alternatively, a term $-R^{-1}k^2\tilde{\omega}$ can be added to the right hand side of Eq. (36)).

Using this model the energy wavenumber power spectrum $\tilde{F}(k, t)$ can be written

$$\tilde{F}(k, t) = k^{-2} \sum_{\ell} \tilde{\omega}_\ell^2(\mathbf{x}_\ell, t) \delta(k - |\mathbf{k}_\ell(t)|) , \tag{37}$$

where the index ℓ runs over all wave packets in existence at time t, $\omega_\ell(t)$ is the wavepacket amplitude, and $\mathbf{k}_\ell(t)$ is the wave vector attached to the ℓth wave packet. Introducing $\lambda = \ell n k/k_0$, we can cast Eq. (37) in the form

$$\tilde{F}(k, t) = k^{-3} \langle |\tilde{\omega}(\mathbf{x}, t)|^2 \rangle_\lambda , \tag{38}$$

where the average $\langle \ldots \rangle_\lambda$ runs over all existing wave packets with fixed exponentiation λ. The quantity λ is the exponentiation of the wave packet's wave vector since its creation by the source, and it is assumed that all new wave packets emitted at the same time have the same initial amplitude and wave vector $\mathbf{k}_0$. Results in the theory of large deviations[12,45] yield the scaling property $e^{-q\Gamma t}\,\langle|\tilde{\omega}(\mathbf{x},t)|^q\rangle_\lambda \sim \exp[\lambda\Lambda(q)]$. Putting this result back in Eq. (38), we obtain $\tilde{F}(k,t) \sim \exp(2\Gamma t)k^\eta$, where the scaling exponent η is given by $\eta = 3-\Lambda(2)$. Thus, as observed in the numerical experiments of Ref. 45 on the flow (29), the theory yields power-law dependence on the wavenumber k. Moreover, by solving the *ordinary* differential equation system (34)–(36) for an ensemble of initial conditions we can numerically perform the average $\langle\ldots\rangle_\lambda$. For the flow numerically studied in Ref. 45 this procedure yields an exponent of $\eta = 1.70$ which compares well with the exponent from full numerical solution of the partial differential equations, Eq. (33), 1.85 ± 0.22. For a theory along similar lines giving the D_q multifractal dimension spectrum of the vorticity measure see Ref. 45.

4.2 Discussion

Due to the freedom to pick the force density $\mathbf{g}(\mathbf{x},t)$ in the momentum equation, our basic state $\mathbf{v}_0(\mathbf{x},t)$ can, in principle, be any arbitrarily chosen divergenceless velocity field, and we have taken $\mathbf{v}_0$ to be smooth and Lagrangian chaotic. The assumption of Lagrangian chaos in $\mathbf{v}_0$ is natural, since Lagrangian chaos is a generic feature of three dimensional flows (the absence of Lagrangian chaos in a three dimensional flow signals the presence of some special property of the flow, such as a symmetry). The assumption that $\mathbf{v}_0$ is smooth (i.e., has significant variation only on large spatial scale) is, however, not so natural. In particular, it is experimentally observed that, for sufficiently large Reynolds number, fluids exist in a turbulent state with intermittency and a power law wavenumber energy spectrum. What are the physical implications of a stability theory for a basic state that does not exist in nature? Our view is that the stability theory tells us something about what to expect of the turbulent state. In particular, we began by assuming the absence of fine spatial scale variations ($\mathbf{v}_0$ is smooth), but then find an instability that produces such variations in the form of a fractal vorticity distribution and a power law energy wavenumber spectrum. We regard this as implying that the absence of power law behavior of the energy wavenumber spectrum and of fractal vorticity are unnatural at large Reynolds number, since they are spontaneously created when initially absent. Thus these properties should be present in turbulence, as indeed they are experimentally observed to be. Note that the vorticity squared, which is a surrogate for the energy dissipation density, yields a fractal measure so that fractality of the vorticity implies fractality of energy dissipation. The latter has been shown by Benzi et al.[46] to imply the presence of intermittency as evidenced by the behavior of longitudinal velocity structure functions. Thus we view our results as complementary to other theories yielding intermittency (e.g., cascade based theories). One advantage of our approach is that it is obtained by actual solution of the Navier-Stokes equations, whereas cascade theories, for example, are not. The mitigating factor is, of course, that the Navier-Stokes equations that we solve are linearized.

Another possible insight from our theory concerns the issue of locality in cascade theories. One view of an inertial cascade is that structures on a given scale are unstable (e.g., Kelvin-Helmholtz type instability) to the production of structures on smaller scale, which are likewise unstable, and so on. In our linear stability analysis, however, we find that a power law and fractality are produced by a single instability. Thus a local cascade of instabilities may not be necessary in order to produce these properties and significant variations down to very small scale can be produced by a single instability at much larger scale.

This work was supported by the United States Office of Naval Research (Physics).

References

[1] F. Varosi, T. M. Antonsen, and E. Ott, Phys. Fluids A **3**, 1017 (1991); E. Ott and T. M. Antonsen, Phys. Rev. Lett. **61**, 2839 (1988).

[2] L. Yu, E. Ott and Q. Chen, Phys. Rev. Lett. **65**, 2935 (1990); Physica D **53**, 102 (1991).

[3] J. C. Sommerer and E. Ott, Science **259**, 335 (1993); J. C. Sommerer, Physica D **76**, 85 (1994).

[4] T. M. Antonsen, Z. Fan, and E. Ott, Phys. Rev. Lett. **75**, 1751 (1995).

[5] T. M. Antonsen, A. Namenson, E. Ott and J. C. Sommerer, Phys. Rev. Lett. **75**, 3438 (1995).

[6] C. Reyl, T. M. Antonsen, and E. Ott, Phys. Rev. Lett. **76**, 2270 (1996).

[7] M. A. F. Sanjuán, J. Kennedy, E. Ott, and J. A. Yorke, Phys. Rev. Lett. **78**, 1892 (1996).

[8] P. Grassberger and I. Procaccia, Physica D **9**, 189 (1983).

[9] E. Ott, *Chaos in Dynamical Systems* (Cambridge University Press, 1994), pp. 78–85.

[10] J. D. Farmer, E. Ott, and J. A. Yorke, Physica D **7**, 153 (1983).

[11] Ref. 9, Sec. 9.4.

[12] R. S. Ellis, *Entropy, Large Deviations, and Statistical Mechanics* (Springer-Verlag, New York, 1985).

[13] R. Ramshakar and J. Gollub, Phys. Fluids A **3**, 1344 (1991).

[14] F. Ladrappier and L. S. Young, Comm. Math. Phys. **117**, 529 (1988).

[15] C. Jung and E. Ziemniak, J. Phys. A **25**, 3929 (1992); C. Jung, T. Tél and E. Ziemniak, Chaos **3**, 555 (1993); A. Péntek, Z. Toroczkai, T. Tél, C. Grebogi, and J. A. Yorke, Phys. Rev. E **51**, 4076 (1995).

[16] J. Jacobs, E. Ott, T. M. Antonsen, and J. A. Yorke, Physica D **110**, 1 (1997).

[17] J. C. Sommerer, H.-C. Ku, and H. E. Gilreath, Phys. Rev. Lett. **77**, 5055 (1966).

[18] Ref. 9, Chapter 5, and references therein; see also E. Ott and T. Tél,Chaos **3**, 417 (1993).

[19] S. I. Vainshtein and Ya. B. Zeldovich, Sov. Phys. Usp. **15**, 159 (1972).

[20] V. I. Arnold, Ya. B. Zeldovich, A. A. Ruzmaikin, and D. D. Sokolov, Sov. Phys. JETP **54**, 1083 (1981).

[21] B. J. Bayly and S. Childress, Geophys. Astrophys. Fluid Dyn. **44**, 211 (1988); ibid. **49**, 23 (1989).

[22] J. M. Finn and E. Ott, Phys. Fluids **31**, 2992 (1988).

[23] B. J. Bayly and S. Childress, Geophys. Astrophys. Fluid Dyn. **44**, 211 (1988).

[24] J. Finn, J. Hanson, I. Kan, and E. Ott, Phys. Rev. Lett. **62**, 2965 (1989).

[25] M. M. Vishik, Geophys. Astrophys. Fluid Dyn. **48**, 151 (1989).

[26] E. Ott and T. M. Antonsen, Phys. Rev. A **39**, 3660 (1989).

[27] J. Finn and E. Ott, Phys. Fluids B **2**, 916 (1990).

[28] J. Finn, J. Hanson, I. Kan, and E. Ott, Phys. Fluids B **3**, 1250 (1991).

[29] D. J. Galloway and M. R. E. Proctor, Nature **356**, 691 (1992).

[30] A. D. Gilbert and B. J. Bayly, J. Fluid Mech. **241**, 199 (1992).

[31] A. D. Gilbert, Philos. Trans. R. Soc. London Ser. A **339**, 627 (1992).

[32] E. Ott, Y. Du, K. R. Sreenivasan, A. Juneja, and A. K. Suri, Phys. Rev. Lett. **69**, 2654 (1992).

[33] Y. Du and E. Ott, Physica D **67**, 387 (1993).

[34] Y. Du and E. Ott, J. Fluid Mech. **257**, 265 (1993).

[35] Y. -T. Lau and J. Finn, Phys. Fluids B **5**, 365 (1993).

[36] A. Soward, Geophys. Astrophys. Fluid Dyn. **73**, 179 (1993).

[37] A. D. Gilbert, N. F. Otani, and S. Childress, in *Theory of Solar and Planetary Dynamics*, edited by M. R. E. Proctor, P. C. Matthews, and A. M. Rucklidge (Cambridge University Press, New York, 1993), pp. 129–136.

[38] N. F. Otani, J. Fluid Mech. **253**, 327 (1993).

[39] Y. Ponty, A. Pouquet, and P. L. Sulem, Geophys. Astrophys. Fluid Dyn. **79**, 239 (1995).

[40] F. Cattaneo, E. Kim, M. Proctor, and L. Tao, Phys. Rev. Lett. **75**, 1522 (1995).

[41] I. Klapper and L. S. Young, Comm. Math. Phys. **173**, 623 (1995).

[42] C. Reyl, E. Ott, and T.M. Antonsen, Phys. Plasmas **3**, 2564 (1996).

[43] D. Galloway and U. Frisch, Geophys. Astrophys. Fluid Dyn. **36**, 53 (1986); V. I. Arnold and E. I. Korkiina, Vestn. Mosk. Univ. Mat. Mekh. **3**, 43 (1983) (in Russian).

[44] E.g., see Ref. 9, p. 143.

[45] C. Reyl, T. M. Antonsen, and E. Ott, Phys. Rev. Lett. **78**, 2559 (1997); Physica D, to be published (1998).

[46] R. Benzi, G. Paladin, G. Parisi, and A. Vulpiani, J. Phys. A.**17**, 3521 (1984).

Trends in Mathematics, © 1999 Birkhäuser Verlag Basel/Switzerland

On the geometry of two-dimensional slices of irregular level sets in turbulent flows

P. E. Dimotakis, H. J. Catrakis, A. W. Cook, J. M. Patton

Abstract

Isoscalar surfaces in turbulent flows are found to be more complex than (self-similar) fractals, in both the far field of liquid-phase turbulent jets and in a realization of Rayleigh-Taylor-instability flow. In particular, they exhibit a scale-dependent coverage dimension, $D_2(\lambda)$, for 2-D slices of scalar level sets, that increases with scale, from unity, at small scales, to 2, at large scales. For the jet flow and Reynolds numbers investigated, the isoscalar-surface geometry is both scalar-threshold- and *Re*-dependent; the level-set (coverage) length decreases with increasing *Re*, indicating enhanced mixing with increasing Reynolds number; and the size distribution of closed regions is well described by lognormal statistics at small scales. A similar $D_2(\lambda)$ behavior is found for level-set data of 3-D density-interface behavior in recent direct numerical-simulation studies of Rayleigh-Taylor-instability flow. A comparison of (spatial) spectral and isoscalar coverage statistics will be discussed.

1. Introduction

Following the introduction of Fourier spectra in the analysis of velocity correlations and the decomposition of turbulent kinetic-energy by G. I. Taylor (1938), descriptions of turbulent flow have largely relied on spectral measurement and analysis methods. Subsequent contributions introduced the notion of *similarity* in the description of small-scale turbulence (Kolmogorov 1941a,b,c; "K41") as discussed by G. K. Batchelor (1953) and J. O. Hinze (1975), with important refinements and extensions subsequently put forth (*e.g.*, Monin & Yaglom 1975, Frisch 1995).

The successes of these proposals, however, must be tempered by the host of turbulent-flow issues that cannot be addressed by correlation/spectral/moment analyses that classical descriptions have provided, that are also typically limited to uniform and isotropic flows. Part of the difficulty can be traced to the fact that information offered by such analyses is not invertible. Given a process, its spectrum, for example, is specified. Knowledge of the spectrum, alone, yields only limited other information about the process.

Such turbulent-flow issues often pose questions regarding the geometrical properties of turbulence-generated fields. Examples of such issues include, heat and mass transfer in turbulent flows; mixing and chemically-reacting turbulent flows, requiring information about the surface-to-volume ratio of scalar level sets;

aerooptics and optical-beam propagation through a turbulent medium, which (absent additional modeling and assumptions) require geometrical information about index-of-refraction gradients; aeroacoustics and weak- and strong-wave propagation through turbulence, which rely on the geometrical properties of both scalar and velocity fields; and many others. While important progress has been made in these phenomena as well, which has derived considerable benefit from classical turbulence theory, in almost all cases, additional, often *ad hoc*, assumptions, variations, and models are employed, often implicitly.

More recently, the realization that Direct Numerical Simulation (DNS) methods cannot hope to represent turbulent phenomena at the high Reynolds numbers of interest, especially when coupled to other physical processes that must be computed concurrently, has led to the quest for subgrid-scale (SGS) models that describe the behavior of scales smaller than those that can potentially be resolved in the simulations. Significantly, classical models do not yield the necessary SGS models, which also require additional structure and assumptions, as would be employed in Large Eddy Simulations (LES) calculations. Geometrical scaling information that would permit an extension of descriptions founded on an underresolved range of scales would facilitate this quest.

An important contribution that addressed some geometric-scaling issues was made by B. Mandelbrot (1975, 1982), who proposed that (power-law/self-similar) fractals could be used to describe level-set behavior of scalar and other turbulence-generated fields. A considerable body of experimental and modeling work by many investigators followed these proposals, as discussed by Sreenivasan (1991, 1994). An extension of these proposals, necessitated by recent experiments and direct numerical simulations, will be discussed below.

2. Fractals and scale-dependent irregular level sets

Mandelbrot (1975, 1977, 1982), proposed to apply the notion of stochastic geometric self-similarity to describe a host of physical phenomena, including turbulence. Level sets in a d-dimensional embedding space conforming to this description he dubbed as *fractal* and can be covered by a number of elements (boxes, tiles, line segments), $N_d(\lambda)$, of size λ, given by,

$$N_d(\lambda) \propto \lambda^{-D_d} , \tag{1a}$$

where,

$$D_d = -\frac{\mathrm{d}\log N_d(\lambda)}{\mathrm{d}\log\lambda} , \tag{1b}$$

must be a constant. For level sets generated by turbulent flow, Eq. 1b must be regarded as potentially applicable over some finite range of scales, only, *i.e.*,

$$\lambda_{\mathrm{i}} \equiv \max\{\lambda_\nu, \lambda_{\mathcal{D}}\} \ll \lambda_{\min} < \lambda < \lambda_{\max} \ll \delta . \tag{1c}$$

Here δ is the outer scale and $\lambda_{\rm i}$ is the inner scale, *e.g.*, viscous-/diffusion-scale maximum, of the flow.

Irregular level sets need not be geometrically self-similar and may be characterized by a coverage, $N_d(\lambda)$, whose logarithmic derivative, $D_d(\lambda)$, may now be a function of scale, *i.e.*,

$$D_d(\lambda) = -\frac{\mathrm{d}\log N_d(\lambda)}{\mathrm{d}\log\lambda}, \tag{2a}$$

which we may dub the (scale-dependent) *coverage dimension*. Inverting Equation 2a leads to,

$$N_d(\lambda) = \exp\left\{\int_\lambda^\delta D_d(\lambda')\,\frac{\mathrm{d}\lambda'}{\lambda'}\right\}, \tag{2b}$$

for a coverage count normalized at $\lambda = \delta$, *i.e.*, for $N_d(\delta) = 1$ (Takayasu 1982, 1992; Miller & Dimotakis 1991; Dimotakis 1991; and Catrakis & Dimotakis 1996a). A more complete discussion of these notions can be found in the review by Dimotakis & Catrakis (1996).

Equations 2a,b are the counterparts of Equations 1b,a, that may be regarded as a special case, with the more relaxed definition of $D_d(\lambda)$ in the former now allowing their application over the whole range of flow scales. They will be used to analyze two-dimensional slices of the scalar field in the far-field of liquid-phase turbulent jets and of the density field in a Rayleigh-Taylor-instability flow.

3. Isoscalar geometry in turbulent jets

In recent liquid-phase jet experiments (Catrakis & Dimotakis 1996a), the fluid Schmidt number ($Sc \simeq 2.0 \times 10^3$) and flow Reynolds number ($Re \simeq 4.5 \times 10^3$, 9.0×10^3, and 18×10^3) resulted in a jet-fluid concentration field with a complex, multiscale geometry. Laser-induced-fluorescence images of the scalar far field in this flow, in a plane normal to the jet axis, are shown in Fig. 1. They span the whole jet-fluid-concentration field, at the $z/d_{\rm j} = 275$ downstream measurement location.

Spatial spectra computed for the jet scalar-field data are shown in Fig. 2. The 2-D spectrum (left) is for a single image realization ($Re = 9 \times 10^3$) and very-nearly axisymmetric. Radial spectra, obtained by azimuthal integration, are also shown (right) for the three Reynolds numbers investigated. As can be seen, other than axisymmetry, very little information is conveyed by the 2-D spectrum; similarly for the radially-averaged spectra, which, however, indicate a decreasing wavenumber content with increasing *Re*, in contrast to classically-expected behavior. The departure occurs at wavenumbers corresponding to scales roughly 1/3 the image extent and are not the consequence of measurement resolution, which is adequate.

The concentration data were thresholded at a level $c(x, y) = c_2$, corresponding to the peak in the scalar pdf at the two lower Reynolds numbers (*cf.* Catrakis

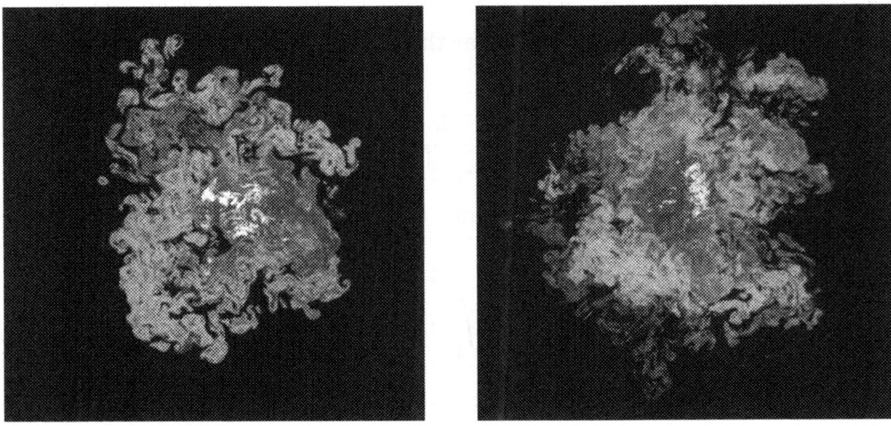

FIG. 1 Laser-induced fluorescence scalar-field data in a liquid-phase turbulent jet at $Re \simeq 4.5 \times 10^3$ (left) and 18×10^3 (right) in a plane normal to the jet axis in the far field ($z/d_{\mathrm{j}} = 275$). Color denotes jet-fluid concentration level.

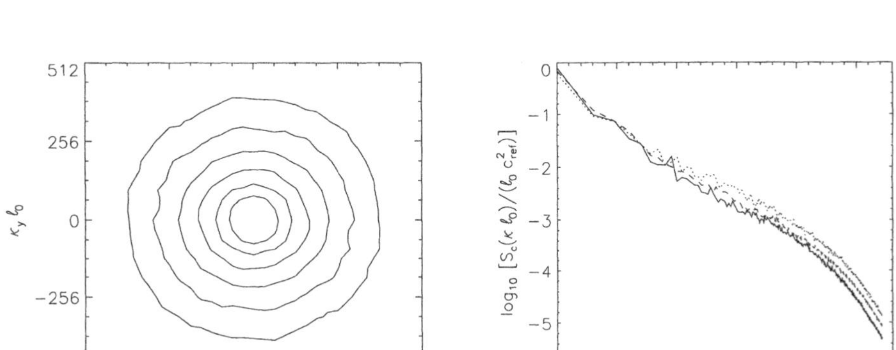

FIG. 2 Scalar spectrum for single image realization in a turbulent jet, at $Re = 9 \times 10^3$ (left) and ensemble-averaged radial scalar spectra (right), for $Re = 4.5 \times 10^3$, 9×10^3, and 18×10^3; lines of increasing solidity denote increasing Re.

& Dimotakis 1996a, Fig. 8) and the δ_b-size bounding box was computed, *i.e.*, the circumscribing rectangle of the level set. At this threshold, δ_b is independent of *Re*, within measurement statistics. Figure 3 depicts selected stages of a typical coverage sequence of a c_2-level set, at $Re \simeq 9.0 \times 10^3$, through binary subdivisions of the bounding box. Size is here defined as the square-root of the box/tile area.

One δ_b-size tile covers the set, *i.e.*, $N_2(\delta_b) = 1$, while four half-size tiles cover the set, *i.e.*, $N_2(\lambda = \delta_b/2) = 4$, yielding $D_2(\lambda) \to 2$, as $\lambda \to \delta_b$. As λ gets smaller, only a fraction of the total number of tiles is needed, *i.e.*, $N_2(\lambda) < N_{2,\mathrm{tot}}(\lambda) = (\delta_b/\lambda)^2$, and the coverage dimension decreases. Finally, at the smallest λ's, the coverage approximates the level-set contour. In this limit, $N_2(\lambda)$, increasing linearly with λ, as $\lambda \to 0$. The latter limit can be assessed by computing the normalized scalar level-set coverage length, $L(\lambda)$, from the coverage count, $N_2(\lambda)$, as a function of the coverage scale, λ, *i.e.*,

$$\frac{L(\lambda)}{\delta_b} = \frac{\lambda}{\delta_b} N_2(\lambda) \;\to\; \text{const.}, \quad \text{as } \lambda \to 0 \,. \tag{3}$$

This is plotted in Fig. 4 (left), for the three Reynolds numbers investigated. As can be seen, the data indicate a scale-independent level-set coverage length, at small scales; a limiting behavior attained for scales higher than the measurement resolution. The data also indicate a level-set coverage length that *decreases* with increasing *Re*.

The coverage dimension, $D_2(\lambda)$, is shown in Fig. 4 (right), for the Reynolds numbers investigated (Eq. 2a). It is a smoothly-increasing function of scale, bounded by its limiting value of unity (topological dimension), at the smallest scales, and 2 (embedding dimension), at the largest scales (*cf.*, also, Miller & Dimotakis 1991). A similar behavior obtains for lower and higher values of the scalar threshold.

Both the coverage length and dimension indicate less-convoluted level sets with increasing *Re*, in accord with the radially-averaged spatial-spectrum sequence (Fig. 2, right). The limiting value of $L(\lambda)$, as $\lambda \to 0$, and the coverage dimension, $D_2(\lambda)$, at medium-to-large scales, both decrease with increasing *Re*. These findings are consistent with enhanced mixing, relative to stirring, as *Re* increases, leading to improved local homogenization of the scalar field and geometrically-simpler scalar level sets. These, in turn, result in lower surface-to-volume ratios, with increasing *Re*. This is manifest in the comparison plot of sample c_2-level sets computed from $Re = 4.5 \times 10^3$ (left) and $Re = 18 \times 10^3$ (right) realizations (Fig. 5).

For such data, scalar level sets consist of individual (disjoint) "islands" and "lakes", depending on whether the interior is at a lower, or higher, scalar level, respectively. It is useful to analyze island/lake statistics, such as size and shape complexity. In the context of combustion, for example, an island would be associated with an unburnt fuel pocket in a non-premixed turbulent-jet flame. Such an analysis indicates that the size distribution of such features is well approximated by a log-normal pdf, at small-to-intermediate scales (Catrakis & Dimotakis 1996b). Size here is defined as $\sqrt{A}$, with A the individual island/lake area.

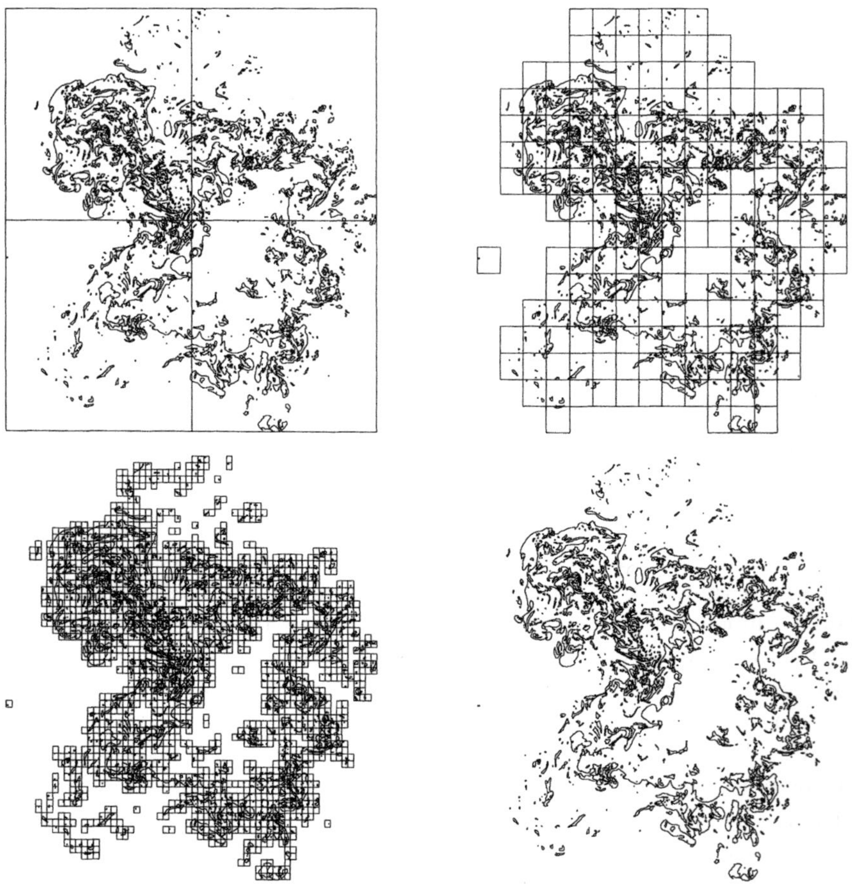

FIG. 3 Coverage sequence stages of a scalar level set in a liquid-phase turbulent jet ($Re \simeq 9.0 \times 10^3$).

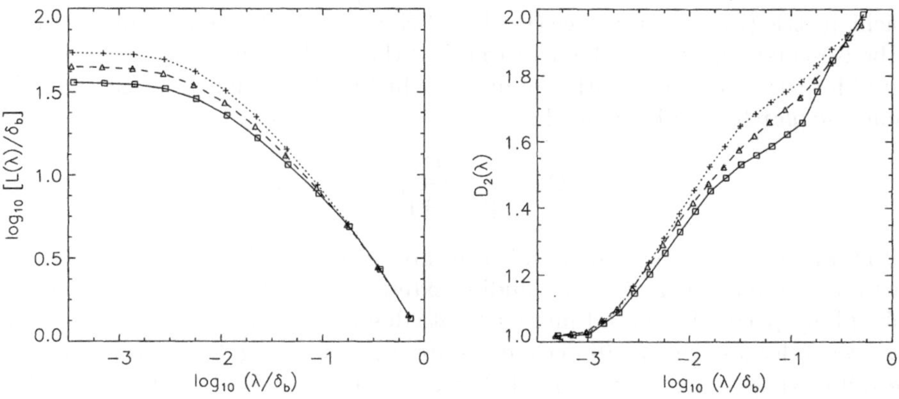

FIG. 4 Coverage length and coverage dimension as a function of (normalized) scale for scalar level sets in a turbulent jet. $Re \simeq 4.5 \times 10^3$: dotted/crosses; $Re \simeq 9.0 \times 10^3$: dashed/triangles; $Re \simeq 18 \times 10^3$: solid/squares.

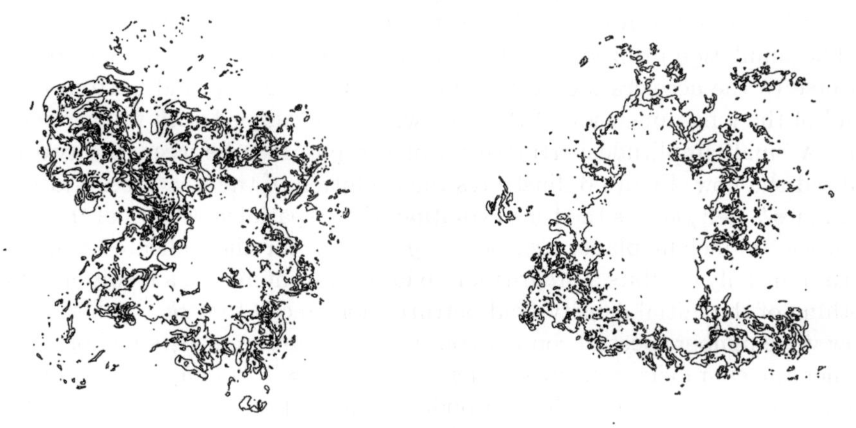

FIG. 5 Scalar c_2 level sets at $Re \simeq 4.5 \times 10^3$ (left) and 18×10^3 (right), indicating simpler topology at higher Reynolds number.

Returning to chemical reactions and combustion in non-premixed hydrocarbon turbulent flames, in which combustion is largely confined to the instantaneous stoichiometric (isoscalar) surface (Burke & Schumann 1928), area-volume measures of the isoscalar surface may be used to relate the local burning rate to the time required for the local consumption of unburnt fuel pockets. Such a measure, dubbed *shape complexity*, can be defined as,

$$1 \leq \Omega_2 \equiv \frac{P}{2\,(\pi A)^{1/2}} \leq \infty\,, \tag{4}$$

in 2-D, where P is the perimeter and A the area of an island or lake, with $(\Omega_2)_{\min} = 1$ attained for a circle, and corresponding extensions for Ω_d, for higher-dimensional embedding spaces. The liquid-phase jet data described above indicate that a power law over 3 decades in size (6 decades in area) provides a good approximation for the pdf of shape complexity. This behavior is equivalent to log-Poisson statistics for Ω_2 (Catrakis & Dimotakis 1998).

4. Isoscalar geometry in Rayleigh-Taylor-instability flow

A coverage analysis was also performed on isodensity data from a Navier-Stokes DNS study of the evolution of a Rayleigh-Taylor-instability flow, of a $Sc \equiv \nu/\mathcal{D} = 1$ fluid. The flow was initialized with a $\rho = 3$ fluid on top and a $\rho = 1$ fluid on the bottom, in a $256^2 \times 512$ rectangular box. The three-dimensional DNS of the evolving flow was terminated when the spatial-resolution requirements could no longer be met by the fixed grid, at $Re_{\mathrm{final}} \simeq 1.1 \times 10^3$, based on the vertical extent and growth rate of the Rayleigh-Taylor mixing region (Cook 1998).

The simulation utilized periodic boundary conditions in the boundary planes transverse to the acceleration vector, and no-slip at the top and bottom faces at the end of the long dimension of the box, which was aligned with the acceleration vector. A small-amplitude perturbation of the interface between the two fluids initialized the flow. Figure 6 illustrates the evolution of the $\rho = 2$ isosurface.

Figure 7 (left) shows the time evolution of the spatial spectrum of the density field in the midheight plane, *i.e.*, of $\rho(x, y, z = 0)$, for the indicated times. The spectrum initially exhibits a progression to lower wavenumbers, a result of diffusive smoothing of the initial density-field perturbation (recall that $Sc = 1$ here). While the low wavenumber spectral content continues to increase, a sustained progression to higher wavenumbers with increasing time (for $t > 1.4$) can be seen, plausibly as the Rayleigh-Taylor and other, secondary, flow instabilities take over, with the spectrum reflecting the growth of small-scale features at the midheight plane.

Figure 7 (right) plots the coverage dimension, $D_2(\lambda)$, for the $\rho(x, y, z = 0) = 2$ isodensity contours, in the same plane, computed by successive binary subdivisions of the midheight slice, as in the jet scalar-data analysis, above. The resulting coverage dimension $D_2(\lambda)$ spans the range of values from unity (the topological dimension), to 2 (the embedding dimension), smoothly transitioning between the two limiting values, at the smallest and the largest scales, respectively, a finding not

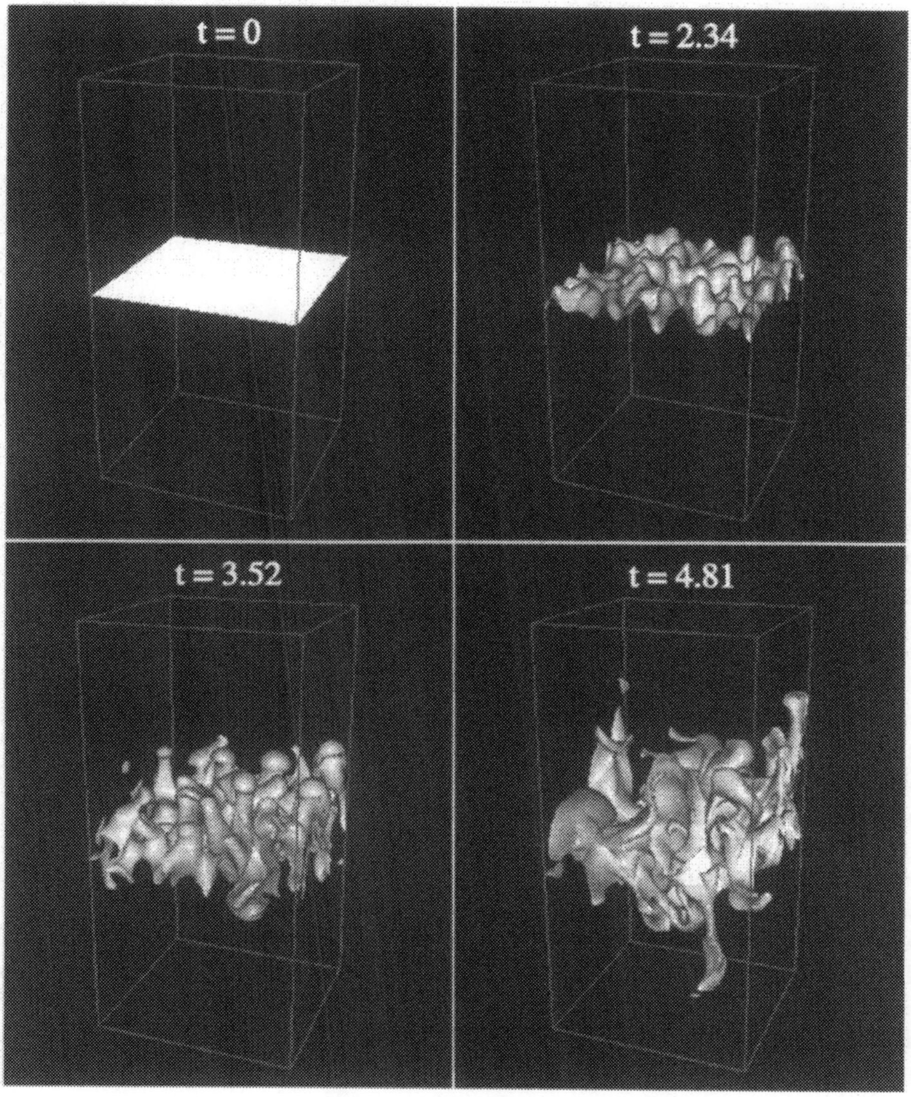

FIG. 6 Rayleigh-Taylor-instability flow. $\rho = 3$ fluid initially on top, $\rho = 1$ fluid on the bottom. Plot of $\rho = 2$ isosurface at indicated times.

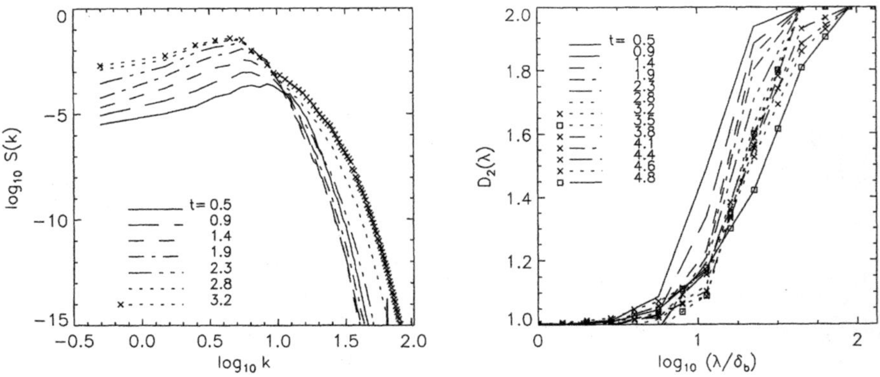

FIG. 7 Temporal evolution of scalar power spectrum (left) and coverage dimension (right) for a 2D slice (at mid-height) of the $\rho = 2$ isoscalar surface in a numerical simulation of the Rayleigh-Taylor-instability flow.

in accord with (constant) fractal-dimension behavior assumed by Timmes (1994) in modeling Rayleigh-Taylor-instability flow, for example. Interestingly, the temporal progression indicated by $D_2(\lambda)$ is from small to large scales, for $t \lesssim 3.5$, *i.e.*, opposite the high-wavenumber trend in the spectral analysis. It is not until near the end of the simulation (for $t \gtrsim 3.5$), where this trend is reversed, at small scales only. The reasons for this apparent disagreement will be discussed below.

The spectrum measures the wavenumber content of the $\rho(x, y, z = 0; t)$ density field, *i.e.*, the density surface in the midheight plane. The level-set coverage analysis interrogates the $\rho(x, y, z = 0; t) = 2$ contours in the midheight plane, *i.e.*, the $(\rho = 2)$-elevation slice of the former. These data are depicted in Fig. 8, left and right, respectively, at the indicated times. Comparison of the two early-time ($t = 0.938$ and 1.875) top figures, indicates increasingly smoother contours (right), while the density surface is developing sharp peaks. The latter are responsible for the spectral progression to higher wavenumbers. This trend continues for a while, with a subsequent transition evident in the behavior of the density surfaces (left). These develop a more complicated topology, characterized by folds between local maxima and minima, at late times (*cf.* Fig. 8, bottom: $t = 4.607$ and 4.810), indicative of mixing. This is also evident in the level-set contours that now develop small-scale features, on top of the larger-scale features that continue to increase in size.

Viewing the flow evolution through the three-dimensional density-field data (not discussed here) indicates that the likely cause of this transition is the development of secondary instabilities, of the Kelvin-Helmholtz type in the high-shear, near-midheight regions generated by the interpenetrating Rayleigh-Taylor fingers, and the formation of mushroom-like structures at their tips (*cf.* Fig. 6). The crossover in the coverage dimension, $D_2(\lambda)$, at small scales for late times revealed

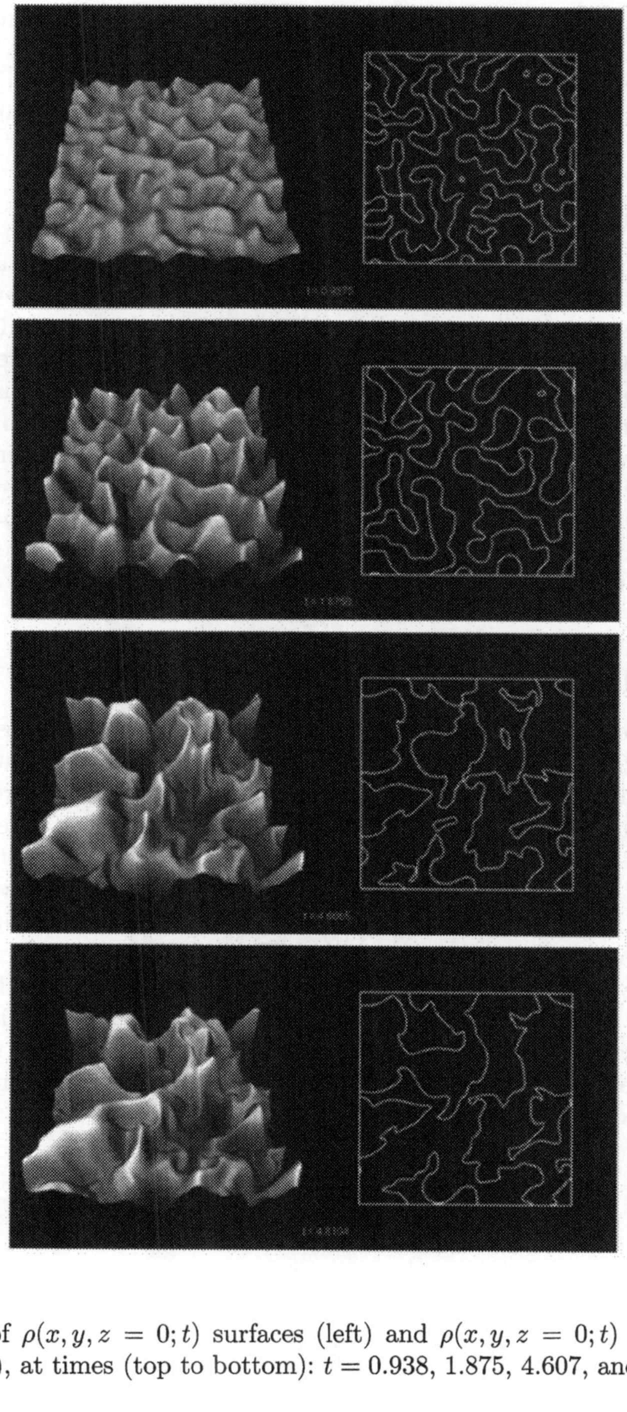

FIG. 8 Plot of $\rho(x, y, z = 0; t)$ surfaces (left) and $\rho(x, y, z = 0; t) = 2$ contours (right), at times (top to bottom): $t = 0.938$, 1.875, 4.607, and 4.810.

this transition, even though there is scant, if any, evidence for it in the spectral data. It is an attestation of the scale-local capability of the coverage analysis that such geometrical properties were clearly registered in those statistics. They were subsequently confirmed by computer-visualization of the corresponding field information.

5. Conclusions

Classical turbulence statistics and theories, while providing important guidance, do not explicitly address geometrical issues. Such issues require new tools, such as the ones derived from the original self-similar fractal ideas, and as extended to accommodate the behavior of scale-dependent irregular level sets. Statistics that derive from the coverage analysis of level sets provide many useful geometric measures that complement information from classical theories and analysis of turbulence. Two turbulent-flow cases, involving experimental data in the far-field of liquid-phase turbulent jets and a moderate-Re realization of a $Sc = 1$ Rayleigh-Taylor-instability flow, confirm the need for such extensions. In particular, the scale-local nature of coverage analysis of irregular level sets makes it a better register of geometrical information, which is difficult to infer from spectral data alone.

Acknowledgements

This work was supported by the Air Force Office of Scientific Research, Grant Nos. F49620–94–1–0353 and F49620–98–1–0052, as part of an effort to study turbulent mixing; by the Academic Strategic Alliance Program from Lawrence Livermore National Laboratory (LLNL), under Department of Energy Contract W–7405–ENG–48; and by Caltech's Center for Advanced Computing Research (CACR). The DNS work on the Rayleigh-Taylor-instability flow was performed on the IBM ASCI-Pacific-Blue computer at Lawrence Livermore National Laboratory, while one of us (AC) was at Los Alamos National Laboratory, as part of an effort to test LES models, under the Department of Energy Contract W–7405–ENG–36. We would like to acknowledge discussions with Jerry Shan and Paul Miller, help by Paul Miller and Fokion Egolfopoulos with the text, as well as the expert assistance by Dan Lang on the digital-imaging, computing, and networking parts of the experimental work.

References

BATCHELOR, G. K. 1953 *The Theory of Homogeneous Turbulence* (Cambridge U.P., London).

BURKE, S. P. & SCHUMANN, T. E. W. 1928 "Diffusion Flames," *Ind. Eng. Chem.* **20**, 998.

CATRAKIS, H. J. & DIMOTAKIS, P. E. 1996a "Mixing in turbulent jets: scalar measures and isosurface geometry," *J. Fluid Mech.* **317**, 369–406.

CATRAKIS, H. J. & DIMOTAKIS, P. E. 1996b "Scale Distributions and Fractal Dimensions in Turbulence," *Phys. Rev. Lett.* **77**, 3795–3798.

CATRAKIS, H. J. & DIMOTAKIS, P. E. 1998 "Shape Complexity in Turbulence," *Phys. Rev. Lett.* **80**, 968–971.

COOK, A. W. 1998 "Large-eddy simulation of variable-density flows," submitted to the *J. Fluid Mech.* .

DIMOTAKIS, P. E. 1991 "Fractals, dimensional analysis and similarity, and turbulence," *Nonlinear Sci. Today* #2/91, pp. 1, 27–31.

DIMOTAKIS, P. E. CATRAKIS, H. J. 1996 "Turbulence, fractals, and mixing," NATO Advanced Studies Institute series, *Mixing: Chaos and Turbulence*, GALCIT Report FM97–1.

FRISCH, U. 1995 *Turbulence. The Legacy of Kolmogorov* (Cambridge U.P., Cambridge U.K.).

HINZE, J. O. 1975 *Turbulence* (2nd ed., McGraw-Hill).

KOLMOGOROV, A. N. 1941a "Interpolation and Extrapolation of Stationary Random Sequences," translated from the Russian by W. Doyle and I. Selin, RAND Corp. Memorandum RM-3090-PR (April 1962).

KOLMOGOROV, A. N. 1941b "On degeneration of isotropic turbulence in an incompressible fluid," *C. R. Acad. Sci. U.R.S.S.* **32**, 538.

KOLMOGOROV, A. N. 1941c "Dissipation of energy in locally isotropic turbulence," *Doklady Akad. Nauk. S.S.S.R.* **66**, 825.

MANDELBROT, B. B. 1975 "On the geometry of homogeneous turbulence, with stress on the fractal dimension of the iso-surfaces of scalars," *J. Fluid Mech.* **72**, 401–416.

MANDELBROT, B. B. 1977 *Fractals. Form, Chance, and Dimension* (W. H. Freeman & Co., San Francisco).

MANDELBROT, B. B. 1982 *The Fractal Geometry of Nature* (W. H. Freeman & Co., NY).

MILLER, P. L. & DIMOTAKIS, P. E. 1991 "Stochastic geometric properties of scalar interfaces in turbulent jets," *Phys. Fluids A* **3**, 168–177.

MONIN, A. S. & YAGLOM, A. M. 1975 *Statistical Fluid Mechanics: Mechanics of Turbulence* **II** (Ed. J. Lumley, MIT Press, Cambridge, MA).

SREENIVASAN, K. R. 1991 "Fractals and Multifractals in Fluid Turbulence," *Ann. Rev. Fluid Mech.* **23**, 539–600.

SREENIVASAN, K. R. 1994 "Fractals in Fluid Mechanics," *Fractals* **2**, 253–263.

TAKAYASU, H. 1982 "Differential Fractal Dimension of Random Walk and Its Applications to Physical Systems," *J. Phys. Soc. Japan* **51**, 3057–3064.

TAKAYASU, H. 1992 *Fractals in the Physical Sciences* (Wiley, Chichester, UK).

TAYLOR, G. I. 1938 "The spectrum of turbulence," *Proc. Roy. Soc.* A **164**, 476.

TIMMES, F. X. 1994 "On the Acceleration of Nuclear Flame Fronts in White-Dwarfs," *Ap. J.* **423**(2), Part 2, L131–L134.

Trends in Mathematics, © 1999 Birkhäuser Verlag Basel/Switzerland

Universal and nonuniversal properties of the passive scalar statistics

G. Falkovich

The main physical problem of the theory of developed turbulence is to understand the degree of universality of the statistics, that is independence of the conditions of excitation and dissipation. Non-equilibrium systems generally do not enjoy the same degree of universality as equilibrium ones. At every case, the starting point is to identify the integral of motion that cascades through the inertial interval. Requiring the flux of that integral to be constant in k-space one usually gets an exact expression for some correlation function. We consider the mixing of the passive scalar generated by some random large-scale pumping ϕ and advected by an incompressible random velocity v:

$$(\partial_t + v_\alpha \nabla_\alpha - \kappa \triangle)\theta = \phi, \quad \nabla_\alpha v_\alpha = 0, \tag{1}$$

Advection preserves θ^2 which brings the flux relation valid under the condition of local isotropy [1]:

$$\langle \delta v (\delta\theta)^2 \rangle = -4\bar{\epsilon} r/3 , \tag{2}$$

where $\bar{\epsilon} = \langle \kappa |\nabla\theta|^2 \rangle$ and δv and $\delta\theta$ are longitudinal velocity and scalar differences taken at the distance r which is assumed to lie in the convective interval (that is much smaller than the pumping scale and much larger than diffusion scale). Apart from this exact relation, the rest of turbulent mixing theory is the set of more or less plausible assumptions except for a few exact results in two simplest cases of the velocity either white in time or smooth in space. Even if the velocity field is completely scale-invariant $\delta v \propto r^a$ the prediction that one may draw from (2) in the spirit of Kolmogorov 41

$$S_{2n}(r) = \langle (\delta\theta)^{2n} \rangle = C_n r^{\zeta_n} \propto \bar{\epsilon}^n r^{2(1-a)n}$$

is generally incorrect. The scale invariance of the scalar statistics is thus broken that is the probability density function (PDF) of scalar differences $\mathcal{P}(\delta\theta, \mathbf{r})$ cannot be presented as a function of the single argument $\delta\theta(\bar{\epsilon} r)^{(1-a)/2}$ and cannot be characterized solely by the flux $\bar{\epsilon}$, that is the scalar statistics is nonuniversal. There are three levels of nonuniversality. First, advection preserves an infinite number of conserved quantities $\int \theta^n dr$ and their fluxes are parameters of the PDFs so that scalar statistics generally depends on pumping statistics (except for the Batchelor case of a smooth velocity – see below). Second, under a given pumping statistics (say, Gaussian) the statistics of the scalar and it's derivatives may depend on the spatial form of pumping variance. And third, the statistics of the scalar may depend on the statistics of the velocity field.

We shall see below that ζ_n are less than those prescribed by a normal scaling hypothesis and flux law; this means, in particular, that even if the excitation produces Gaussian statistics at large scale, the statistics is getting more and more non-Gaussian as one goes to smaller scales in the inertial interval: $S_{2n}/S_2^n \propto (r/L)^{\zeta_{2n}-n\zeta_2} = (r/L)^{-\Delta_{2n}}$. When r is getting smaller than the diffusion scale, the statistics of velocity differences saturates (and coincides with the statistics of velocity derivatives) on some strongly non-Gaussian distribution, usually stretched exponential $\exp[-|\nabla\theta|^b]$. Both anomalous scaling and the strong non-Gaussianity are manifestations of spatial and temporal intermittency of developed turbulence. The main target is to find the numbers Δ_n and b, to explain the physical mechanism of intermittency and anomalous scaling in turbulent mixing, and to understand what is really universal in the convective interval statistics.

Most of the rigorous results (including the recent breakthrough in the study of anomalous scaling and stretched-exponential PDF) have been obtained hitherto within the framework of the Kraichnan's model of white-in-time Gaussian velocity [2]. The pair correlation function $\langle v^\alpha(t_1,\mathbf{r}_1)v^\beta(t_2,\mathbf{r}_2)\rangle = \delta(t_1-t_2)[V_0\delta^{\alpha\beta} - \mathcal{K}^{\alpha\beta}(\mathbf{r}_{12})]$ is expressed via the so-called eddy diffusivity

$$\mathcal{K}^{\alpha\beta} = \frac{D}{r^\gamma}(r^2\delta^{\alpha\beta} - r^\alpha r^\beta) + \frac{D(d-1)}{2-\gamma}\delta^{\alpha\beta}r^{2-\gamma} ,$$

where $0<\gamma<2$ and isotropy is assumed. The source is spatially correlated on a scale L i.e. the pair correlation function $\langle\phi(t_1,\mathbf{r}_1)\phi(t_2,\mathbf{r}_2)\rangle = \delta(t_1-t_2)\chi(r_{12})$ as a function of the argument $r_{12}\equiv|\mathbf{r}_1-\mathbf{r}_2|$ decays on the scale L. The value $\chi(0)=\bar{\epsilon}$ is the production rate of θ^2. Considering steady state and averaging (1) over the statistics of $\mathbf{v}$ and ϕ [3], one gets the closed balance equation for the simultaneous correlation function of the scalar $F_{1\ldots 2n} = F(\mathbf{r}_1,\ldots,\mathbf{r}_{2n}) = \langle\theta(\mathbf{r}_1),\ldots,\theta(\mathbf{r}_{2n})\rangle$:

$$-\hat{\mathcal{L}}F_{1\ldots 2n} = F_{1\cdots 2n-2}\chi_{2n-1,2n} + \text{permutations}. \tag{3}$$

The operator $\hat{\mathcal{L}} \equiv \sum_{i,j}\mathcal{K}^{\alpha\beta}(\mathbf{r}_{ij})\nabla_i^\alpha\nabla_j^\beta/2 + \kappa\sum\Delta_i$ describes both turbulent and molecular diffusion. For $n=1$, (3) is the analog of Yaglom equation for the white case, it is an ordinary differential equation which can be readily solved. There is an equilibrium equipartition at $(V_0/D)^{1/(2-\gamma)} \gg r \gg L$ [4]:

$$F_2(r) = \bar{\chi}r^{\gamma-d}(2-\gamma)/d(d-1)(d-\gamma)D ,$$

which can be thought of as Rayleigh-Jeans distribution $\langle\theta_k\theta_{k'}\rangle = \delta(k+k')\bar{\chi}/\omega_k$ with the "temperature" $\bar{\chi} = \int\chi(x)x^{d-1}dx$ and $\omega_k = k^\gamma\gamma d(d-1)(d-\gamma)D/(2-\gamma)$ being inverse stretching time. We introduce now the diffusion scale $r_d^{2-\gamma} = 2\kappa(2-\gamma)/(D(d-1)$ and consider the convective interval $L \gg r_{ij} \gg r_d$ where $\kappa\Delta$ can be neglected and the operator $\hat{\mathcal{L}}$ is scale invariant. Since we are interesting in r-dependencies then to avoid constant contributions we study products at $r \gg L$ and differences at $r \ll L$. At $r_d \ll r < L$ one gets

$$S_2(r) = \bar{\epsilon}\frac{2-\gamma}{\gamma(d-1)D}\frac{r^\gamma}{d} .$$

We see that $\zeta_2 = \gamma$. The scaling exponent of $\hat{\mathcal{L}}$ is $-\gamma$; the solution of (3) may thus be presented in the form $F = F_{\text{forc}} + \mathcal{Z}$, where we separated the so-called "forced" part of the solution (with the scaling exponent $\zeta_{2n-1} + \gamma$ prescribed by the rhs) from the zero mode $\mathcal{Z}$ that may have a different scaling. It has been recognized independently by the authors of [4, 5, 6] that they are the zero modes of the operator $\hat{\mathcal{L}}$ that are responsible for the anomalous scaling. Even though the equation $\mathcal{L}\mathcal{Z} = 0$ is a linear partial differential equations of the second order, we cannot yet solve it for general n, γ, d. To learn about the physics of turbulent mixing, it is natural for a theorist to explore the whole admissible range of the parameters: $0 \le \gamma \le 2$ and $2 \le d \le \infty$. In this way, it has been understood that there are three limits where the statistics of the scalar is getting Gaussian: i) the limit of infinite dimensionality [2, 4, 7, 8], ii) extremely irregular velocity $\gamma \to 2$ [5, 9, 10] and iii) the (Batchelor) limit of smooth velocity $\gamma \to 0$ [2, 3, 6, 11, 12, 13, 14]. The perturbation theory is regular in the first two cases and the exponents at $\Delta_n \ll n\gamma$ are given by [7, 9]

$$\zeta_n = \gamma n/2 - 2n(n-1)(2-\gamma)/(d+2) \ . \tag{4}$$

The perturbation theory around the Batchelor limit is singular [6], only the third structure function has been found so far [15, 16]. It is instructive to discuss the significant difference between the first two limits on the one hand and the third limit on the other. Since the scalar field at any point is the superposition of fields brought from d directions then it follows from a central limit theorem that scalar's statistics approaches Gaussian when space dimensionality d increases. In the case $\gamma = 2$, an irregular velocity field acts like Brownian motion so that turbulent diffusion is much like linear diffusion: statistics is Gaussian provided the input is Gaussian. What is general in both limits $d = \infty$ and $\gamma = 2$ is that the degree of Gaussianity (say, flatness) is independent of the ratio r/L. Quite contrary, $\ln(L/r)$ is the parameter of Gaussianity in the Batchelor limit [13, 17] so that statistics is getting Gaussian at small scales whatever the input statistics. At $\gamma = 0$ the mechanism of Gaussianity is temporal rather than spatial: since the stretching is exponential in a smooth velocity field then the cascade time grows logarithmically as the scale decreases. That leads to the essential difference: at small yet nonzero $(2-\gamma)/d$, the degree of non-Gaussianity increases downscales while at small γ the degree of non-Gaussianity first decreases downscales until $\ln(L/r) \simeq 1/\gamma$, and then starts to increase, the first region grows with diminishing γ. Already that simple reasoning shows that the way from the Batchelor limit towards an anomalous scaling at non-smooth velocity field is not to be easy. The formal reason for this perturbation theory to be singular is that, at the limit $\gamma = 0$, the many-point correlation functions have singularity (smeared by molecular diffusion only) at the collinear geometry – smooth velocity provides for homothetic transformation that does not break collinearity [14]. Even weak non-smoothness of the velocity smears the singularity i.e. strongly influences the solution in the narrow region near collinearity; such a situation calls for a boundary layer approach [6, 11].

Describing the behavior at small γ is important for our understanding of the whole picture not only because 3d Kolmogorov scaling corresponds to $\gamma = 2/3$ yet also because the dependence $\Delta_n(\gamma)$ has to be nonmonotonic since $\Delta_n(0) = \Delta_n(2) = 0$. There is a transparent physics behind the nonmonotonic dependence $\Delta(\gamma)$ because the influence of velocity nonsmoothness (measured by γ) on scalar intermittency is twofold: if one considers scalar fluctuation of some scale then velocity harmonics with comparable scales produce intermittency while small-scale velocity harmonics act like diffusivity and smooth it out. There exists some γ_* such that at $0 < \gamma < \gamma_*$ the first mechanism is stronger while at $\gamma_* < \gamma < 2$ the second one takes over. Still, our understanding is only qualitative here, we don't know how the maximum position γ_* depends on n and d. The only data are due to numerics [18]. The mechanism of d-dependence is probably due to the fact that collinear singularity is getting less important as d grows. Since large-d expansion has actually γd as an expansion parameter it is likely that $\gamma_* \propto 1/d$ at large d.

High moments (at $n \geq d$) cannot be found perturbatively since Δ_n is already of order $n\gamma$. Another approach can be implemented in this case: using n as large parameter one can apply an instanton formalism [19, 20] which shows that at $d\gamma \gg 1$ the expression (4) is valid for all $n < n_c = d\gamma/2(2-\gamma)$ and then the exponents saturate at $n > n_c$ [20]:

$$\zeta_n = \zeta_c = \frac{d\gamma^2}{8(2-\gamma)}. \tag{5}$$

The exponents ζ_n are thus universal within the Kraichnan model, they depend only on γ, d. Zero modes enter the solution of (3) due to boundary conditions at $r = L$ so that their contributions depend generally on pumping statistics and the form of pumping correlation functions. Let us ask: if the pumping statistics is given (say, Gaussian) how many constants one needs to describe the PDF $\mathcal{P}(\delta\theta, r)$? The answer is given in [20]:

$$S_n = \left(\frac{n}{\gamma}\frac{\bar{\epsilon}L^\gamma}{D}C_1\right)^{n/2}\left(\frac{r}{L}\right)^{\gamma n/2 - 2n(n-1)(2-\gamma)/(d+2)} \qquad \text{at } n < n_c\,, \tag{6}$$

$$S_n = \left(\frac{n}{\gamma}\frac{\bar{\epsilon}L^\gamma}{D}C_2\right)^{n/2}\left(\frac{r}{L}\right)^{\zeta_c} \quad \text{at } n \gg n_c\,. \tag{7}$$

The quantities C_1 and C_2 in (6) are non-universal constants of order unity, whose values depend on the shape of χ (that is on the spatial structure of pumping covariance). The physical meaning of C_1 and C_2 is unclear at the moment. Note that the r-independent factor in (6,7) is determined by the single-point rms value $\langle\theta^2\rangle = \bar{\epsilon}L^\gamma/D\gamma$. The expression (7) corresponds to the factorized Gaussian PDF valid at $(\delta\theta)^2 > \langle\theta^2\rangle$

$$\mathcal{P}(\delta\theta) \propto \left(\frac{r}{L}\right)^{\zeta_c} \exp\left[-\frac{(\delta\theta)^2}{2C_2\langle\theta^2\rangle}\right]. \tag{8}$$

At $S_2 \ll (\delta\theta)^2 \ll \langle\theta^2\rangle$ the PDF is strongly non-Gaussian due to the n-dependence of the last factor in (6). Note that the single-point $\mathcal{P}(\theta)$ has not been yet found for nonzero γ, it is likely to be close to Gaussian for a Gaussian pumping.

For $\gamma = 0$, the PDF $\mathcal{P}(\delta\theta)$ is Gaussian until $\delta\theta \simeq \bar{\epsilon}\ln^2(r/r_d)/D$ and then it is exponential [19, 21], the single-point $\mathcal{P}(\theta)$ has a similar form [3, 13, 21]; the scalar PDFs at $\gamma = 0$ are universal. Remarkably, those results are valid for an arbitrary velocity statistics (as long as the velocity is ergodic and has finite temporal correlation) and for an arbitrary pumping statistics [13]. Since it is a logarithmic regime $\langle\theta_1^n\theta_2^n\rangle \propto \ln^n(L/r_{12})$ then single-point moments are determined by the whole convective interval (not by the pumping scale as for power-law case) and the single-point statistics is universal as well. High-order fluxes are irrelevant in this case due to the phenomenon of "distributed pumping" [22]. Note that the correlation functions depend not only on the flux yet also on pumping scale L yet it is logarithmic dependence so intrinsic ambiguity in L definition (up to order-unity constant) doesn't matter.

Let us discuss now the statistics of scalar gradients, in particular, the PDF of dissipation $\epsilon = \kappa|\nabla\theta|^2$. Considering simple fusion rules for $2n$ points one can express dissipation moments via anomalous exponents of the scalar: $\langle\epsilon^n\rangle = c_n\bar{\epsilon}^n(L/r_d)^{\Delta_{2n}}$ [4, 7]. Here the constants c_n are due to fluctuations of dissipation scale, most likely they are of the form n^{qn} with yet unknown q. In the perturbative domain $n \ll d\gamma/(2-\gamma)$ when $\Delta_{2n} \propto n(n-1)$, the main factor is $(L/r_d)^{\Delta_{2n}}$ and $\mathcal{P}(\epsilon)$ is close to lognormal [7], the form of the more distant tail is yet unknown. One may think that finding $\mathcal{P}(\epsilon)$ in developed turbulence is so difficult because of non-analyticity with respect to κ, indeed, $\bar{\epsilon}$ stays finite while $\langle\epsilon^n\rangle$ diverges (at nonzero γ) as $\kappa \to 0$. While nonanalyticity is indeed the fact so that diffusion term cannot be treated by a straightforward perturbative theory, the drastic difference between the diffusive and advective timescales can be exploited in a dynamical Lagrangian formalism by a time separation method [23, 24]. Such an approach reveals (not surprisingly) that the main difficulty is in treating advection, as long as it can be effectively described, the dissipation PDF can be found. The case in question is Kraichnan model at $\gamma = 0$, velocity field is spatially smooth which corresponds to a large Prandtl number. In this case, we can introduce $\sigma_{\alpha\beta} \equiv \nabla_\beta v_\alpha$ and passing into the comoving reference frame we write (1) as $\partial_t\theta + r_j\sigma_{jl}(t)\partial_l\theta - \kappa\Delta\theta = \phi(t, \mathbf{r})$. Making spatial Fourier transform and introducing $\dot{\mathcal{W}} = \hat{\sigma}\mathcal{W}$, one can write $\theta_k(t)$ as an integral over Lagrangian trajectories

$$\int_0^t dt'\phi\left(t-t', \hat{\mathcal{W}}^{-1,T}(t')\mathbf{k}\right)\exp\left[-\kappa k_\mu\int_0^{t'}\left[\hat{\mathcal{W}}^{-1}(\tau)\hat{\mathcal{W}}^{-1,T}(\tau)\right]_{\mu\nu}d\tau k_\nu\right].$$

Averaging over ϕ a product of $2n$ replicas of the inverse Fourier of $\mathbf{k}\theta_k$, we get the n-th moment of $\epsilon = \kappa\left(\nabla\theta\right)^2$ and restore the PDF

$$\mathcal{P}(\epsilon) = \int_{0^+-i\infty}^{0^++i\infty}\frac{dse^{s\epsilon}}{2\pi^2 i}\int d\mathbf{m}e^{-m^2}\left\langle e^{-s\int_0^\infty dtQ}\right\rangle_\sigma, \tag{9}$$

$$Q = \int d\mathbf{q}\chi_q \left[\frac{\mathbf{q}\hat{\mathcal{W}}(t)\mathbf{m}}{2\pi Pe}\right]^2 \exp\left[-\frac{\mathbf{q}\hat{\Lambda}(t)\mathbf{q}}{Pe^2}\right], \tag{10}$$

where $\hat{\Lambda}(t) \equiv D\hat{\mathcal{W}}(t)\int^t dt'\hat{\mathcal{W}}^{-1}(t')\hat{\mathcal{W}}^{-1,T}(t')\hat{\mathcal{W}}^T(t)$ and $\mathbf{q} = \mathbf{k}L$, an extra integration over the auxiliary vector $\mathbf{m}$ takes care of combinatorics. The next step is Kolokolov transformation [25] to replace time-ordered exponent by a regular function, the natural price to pay is measure nonlocality in the path integral. Then we implement time separation between two different time scales which are clearly seen in (9–10) for $\epsilon \gg \langle\epsilon\rangle/Pe^2$. Indeed, $\int Q dt$ has to be large, which is achieved on such realizations of $\hat{\sigma}$ where the integrand grows exponentially due to $\mathcal{W}(t)$. The last exponential factor restricts the growth at the time of order $D^{-1}\ln \mathrm{Pe}$. Well before, on a time scale of the inverse stretching rate D^{-1}, the exponentially growing $\mathcal{W}(t')$ saturates the integral over dt' in $\hat{\Lambda}(t)$. A dynamical picture in r-space is as follows: For the Lagrangian trajectories giving the main contribution into $\langle\epsilon^n\rangle$, fluid particles start very close. Diffusion remains dominant until the particles separate by a distance comparable to the diffusion scale r_d. This phase of the dynamics takes place on times of order D^{-1}, this time is diffusion-independent since r_d is a scale where diffusion and stretching are comparable. Once the distance between particles outgrows r_d, random multiplicative stretching due to velocity becomes dominant, the time to go from r_d to L is proportional to $D^{-1}\ln \mathrm{Pe}$. Time separation provides for a reduction to two subsequent path integrations both local in time: one describing the long evolution (since the stretching effectively goes along one direction it is reduced to ODE that can be solved for any given χ) and another describing the fast fluctuations which can be integrated exactly due to strain Gaussianity [24]. The tail of distribution at $\epsilon \gg \bar{\epsilon}$ is a stretched exponent

$$\mathcal{P}(\epsilon) \sim \epsilon^{-1/2} \exp\left[-C_3(\epsilon/\bar{\epsilon})^{1/3}\right] , \tag{11}$$

where C_3 is order-unity constant depending on the form of pumping covariance χ. The difference between $\ln \mathcal{P}(\nabla\theta) \propto |\nabla\theta|^{2/3}$ and the exponential tail of $\mathcal{P}(\delta\theta)$ appears due to fluctuations of the dissipation scale r_d. Indeed, assuming that the fluctuations of $\delta\theta$ and r_d are independent one gets 2/3-tail for $w = \delta\theta/r_d$ integrating together exponential $\mathcal{P}(\delta\theta)$ and Gaussian $\mathcal{P}(1/r_d)$: $\mathcal{P}(w) = \int \mathcal{P}(\delta\theta)\mathcal{P}(1/r_d)\delta(w - \delta\theta/r_d)d\delta\theta dr_d^{-1}$. Note that the index 1/3 of the stretched exponent in (11) is determined by the velocity statistics that determines the statistics of the dissipation scale. It is likely that the same law is correct for the Prandtl number of order unity as well because the tail is determined by the fluctuations with r_d smaller than the mean value (and the viscous scale), velocity can be considered smooth at such scales. At $\epsilon \ll \bar{\epsilon}$, we find $\mathcal{P} \propto 1/\sqrt{\epsilon}$. We see that in a spatially smooth and temporally random Gaussian flow ($\gamma = 0$) the gradient PDF depends on velocity statistics and pumping covariance (while scalar PDFs are completely universal).

Finally, let me discuss how difficult it is to introduce any meaningful model of the velocity statistics with finite temporal correlations into the studies of passive scalar. If one is interested in the dispersion of Lagrangian trajectories $\rho(t)$ given by $\partial_t \rho = \mathbf{v}(t,\rho)$ then the statistics of the separation $\mathbf{R}(t) = \rho_1 - \rho_2$ is determined by the velocity statistics. If the velocity statistics were Gaussian in the Lagrangian frame it would be characterized by the variance

$$\mathcal{K}_{\alpha\beta}(t,\mathbf{R}_1,\mathbf{R}_2) = \Big\langle \Big(v_\alpha[0,\rho(0)+\mathbf{R}_1] - v_\alpha[0,\rho(0)] \Big) \Big(v_\beta[t,\rho(t)+\mathbf{R}_2] - v_\beta[t,\rho(t)] \Big) \Big\rangle$$

which has to satisfy "triangle identity" (the symmetry of the statistics with respect to an interchange of the three Lagrangian separations $\mathbf{R}_1, \mathbf{R}_2$ and $\mathbf{R}_1 - \mathbf{R}_2$) [26]

$$\begin{aligned} &\mathcal{K}[t-t',\mathbf{R}_1(t),\mathbf{R}_1(t')] + \mathcal{K}[t-t',\mathbf{R}_2(t),\mathbf{R}_2(t')] - \mathcal{K}[t-t',\mathbf{R}_1(t),\mathbf{R}_2(t')] \\ &-\mathcal{K}[t-t',\mathbf{R}_2(t),\mathbf{R}_1(t')] = \mathcal{K}[t-t',\mathbf{R}_1(t)-\mathbf{R}_2(t),\mathbf{R}_1(t')-\mathbf{R}_2(t')] \,. \end{aligned}$$

The relation is automatically satisfied only in a delta-correlated case (for any γ) or for $\gamma = 0$ (arbitrary temporal correlations). Spatially non-smooth and temporally correlated velocity cannot be described by the action of the second order in $\mathbf{R}$ that is Gaussian velocity is meaningless in this case. That shows that the approach usually adopted in studying passive scalar or Lagrangian dispersion, namely "take some (usually Gaussian) velocity statistics and study particle dispersion", does not work for a finite correlated velocity with nonzero γ. In such a general case, different-time velocity statistics has to be restored from single-time statistics. Probably the only way to avoid unphysical results is to use Navier-Stokes equation.

To conclude, let me briefly list the open problems mentioned above:

1. Within the Kraichnan model, to understand the behaviour of $\Delta_n(\gamma, d)$, in particular, the behavior at small γ and the position of the maximum.

2. To find the dissipation PDF at $\gamma \neq 0$, some combination of instanton and time-separation methods may give the PDF tail.

3. To find high-order correlation functions at large scales and to understand how universal the equilibrium is. What is the extremum principle that governs such an equilibrium? Is there something special in two dimensions? Is the statistics of decaying scalar turbulence universal at large scales?

4. To go beyond Kraichnan model considering finite-correlated velocity. Here we don't know neither the pair correlation function of the scalar nor the law of two-particle dispersion (Richardson law most likely is not exact). In my opinion, the only chance here to develop a consistent theory is to exploit large d approach.

I'm grateful to A. Zamolodchikov and V. Lebedev for the useful remarks.

References

[1] A. M. Yaglom, Dokl. Akad. Nauk **69**, 743 (1949).

[2] R. H. Kraichnan, Phys. Fluids **11**, 945 (1968); J. Fluid Mech. **64**, 737 (1974).

[3] B. Shraiman and E. Siggia, Phys. Rev. E**49**, 2912 (1994).

[4] M. Chertkov, G. Falkovich, I. Kolokolov and V. Lebedev, Phys. Rev. E**52**, 4924 (1995).

[5] K. Gawedzki and A. Kupiainen, Phys. Rev. Lett. **75**, 3608 (1995).

[6] B. Shraiman and E. Siggia, CRAS **321**, Ser. II, 279 (1995).

[7] M. Chertkov and G. Falkovich, Phys. Rev. Lett. **76**, 2706 (1996).

[8] D. Gutman and E. Balkovsky, Phys. Rev. E**54**, 4435 (1996).

[9] D. Bernard, K. Gawedzki and A. Kupiainen, Phys. Rev. E**54**, 2564 (1996).

[10] A. Pumir, Europh. Lett. **34**, 25 (1996).

[11] B. I. Shraiman and E. D. Siggia, Phys. Rev. Lett, **77**, 2463, (1996).

[12] G.K. Batchelor, J. Fluid Mech. **5**, 113 (1959).

[13] M. Chertkov, G. Falkovich, I. Kolokolov, and V. Lebedev, Phys. Rev. E**51**, 5068 (1995).

[14] E. Balkovsky, M. Chertkov, I. Kolokolov, and V. Lebedev, JETP Letters **61**, 1049 (1995).

[15] A. Pumir, B.I. Shraiman, E.D. Siggia, preprint cond-mat/9609272.

[16] E. Balkovsky, G. Falkovich and V. Lebedev, Phys. Rev. E**55**, 4881 (1997).

[17] G. Falkovich and V. Lebedev, Phys. Rev. E **49**, R1800 (1994).

[18] U. Frisch, A. Mazzino and M. Vergassola, chao-dyn/9802192.

[19] G. Falkovich, I. Kolokolov, V. Lebedev and A. Migdal, Phys. Rev. E**54**, 4896 (1996).

[20] E. Balkovsky and V. Lebedev, chao-dyn/9803018.

[21] D. Bernard, K. Gawedzki and A. Kupiainen, cond-mat/9706035; J. Stat. Phys. (1998).

[22] G. Falkovich, Phys. Rev. E**49**, 2468 (1994).

[23] M. Chertkov, I. Kolokolov, and M. Vergassola, Phys. Rev. E**56**, 5483 (1997).

[24] M. Chertkov, G. Falkovich and I. Kolokolov, Phys. Rev. Lett. **79**, 2121 (1998).

[25] I. Kolokolov, Ann. of Phys. **202**, 165 (1990).

[26] G. Falkovich, V. Kazakov and V. Lebedev, Physica A (1998).

Trends in Mathematics, © 1999 Birkhäuser Verlag Basel/Switzerland

Multi-scalar mixing and Lagrangian approaches

P. K. Yeung

Abstract

Fundamental challenges in the statistical description of multi-scalar mixing in homogeneous turbulence are considered, with particular emphasis on the differential diffusion of multiple passive scalars with different molecular diffusivities. New concepts and difficulties arising in addition to those in single-scalar mixing are discussed using typical results from direct numerical simulations of isotropic turbulence. The latter include both Eulerian and Lagrangian quantities, such as multi-variate conditional expectations of molecular diffusion, the time series of scalar fluctuations following fluid particle trajectories, and the dispersion between different species molecules which diffuse relative to the fluid. Knowledge of small-scale processes is important for improvements in Lagrangian statistical modeling.

1 Introduction

Many applications of turbulent mixing require the simultaneous consideration of multiple diffusing entities such as scalars representing temperature or concentration fluctuations. Statistical and spatial correlations between different scalars in the same flow are often important: for example combustion efficiency depends strongly on how well the reactants and oxidants are mixed together (and down to the smallest scales). Although equal-diffusivity assumptions can bring appealing simplifications based on linearity properties of the scalar transport equations, in general they do miss an important part of the essential physics. Our motivation in this paper is thus to review and explore some of the significant challenges in studying multi-scalar mixing, in the context of (as a prototypical problem) the differential diffusion of multiple passive scalars with different molecular diffusivities.

In general it is instructive to describe turbulent mixing using both Eulerian and Lagrangian viewpoints (McComb 1990). A fundamental difficulty in the prediction of multi-scalar statistics is that knowledge of joint probability distributions is required. Exact Eulerian transport equations for probability density functions (PDF) and the associated closure issues are well known (Pope 1985). However, because of the high dimensionality of PDF transport equations in sample space, computational tractability requires the use of Monte-Carlo methods where the modeling must be formulated in Lagrangian terms (Pope 1994). The Lagrangian viewpoint is similarly crucial in stochastic models of contaminant dispersion based on fluid particle motion, where (e.g. Borgas & Sawford 1996) the effects of molecular diffusion may be treated via the motion of species "molecules" which undergo Brownian motion relative to the fluid.

In the following we discuss both Eulerian and Lagrangian data obtained from direct numerical simulations (DNS) with a particle tracking algorithm. The flow studied is stationary isotropic turbulence, which is idealized but allows new concepts to be addressed with greater clarity. Eulerian scalar fields are driven by uniform mean gradients and also stationary in time. Typical results at Taylor-scale Reynolds numbers (R_λ) 90 and 160 are given below for discussion.

2 Differential diffusion: Eulerian measures

For our purposes here we assume that the important statistics (mean, variance, dissipation, or the probability density function) of each scalar *taken separately* are known (or otherwise modeled satisfactorily), and focus on quantities that describe the statistical relationships between the fluctuations of different scalars. In particular, quantities that have been used as statistical measures of differential diffusion have included correlation coefficients in physical and Fourier space (as a function of scale size), the mean-squared concentration difference between two scalars, and the conditional fluctuations (Bilger 1993) of one scalar relative to another. Recent evidence from DNS (Yeung 1996, 1998) on these Eulerian measures has indicated that molecular diffusivity differences remain important in the small scales even at high Reynolds number.

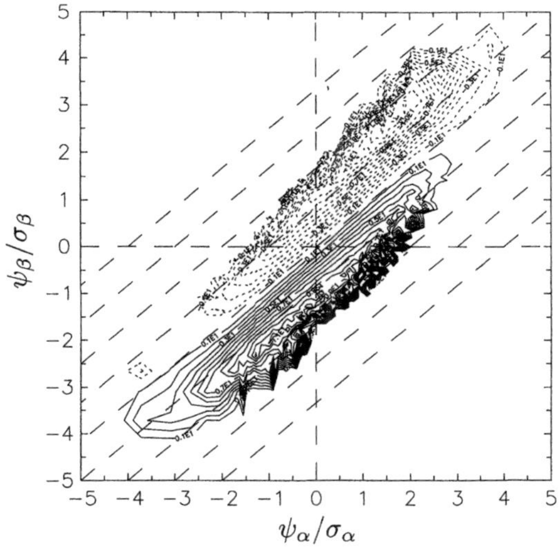

Figure 1: Contours of normalized conditional diffusion of scalar ϕ_α with Schmidt number (Sc) 1/4 given both scalars ($Sc_\alpha = 1/4, Sc_\beta = 1$) normalized by r.m.s values, from 256^3 simulation at R_λ 160. Dashed lines represent slope given by a linear regression model.

From a statistical point of view, for a known velocity field the joint PDF of the scalars may be considered to contain a complete (one-point, Eulerian) description of multi-scalar mixing. For scalar fluctuations driven by a mean gradient of magnitude G the transport equation for the joint PDF of a set of scalars $\underline{\phi} = (\phi_1, \phi_2, ...\phi_n)$ is given by

$$\frac{\partial f_{\underline{\phi}}}{\partial t} = G \sum_{\alpha} \frac{\partial}{\partial \psi_\alpha} [\langle u_1 | \underline{\psi} \rangle f_{\underline{\phi}}] - \sum_{\alpha} \frac{\partial}{\partial \psi_\alpha} [\langle D_\alpha \nabla^2 \phi_\alpha | \underline{\psi} \rangle f_{\underline{\phi}}] \,, \tag{1}$$

where summation is taken all the scalars, and $\underline{\psi}$ is the sample space variable corresponding to the "composition vector" ($\underline{\phi}$). The fundamental quantity to model is the conditional expectation of the molecular diffusion of each scalar given all the scalars. Figure 1 shows contours of $\langle D_\alpha \nabla^2 \phi_\alpha | \psi_\alpha, \psi_\beta \rangle$ in a normalized form that is seen to compare well with a linear regression model (Yeung 1998). However, sampling limitations inherent in conditional averages are also evident in low-probability regions (away from the diagonal) where relatively few conditional samples are available.

Whereas the agreement demonstrated in Fig. 1 suggests that a linear regression approach may be viable for the modeling of conditional diffusion, extension to more general situations is still difficult. In particular the parameters required in the model include correlation coefficients between multiple scalars and their corresponding gradients, which may themselves be difficult to predict. Furthermore, consideration of more than two scalars or of the full velocity-scalar joint PDF would require the use of additional conditioning variables in sample space. The latter is expected to further aggravate data analysis difficulties (Fox 1996) in the statistical estimation of conditional quantities.

3 Lagrangian descriptions

In general, the set of independent variables in an Eulerian PDF consists of time, three space coordinates as well as the velocity components and scalars in sample space. In practice, and especially for inhomogeneous flows, the modeled equations are usually solved in a Lagrangian Monte-Carlo formulation. Essentially, the fluid medium is replaced by a collection of discrete fluid particles, and hence we are interested in the evolution of Lagrangian time series of the velocity, scalar or other flow properties evaluated along the particle trajectories. Here a problematic issue for scalar transport is the development of stochastic mixing models, many of which currently exhibit significant limitations (see Pope 1994). Direct testing of these models requires detailed knowledge of the Lagrangian statistics of scalar fluctuations (following fluid particle paths), which are currently little known from either DNS or experiment.

In the present simulations the instantaneous Lagrangian scalar value $\tilde{\phi}^+(t)$ can be written as the sum of two contributions: one due to the scalar fluctuation field evaluated at the instantaneous fluid particle position (as $\phi^+(t)$) and the other

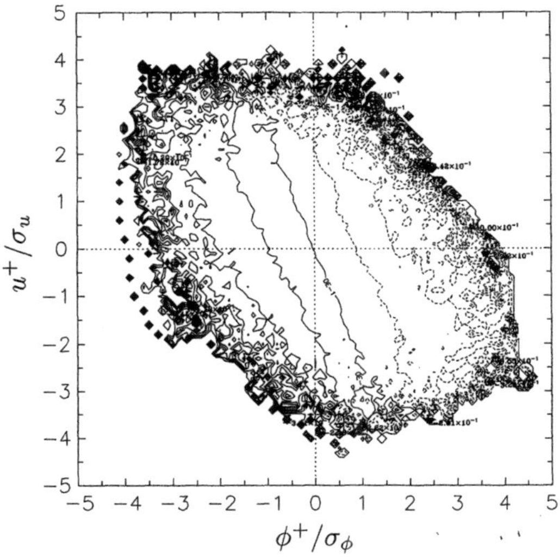

Figure 2: Contours of small-time Lagrangian scalar increment $[\Delta_\tau \tilde{\phi}^+(t)]$ conditioned on the scalar ($Sc = 1$) and velocity. Data at R_λ 90, 128^3.

due to the mean scalar field evaluated also at the particle location. The rate of change of $\tilde{\phi}^+(t)$ is, of course, the material derivative

$$\frac{D\tilde{\phi}}{Dt} = D_\alpha \nabla^2 \phi \; . \tag{2}$$

Consequently, the rate of molecular diffusion is again an important quantity, especially since most practical stochastic models are Markovian—that is, aimed at attempting to predict the "future" value $\tilde{\phi}^+(t+\tau)$ (with $\tau > 0$) using only information from the "current" state $\tilde{\phi}^+(t)$.

From the discussion above, an important quantity for stochastic modeling is the Lagrangian conditional expectation of the diffusion, namely $\langle D_\phi \nabla^2 \phi^+ | \phi^+ \rangle$, where derivatives are taken in space but evaluated at the current particle position. Whereas the averaging is formally taken over an ensemble of fluid particles, for homogeneous turbulence this is the same as Eulerian averaging since particle positions then represent an unbiased sample of observation points. It is important, however, to bear in mind that the dominance of advective transport in the Lagrangian view implies that (Fox 1996, Pope 1998) conditioning should be performed on the fluid particle velocity $u^+(t)$ as well. Effectively this also implies that a fully consistent probabilistic description of scalar transport requires the use of the full velocity-scalar joint PDF equations.

In Figure 2 we present contours of the conditional average of the temporal increment $\Delta_\tau \tilde{\phi}^+(t) \equiv \tilde{\phi}^+(t+\tau) - \tilde{\phi}^+(t)$ of a scalar with Schmidt number unity, over a time interval which is less than the Kolmogorov time scale, and conditioned

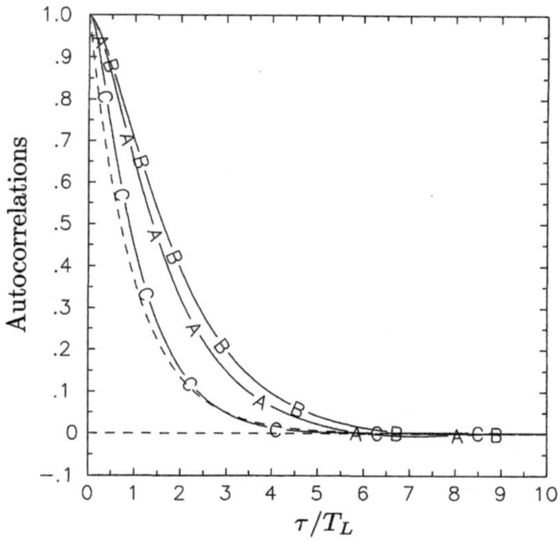

Figure 3: Lagrangian autocorrelations of scalar fluctuation (ϕ^+) with $Sc = 1/8$ (A), 1/4 (B) and their difference (C), compared with exponential approximation for velocity autocorrelation (dashed curve). Data at R_λ 90, 128^3.

on the scalar itself and the velocity component in the mean gradient direction. For reasons stated above this is approximately equivalent to the Eulerian conditional diffusion given the scalar and velocity, which has been studied by Overholt & Pope (1996). A linear dependence on both conditioning variables appears plausible in the interior of the sampling envelope (which is relatively free of statistical noise).

In addition to the temporal increment, information on Lagrangian time scales for scalars and scalar dissipation is important for stochastic modeling (Fox 1997). If a given fluid particle property is statistically stationary in time, then in a Markovian model the Lagrangian autocorrelation between property values at times t and $t + \tau$ would be completely specified by the integral time scale (which is the area under the autocorrelation curve). Figure 3 shows the autocorrelations of scalar fluctuations (ϕ^+) at different Schmidt numbers and of their difference. For comparison purposes, the time lag is normalized by the Lagrangian integral time scale (T_L), and the velocity autocorrelation is represented approximately (except at small times) by the exponential form $\exp(-\tau/T_L)$. It is clearly seen that, especially at higher Schmidt numbers, the scalars are correlated for longer times than for the velocity. On the other hand, the autocorrelation for the difference apparently decays faster than the autocorrelations of both scalars. This is perhaps not a surprise, since the difference $\phi_\alpha - \phi_\beta$ is driven by processes at smaller scales. The difference autocorrelation appears to be remarkably close to that of the velocity, but further results at other Reynolds numbers are needed to ascertain whether this is a general feature.

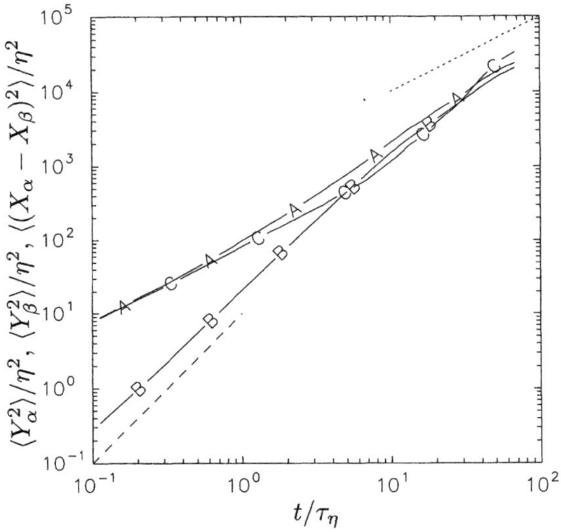

Figure 4: Growth of mean-square displacement of molecules with $Sc = 0.02618$ (A), 2.618 (B), and inter-species dispersion (C), normalized by Kolmogorov variables. Dashed lines show asymptotic slopes of 2 and 1. Data at R_λ 90, 128^3.

In the study of turbulent dispersion it is usual to take a Lagrangian viewpoint, based on the idea that passive contaminants are dispersed primarily by being carried along with the fluid velocity fluctuations. However, it is also important (Sawford & Hunt 1986) to recognize the fact that the contaminant material does diffuse relative to the fluid (and hence do not follow fluid particle paths exactly), at a rate determined by the molecular diffusivity. Both DNS and stochastic modeling results have recently been obtained and compared (Yeung & Borgas 1997), for the statistics of molecular path displacements, which are consistent with the small-time theoretical result by Saffman (1960) but is applicable for arbitrary diffusion times and diffusivity values.

In the molecular path description certain aspects of differential diffusion can be studied in terms of the dispersion between species molecules which are initially coincident but subsequently diffuse relative to the fluid (in a Brownian manner) at different rates. Figure 4 shows such a result, where the mean-squared separation $\langle (X_\alpha - X_\beta)^2\rangle$ between different species is compared with individual-species mean-squared displacements ($\langle Y_\alpha^2\rangle$, $\langle Y_\beta^2\rangle$). It may be seen that (depending on the Reynolds number), whereas molecular diffusivity effects on displacements at large times are fairly weak, the dispersion between species molecules of different diffusivities is of considerable magnitude. It is expected that the theoretical description of these effects will require a "cross-species" generalization of the "substance autocorrelation" concept introduced by Saffman (1960).

4 Concluding remarks

In summary, we stress that multi-scalar mixing in turbulence presents a number of problematic issues which remain to be addressed satisfactorily in the literature. Understanding of small-scale processes is important, in part because molecular diffusivity differences remain important in the smallest scales even at high Reynolds number, and in part because (for example) many combustion phenomena are sensitive to small-scale mixing. Lagrangian approaches are seen to play an important role, but progress in stochastic modeling is often limited by extreme and well-known difficulties in Lagrangian measurements in the laboratory. Recent advances in parallel computing have improved the prospects of extracting Lagrangian data from DNS at Reynolds numbers that are just sufficient to capture important characteristics of high Reynolds number turbulence. However, sampling difficulties for multi-scalar conditional averages are great, and are even more so for inhomogeneous flows.

This work is supported by NSF Grant CTS-9705678, with computer time provided by Cornell Theory Center and San Diego Supercomputing Center. Helpful discussions with R.O. Fox and M.S. Borgas are also acknowledged.

References

BILGER, R.W. (1993) Conditional moment closure for turbulent reacting flow. *Phys. Fluids A* **4**, 436–444.

BORGAS, M.S. & SAWFORD, B.L. (1996) Molecular diffusion and viscous effects on concentration statistics in grid turbulence. *J. Fluid Mech.* **324**, 25–54.

FOX, R.O. (1996) On velocity-conditioned scalar mixing in homogeneous turbulence. *Phys Fluids* **8**, 2678–2691.

FOX, R.O. (1997) The Lagrangian spectral relaxation model of scalar dissipation in homogeneous turbulence. *Phys Fluids* **9**, 2364–2386.

MCCOMB, W.D. (1990) *The Physics of Fluid Turbulence*, Clarendon Press, Oxford, U.K.

OVERHOLT, M.R. & POPE, S.B. (1996) Direct numerical simulation of a passive scalar with imposed mean gradient in isotropic turbulence. *Phys Fluids* **8**, 3128–3148.

POPE, S.B. (1985) PDF methods for turbulent reactive flows. *Prog. Energy Combust. Sci* **11**, 119–192.

POPE, S.B. (1994) Lagrangian pdf methods for turbulent flows. *Ann. Rev. Fluid Mech.* **26**, 23–63.

POPE, S.B. (1998) The vanishing effect of molecular diffusivity on turbulent dispersion: implications for turbulent mixing and the scalar flux. *J. Fluid Mech.* (in press).

SAFFMAN, P.G. (1960) On the effect of molecular diffusivity in turbulent diffusion. *J. Fluid Mech.* **8**, 273–283.

SAWFORD, B.L. & HUNT, J.C.R. (1986) Effects of turbulence structure, molecular diffusion and source size on scalar fluctuations in homogeneous turbulence. *J. Fluid Mech.* **165**, 373–400.

YEUNG, P.K. (1996) Multi-scalar spectral transfer in differential diffusion with and without mean scalar gradients. *J. Fluid Mech.* **321**, 235–278.

YEUNG, P.K. (1998) Correlations and conditional statistics of differential diffusion: scalars with uniform mean gradients. Submitted to *Phys. Fluids* (under review).

YEUNG, P.K. & BORGAS, M.S. (1997) Molecular path statistics in turbulence: simulation and modeling. in *Advances in DNS/LES*, eds. C. Liu, Z. Liu & L. Sakell, Greyden Press, Columbus, Ohio, USA. (Proc. First AFOSR International Conference on Direct Numerical Simulation and Large Eddy Simulation Louisiana Tech University, Ruston, LA, August 1997)

Trends in Mathematics, © 1999 Birkhäuser Verlag Basel/Switzerland

Turbulent transport of vector (magnetic) field: intermittency and anomalous scalings

N. Kleeorin, I. Rogachevskii

Abstract

Numerous studies of magnetic fluctuations with a zero mean-field for small magnetic Prandtl numbers ($\mathrm{Pr}_m \ll 1$) show that magnetic fluctuations cannot be generated by turbulent fluid flow with the Kolmogorov energy spectrum. In addition, the generation of magnetic fluctuations with a zero mean-field for $\mathrm{Pr}_m \ll 1$ were not observed in numerical simulations. However, in astrophysical applications the magnetic Prandtl numbers are small and magnetic fluctuations are observed. Thus the mechanism of generation of magnetic fluctuations for $\mathrm{Pr}_m \ll 1$ still remains poorly understood.

The generation of magnetic fluctuations by turbulent flow of a conducting fluid with a zero mean magnetic field for $\mathrm{Pr}_m \ll 1$ is studied. The equation for the high-order correlation functions of the magnetic field is derived. It is shown that the growth rates of the higher moments of the magnetic field are larger than those of the lower moments, i.e., the spatial distribution of the magnetic field is intermittent. The turbulent fluid velocity field is assumed to be homogeneous and isotropic with a power law energy spectrum ($\propto k^{-p}$) and with a very short scale-dependent correlation time. It is found that magnetic fluctuations can be generated when the exponent $p > 3/2$. In addition, the effect of compressibility (i.e., $\nabla \cdot \mathbf{u} \neq 0$) of the low-Mach-number turbulent fluid flow $\mathbf{u}$ is studied. It is shown that the threshold for the generation of magnetic fluctuations by turbulent fluid flow with $\nabla \cdot \mathbf{u} \neq 0$ is higher than that for incompressible fluid. This implies that the compressibility impairs the generation of magnetic fluctuations.

Problems of anomalous scalings for vector (magnetic) field is discussed as well. The anomalous scaling for the second moment of the magnetic field passively advected by an turbulent fluid flow with very short scale-dependent correlation time is found for the case $3/2 < p < 3$. This anomalous scaling has the complex exponent which depends on degree of compressibility of a low-Mach-number turbulent flow. When $1 < p < 3/2$ the anomalous exponent is real.

Nonlinear effects result in saturation of growth of the magnetic fluctuations. Asymptotic properties of the steady state solution for the second moment of the magnetic field in the case of the Hall nonlinearity for the low-Mach-number compressible turbulent flow are studied.

I. Introduction

The generation of magnetic fluctuations by homogeneous, isotropic and incompressible turbulence with zero mean magnetic field was mainly studied for magnetic

Prandtl numbers $\mathrm{Pr}_m \geq 1$ (see, e.g., [1, 2, 3, 4, 5, 6]). However, in astrophysics the magnetic Prandtl numbers is small. On the other hand, generation of magnetic fluctuations with zero mean field for $\mathrm{Pr}_m \ll 1$ were not observed in numerical simulations (see, e.g., [7]). In addition, the turbulent velocity field cannot be considered as a divergence-free in astrophysics (e.g., accretion disks, solar and stellar convection zones, galaxies).

We study here the generation of magnetic fluctuations with zero mean magnetic field for $\mathrm{Pr}_m \ll 1$. Turbulent fluid velocity field is assumed to be homogeneous and isotropic with a very short scale-dependent correlation time. We have found that magnetic fluctuations can be generated by turbulent motions of conducting fluid flow even with Kolmogorov energy spectrum $\mathrm{Pr}_m \ll 1$. Equation for the high-order correlation functions of the magnetic field is derived. It is shown that spatial distribution of the magnetic field is intermittent. In addition, we study the effect of compressibility (i.e., $\nabla \cdot \mathbf{u} \neq 0$) of the low Mach numbers turbulent fluid flow $\mathbf{u}$ on the generation of magnetic fluctuations.

Nonlinear effects (the influence of the Ampere force on the motion of the conducting fluid and the Hall effect) result in saturation of growth of the magnetic field. Asymptotic properties of the steady state solution for the second moment of the magnetic field in the case of the Hall nonlinearity for the low-Mach-number compressible turbulent flow are studied.

II. Mathematical formulation of the problem

Consider dynamics of magnetic fluctuations. The induction equation in a compressible (i.e., $\nabla \cdot \mathbf{v} \neq 0$) fluid flow reads

$$\frac{\partial \mathbf{H}}{\partial t} + (\mathbf{v} \cdot \nabla)\mathbf{H} = (\mathbf{H} \cdot \nabla)\mathbf{v} - \mathbf{H}(\nabla \cdot \mathbf{v}) + \eta \Delta \mathbf{H} \,, \tag{1}$$

where $\eta = c^2/4\pi\sigma$ is the magnetic diffusion, σ is the electrical conductivity, and c is the light speed. We derive equations for the high-order correlation functions of the magnetic field. To this purpose we use the metod of path integrals (see, e.g., [4, 5, 6, 8, 9, 10]). The equation for the second-order correlation function $W(r,t) = \langle h_\tau(\mathbf{x},t) h_\tau(\mathbf{y},t) \rangle$ is given by

$$\frac{\partial W}{\partial t} = \frac{1}{m} W'' + \mu W' - \frac{\kappa}{m} W \,, \tag{2}$$

where h_τ is the projection of magnetic field $\mathbf{h}(t,\mathbf{x}) = \mathbf{H}(t,\mathbf{x}) - \mathbf{B}(t,\mathbf{x})$ on the direction $\mathbf{r} = \mathbf{x} - \mathbf{y}$, and

$$\begin{aligned} \frac{1}{m} &= \frac{2}{\mathrm{Rm}} + \frac{2}{3}[1 - F - (rF_c)'] \,, \\ \mu &= \frac{4}{mr} + \left(\frac{1}{m}\right)' , \quad \kappa = \frac{2m}{r}(f' + 2f_c') \,, \\ f &= F + rF'/3 \,, \quad f_c = F_c + rF_c'/3 \,, \end{aligned}$$

$\mathrm{Rm} = u_0 l_0/\eta \gg 1$ is the magnetic Reynolds number, u_0 is the characteristic velocity in the maximum scale l_0 of turbulent motions, and $F' = dF/dr$. Here we consider a homogeneous, isotropic and reflectionally invariant (with zero mean helicity) compressible turbulent fluid velocity field. In this case the correlation function $\langle \tau u_m u_n \rangle$ is given by

$$\langle \tau u_m(\mathbf{x}) u_n(\mathbf{x}+\mathbf{r}) \rangle = \eta_T \left[[F(r)+F_c(r)]\delta_{mn} + \frac{rF'}{2}\left(\delta_{mn} - \frac{r_m r_n}{r^2}\right) + rF_c' \frac{r_m r_n}{r^2} \right] \quad (3)$$

(for details see [8]), where $F' = dF/dr$, $\eta_T = u_0^2 \tau_0/3$ is the turbulent magnetic diffusion, $\tau_0 = l_0/u_0$, $F(0) = 1 - F_c(0)$. The function $F_c(r)$ describes the compressible (potential) component whereas $F(r)$ corresponds to the vortical part of the turbulence. We used that the second moment of the magnetic field is given by

$$\langle h_m(\mathbf{x}) h_n(\mathbf{x}+\mathbf{r}) \rangle = W(r)\delta_{mn} + \frac{rW'}{2}\left(\delta_{mn} - \frac{r_m r_n}{r^2}\right) \quad (4)$$

This form of the second moment corresponds to the condition $\vec{\nabla} \cdot \mathbf{h} = 0$ and an assumption of the homogeneous and isotropic magnetic fluctuations. Equation (2) is written in dimensionless variables: coordinates and time are measured in the units l_0 and τ_0, the velocity is measured in the units u_0, the magnetic field is measured in the units B_0. By means of Eq. (2) we will study the generation of magnetic fluctuations (see Section III).

The technique of path integrals allows also to derive the equation for the high-order correlation function of the magnetic field [10]. The solution of this equation yields the growth rate of the s-order correlation function of the magnetic field: $\gamma_s = s(s-1)\gamma_2/2$ [10]. This implies that if the second-order correlation function of the magnetic field grows ($\gamma_2 > 0$), than all high-order correlation functions grow. It is shown in Section III that under certain conditions $\gamma_2 > 0$. Note that the higher moments grow faster than the lower moments of magnetic field. Therefore spatial distribution of magnetic fluctuations is intermittent (i.e., $\gamma_s > s\gamma_2/2$). This is in agreement with a dynamo theorem [4, 5, 11].

We use in the present paper the δ-correlated in time random process to describe a turbulent velocity field. The results remain valid also for the velocity field with a finite correlation time, if the high-order correlation functions of the magnetic field vary slowly in comparison with the correlation time of the turbulent velocity field (see, e.g., [4, 5, 12]). We also take into account the dependence of the momentum relaxation time on the scale of turbulent velocity field: $\tau(\mathbf{k}) = \tau_0 (k/k_0)^{1-p}$, where p is the exponent in spectrum of kinetic turbulent energy, k is the wave number, $k_0 = l_0^{-1}$.

III. Generation of Magnetic Fluctuations

In this section we study the generation of magnetic fluctuations in a low-Mach-number compressible turbulent fluid flow for small magnetic Prandtl numbers. A mechanism of amplification of magnetic fluctuations with zero mean magnetic field comprises stretching, twisting and folding of the original loop of a magnetic field (see, e.g., [4, 5]). These non-trivial motions are three-dimensional and result in an amplification of the magnetic field. Magnetic diffusion leads to reconnection of the field at an X-point. The generation of magnetic fluctuations can be described by Eq. (2) for the second moment of magnetic field. We seek a solution of Eq. (2) in the form $W(t,r) = \Psi(r)\sqrt{m}\exp(\gamma t)/r^2$, where the unknown function $\Psi(r)$ is determined by equation

$$\frac{1}{m(r)}\frac{d^2\Psi}{dr^2} - [\gamma + U(r)]\Psi = 0 , \tag{5}$$

and $U(r) = (\chi^2 + 2\chi' + 4\kappa)/4m(r)$, and $\chi(r) = \mu(r)m(r)$. We consider the case of small magnetic Prandtl numbers $\mathrm{Pr}_m = \nu/\eta \ll 1$ which is typical for many astrophysical and geophysical applications (where ν is the kinematic viscosity). The latter allows to consider magnetic fluctuations only in the inertial range of the turbulent velocity field. We choose the following model of turbulence. Incompressible $F(r)$ and compressible $F_c(r)$ components in inertial range of turbulence $r \ll 1$ are given by $F(r) = (1-\varepsilon)(1-r^{q-1})$ and $F_c(r) = \varepsilon(1-\beta r^{q-1})$, where $r_d < r \ll 1$, q is the exponent in spectrum of the function $\langle \tau u_m u_n \rangle$, and $r_d = \mathrm{Re}^{-1/(3-p)}$, p is the exponent in spectrum of kinetic turbulent energy, $\mathrm{Re} = u_0 l_0/\nu \gg 1$ is the Reynolds number. Note that the exponent p in the spectrum of kinetic turbulent energy differs from that of the function $\langle \tau u_m u_n \rangle$ due to the scale dependence of the momentum relaxation time τ of turbulent velocity field $\mathbf{u}$. The relation between p and q is given by $q = 2p-1$ [10]. Note that when $p > 3/2$, the exponent $q > 2$. We will see in this section that the magnetic fluctuations can be generated if $q > 2$.

The solution of Eq. (5) can be obtained using an asymptotic analysis (see, e.g., [4, 5, 6, 8, 9, 10]). There are three characteristic regions. In the region I, i.e., for $0 \le r \le \mathrm{Rm}^{-1/(q-1)}$, the correlation function $W(r)$ is given by $W(r) = A_1 m^{1/2} r^{-3/2} J_\lambda[(2\lambda/3)\sqrt{\kappa_0 \mathrm{Rm}}\, r^{3/2\lambda}]$, where $1/m(r) \sim 2[1+\beta_m \mathrm{Rm}\, r^{q-1}]/\mathrm{Rm}$, and $\lambda = 3/(q-1)$, and J_λ is the Bessel function of the first kind, and $\beta_m = (1-\varepsilon)(1+q\sigma)/3$, and $\sigma = \varepsilon\beta/(1-\varepsilon)$, and $\kappa_0 = (1-\varepsilon)(1+2\sigma)(2+q)(q-1)/3$. Note that the correlation function $W(r)$ for $r \ll \mathrm{Rm}^{-1/(q-1)}$ is given by $W(r) = 1 - \beta_0 \mathrm{Rm}\, r^{q-1}$, where $\beta_0 = \beta_m + \kappa_0(q-1)/(q+2)$.

In the region II, i.e., for $\mathrm{Rm}^{-1/(q-1)} \ll r \ll 1$ the correlation function $W(r)$ is given by

$$W(r) = A_2 m^{1/2} r^{-3/2} \cos(b \ln r + \varphi_0) . \tag{6}$$

$$b^2 = \left(\frac{q^2-4}{4}\right)\left(\frac{3+\sigma(4-q)}{1+q\sigma}\right) , \tag{7}$$

where $1/m(r) \sim 2\beta_m r^{q-1}[1 + (\beta_m \mathrm{Rm}\, r^{q-1})^{-1}]$. In the region III ($r \gg 1$) the correlation function $W(r)$ is given by $W(r) = A_3 r^{-2}(\sqrt{|\gamma|} + r^{-1})\exp(-\sqrt{|\gamma|}r)$. Matching functions $W(r)$ and $W'(r)$ at the boundaries of these regions yields the constant A_k, φ_0 and the growth (or damping) rate $\gamma_2 \equiv \gamma$ of the magnetic fluctuations. The latter is given by

$$\gamma \simeq \frac{4b^2 + (4-q)^2}{4(q-1)} \ln\left(\frac{\mathrm{Rm}}{\mathrm{Rm}^{\mathrm{cr}}}\right) ,$$

where the critical magnetic Reynolds number $\mathrm{Rm}^{\mathrm{cr}}$ is given by

$$\mathrm{Rm}^{(\mathrm{cr})} \simeq \exp\left[\frac{q-1}{b}\left(\pi k + \arctan\frac{4-q}{2b} + \arctan S\right)\right] , \tag{8}$$

and

$$S = \frac{\sqrt{\kappa_0}}{b} \frac{J'_\lambda[2\lambda\sqrt{\kappa_0}/3]}{J_\lambda[2\lambda\sqrt{\kappa_0}/3]} ,$$

and $k = 1, 2, 3, \ldots$ This analysis shows that the characteristic scales of localization of the magnetic fluctuations is of order

$$l_f \sim l_0 \mathrm{Rm}^{-\frac{1}{2(p-1)}} \exp\left[\frac{1}{b}(\arctan S + \pi n + \pi/2)\right] , \tag{9}$$

where $n \leq k$.

The critical magnetic Reynolds strongly depends on the parameter compressibility σ. Indeed, for incompressible fluid $\sigma = 0$ the critical magnetic Reynolds number $\mathrm{Rm}^{(\mathrm{cr})} = 412$. For compressible fluid flow, i.e., $\sigma = 0.1$ the value $\mathrm{Rm}^{(\mathrm{cr})} = 740$. For larger parameter of compressibility the critical magnetic Reynolds number increases sharply. For $\sigma \gg 1$ the value $\mathrm{Rm}^{(\mathrm{cr})} \to 10^6$. This implies that the compressibility impairs generation of magnetic fluctuations. Here we use $p = 5/3$ (Kolmogorov turbulence) and $\beta = 1$.

Now we discuss the effect of compressibility on the generation of magnetic fluctuations. In compressible ideal conducting fluid flow the vector $\mathbf{H}/\rho$ is frozen into motion of fluid. In incompressible flow at any time mass of fluid flowing into a small volume exactly equals to a mass outflow from this volume. On the other hand, in compressible flow ($\nabla \cdot \mathbf{u} \neq 0$) a mass of fluid flowing into a small volume does not equal to a mass outflow from the volume at any instance. Therefore at times smaller than a characteristic time of the turbulent velocity field the fluid density ρ and the magnetic field $\mathbf{H}$ increase (or decrease) when $\nabla \cdot \mathbf{u} < 0$ (or $\nabla \cdot \mathbf{u} > 0$). Note that the increase and decrease of the magnetic field in a small control volume are separated in time and small molecular magnetic diffusion breaks a reversibility in time. The latter may cause an additional increase $\langle \mathbf{h}^2 \rangle$ caused by the compressibility of fluid flow.

On the other hand, the compressibility affects on the turbulent magnetic diffusion as well since it increases the turbulent magnetic diffusion in small scales. When $q > 2$ the increase of the turbulent magnetic diffusion is stronger than the increase of the growth of the magnetic fluctuations due to the compressibility of the fluid flow (see below). Note also the scale of localization of magnetic fluctuations increases when the degree of the compressibility σ increases [see Eq. (9)]. On the other hand, the scale-dependent turbulent magnetic diffusion increases with the increase of scale.

The generation of magnetic fluctuations depends on the form of solution for the correlation function $W(r)$ in the second region $\mathrm{Rm}^{-1/(q-1)} \ll r \ll 1$ (i.e., in the inertial range of the turbulence). When $b^2 > 0$ magnetic fluctuations are excited. It follows from Eq. (7) that $b_c^2 - b_{\rm inc}^2 = -\sigma(q^2-4)(q-1)$, where $b_{\rm inc} = b(\sigma = 0)$ is the value of parameter b for incompressible fluid flow, and $b_c = b(\sigma > 0)$ is the value of parameter b for compressible fluid flow. It is seen from here that for $q > 2$ (i.e., $b^2 > 0$) we obtain that $0 < b_c^2 < b_{\rm inc}^2$. The latter implies that the compressibility impairs generation of magnetic fluctuations for $q > 2$ [see Eqs. (7) and (8)].

Now we discuss effect of the exponent of the energy spectrum of the turbulent velocity field on the generation of magnetic fluctuations. The solution (6) for the correlation function $W(r)$ is valid for $b^2 > 0$ (i.e., for $q > 2$). In this case ($q > 2$) magnetic fluctuations can be excited. Since $q = 2p - 1$ the necessary condition for the excitation of magnetic fluctuations is $p > 3/2$. On the other hand, when $p < 3/2$ (i.e., for $q < 2$) the solution for $W(r)$ is given by $W(r) = m^{1/2} r^{-3/2}(A_2 r^{-b} + A_5 r^b)$. In this case magnetic fluctuations are not excited.

Note that magnetic fluctuations in a delta-correlated in time incompressible turbulent fluid flow were studied in [2] for $\mathrm{Pr}_m \ll 1$. In the latter model the correlation time is assumed to be independent of the scale of turbulent motions and therefore $q = p$ and the necessary condition for the excitation of magnetic fluctuations is $p > 2$. The Kolmogorov turbulence with $p = 5/3$ and for $\mathrm{Pr}_m \ll 1$ cannot generate magnetic fluctuations in the model with a scale-independent correlation time of the turbulent velocity field [2]. On the other hand, in the model with a scale-dependent correlation time of the turbulent velocity field considered in the present study the magnetic fluctuations are excited for $p > 3/2$.

Remarkably, a spectrum of magnetic fluctuations $\propto k^{-3/2}$ (the Kraichnan spectrum) occur when magnetic energy equals hydrodynamic energy $M(k) = E(k) \propto k^{-3/2}$. Here $M(k)$ and $E(k)$ spectral functions of magnetic and hydrodynamic energy, respectively. We have shown here that if the energy spectrum of hydrodynamic turbulent motions is more steep than $k^{-3/2}$-spectrum, the magnetic fluctuations can be generated.

Note that the condition of the validity of the assumption of the very short correlated velocity field is as follows. The characteristic time of variations of the magnetic fluctuations $\tau_B \sim \gamma_s^{-1}$ is very large in comparison with the momentum relaxation time τ of the velocity field in the scale l_f. The scale l_f determines the conditions of generation and localization of the magnetic fluctuations. This allows us to estimate the maximum number s of the high-order correlation function of the magnetic field:

$$s < \mathrm{Rm}^{\frac{1}{2}} \Big/ \left| \ln\left(\frac{\mathrm{Rm}}{\mathrm{Rm}^{\rm cr}}\right)\right|^{\frac{1}{2}}, \tag{10}$$

when the theory is valid. For the Kolmogorov turbulence ($p = 5/3$) and $\mathrm{Rm} = 10^6$ (e.g., the convection zone of the sun) we obtain that $s < 10^3$.

IV. Anomalous scaling

Problems of anomalous scalings for vector (magnetic) and scalar (particles number density or temperature) fields passively advected by an turbulent fluid flow are a subject of active research in the last years (see, e.g., [13, 14, 15, 10], and references therein). The anomalous scaling means the deviation of the scaling exponents of the correlation function of a vector (scalar) field from their values obtained by the dimensional analysis. For incompressible turbulent flow the anomalous scalings for scalar field can occur beginning only with a forth-order correlation function (see e.g., [14]) while for the vector (magnetic) field the anomalous scalings appear already in the second moment [15].

The anomalous scalings of the magnetic fluctuations were considered in [15] using the model of the delta-correlated in time turbulent velocity field. The correlation time in this model is independent of the scale of turbulent motions. Note that in this model the generation of magnetic fluctuations is possible only for $q = p > 2$. Taking into account the dependence of the correlation time on the scale of turbulent motions we have shown (see Section III) that the magnetic fluctuations can be generated for $3/2 < p < 3$.

Now we discuss the anomalous scaling for the model with the scale-dependent correlation time of turbulent motions. Consider the case when the magnetic Reynolds number $\mathrm{Rm} < \mathrm{Rm}^{\mathrm{cr}}$ and the magnetic fluctuations are caused by an external source. The condition $\mathrm{Rm} < \mathrm{Rm}^{\mathrm{cr}}$ implies that there is no self-excitation (i.e., exponential growth) of the magnetic fluctuations (see Section III). Solution for the correlation function $W(r)$ in the inertial range is given by $W(r) \sim r^{-(2p+1)/2} \cos(b \ln r + \varphi_0)$ [see Eq. (6)]. This corresponds to the anomalous scaling of the magnetic fluctuations. The normal scaling for the second moment of the magnetic fluctuations is given by $W(r) \sim r^{4-2p} \propto r^{3-q}$, where $q = 2p - 1$.

The general solution of the equation with an external source for the correlation function W includes the solution describing the anomalous $W(r) \sim r^{-(2p+1)/2} \cos(b \ln r + \varphi_0)$ and normal $W(r) \propto r^{4-2p}$ scalings. The obtained anomalous scaling $W(r) \sim r^{-(2p+1)/2} \cos(b \ln r + \varphi_0)$ can be presented as the real part of the power-law function r^{ζ} with the complex exponent $\zeta = -1/2 - p - ib$. This anomalous scaling is significantly different from the normal scaling $W(r) \propto r^{4-2p}$, and corresponds to the deviation from the condition of the constant flux of magnetic fluctuations over the spectrum. The obtained anomalous scaling is valid when the exponent of the energy spectrum of the turbulent velocity field is $3/2 < p < 3$. When $1 < p < 3/2$ the anomalous exponent in a low-Mach-number compressible turbulent flow is real, i.e., $\zeta = -1/2 - p + |b(\sigma, q)|$. In the case of incompressible turbulent flow ($\sigma = 0$) this result coincides with that obtained in [15].

V. Nonlinear analysis

Nonlinear effects results in saturation of growth of the magnetic field. There are two kind of nonlinearities: the influence of the Ampere force on the motion of the conducting fluid and the Hall effect. The Hall nonlinearity is dominated if $\Omega_i \tau_0 \ll 1$, where $\Omega_i = e h_0/(m_i c)$ is the ion gyrofrequency, e and m_i are the charge and mass of ion (see, e.g., [6, 16]). The induction equation with taking into account the Hall effect is given by

$$\frac{\partial \mathbf{h}}{\partial t} = \nabla\times[\mathbf{u}\times\mathbf{h} + \Lambda\,\mathbf{h}\times(\nabla\times\mathbf{h})] + \mathrm{Rm}^{-1}\Delta\mathbf{h} ,$$

where the magnetic field h is measured in units of $h_0 = u_0(4\pi m_i n)^{1/2}$, and n is the number density of ions, and $\Lambda = (\Omega_i \tau_0)^{-1}$. The equation for the function Ψ with taking into account the Hall effect is given by

$$\frac{\partial \Psi}{\partial t} = \frac{1}{m_N}\Psi'' - U_N \Psi ,$$

where $m_N^{-1} = m^{-1}(r) + 2(N - N_*)$, the nonlinearity N is defined by $N = W'' + 4W'/r$ and $U_N = U(m_N) - (r^3 N)'/(3r^2)$, and the formula for $U(m_N)$ is obtained from that for the potential of the linear problem U after the change $m \to m_N$. The nonlinear term N is due to the Hall effect. The function N corresponds to a correlation function of the electric current: $N = -\langle j(t,\mathbf{x}) j(t,\mathbf{y})\rangle$, where j is the projection of the electric current $\mathbf{j}$ on the direction $\mathbf{r}$ and is measured in the units of enu_0.

Near the threshold of self-excitation of the magnetic fluctuations, the system tends to a steady state after several correlation times. This is also in agreement with the direct 3D numerical simulation of the magnetic field generation by a prescribed ABC flow of the conducting fluids (see [16]). In particular, the Hall nonlinearity results in a saturation of the growth of the magnetic field.

Now we study asymptotic properties of the steady state solution for the second moment of the magnetic field in the case of the Hall nonlinearity for the low-Mach-number compressible flow. For simplicity the case $q = 3$ is considered. The correlation function W satisfys the following boundary conditions: $W(r=0) = W_*$ and $W(r=\infty) = 0$. Let us study the properties of the solution for small $r \ll 1$. The function Ψ and the nonlinearity N in this range can be reduced to $\Psi \sim \Psi_d r^2 + O(r^4)$, and $N \sim N_* + (N_d - \beta_m) r^2 + O(r^4)$. The boundary condition $W(r=0) = W_*$ and the definition $\Psi = r^2 W/\sqrt{m}$ yield $\Psi_d = W_*\sqrt{2/\mathrm{Rm}}$. Analysis in the vicinity of $r = 0$ provides a relationship between N_* and W_* : $N_* = -5\mathrm{Rm} W_* N_d$. Note that for $r \ll 1$ the function $m_N^{-1} \sim 2/\mathrm{Rm} + 2N_d r^2$.

The asymptotic analysis yields the correlation function W in three regions: when $0 \le r \le \mathrm{Rm}^{-1/2}$ the function W is given by $W \sim W_*(1 - \mathrm{Rm} N_d r^2/2)$; when $\mathrm{Rm}^{-1/2} < r < 1$ the function W is given by $W(r) = C_2 m_N^{1/2} r^{-3/2}\cos(b_N \ln r + \varphi_N)$; and when $r \gg 1$ the function W is given by $W \sim C_3/r^3$, where $b_N^2 = b^2(\beta_m/N_d) + 55(\beta_m/N_d - 1)/12$. Matching functions $W(r)$ and $W'(r)$ at the boundaries of these regions yields the constant C_k, φ_N and N_d. The latter is given by $N_d = \beta_m(1 + 12b^2/55)/(1+s)$, where $s = 36(\pi k + g)^2/[55\ln^2(\mathrm{Rm})]$, and

$$g = \arctan\frac{1}{2b_N} + \arctan\frac{(2N_d^2 - 3N_d - 3)}{2b_N(1+N_d)} .$$

The nonlinear solution is much more strongly localized than the linear one. This is a result of the strong dependence of the mass m on the distance r in the nonlinear case. In the vicinity of $r \sim \mathrm{Rm}^{-1/2}$ the mass drastically decreases from $\mathrm{Rm} \gg 1$ to the value $\mathrm{Rm}/|N_*|$. For $r \gg 1$ the mass is of the order of $|N_*|^{-1} \ll 1$. The sharp drop of the mass in the nonlinear case determines the strong localization of the correlation function W.

VI. Discussion

In this study we have shown that the magnetic fluctuations can be generated for small magnetic Prandtl numbers which occur in many astrophysical and geophysical applications. We have studied the effect of compressibility of the low Mach numbers turbulent fluid flow. The compressibility impairs generation of magnetic fluctuations (i.e., the threshold for generation of magnetic fluctuations by turbulent fluid flow with div $\mathbf{u} \neq 0$ is higher than that for the case of divergence-free fluid flow). The reason is that the compressibility results in increase of the scale of localization of the magnetic fluctuations. In larger scales the scale-dependent turbulent magnetic diffusion is larger. However, the contribution of the compressibility to the generation of the magnetic fluctuations $\propto (\vec{\nabla} \cdot \mathbf{u})^2$ is smaller in the larger scales. This results in increase of the threshold for generation of magnetic fluctuations in a compressible fluid flow.

The model of very short scale-dependent correlation time of the turbulent velocity field is considered. It is shown that the magnetic fluctuations can be generated even in the Kolmogorov turbulence (i.e., for $p = 5/3$). It is shown that the growth rates of the higher moments of the magnetic field is higher than those of the lower moments, i.e., the spatial distribution of the magnetic field is intermittent.

The presented analysis explains why in the numerical simulations by [7] the magnetic fluctuations with zero mean magnetic field were not generated. The parameters in the numerical simulation [7] with zero mean field ($\mathbf{B} = 0$) are $\mathrm{Rm} \leq 200$, $\sigma = 0.01$. We have shown that even for incompressible turbulent fluid flow ($\sigma = 0$) the threshold of the excitation of the magnetic fluctuations $\mathrm{Rm}^{\mathrm{cr}} = 412$ in the case of $\mathrm{Pr}_m \leq 1$. Thus the magnetic fluctuations cannot be generated for the parameters used in [7]. Note that the value $\mathrm{Rm}^{\mathrm{cr}} = 412$ also cannot be achieved in laboratory experiments. On the other hand, in astrophysical conditions $\mathrm{Rm} \gg \mathrm{Rm}^{\mathrm{cr}}$ and $\mathrm{Pr}_m \ll 1$, and therefore the magnetic fluctuations can be generated in the astrophysical conditions.

References

[1] R. H. Kraichnan, and S. Nagarajan, Phys. Fluids, **10**, 859 (1967).

[2] A. P. Kazantsev, Sov. Phys. JETP, **26**, 1031 (1968).

[3] V. G. Novikov, A. A. Ruzmaikin and D. D. Sokoloff, Sov. Phys. JETP. **58**, 527 (1983).

[4] Ya. B. Zeldovich, S. A. Molchanov, A. A. Ruzmaikin and D. D. Sokoloff, Sov. Sci. Rev. C. Math Phys. **7**, 1 (1988).

[5] Ya. B. Zeldovich, A. A. Ruzmaikin, and D. D. Sokoloff, *The Almighty Chance* (Word Scientific Publ., London, 1990), and references therein.

[6] N. Kleeorin and I. Rogachevskii, Phys. Rev. E **50**, 493 (1994); N. Kleeorin, I. Rogachevskii, and A. Eviatar, J. Geophys. Res. A. **99**, 6475 (1994).

[7] A. Brandenburg, R. L. Jennings, A. Nordlund, M. Rieutord, R. F. Stein and I. Tuominen, J. Fluid Mech. **306**, 325 (1996).

[8] T. Elperin, N. Kleeorin and I. Rogachevskii, Phys. Rev. E **52**, 2617 (1995); Phys. Rev. E **55**, 2713 (1997).

[9] T. Elperin, N. Kleeorin and I. Rogachevskii, Phys. Rev. Lett. **77**, 5373 (1996); **76**, 224 (1996).

[10] I. Rogachevskii and N. Kleeorin, Phys. Rev. E **56**, 417 (1997).

[11] S. A. Molchanov, A. A. Ruzmaikin and D. D. Sokoloff, Geophys. Astrophys. Fluid Dyn. **30**, 242 (1984).

[12] P. Dittrich, S.A. Molchanov, A.A. Ruzmaikin and D.D. Sokoloff, Astron. Nachr. **305**, 119 (1984).

[13] R. Kraichnan, Phys. Rev. Lett. **72**, 1016 (1994).

[14] M. Chertkov, G. Falkovich, I. Kolokolov and V. Lebedev, Phys. Rev. E **52**, 4924 (1995).

[15] M. Vergassola, Phys. Rev. E **53**, 3021 (1996).

[16] B. Galanti, N. Kleeorin and I. Rogachevskii, Physics of Plasmas **1**, 3843 (1994); **2**, 4161 (1995).

Trends in Mathematics, © 1999 Birkhäuser Verlag Basel/Switzerland

Intermittency and anomalous scalings of inertial particles distribution in turbulent flow

T. Elperin, N. Kleeorin, I. Rogachevskii

Abstract

Small-scale fluctuations of the number density of inertial particles in turbulent fluid flow are studied. Anomalous scaling for the second-order correlation function of the number density of inertial particles is found. The mechanism for the anomalous scaling is associated with the inertia of particles which results in a divergent velocity field of particles. The anomalous scaling appears already in the second moment when the degree of compressibility $\sigma > 1/27$ (where σ is the ratio of the energies in compressible and incompressible component of the particles velocity). It is also demonstrated that inertia of particles in a homogeneous and isotropic turbulent fluid flow causes a self-excitation (exponential growth) of fluctuations of particles concentration. The growth rates of the higher moments of particle concentration is higher than those of the lower moments, i.e., the particle spatial distribution is intermittent.

I. Introduction

Turbulent transport of passive scalar (e.g., number density of particles) advected by an incompressible fluid flow in the case $\mathbf{U} = \mathbf{v}$ was studied in a large number of publications. Here $\mathbf{U}$ is the particle velocity and $\mathbf{v}$ is the velocity of surrounding fluid.

Interesting features in turbulent transport of passive scalar appear when $\mathbf{U} \neq \mathbf{v}$. In this case new phenomena, e.g., turbulent thermal diffusion [1], turbulent barodiffusion [2], and self-excitation, i.e., exponential growth of fluctuations of the number density of inertial particles [3] occur. These effects are caused by inertia of particles which results in a divergent velocity field of particles. The self-excitation of fluctuations of the number density of particles results in the intermittency in spatial distribution of inertial particles [3].

There are two types of intermittency of passive scalar distribution: the intermittency in the systems with and without external pumping. Intermittency in the systems without external pumping was predicted by Zeldovich, Molchanov, Ruzmaikin and Sokoloff in 1985 (see, e.g., [4, 5, 6]). In these systems under certain conditions there is a self-excitation of fluctuations of passive scalar or vector (magnetic) fields. The growth rate γ_s of the s-order correlation function of passive scalar (see [3]) and vector (see [6, 7]) fields is given by $\gamma_s = s(s-1)\gamma_2/2$ [where γ_2 is the growth rate of the second-order correlation function of passive scalar or vector (magnetic) fields]. This implies that when $\gamma_2 > 0$ (i.e., fluctuations of passive fields are excited), higher moments grow faster than lower moments, i.e., $\gamma_s > \gamma_{s-1}$ and $\gamma_s > s\gamma_2/2$. This results in intermittency, i.e., the appearance of sharp peaks in which the main part of the field intensity is concentrated. Passive scalar fluctuations (e.g., fluctuations of the particles number density) can be excited when $\mathbf{U} \neq \mathbf{v}$ (e.g., for inertial particles) or when $\mathbf{U} = \mathbf{v}$ but div $\mathbf{v} \neq 0$ (e.g., for a low-Mach-number compressible turbulent fluid flow) [3].

Intermittency in the systems with external pumping was predicted by Kraichnan in 1994 (see [8]). Here the fluctuations of passive scalar are sustained by an external source. In these systems a problem of anomalous scaling arises. The anomalous scaling means the deviation of the scaling exponents of the correlation function of a passive field from their values obtained by the dimensional analysis. For incompressible turbulent flow and when $\mathbf{U} = \mathbf{v}$ the anomalous scalings for scalar field can occur only for a forth-order correlation function (see e.g., [9, 10, 11]) while for the vector (magnetic) field the anomalous scalings appear already in the second moment [7, 12].

In the present paper we show that the anomalous scalings appear already in the second-order correlation function of the number density of inertial particles when the degree of compressibility $\sigma > 1/27$ (where σ is the ratio of the energies in compressible and incompressible component of the particles velocity). In this case there is no self-excitation of fluctuations of the number density of inertial particles and these fluctuations are maintained by an external source.

Which type of intermittency can occur in a system? It depends on Reynolds number in the case of inertial particles and on magnetic Reynolds number in the case of magnetic fluctuations. When Reynolds number is larger than a certain critical value the first type of intermittency (without external pumping) occurs [3]. On the other hand, when Reynolds number is smaller than the certain critical value the second type of intermittency (with external pumping) appears.

II. Small-scale fluctuations

Number density $n_p(t, \mathbf{r})$ of small particles in a turbulent flow is determined by the equation:

$$\frac{\partial n_p}{\partial t} + \vec{\nabla} \cdot (n_p \mathbf{U}) = D \Delta n_p \ , \tag{1}$$

where $\mathbf{U}$ is a random velocity field of the particles which they acquire in a turbulent fluid velocity field, D is the coefficient of molecular diffusion. We consider the case of large Reynolds and Peclet numbers.

To study the fluctuations of inertial particles concentration we derive equation for the second-order correlation function of particles concentration. For this purpose we use a method of path integrals (Feynman-Kac formula) which has been previously employed in magnetohydrodynamics [5, 6, 7, 13] and in the problems of passive scalar transport in incompressible [5, 6] and compressible [1, 2, 3, 14] turbulent flows. The use of this technique allows to derive an equation for the second-order correlation function $\Phi = \langle \Theta(\mathbf{x})\Theta(\mathbf{y}) \rangle$:

$$\frac{\partial \Phi}{\partial t} = -2T_{mn}\frac{\partial^2 \Phi}{\partial x_m \partial y_n} + 2\langle \tau b(\mathbf{x}) b(\mathbf{y}) \rangle \Phi - 4\langle \tau u_m(\mathbf{x}) b(\mathbf{y}) \rangle \frac{\partial \Phi}{\partial x_m} + I \, , \tag{2}$$

where $\Theta = n_p - N$, $T_{mn} = D_{mn}(0) - D_{mn}(\mathbf{r})$, $\mathbf{r} = \mathbf{y} - \mathbf{x}$, and $I = 2\langle \tau b(\mathbf{x}) b(\mathbf{y}) \rangle N^2$, and $D_{pm} = D\delta_{pm} + \langle \tau u_p u_m \rangle$, and $N = \langle n_p \rangle$ is the mean number density of particles, $\mathbf{U} = \mathbf{V}_p + \mathbf{u}$, and $\mathbf{V}_p = \langle \mathbf{U} \rangle$ is the mean particles velocity, $b = \vec{\nabla} \cdot \mathbf{u}$, and τ is the momentum relaxation time of random velocity field $\mathbf{u}$, which depends on the scale of turbulent motion. We use here for simplicity the δ-correlated in time random process to describe a turbulent velocity field. However, the results remain valid also for the velocity field with a finite correlation time, if all moments of the number density of the particles vary slowly in comparison with the correlation time of the turbulent velocity field (see, e.g., [15]).

Equation (2) for $b = 0$ was first derived by Kraichnan (see [16]). In this particular case, $b = 0$, this equation describes a relaxation of the second moment of particles number density. On the other hand, when $b \neq 0$, i.e., when the velocity of particle is divergent, Eq. (2) implies both, an effect of self-excitation (exponential growth) of fluctuations of particles number density caused by the second term in (2) (see [3]) and anomalous scalings for the fluctuations (see Section III). Another interesting feature of Eq. (2) is the emergence of the "internal" source term $I = 2\langle \tau b(\mathbf{x}) b(\mathbf{y}) \rangle N^2$. The latter means that external pumping is not required in order to sustain the fluctuations even when there is no self-excitation of the fluctuations of particles number density [3].

We consider a homogeneous and isotropic turbulent velocity field of fluid. In this case the particles velocity field is also homogeneous and isotropic, and it is compressible, i.e., $\vec{\nabla} \cdot \mathbf{U} \neq 0$. Indeed, the velocity of particles $\mathbf{U}$ depends on the velocity of the surrounding fluid $\mathbf{v}$, and it can be determined from the equation of motion for a particle: $d\mathbf{U}/dt = (\mathbf{v} - \mathbf{U})/\tau_p$, where τ_p is the characteristic time of coupling between the particle and surrounding fluid (Stokes time). A solution of the equation of motion for particles can be written in the form $\mathbf{U} = \mathbf{v} + \tau_p \mathbf{f}(\mathbf{v}, \tau_p)$. The second term in this solution describes the difference between the local fluid velocity and particle velocity arising due to the small but finite inertia of the particle. Calculate the divergence of the equation of motion for particles, and after simple manipulation we obtain

$$\vec{\nabla} \cdot \mathbf{U} = -\tau_p \vec{\nabla} \cdot [(\mathbf{v} \cdot \vec{\nabla})\mathbf{v}] - \tau_p^2 \vec{\nabla} \cdot \left[\frac{\partial \mathbf{f}}{\partial v_k} \frac{dv_k}{dt} \right] . \tag{3}$$

When τ_p is very small Eq. (3) coincides with the results obtained in [17]. Navier-Stokes equation for the fluid yields $\vec{\nabla} \cdot [(\mathbf{v} \cdot \vec{\nabla})\mathbf{v}] = -\Delta P_f / \rho$, where P_f is the pressure of a fluid. From the latter equation and Eq. (3) it is seen that $\vec{\nabla} \cdot \mathbf{U} \neq 0$. The correlation function of a compressible homogeneous and isotropic random velocity field was derived in [18]. The second moment for the particle velocity can be chosen in the same form (see below):

$$\langle \tau u_m(\mathbf{x}) u_n(\mathbf{x}+\mathbf{r})\rangle = D_T\left[[F(r)+F_c(r)]\delta_{mn} + \frac{rF'}{2}\left(\delta_{mn} - \frac{r_m r_n}{r^2}\right) + rF_c'\frac{r_m r_n}{r^2}\right] \quad (4)$$

(for details see [18]), where $F' = dF/dr$, $F(0) = 1 - F_c(0)$, and $D_T = u_0 l_0/3$, and l_0 is the maximum scale of turbulent motions, u_0 is the characteristic velocity in this scale, The function $F_c(r)$ describes the potential component whereas $F(r)$ corresponds to the vortical part of the turbulent velocity of particles.

We seek a solution to the equation for Φ without the source I in the form:

$$\Phi(t,r) = \frac{\Psi(r)}{r}\exp\left[-\int_0^r \chi(x)\,dx\right]\exp(\gamma t)\,. \quad (5)$$

Substitution (5) into Eq. (2) yields an equation for unknown function $\Psi(r)$:

$$\frac{1}{m(r)}\frac{d^2\Psi}{dr^2} - [\gamma + U_0(r)]\Psi = 0\,, \quad (6)$$

where

$$U_0(r) = \frac{1}{m(r)}\left(\frac{2\chi}{r} + \chi^2 + \chi'\right) - \kappa(r), \qquad \frac{1}{m(r)} = \frac{2}{\mathrm{Pe}} + \frac{2}{3}[1 - F - (rF_c)']\,,$$

and $\chi(r) = m(r)(10F_c' - F' + 2rF_c'')/3$, and $\kappa(r) = -2(8F_c'/r + 7F_c'' + rF_c''')/3$, and distance r is measured in units of l_0, time t is measured in units of $\tau_0 = l_0/u_0$, and $\mathrm{Pe} = l_0 u_0/D \gg 1$ is the Peclet number.

Now we discuss the above model of a random velocity field of inertial particles. Consider a case when $\tau_\nu \ll \tau_p \ll \tau_0$, and the particle radius $a_* \ll l_\nu$, where τ_ν is the correlation time in the viscous dissipation scale l_ν of a fluid flow. The viscous scale is $l_\nu \sim \mathrm{Re}^{-1/(3-p)}$, where $\mathrm{Re} = l_0 u_0/\nu_0 \gg 1$ is the Reynolds number, ν_0 is the kinematic viscosity of the fluid, p is the exponent in the spectrum of the turbulent kinetic energy of fluid. Consider the case when the material density ρ_p of particles is much larger than the density ρ of fluid. Introduce a scale r_a in which $\tau_p = \tau(r = r_a)$, where $l_\nu \ll r_a \ll 1$, and $\tau(r)$ is the correlation time of the turbulent fluid velocity field in the scale r. In the range $r_a \ll r < 1$ the effect of inertia of particles is very small and particles velocity is close to the fluid velocity. In this case $F = 1 - r^{q-1} + O(\tau_p^2/\tau_0^2)$ and $F_c = O(\tau_p^2/\tau_0^2)$, where $q = 2p - 1$. Note that the exponent p in the spectrum of kinetic turbulent energy is different from that of the function $\langle \tau u_m u_n\rangle$ due to the scale dependence of the momentum relaxation time τ of turbulent velocity of fluid [7]. Thus in the scales $r_a \ll r < 1$ the effects of compressibility of the particles velocity field is negligible.

On the other hand, in scales $l_\nu \ll r < r_a$ the effect of inertia is important so that $\vec{\nabla}\cdot\mathbf{U} \neq 0$. In these scales incompressible $F(r)$ and compressible $F_c(r)$ components of the turbulent velocity field of particles can be chosen as $F(r) = (1-\varepsilon)(1-r^{q-1})$, and $F_c(r) = \varepsilon(1-r^{q-1})$. We take into account that in the equation of motion for particles $d\mathbf{U}/dt = (\mathbf{v} - \mathbf{U})/\tau_p$ the last term $|\mathbf{U}/\tau_p| \ll |d\mathbf{U}/dt|$ in the scales $l_\nu \ll r < r_a$. In this case the equation of motion for particles coincides with Navier-Stokes equation for fluid in the inertial range (where viscous term is dropped out) except for the term $\propto \vec{\nabla} P$. In the latter equation for particles motion the term $\mathbf{v}/\tau_p$ can be interpreted as a stirring force. Thus in this case it is plausible to suggest that the exponent in the spectrum of the second moment of particle velocity coincides with that of the turbulent fluid velocity. However, $|\vec{\nabla} \cdot \mathbf{U}| \propto |\vec{\nabla} \cdot [(\mathbf{U} \cdot \vec{\nabla})\mathbf{U}]| \neq 0$.

In scales $0 \ll r < l_\nu$ incompressible $F(r)$ and compressible $F_c(r)$ components of the turbulent velocity field of particles are given by $F(r) = (1-\varepsilon)(1-\alpha r^2)$, and $F_c(r) = \varepsilon(1-\alpha r^2)$, where $\alpha = \mathrm{Re}^{(3-q)/(3-p)}$.

We consider the case of large Schmidt numbers, $\mathrm{Sc} = \nu/D \gg 1$. This condition is always satisfied for Brownian particles. Solution of Eq. (6) can be obtained using an asymptotic analysis (see, e.g., [3, 5, 6, 7, 13, 18]). This analysis is based on the separation of scales. In particular, the solution of the Schrödinger equation (6) with a variable mass has different regions where the form of the potential $U_0(r)$, mass $m(r)$ and, therefore, eigenfunctions $\Psi(r)$ are different. Solutions in these different regions can be matched at their boundaries. Note that the most important part of the solution is localized in small scales (i.e., $r \ll 1$). The results obtained by this asymptotic analysis are presented below. The solution of Eq. (6) has several characteristic regions. In the region I, i.e., for $0 \le r \ll l_\nu$ the mass $m(r)$, the potential $U_0(r)$ and the functions $\Psi(r)$ and $\Phi(r)$ are given by

$$
\begin{aligned}
\frac{1}{m} &\sim \frac{2}{\mathrm{Pe}}(1+X^2)\,, \\
U_0 &\sim 2\alpha\beta_m\left[\zeta(\zeta+1) + \frac{1-\mu^2}{1+X^2}\right]\,, \\
\Psi &= (1+X^2)^{\frac{1}{4}} S(X)\,, \\
\Phi(r) &= \frac{1}{X}(1+X^2)^{\frac{\mu}{2}} S(X)\,,
\end{aligned}
$$

where $X = (\alpha\beta_m \mathrm{Pe})^{1/2} r$, $S(X) = Re\{A_1 P_\zeta^\mu(iX) + A_2 Q_\zeta^\mu(iX)\}$ is a real part of the complex function, $P_\zeta^\mu(Z)$ and $Q_\zeta^\mu(Z)$ are the Legendre functions with imaginary argument $Z = iX$, and $\beta_m = (1+3\sigma)/3(1+\sigma)$, and

$$
\mu = 15\sigma/(1+3\sigma)\,, \tag{7}
$$

$$
\zeta(\zeta+1) = \mu^2 - 5\mu + 2\,, \tag{8}
$$

and the parameter of compressibility $\sigma = \varepsilon/(1-\varepsilon)$.

Condition $\Phi(r=0) = \mathrm{const}$ yields the ratio A_1/A_2. The correlation function has a global maximum at $r = 0$ and therefore it satisfies the conditions: $\Phi'(r = 0) = 0$, and $\Phi''(r=0) < 0$, and $\Phi(r=0) > |\Phi(r > 0)|$. The function Φ for $X \ll 1$

[i.e., for $0 \le r \ll (\alpha \text{Pe})^{-1/2}$] is given by

$$\Phi \sim B_1 \left[1 - \frac{\mu}{3} \left(X^2 - \frac{(3-\zeta-\mu)(4+\zeta-\mu)}{20} X^4 + O(X^6) \right) \right] , \tag{9}$$

where $B_1 \sim A_2$ and we use Eqs. (7) and (8). It follows from Eq. (9) that for incompressible velocity field $\sigma = 0$ (i.e., $\mu = 0$) the correlation function $\Phi = B_1 =$ constant. The same result follows from the solution of Eq. (2) for $b = 0$ (i.e., $\sigma = 0$):

$$\Phi = B_1 + B_2 \int [m(r)/r^2] \, dr \ . \tag{10}$$

The integral in Eq. (10) has a singularity at $r = 0$, and therefore an unique nonsingular solution for the correlation function $\Phi(r)$ for incompressible velocity field is a trivial solution $\Phi = B_1$ (i.e., $B_2 = 0$). On the other hand, when $\sigma \neq 0$ there is no singularity in the function $\Phi(r)$.

Now we consider an asymptotic solution of Eq. (6) for $X \gg 1$ [i.e., for $(\alpha \text{Pe})^{-1/2} \ll r \ll l_\nu$]. A form of the solution depends on the parameter $\zeta = -[1 \pm (4\mu^2 - 20\mu + 9)^{1/2}]/2$. When $\sigma > 1/27$ (i.e., $\mu > 1/2$) the parameter $\zeta = -1/2 \pm i\zeta_i$ is a complex number, and the correlation function is given by

$$\Phi = A X^{\mu - 3/2} \cos(\zeta_i \ln r + \varphi_0) \ , \tag{11}$$

where $\zeta_i = |4\mu^2 - 20\mu + 9|^{1/2}/2$. This solution is valid for $1/27 < \sigma < 1/7$ (i.e., $1/2 < \mu < 3/2$). The parameters A and φ_0 are determined by the coefficients A_1 and A_2. When $\sigma < 1/27$ (i.e., $\mu < 1/2$) the parameter $\zeta = -1/2 \pm \zeta_i$ is a real number, and the correlation function is given by

$$\Phi = A X^{-(3/2 - \mu - \zeta_i)} \ ,$$

where $3/2 - \mu - \zeta_i < 0$ and for small σ the exponent $3/2 - \mu - \zeta_i \sim 10\sigma$.

In the region II, i.e., for $l_\nu \le r \ll r_a$ the mass $m(r)$ and the potential $U_0(r)$ are given by

$$\frac{1}{m} \sim 2\beta r^{q-1}, \quad U_0 \sim -\frac{1+4c^2}{4mr^2} \ ,$$

where $c = \sqrt{M(q,\sigma)}/2(q+3)(1+q\sigma)$, and $M(q,\sigma) = b_1 B^2 + b_2 B + b_3$, and $B(q,\sigma) = 2\sigma(q+3) - 1$, and $b_1(q) = 4(q-1)(9-2q^2) - q^2$, and $b_2(q) = 2(q+2)[2(q-1)(4q^2+9q+9) - 3q]$, and $b_3(q) = 3(q+2)^2(2q-3)(2q+1)$, and $a = [q - \sigma(2q^2+3q-6)]/2(1+q\sigma)$, and $\beta = (1+q\sigma)/3(1+\sigma)$. The solution of Eq. (6) in this region depends on the parameter compressibility σ. When σ is in the range: $\sigma_1 < \sigma < \sigma_2$, the functions $\Psi(r)$ and $\Phi(r)$ are given by

$$\begin{aligned} \Psi &= A_3 r^{1/2} \cos(c \ln r + \varphi_1) \\ \Phi(r) &= A_3 r^{-a} \cos(c \ln r + \varphi_1) \ , \end{aligned}$$

where $\sigma_i = (B_i + 1)/2(q+3)$, and B_i are the roots of the equation $M = 0$. When $0 < \sigma < \sigma_1$ the functions $\Psi(r)$ and $\Phi(r)$ are given by

$$\Psi = r^{1/2}(A_3 r^{|c|} + A_4 r^{-|c|}), \quad \Phi(r) = \Psi / r^a \; .$$

In the region III $r_a \ll r \ll 1$ the mass $m(r)$, the potential $U_0(r)$ and the functions $\Psi(r)$ and $\Phi(r)$ are given by

$$\begin{aligned} \frac{1}{m} &\sim 2r^{q-1}/3 \,, \quad U_0 \sim -\frac{1-q^2}{4mr^2} \,, \\ \Psi &= r^{1/2}(A_5 r^{q/2} + A_6 r^{-q/2}) \,, \quad \Phi(r) = A_5 + A_6 r^{-q} \,, \end{aligned} \tag{12}$$

When $r_a < l_\nu$ (i.e., $\text{Re} < 1/r_a^{3-p}$) the region II vanishes and the solution (11) for $0 < r < l_\nu$ borders on the solution (12) for $l_\nu < r < 1$.

In large scales, i.e., $r \gg 1$, functions $F(r)$ and $F_c(r)$ tend to zero and, therefore $1/m(r) \to 2/3$ and $U_0 \to 0$. The correlation function in this range is given by $\Phi(r) = A_7 r^{-1} \exp(-r\sqrt{3|\gamma|/2})$. Parameters A_k, φ_1 and the growth rate of fluctuations γ are determined by matching functions $\Phi(r)$ at the boundaries of these regions. In particular, the growth rate of fluctuations of particles concentration is given by

$$\gamma = \frac{2[c^2 + (q-a)^2]^2}{3q^4(3-p)^2 r_a^{2q}} \ln^2\left(\frac{\text{Re}}{\text{Re}^{(\text{cr})}}\right) , \tag{13}$$

where $r_a = (\tau_p/\tau_0)^{1/(p-1)}$, $\text{Re} > \text{Re}^{(\text{cr})}$ and the critical Reynolds number $\text{Re}^{(\text{cr})}$

$$\text{Re}^{(\text{cr})} \simeq r_a^{p-3} \exp\left[\frac{3-p}{c}\left(\pi k + \arctan\frac{q-a}{c} + \arctan(3/2 - \mu - a + c_* \zeta_i)\right)\right] , \tag{14}$$

where $k = 1, 2, 3, \ldots$ and $c_* = \tan(\zeta_i \ln l_\nu + \varphi_0)$. Therefore, the fluctuations of particle concentration can be excited without an external source.

Divergent velocity field of inertial particles is the main reason of the self-excitation (exponential growth) of fluctuations of concentration of small particles in a turbulent fluid flow. Indeed, multiplication of Eq. (1) by n_p and simple manipulations yield

$$\frac{\partial n_p^2}{\partial t} + (\vec{\nabla} \cdot \mathbf{S}) = -n_p^2(\vec{\nabla} \cdot \mathbf{U}) - 2D(\vec{\nabla} n_p)^2 \,, \tag{15}$$

where $\mathbf{S} = n_p^2 \mathbf{U} - D\vec{\nabla} n_p^2$. The latter equation implies that if $\vec{\nabla} \cdot \mathbf{U} < 0$, a perturbation of the equilibrium homogeneous distribution of inertial particles can grow in time, i. e., $(\partial/\partial t) \int n_p^2 \, d^3r > 0$. However, the total number of particles is conserved. Averaging Eq. (15) over a volume V_* we obtain

$$\frac{\partial \langle n_p^2 \rangle}{\partial t} \sim -\langle n_p^2 (\vec{\nabla} \cdot \mathbf{U}) \rangle - 2D\langle (\vec{\nabla} n_p)^2 \rangle \,. \tag{16}$$

Here we used $\int(\vec{\nabla}\cdot\mathbf{S})\,dV_* = \int \mathbf{S}\cdot d\mathbf{A} \ll \int n_p^2(\vec{\nabla}\cdot\mathbf{U})\,dV_*$, and $\mathbf{A}$ is a closed surface. Equation (16) implies that variation of particle concentration during the time interval $\tau_0 = l_0/u_0$, around the value $n_p^{(0)}$ is of the order of $\delta n_p \sim -n_p^{(0)}\tau_0(\vec{\nabla}\cdot\mathbf{U})$, where u_0 is the characteristic velocity in the energy containing scale l_0. Substitution $n_p = n_p^{(0)} + \delta n_p$ into Eq. (16) yields $\partial\langle n_p^2\rangle/\partial t \sim 2\tau_0\langle n_p^2(\vec{\nabla}\cdot\mathbf{U})^2\rangle$. Therefore, the growth rate of fluctuations of particles concentration $\gamma \sim 2\tau_0\langle(\vec{\nabla}\cdot\mathbf{U})^2\rangle$. This estimate in a good agreement with the analytical results obtained above.

The physics of self-excitation (exponential growth) of fluctuations of particle concentration is as follows. The inertia causes particles inside the turbulent eddy to drift out to the boundary regions between eddies (the regions with maximum pressure of the fluid). Indeed, Eq. (3) shows that particles inertia results in $\vec{\nabla}\cdot\mathbf{U} \propto \tau_p\Delta P/\rho$. On the other hand, for large Peclet numbers $\vec{\nabla}\cdot\mathbf{U} \propto -dn_p/dt$ [see Eq. (1)]. Therefore, $dn_p/dt \propto -\tau_p\Delta P/\rho$. Thus there is accumulation of inertial particles (i.e., $dn_p/dt > 0$) in regions with the maximum pressure of a turbulent fluid (i.e., where $\Delta P < 0$). Similarly, there is an outflow of inertial particles from the regions with the minimum pressure of fluid.

This mechanism acts in a wide range of scales of a turbulent fluid flow. Turbulent diffusion results in relaxation of fluctuations of particle concentration in large scales. However, in small scales where turbulent diffusion is small, the relaxation of fluctuations of particle concentration is very weak. Therefore the fluctuations of particle concentration are localized in the small scales.

This phenomenon is considered for the case when density of fluid is much less than the material density of particles ($\rho \ll \rho_p$). However, the results of this study can be easily generalized to include the case $\rho \geq \rho_p$ using equation of motion of particles in fluid flow presented in [19]. This equation of motion takes into account contributions due to the pressure gradient in the fluid surrounding the particle (caused by acceleration of the fluid) and the virtual ("added") mass of the particles relative to the ambient fluid. The results for $\rho \geq \rho_p$ coincide with those obtained for the case ($\rho \ll \rho_p$) except for the transformation $\tau_p \to \beta_*\tau_p$, where

$$\beta_* = \left(1 + \frac{\rho}{\rho_p}\right)\left(1 - \frac{3\rho}{2\rho_p + \rho}\right) .$$

For $\rho \geq \rho_p$ the value $dn_p/dt \propto -\beta_*\tau_p\Delta P/\rho$. Thus there is accumulation of inertial particles (i.e., $dn_p/dt > 0$) in regions with the minimum pressure of a turbulent fluid since $\beta < 0$.

III. Anomalous scaling for fluctuations of particles concentration

Problems of anomalous scalings for vector (magnetic) and scalar (particles number density or temperature) fields passively advected by an turbulent fluid flow are a subject of active research in the last years (see, e.g., [7, 8, 9, 10, 11, 12, 14]). For an incompressible turbulent flow and when $\mathbf{U} = \mathbf{v}$ the anomalous scalings for scalar field can occur only for a forth-order correlation function (see e.g., [9, 10, 11]). In this section we show that the anomalous scaling appears already in the second moment of the number density of inertial particles when the degree of compressibility $\sigma > 1/27$.

Consider fluctuations of inertial particles concentration in the presence of a source $I(r)$ in Eq. (2). We study the case when there is no self-excitation of the fluctuations of the number density of inertial particles, i.e., the case when $\mathrm{Re} < \mathrm{Re}^{(\mathrm{cr})}$. Substituting (5) in Eq. (2) yields equation for the unknown function $\psi(t,r)$

$$\frac{\partial \psi}{\partial t} = \frac{1}{m}\frac{\partial^2 \psi}{\partial r^2} - U_0(r)\psi + \tilde{f}(r)\,, \tag{17}$$

where $\tilde{f}(r) = rI(r)\exp[\int_0^r \chi(x)\,dx]$. Determine the stationary solution of Eq. (17). The external source in these scales is chosen as follows; $I(r) = I_0(1 - r^s)$, where $s > 0$, and for $r > 1$, $I(r) = 0$. General solution of Eq. (17) reads

$$\psi(r) = A\Psi_1 + B\Psi_2 + \int_0^\infty G(r,\xi)\tilde{f}(\xi)\,d\xi\,, \tag{18}$$

where Ψ_1 and Ψ_2 are solutions of Eq. (17) with $I = 0$, and Green function $G(r,\xi)$ is given by

$$G(r,\xi) = m(\xi)H(r-\xi)\frac{\Psi_1(r)\Psi_2(\xi) - \Psi_2(r)\Psi_1(\xi)}{\Psi_1(\xi)\Psi_2'(\xi) - \Psi_2(\xi)\Psi_1'(\xi)}\,,$$

and $H(y)$ is a Heviside function. Equation (18) yields the formula for the second moment $\Phi(r)$ in different regions. When $l_\nu \le r < r_a$

$$\Phi = r^{1/2-a}(A_3 r^{|c|} + A_4 r^{-|c|}) - \frac{I_0}{2\beta_m[(3-q+a)^2 - c^2]} r^{3-q}$$

(for $1/27 < \sigma < \min(\sigma_1, 1/7)$, and

$$\Phi = A_3 r^{-a}\cos(c\ln r + \varphi_1) - \frac{I_0}{2\beta_m[(3-q+a)^2 + c^2]} r^{3-q}$$

(for $\sigma_1 < \sigma < 1/7$). When $r_a < r \ll 1$

$$\Phi(r) = A_5 + A_6 r^{-q} - \frac{I_0}{2(3-q)} r^{3-q}\,,$$

and when $r \gg 1$ the function $\Phi(r) = A_7/r$. Matching functions $\Phi(r)$ and $\Phi'(r)$ at the boundaries of these regions yields the constant A_k. The term $\propto r^{3-q}$ in these equations corresponds to a normal scaling for the second moment of inertial particles concentration, whereas the term $\propto r^{-q}$ in the range $r_a < r \ll 1$ corresponds to the anomalous scaling. When $\sigma_1 < \sigma < 1/7$ the anomalous scaling in the range $l_\nu \le r < r_a$ is complex ($\propto r^{-a \pm i|c|}$). Note that when $\sigma > 1/7$ a zero mode (i.e. $\gamma = 0$) does not exist.

VI. Conclusions

Fluctuations of number density of inertial particles in turbulent fluid flow are investigated. It is shown that the anomalous scaling appears already in the second moment of the number density of inertial particles when the degree of compressibility of the particles velocity $\sigma > 1/27$. It is demonstrated that inertia of particles in a homogeneous and isotropic turbulent fluid flow causes a self-excitation (exponential growth) of fluctuations of particles concentration. The growth rates of the higher moments of particles concentration is higher than those of the lower moments, i.e., particles spatial distribution is intermittent. This process can be damped by the nonlinear effects (e.g., two-way coupling between fluctuations of particles concentration and turbulent fluid flow). Note that when the particles velocity field is divergence free, i.e., $\vec{\nabla} \cdot \mathbf{U} = 0$ all the moments of the concentration field do not grow and there is no intermittency without an external source of fluctuations of particles concentration. When the inertia effect is negligible (e.g., for small size of particles or gaseous admixture) but the fluid velocity field is divergent, i.e., $\vec{\nabla} \cdot \mathbf{u} \neq 0$, the moments of the concentration field grow and there is intermittency without an external source of fluctuations of particles concentration. In this case Eqs. (13) and (14) with $r_a = 1$ determine the growth rate of fluctuations of particles concentration and the critical Reynolds number, respectively.

The analyzed effect of self-excitation (exponential growth) of fluctuations of particles concentration is important in turbulent fluid flows of different nature with inertial particles or droplets (e.g., in atmospheric turbulence, combustion and in a laboratory turbulence). In particular, this effect causes formation of inhomogeneities in spatial distribution of fuel droplets in internal combustion engines. The self-excitation of fluctuations of particles concentration is observed in atmospheric turbulence, e.g., this effect causes formation of small-scale inhomogeneities in droplet clouds ("inch clouds") which were discovered recently [20]. Small-scale inhomogeneities in spatial distribution of inertial particles were observed also in laboratory [21].

References

[1] T. Elperin, N. Kleeorin and I. Rogachevskii, Phys. Rev. Lett. **76**, 224 (1996).

[2] T. Elperin, N. Kleeorin and I. Rogachevskii, Phys. Rev. E **55**, 2713 (1997).

[3] T. Elperin, N. Kleeorin and I. Rogachevskii, Phys. Rev. Lett. **77**, 5373 (1996); Phys. Rev. E **58**, No. 3 (1998).

[4] Ya. B. Zeldovich, S. A. Molchanov, A. A. Ruzmaikin, and D. D. Sokoloff, Zh. Eksp. Teor. Fiz. **89**, 2061 (1985) [Sov. Phys. JETP **62**, 1188 (1985)].

[5] Ya. B. Zeldovich, S. A. Molchanov, A. A. Ruzmaikin and D. D. Sokoloff, Sov. Sci. Rev. C. Math Phys. **7**, 1 (1988), and references therein.

[6] Ya. B. Zeldovich, A. A. Ruzmaikin, and D. D. Sokoloff, *The Almighty Chance* (Word Scientific Publ., London, 1990), and references therein.

[7] I. Rogachevskii and N. Kleeorin, Phys. Rev. E **56**, 417 (1997).

[8] R. Kraichnan, Phys. Rev. Lett. **72**, 1016 (1994).

[9] M. Chertkov, G. Falkovich, I. Kolokolov and V. Lebedev, Phys. Rev. E **52**, 4924 (1995); M. Chertkov and G. Falkovich, Phys. Rev. Lett. **76**, 2706 (1996); M. Chertkov, G. Falkovich and V. Lebedev, Phys. Rev. Lett. **76**, 3707 (1996).

[10] K. Gawedzki and A. Kupiainen, Phys. Rev. Lett. **75**, 3834 (1995).

[11] B. I. Shraiman and E. D. Siggia, C. R. Acad. Sci. **321**, 279 (1995).

[12] M. Vergassola, Phys. Rev. E **53**, 3021 (1996).

[13] N. Kleeorin and I. Rogachevskii, Phys. Rev. E **50**, 493 (1994).

[14] T. Elperin, N. Kleeorin and I. Rogachevskii, Phys. Rev. E **55**, 7043 (1997).

[15] P. Dittrich, S.A. Molchanov, A.A. Ruzmaikin and D.D. Sokoloff, Astron. Nachr. **305**, 119 (1984).

[16] R. H. Kraichnan, Phys. Fluids **11**, 945 (1968).

[17] M. R. Maxey, J. Fluid Mech. **174**, 441 (1987).

[18] T. Elperin, N. Kleeorin and I. Rogachevskii, Phys. Rev. E **52**, 2617 (1995).

[19] W. D. McComb, *The Physics of Fluid Turbulence* (Clarendon, Oxford, 1990).

[20] B. Baker, J. Atmosph. Sci. **49**, 387 (1992).

[21] J. R. Fessler, J. D. Kulick, and J. K. Eaton, Phys. Fluids **6**, 3742 (1994).

Trends in Mathematics, © 1999 Birkhäuser Verlag Basel/Switzerland

Scalar fluctuation PDF's and kinetics of turbulent mixing

E. Villermaux, C. Innocenti, J. Duplat

Abstract

Experiments are conducted aiming at understanding the kinetics of turbulent mixing and its sensivity to the intrinsic diffusive molecular properties of the scalar being mixed. The concentration fluctuations PDF $P(\frac{C}{C_0})$ resulting from the transient turbulent mixing of a scalar blob exhibits a rapidly decreasing exponential shape, sharpening in time, i.e. $P(\frac{C}{C_0}) \sim (\gamma t_s)^{-1/2}\exp(-\frac{t}{t_s}\frac{C}{C_0})$. It is parametrised by a mixing time $t_s = 0.12\frac{d}{u}\ln(5Sc)$ which depends on the injection time of the blob in the medium $\frac{d}{u}$, on the Schmidt number Sc and is found to be independent of the Reynolds number. The transient turbulent mixing of scalars is thereby shown to "bypass" the hierarchy of scales of the Kolmogorov cascade, the reduction process of the transverse scalar sheet thickness peeled off from the blob just after its formation and pertaining initially to the dissipative range of scales involving a strong, but constant stretching rate.
Key words: Mixing, passive scalar, turbulence.

Mixing, as suggested by our common sense, is the operation by which a system evolves from one state of simplicity (the initial segregation of the constituents) to another state of simplicity (the complete uniformity of the mixture). Between these two extremes, complex patterns birth, and die. Questions then arise: how can the complex patterns geometry be characterized, what is the clock, the time-scale of the process and what are the structures involved in the flow ? Indeed mixing is, in the strict sense, a transient process from the initial segregation to the ultimate homogeneity. Transients call for the understanding of the kinetics, and timescales. This paper focuses on this aspect.

1 Principles, methods and results

We inject continuously in the far field and on the axis of a turbulent jet the scalar to be mixed via a small tube whose diameter d is smaller than the local integral scale L (typically $d/L \approx 1/10$, see Figure 1) and we record the concentration signal at different locations x downstream of the injection point, up to a distance of the order of the integral scale.

The exit velocity of the injection tube is maintained constant and equal to the velocity of the main jet carrying the turbulence at the injection point so that the tube behaves neither as a source, nor as a sink of momentum, in the mean. The choice of a jet as a turbulence generator is not, in principle, limitative. One wishes the properties of the turbulence (integral scale, Reynolds number,

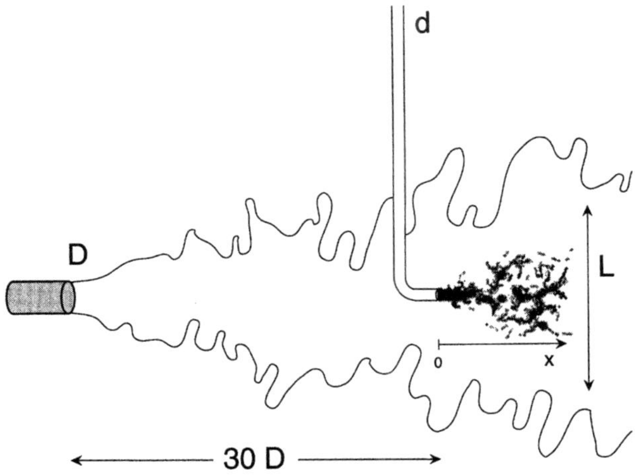

Figure 1: Sketch of the experiment ($Re = u'L/\nu \approx 10^4$). The injection scale of the scalar ($d/L = 0.05, 0.1, 0.16$) lies in the inertial range of the turbulent flow field i.e. $\eta = LRe^{-3/4} < d < L$. The injection velocity equals the mean velocity of the turbulent flow at the injection point. $D = 2.7$ cm, $d = 0.3, 0.6, 1$ cm ($d/L = 0.05, 0.1, 0.16$).

turbulence intensity, Kolmogorov scale) to remain constant within the observation range. The turbulent jet is well documented (Antonia et al. 1980) and allows to reach, in addition, high Reynolds numbers.

These experiments involve three types of scalars: temperature in air ($Sc = 0.7$), temperature in water ($Sc = 7$) and the concentration of disodium fluorescein in water ($Sc = 2000$). The injection point is located 30 diameters downstream of the exit of the main jet and the turbulence Reynolds number $\frac{u'L}{\nu}$ was varied from 6000 to 45000. u' denotes the local r.m.s. velocity ($u'/u \approx 0.25$). The fluctuations of concentrations are measured in the three different experiments by a cold wire thermometer, a cold film thermometer and a fiber optics probe constructed at the laboratory, respectively. The resolution of the probes matches the Kolmogorov scale, in all cases.

When the measurement point is moved on the jet axis farther from the injection location, one observes that the scalar concentration probability density function (PDF) evolves progressively from a bimodal shape, characteristic of a segregated field, towards a continuous, decreasing shape from the lowest concentration levels to the injection concentration C_0. The height of the zero concentration bin (concentration of the diluting medium) is all the more large that the scalar plume is dispersed in the medium, and therefore increases with the downstream distance x/d. The PDF's present an exponential decrease for the intermediary probability

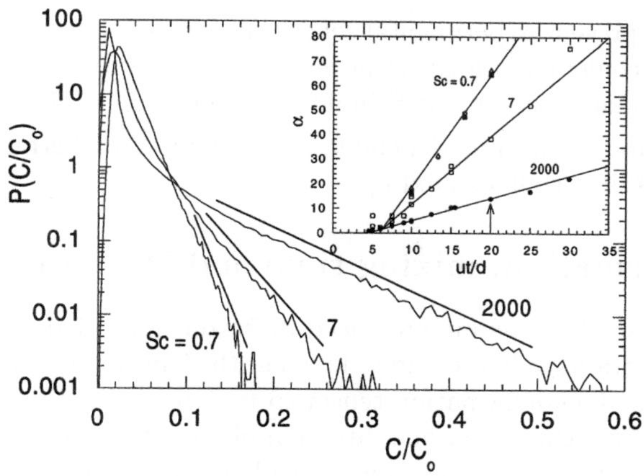

Figure 2: Fluctuations PDF's, normalized by the initial concentration (temperature), recorded for the three Schmidt numbers 20 diameters d downstream from the injection point. The PDFs exhibit an exponential decay (eq. (1)) of the form $P(\frac{C}{C_0}) \sim \exp(-\alpha\frac{C}{C_0})$. Insert: The argument of the exponentials for different distances $\frac{x}{d} = \frac{ut}{d}$ and three Schmidt numbers is such that $\alpha = \frac{ut/d}{0.12\ln(5Sc)}$. • : $Sc = 2000, Re = 6000$ and $12000, d/L = 0.05, 0.1, 0.6$. □ : $Sc = 7, Re = 6000, d/L = 0.05, 0.1, 0.16$. ■ : $Sc = 0.7, d/L = 0.08, Re = 23000$; ○, △ : $Sc = 0.7, d/L = 0.08, Re = 45000$.

levels, exponential whose argument is all the more large that x/d is large for a given Schmidt number, and all the more large, for a given distance x/d, that the Schmidt number is small (Figure 2).

The shape of the PDF depends solely on the distance x measured in units of the injection diameter d (i.e. x/d), and on the Schmidt number. The shape is, in the limit of these high Reynolds numbers, independent of the Reynolds number Re, at the precision of these experiments and on the range of Reynolds numbers we were able to investigate (a factor of 7 among the different cases, roughly).

These data thus suggest that the tail of the concentration PDF's can be represented by

$$P\left(\frac{C}{C_0}\right) \sim \exp\left(-\frac{x/d}{A\ln(BSc)}\frac{C}{C_0}\right), \tag{1}$$

for $x/d > 6$ with $A = 0.12$ and $B = 5$ (Figure 2). The dependence to the Schmidt number is weak (logarithmic), but nevertheless, owning to the broad range covered by our experiments (0.7–2000), clearly noticeable, and quantifiable.

The measurement of the molecular diffusivity of a solute in a fluid substrate is a difficult operation which generally requires the strict immobility of the medium (see for instance Bird et al. 1960, Cussler 1984, Jost 1960). Buoyant motions are undesirable. The present experimental scheme is, on the opposite side, completely dominated by turbulent motions. It nevertheless allows the determination of at least an order of magnitude of the Schmidt number $Sc = \nu/D$.

The shape and the scaling dependencies of (1) are, furthermore, indicative of the fine processes of turbulent mixing.

2 Sheets formation, mixing time and PDF shape

Mixing is a matter of stretching enhanced diffusion. Except in very peculiar and rare situations, motions in flows tend to increase the length and the area of material lines and surfaces. The deformation tensor in turbulent flows bears, in the mean, two directions of stretching and one direction of compression (Betchov 1956) and it has been recognized that this property leads to the increase of the length and area of material lines and surfaces (Batchelor and Townsend 1956). The condition for the net increase of lengths and area was later recognized to be even weaker, and such that only isotropy of the flow is required (Cooke 1969, Orzag 1969, Duplat and Villermaux 1998).

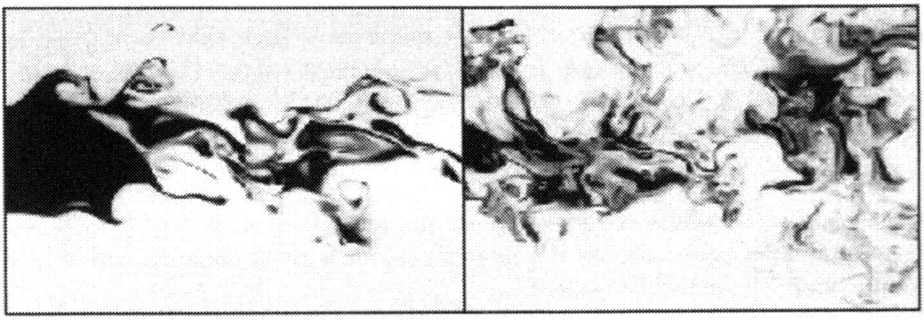

Figure 3: Instantaneous planar cuts of the scalar field downstream of the injection point illustrating how a scalar blob, initially compact and smooth, is progressively converted into disjointed sheets with broad fluctuations in thickness and concentration. $d = 1$ cm. Left: the picture covers the region just downstream of the injection tube $0 < x/d < 4$. Right: farther downstream $4 < x/d < 8$.

As illustrated on Figure 3, a scalar blob, if initially smooth and compact is progressively "laminated" into disjointed sheets whose transverse thickness gets thinner and thinner in the course of time. This process of thickness reduction goes on until the rate of compression of the concentration gradient transverse to the sheet balances the rate of diffusive spreading of the concentration profile across the sheet. One readily finds that the time t_s necessary to form a sheet whose thickness has shrunk to the equilibrium value $\sqrt{D/\gamma}$ from an initial size s_0 and under the action of a permanent stretching rate γ is

$$t_s \sim \frac{\gamma^{-1}}{2} \ln \left(\frac{s_0^2 \gamma}{D} \right), \tag{2}$$

after what the maximal concentration at the center of the sheet vanishes according to $C_M(t) = C_0 \mathrm{e}^{-\gamma(t-t_s)}$, in a timescale of the order of $1/\gamma$.

In a disordered, turbulent flow, the stretching rate depends on the initial separation scale and, at a given scale, presents in addition broad fluctuations around the mean γ. The correlation time τ_i of a stretching event of intensity γ_i is, on dimensional grounds, inversely proportional to γ_i so that

$$\gamma_i \tau_i \approx 1. \tag{3}$$

The value of the constant is discussed by Duplat and Villermaux (1998). At time $t = \sum_{i=1}^{n} \tau_i$ after its formation, an element of scalar of initial size s_0 has thus experienced a number of "stretching sollicitations" equal to $n = \sum_{i=1}^{n} \gamma_i \tau_i$. Parameterized by n, the transverse thickness of the element of scalar is $s(n) = s_0 \exp(-n)$ and the maximal concentration it carries evolves as

$$C_M(n) = C_0 \exp[-(n - n_s)] \qquad \text{for} \qquad n > n_s, \tag{4}$$

the number of sollicitations n_s necessary to reach the dissipation thickness $s(n_s) = \sqrt{D/\gamma}$ which fixes the steepness of the concentration gradient being given by

$$n_s = \gamma t_s. \tag{5}$$

For a given travel time t between the region where the initial structure have been formed (close to the injection point) and a given point downstream (for instance the measurement location), the number n is likely to fluctuate from a structure to another: the elements of scalar *do not* all have the same history, some have experienced a large number of sollicitations, some others only a few. In order to understand how the maximal concentrations (4) are distributed because of the distribution of the number of sollicitations n around n_s, one has to make an assumption about the shape of the probability density function of the number of sollicitations $\mathcal{P}(n)$ for a given mean number of sollicitations $\overline{n} = \gamma t$. It is easy to check that the overall concentration PDF $P(\frac{C}{C_0})$ amounts to the PDF of the maximal concentration C_M across the sheets given by (4), and also that the most probable events of $\mathcal{P}(n)$ dominate the statistics of the resulting concentration PDF $P(\frac{C}{C_0})$. The mean number of stretchings n_s is large enough for $\mathcal{P}(n)$ to be approximated by a Gaussian shape around its maximum (Villermaux et al 1998, Duplat and Villermaux 1998) such that

$$\mathcal{P}(n) \sim \sigma^{-1} \exp \left(-\frac{(n - \overline{n})^2}{\sigma^2} \right), \tag{6}$$

the variance σ^2 being itself proportional to the mean $\overline{n}$. Indeed, by imposing the proportionality of σ^2 to $\overline{n}$, we account for the fact that the existence of different histories among a population of structures at a given finite time results from the finiteness of $\overline{n}$. The elements of scalar will all have the same history in the limit of a very large number of sollicitations, when complete uniformity has been reached: $\mathcal{P}(n) \stackrel{\overline{n}\to\infty}{\longrightarrow} \delta(n-\overline{n})$. Writing $n = n_s + \Delta n$ for $\Delta n \ll n_s$, (4) and (6) give

$$P\left(\frac{C}{C_0}\right) \sim (\gamma t_s)^{-1/2} \exp\left(-\frac{t-t_s}{t_s}\frac{C}{C_0}\right). \tag{7}$$

The exponential form (7) is very general and encountered in several different contexts (see e.g. Jayesh and Warhaft 1992, Castaing et al. 1989) and has prompted some interpretations (Pumir et al. 1991, Shraiman and Siggia 1994). We suggest here that this shape is a consequence of the distribution of the cumulated stretchings experienced by the different fluid particles in the flow at a given instant of time. The mixing time t_s incorporates the delay to the dissipation prescribed by the molecular diffusivity D, and the relevant stretching rate. The present experiments indicate that (see (1) with $x = ut$)

$$t_s = 0.12\frac{d}{u}\ln(5Sc), \tag{8}$$

independently of the Reynolds number. Note that the molecular diffusivity shows up inside a logarithm only, and that the timescale is prescribed by the injection timescale d/u.

In the spirit of (2), this result means that the stretching rate which prescribes the chronology of reduction of transverse scales of the scalar elements from their initial size s_0 down to the dissipation scale $\sqrt{D/\gamma}$ is constant through the scales and depends solely on d. Indeed, (8) indicates that $\frac{\gamma^{-1}}{2} \approx 0.12\frac{d}{u}$, that is $\gamma \approx \frac{4u}{d}$. The "sheet formation time", or mixing time t_s is observed to be essentially independent of the Reynolds number (consistently with an old, well known fact see e.g. Nagata 1975); the initial size of the structures is thus of the order of $s_0 \approx \sqrt{5\frac{\nu}{\gamma}}$ since in that case $\frac{s_0^2\gamma}{D} \approx 5Sc$, accordingly to (8). This scale is substantially smaller than the injection scale d, proportional to the Taylor scale λ, i.e. $s_0/\lambda = \sqrt{\frac{1}{15}\frac{d}{L}\frac{u'}{u}} \approx 0.04$ and is in practice of the order of the Kolmogorov scale η, i.e. $s_0/\eta = \sqrt{\frac{d}{L}\frac{u'}{u}}Re^{1/4} \approx 2$. These structures are "peeled off" at the scalar blob frontier just after its formation (Villermaux and Gagne 1994) in a timescale of the order of $\frac{d}{u}Re^{-1/2}$ (this timescale has a Reynolds number dependence identical to the one of the Kolmogorov turnover time), short compared to the mixing time t_s.

Interestingly, the dependencies of t_s and s_0 show how the mixing of scalars, at least in a situation like ours, proceeds from the reduction of scale, by a strong and permanent stretching rate ($\gamma \approx \frac{4u}{d}$), of structures formed initially in the dissipative range of scales ($s_0 \sim \lambda$).

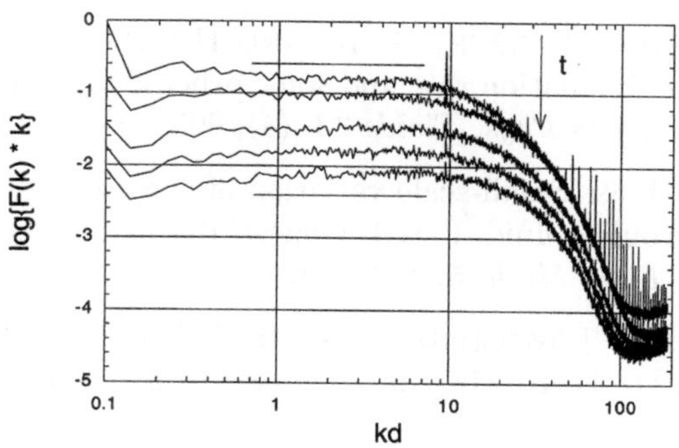

Figure 4: Compensated power spectra $kF(k)$ of the concentration field at different successive times of the mixing process ($kd = d/r$). In the direction of the arrow: $x = ut = 1.5$, 3, 4.5, 6 and 7.5 cm. $d/L = 0.1$, $L = 6$ cm. $Sc = 2000$.

The travel time through the "Kolmogorov cascade", starting from a blob of size d embedded in a turbulence of integral scale L is $t_c(\frac{d}{L}) = \frac{L}{u'}\left(\frac{d}{L}\right)^{2/3}$. The ratio of the cascade time t_c to the mixing time is

$$\frac{t_s}{t_c(\frac{d}{L})} = 0.12\frac{u'}{u}\left(\frac{d}{L}\right)^{1/3} \ln(5Sc). \tag{9}$$

This ratio is notably smaller than unity (of the order of 0.1), even for large Schmidt numbers.

This means that the process of reduction of scale by sheet thinning in turbulent flows *does not* follow *step by step* the cascade prescribed by the pre-existing hierarchy of structures in the flow, even at large Reynolds numbers, but instead "bypasses" the cascade by a permanent, and efficient stretching rate associated with the initial size of the blob and the maximal velocity in the flow. The spectrum of the scalar fluctuations reflects this phenomenon and decays self-similarly from the injection scale d to the resolution scale of the probes with a k^{-1} shape (Figure 4).

The experimentally observed "extension" of this "Batchelor-like" regime (Batchelor 1959) to the inertial range of scales has to be contrasted with the traditional description of scalars being advected in a turbulent flow (Oboukhov 1949, Corrsin 1951) and is reminiscent of the systematic deviation from $-5/3$ of the scalar spectrum exponent in various turbulent flows (Miller and Dimotakis 1996, Sreenivasan 1996).

References

ANTONIA, R. A., SATYAPRAKASH, B. R. AND HUSSAIN, A. K. M. F. 1980 Measurements of dissipation rate and some other characteristics of turbulent plane and circular jets *Phys. Fluids* **23**, (4), 695–699.

BATCHELOR, G. K. 1959 Small-scale variation of convected quantities like temperature in a turbulent fluid. Part 1. General discussion and the case of small conductivity. *J. Fluid Mech.* **5**, 113–133.

BATCHELOR, G. K. AND TOWNSEND, A. A. 1956 Turbulent diffusion. In Batchelor, G. K. and Davis, R. M. (eds), *Surveys in Mechanics*, Cambridge University Press, 352–399.

BETCHOV, R. 1956 An inequality concerning the production of vorticity in isotropic turbulence *J. Fluid Mech.* **1**, 497–504.

BIRD, R. B., STEWART, W. E. AND LIGHTFOOT, E. N. 1960 *Transport Phenomena*, John Wiley & Sons, Inc.

CASTAING, B., GUNARATNE, G., HESLOT, F., KADANOFF, L., LIBCHABER, A., THO-MAE, S., WU, X. Z., ZALESKI, S. AND ZANETTI, G. 1989 Scaling of hard thermal turbulence in Rayleigh-Bénard convection *J. Fluid Mech.* **204**, 1–30.

COCKE, W., J. 1969 Turbulent hydrodynamic line stretching: Consequences of isotropy. *Phys. Fluids* **12**, (12), 2488–2492.

CORRSIN, S. 1951 On the spectrum of isotropic temperature fluctuations in an isotropic turbulence. *J. Appl. Phys.* **22**, 469–473.

CUSSLER, E. L. 1984 *Diffusion*, Cambridge University Press.

DUPLAT, J., & VILLERMAUX, E. 1998 Persistency of material element deformation in isotropic flows and growth rate of lines and surfaces. *Submitted to J. Fluid Mech.*.

JAYESH AND Z. WARHAFT 1992 Probability distributions, conditional dissipation, and transport of passive temperature fluctuations in grid-generated turbulence *Phys. Fluids A* **4**, (10), 2292–2307.

JOST, W. 1960 *Diffusion*, Academic Press Inc.

MILLER, P. L. AND DIMOTAKIS, P. E. 1996 Measurements of scalar power spectra in high Schmidt number turbulent jets *J. Fluid Mech.* **308**, 129–146.

NAGATA, S. 1975 *Mixing, principles and applications.* John Wiley & sons, New York.

OBOUKHOV, A. M. 1949 Structure of the temperature field in a turbulent flow. *Izv. Acad. Nauk SSSR, Geogr. i Geofiz* **13**, 58–69.

ORZAG, S., A. 1970 Comments on "Turbulent hydrodynamic line stretching: Consequences of isotropy". *Phys. Fluids* **12**, (12), 2203–2204.

PUMIR, A., SHRAIMAN, B. I. AND SIGGIA, E. D. 1991 Exponential tails and random advection *Phys. Rev. Letters* **66**, (23), 2984–2987.

SHRAIMAN, B. I. AND SIGGIA, E. D. 1994 Lagrangian path integrals and fluctuations in random flow *Phys. Rev. E* **49**, (4), 2912–2927.

SREENIVASAN, K. R. 1996 The passive scalar spectrum and the Obukhov-Corrsin constant *Phys. Fluids* **8**, (1), 189–196.

VILLERMAUX, E. AND GAGNE, Y. 1994 Line dispersion in homogeneous turbulence: stretching, fractal dimensions, and micromixing *Phys. Rev. Letters* **73**, (2), 252–255.

VILLERMAUX, E., INNOCENTI, C. AND DUPLAT, J. 1998 Scalar fluctuations PDF's in transient turbulent mixing. *C. R. Acad. Sci. Paris, t. 326, Serie IIb*, 21–26.

Trends in Mathematics, © 1999 Birkhäuser Verlag Basel/Switzerland

Magnetohydrodynamics in turbulent swirling flow

J.-F. Pinton, P. Odier, S. Fauve

We report an experimental study of the magnetic field fluctuations generated in a turbulent flow of liquid gallium, in the presence of an externally applied field. We consider the case of a weak 'seed' field $\vec{B_0}$, so that the Lorentz forces do not modify the flow [1]. The velocity gradients induce magnetic field fluctuations $\vec{b}$ at all scales, the description of which pertains to the dynamics of a 'passive vector' in turbulence, in analogy to the passive scalar case [2]. However this passive vector dynamics involves stretching of magnetic field lines by velocity gradients, analogous to stretching of vorticity lines, and is thus at an intermediate level of complexity between passive scalar advection and fully developed turbulence. In particular, stretching of magnetic field lines by velocity gradients may overcome Joule dissipation and generate a large scale magnetic field by amplification of weak initial disturbances – this is the dynamo effect.

We use the flow created in the gap between two coaxial rotating disks, the von Kármán swirling flow, as it is known to produce a very intense turbulence in a compact region of space [3]. In addition, this flow possesses many features, such as differential rotation or poloidal and toroidal mean flow components, which are known to favor dynamo action [4]. Two 11 kW ac-motors are used to drive the disks at a constant frequency Ω, adjustable in the range 5 to 50 Hz. The enclosing cylindrical vessel has a volume of 5.5 liters. It is filled with liquid gallium (density $\rho = 6.09 \times 10^3$ kg.m^{-3}), chosen for its high electrical conductivity ($\sigma = 3.68 \times 10^6$ ohm.m^{-1}). Its kinematic viscosity is $\nu = 3.1 \times 10^{-7}$ m^{-2}.s^{-1}. The integral kinematic and magnetic Reynolds numbers of the flow are defined as: $R_e = 2\pi R^2\Omega/\nu \in [10^6,\ 10^7]$ and $R_m = 2\pi\mu_0\sigma R^2\Omega \in [1.3,\ 14]$. Two pairs of Helmholtz coils are set to produce an external field $B_0 \sim 40$ gauss, either parallel or perpendicular to the rotation axis. Magnetic measurements are performed inside the vessel using directional and temperature compensated Hall probes.

Figure 1(a) shows typical power spectra of magnetic field fluctuations; they start with a flat frequency content followed by a steep cut-off region. The transition between the two regimes happens for a frequency of the order of Ω. The high frequency regions are quite similar and show an algebraic decay $\tilde{b}^2(f) \propto f^{-\alpha}$. Our measurements yield $\alpha = 3.8 \pm 0.2$ for all orientations and accessible values of R_m. This scaling is observed in a range of frequencies where the pressure fluctuations also follow a power law (Figure 1(b)), i.e. in the inertial range of the velocity field [5]. These results can be understood as follows: in the presence of a uniform and constant applied field $\vec{B_0}$, magnetic field perturbations, $\vec{b}$, are governed by the

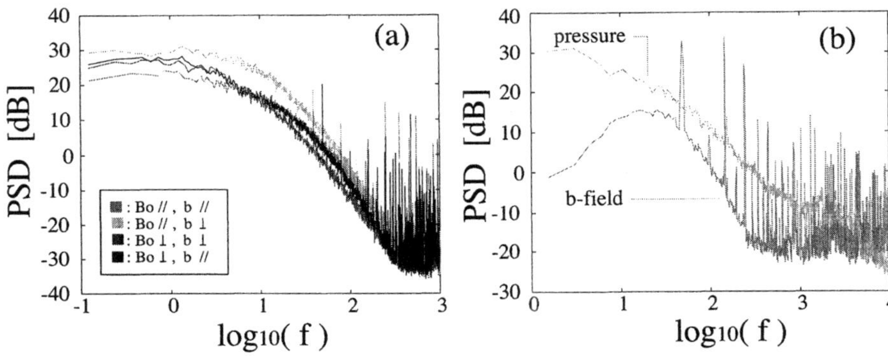

Figure 1: (a) magnetic spectra for different orientations $(\vec{B}_0, \vec{b})$ with respect to the rotation axis, of the applied and induced fields; $\Omega = 40$ Hz, $R_m = 10.8$; probe at $d = 1$ cm from the wall. (b) comparison with a spectrum of pressure fluctuations at the wall, in the same conditions. The low frequency cut-off of the magnetic spectrum is due to the AC filtering of the gaussmeter. These features are found for every value of R_m and d.

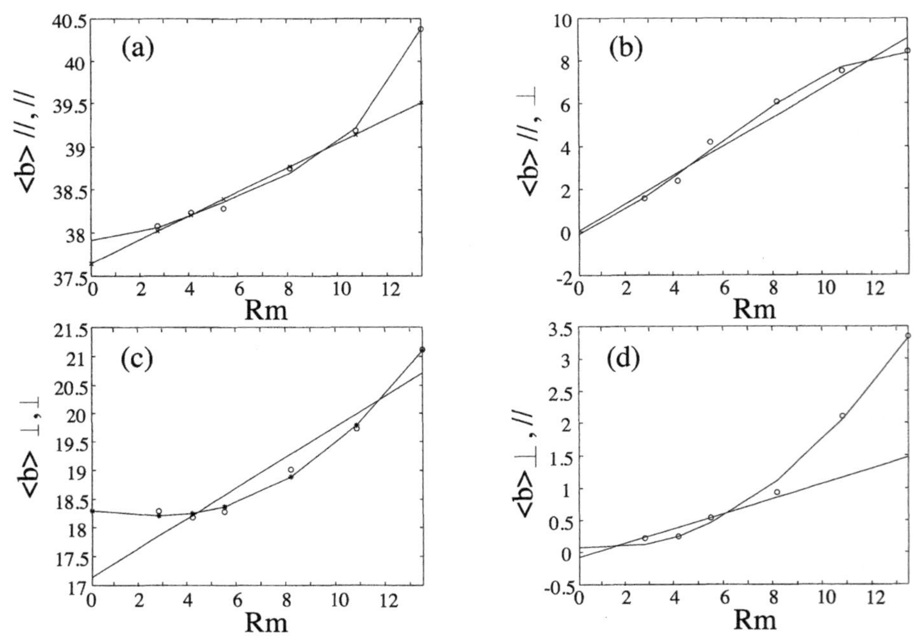

Figure 2: Average value of magnetic for different orientations $(\vec{B}_0, \vec{b})$ with respect to the rotation axis, of the applied and induced field.

equations [4]:

$$(\partial_t + \vec{u}.\vec{\nabla})\vec{b} = \left[(\vec{b} + \vec{B}_0).\vec{\nabla}\right]\vec{u} + \lambda\Delta\vec{b}\,,\ \vec{\nabla}.\vec{b} = 0. \tag{1}$$

Since the magnetic diffusivity is orders of magnitude larger than the kinematic viscosity, $\vec{b}$ adiabatically follows $\vec{u}$. In this "quasistatic" approximation, one has for the leading order magnetic field perturbation $\vec{b}_0$:

$$\lambda\Delta\vec{b}_0 \approx -(\vec{B}_0.\vec{\nabla})\vec{u}, \tag{2}$$

provided that $b \ll B_0$, as is well verified in our experiment. Thus, $\vec{b}_0$ obeys a Poisson equation – this is analogous to the pressure field albeit second order derivatives of the velocity field are involved in that case. Keeping in mind that the flow is not modified by $\vec{B}_0$, dimensional analysis in the framework of Kolmogorov phenomenology then leads to [6]:

$$\tilde{b}_0^2(k) \propto k^{-2}\tilde{u}^2(k) \sim k^{-11/3}\ , \tag{3}$$

in the inertial range. This is consistent with our measurements, providing Taylor's hypothesis may be used. Note that, when EDQNM closures are applied to MHD equations, the relation $\tilde{b}^2(k) \propto k^{-2}\tilde{u}^2(k)$ between kinetic and magnetic energy subsists when the magnetic field is generated by the dynamo effect, although both spectra are steeper because of the effect of the Lorentz force [7].

We now consider the temporal average, $\overline{b}$ of the magnetic field $\vec{b}$. Equation (2) yields:

$$\overline{b}_0(\vec{r}) \approx g \star (\vec{B}_0.\vec{\nabla})\vec{u}\ , \tag{4}$$

where g is a Green function (equal to the free space one only for insulating boundary conditions at the vessel walls). $\overline{b}_0(\vec{r})$ is the leading order distorsion of the magnetic field lines; it is due to the non uniformity of the flow along $\vec{B}_0$ and varies linearly with R_m at small R_m [1,4]. This is evidenced in Figure 2(b), where the azimutal component of the induced field is plotted against R_m, for an applied field parallel to the rotation axis z. Here an azimutal component b_θ is induced by the stretching term $\partial_z u_\theta$, i.e. by differential rotation (the disks rotate in opposite directions). Thus, one of the important source mechanisms for dynamos [4] occurs in our experimental flow configuration.

For higher values of R_m, we observe that the growth of the induced field is no longer linear – see Figure 2; the change being quite clear for $R_m \geq 6$, particularly when the applied field is perpendicular to the rotation axis. We interpret this as follows: from (1), the next order correction $\overline{b}_1$ to $\overline{b}_0$ is governed by the equation

$$\lambda\Delta\overline{b}_1 \approx \overline{(\vec{u}.\vec{\nabla})\vec{b_0}} - \overline{(\vec{b}_0.\vec{\nabla})\vec{u}}\ , \tag{5}$$

showing that $\overline{b}_1$ is quadratic in R_m. Its sign depends on the direction of $\vec{B}_0$ and on the sign of the velocity gradients: our measurements show a saturation effect in the perpendicular direction when $\vec{B}_0$ is parallel to the rotation axis (Figure 2(b)), but a quadratic growth in the direction of the applied field (Figure 2(a,c)).

As regards to dynamo action, this non linear growth may lead to an increase of the net magnetic energy. The threshold for dynamo effect corresponds to a vertical asymptote for $< b^2 > /B_0^2$ (spatial averages here) as a function of R_m since a finite b would be obtained for $B_0 \to 0$. We cannot compute $< b^2 >$ without a complete 3D measurement of $\vec{b}$ throughout the flow. We can however try to get an estimate of the threshold R_m^c for the onset of dynamo action by interpolation of our results for $\overline{b}$ at higher magnetic Reynolds numbers. We compute $B_0 R_m / \overline{b}$ as a function of R_m: in the limit of a week seed field, $B_0 \to 0$, this quantity is the inverse of a "susceptibility" that characterizes the magnetic response of the turbulent fluid to the applied field $\vec{B}_0$. We observe that $B_0 R_m / \overline{b}$ decreases with R_m and yields $R_m^c \sim 25$ as an estimate for the onset of dynamo action in this flow. We note that numerical studies of flows with similar geometries [7, 8, 9] give comparable values for R_m^c. Work is underway to operate the same flow using liquid sodium as a working fluid; magnetic Reynolds numbers exceeding 50 will be reached.

References

[1] Roberts P.H. (1967), *An introduction to MHD*, Longmans.

[2] Monin A.S., Yaglom A.M. (1975), *Statistical fluid mechanics*, MIT Press.

[3] Pinton J.-F., Labbé R. (1994), *J. Phys. II France*, **4**, pp. 1461–1468.

[4] Moffatt H. K. (1978), *Magnetic field generation in conducting fluids*, Cambridge U. P.

[5] Fauve S., Laroche C., Castaing B. (1993), *J. Phys. II France*, **3**, p. 271.

[6] Moffat H. K. (1961), *J. Fluid Mech.*, **11**, pp. 625–635.

[7] Léorat J., Pouquet A., Frisch U. (1981), *J. Fluid Mech.*, **104**, pp. 419–443.

[8] Dudley N.L., James R.W. (1989), *Proc. Roy. Soc. Lond.*, **A425**, p. 407.

[9] Nore C., Brachet M.-E., Politano H., Pouquet A. (1997), *Phys. Plasmas*, **4**, pp. 1–3.

List of participants and authors[1]

Nikolaus Adams*
Fluiddynamik
ETHZ
CH-8092 Zürich, Schweiz
adams@ifd.mavt.ethz.ch

G. Amati†
CASPUR
Università "La Sapienza",
P.le Aldo Moro 5, I-00185, Roma,
Italy

Bruno Andreotti*
ENS Paris,
Lab. de Physique Statistique
24, rue Lhomond
F-75231 Paris Cedex 05, France

R. A. Antonia†
Department of Mechanical
Engineering, University of
Newcastle
NSW 2308 Australia

Hassan Aref*†
Dept. Theor. & Appl. Mech.
Univ. of Illinois at
Urbana-Champaign
104 South Wright Street
Urbana, IL 61801-2935, USA
h-aref@uiuc.edu

Sabine Attinger*
IHW
ETHZ
CH-8093 Zürich, Schweiz
attinger@ihw.baum.ethz.ch

Claude Bardos*†
ENS, LMD-CNRS
24, rue Lhomond
F-75231 Paris Cedex 05, France
Claude.Bardos@cmla.ens-cachan.fr

C. Baudet†
CNRS, Laboratoire de Physique,
E.N.S. de Lyon
46 Allée d'Italie, 69364 Lyon,
France

P. Bayer*
Equipe Turbulence Plasma
UMR 633 CNRS-Université de
Provence
Centre de Saint Jérôme-case 321
F-13397 Marseille Cedex 20

F. Belin†
Lab. de Phys. Stat.
École Normale Supérieure
24 rue Lhomond, 75231 Paris,
France

S. Benkadda*
Equipe Turbulence Plasma
UMR 633 CNRS-Université de
Provence
Centre de Saint Jérôme-case 321
F-13397 Marseille Cedex 20
benkadda@newsup.univ-mrs.fr

R. Benzi†
AIPA
Via Po 14, I-00100, Roma, Italy

[1] Participants are indicated by *, whereas authors of a paper are marked with †.

Luca Biferale*
Università di Roma, Tor Vergata
Dep. of Physics
Via della Ricerca Scientifica 1
I-00133 Roma, Italia
biferale@roma2.infn.it

Stefan Blaser*
IHW
ETHZ
CH-8093 Zürich, Schweiz
blaser@ihw.baum.ethz.ch

P. N. Blossey†
Sibley School of Mechanical &
Aerospace Engineering
Cornell University, Ithaca, NY
14853 USA

L. Blumenfeld†
DRE/SPCP, CEA Saclay
91191, Gif-sur-Yvette Cedex,
France

O. N. Boratav†
Cornell University, Ithaca, NY

P. L. Boyland†
Department of Mathematics,
University of Florida

Marc-Etienne Brachet*
ENS Paris
Lab. de Physique Statistique
24, rue Lhomond
F-75231 Paris Cedex 05, France
brachet@physique.ens.fr

James G. Brasseur*†
Dept. Mech. Engineering
Pennsylvania State University
University Park, PA 16802-1413,
USA
brasseur@jazz.me.psu.edu

P. Caillol†
Laboratoire de Météorologie
Dynamique,
ENS - Université P. et M. Curie,
France

Claude Cambon*†
Ecole central de Lyon
L.M.F.A., U.R.A. 5509 CNRS
BP 163
F-69131 Ecully Cedex, France
cambon@mecaflu.ec-lyon.fr

H. J. Catrakis†
Graduate Aeronautical
Laboratories
California Institute of Technology
Pasadena, California 91125

S. Cerutti†
Department of Mechanical
Engineering
The Johns Hopkins University,
Baltimore, Maryland 21218, USA

J. Chasnov†
The Hong Kong University of
Science and Technology
Clear Water Bay
Kowloon, Hong Kong

F. Chillá†
ENS, Laboratoire de Physique,
F-69364 Lyon

Sergio Ciliberto*†
Lab. de Physique
ENS de Lyon (URA 1325 CNRS)
46, Allée d'Italie
F-69364 Lyon Cedex 07, France
cilibe@physique.ens-lyon.fr

A. W. Cook†
Lawrence Livermore National Laboratories
Livermore, California 94550

Yves Couder*
ENS Paris
Lab. de Physique Statistique
24, rue Lhomond
F-75231 Paris Cedex 05, France
Yves.couder@peterpan.ens.fr

Paul E. Dimotakis*†
John K. Northrop Aeronautics
Appl. Physics
Caltech 301-46
Pasadena, CA 91125, USA
pxd@anemos.caltech.edu

Charles R. Doering*†
Dept. Mathematics
University of Michigan
Ann Arbor, MI 48109-1109, USA
doering@math.lsa.umich.edu

D. G. Dommermuth†
Science Application International Corporation
San Diego, CA 92121

Jérôme Duplat*†
LEGI-CNRS
Institut de Mécanique de Grenoble
BP 53X
F-38041 Grenoble Cedex, France

F. Dupont†
McGill University, Montreal, Canada

Tov Elperin*†
Dept. Mech. Engineering
Ben-Gurion Univ. of the Negev
P.O.B. 653
IL-84105 Beer-Sheva, Israel
elperin@menix.bgu.ac.il

P. G. Esposito†
INSEAN
Via di Vallerano 139,
00128 Rom, Italy

Gregory L. Eyink*†
Dept. of Mathematics
University of Arizona
Tucson, AZ 85721, USA
eyink@math.arizona.edu

Gregory Falkovich*†
Weizmann Institute
IL-76100 Rehovot , Israel
fnfal@wicc.weizmann.ac.il

Marie Farge*†
ENS, LMD-CNRS
24, rue Lhomond
F-75231 Paris Cedex 05, France
farge@lmd.ens.fr

Brian F. Farrell*†
Harvard University, Meteorology
Dept. Earth Planetary Sciences
29 Oxford St. Pierce Hall
Cambridge, MA 02138, USA
farrell@io.harvard.edu

S. Fauve†
Ecole Normale Supérieure
CNRS URA 1306, 75231 Paris,
France

H.-H. Fernholz*
HFI, TU Berlin
Strasse des 17. Juni 135
D-10623 Berlin, Deutschland
fernholz@pi.tu-berlin.de

R. Friedrich†
Institut für Theo. Physik und
Synergetik
Universität Stuttgart
D-70550 Stuttgart

Andrei V. Fursikov*†
Dept. Mech. & Mathematics
Moscow State University
119899 Moscow, Russia
fursikov@sci.lpi.ac.ru

E. Gaudin*
Lab. de Physique & Méc. des
Mil. Hét.
URA CNRS 857
ES de Phy. & de Chimie Indust.
10, rue Vauquelin
F-75005 Paris, France

William K. George*†
Turbulence Research Laboratory
State University of New York
Buffalo, NY 14260, USA
trlbill@eng.buffalo.edu

Massimo Germano*†
Politecnico di Torino
Dep. Ing. Aeronautica e Spaziale
C. so Duca degli Abruzzi 24
I-10129 Torino, Italia
germano@polito.it

John D. Gibbon*†
Dept. Mathematics
Imperial College
London SW7 2BZ, Great Brittain
j.gibbon@ic.ac.uk

J. M. Greene†
General Atomics, San Diego, CA

Y. Gulak†
Rutgers University
Piscataway, NJ

Albert Gyr*
IHW
ETHZ
CH-8093 Zürich, Schweiz
gyr@ihw.baum.ethz.ch

Jackson R. Herring*†
NCAR
P.O. Box 3000
Boulder, CO 80307, USA
herring@oak.mmm.ucar.EDU

Mei-Jiau Huang*†
Mech. Engineering Dept.
National Taiwan University
Taiwan, ROC
mjhuang@w3.me.ntu.edu.tw

C. Innocenti†
LEGI-CNRS
Institut de Mécanique de Grenoble
BP 53X
38041 Grenoble Cedex, France

Petros J. Ioannou*†
Harvard University, Meteorology
Dept. Earth & Planetary Sc.
29 Oxford St. Pierce Hall
Cambridge, MA 02138, USA
pji@io.harvard.edu

Tsutomu Kambe*
Dept. of Physics
University of Tokyo
7-3-1 Hongo, Bunkyo-ku
Tokyo 113-0033, Japan
kambe@phys.s.u-tokyo.ac.jp

Robert M. Kerr*†
NCAR
P.O. Box 3000
3450 Mitchell Lane
Boulder, CO 80307-3000, USA
kerr@oak.mmm.ucar.EDU

Nicholas K.-R. Kevlahan†
ENS, LMD-CNRS
24, rue Lhomond
F-75231 Paris Cedex 05, France

Yoshifumi Kimura*†
Grad. School of Mathematics
Nagoya University
Nagoya 464-01, Japan
kimura@math.nagoya-u.ac.jp

Wolfgang Kinzelbach*
IHW
ETHZ
CH-8093 Zürich, Schweiz
kinzelbach@ihw.baum.ethz.ch

N. Kleeorin*†
Dept. Mech. Engineering
Ben-Gurion Univ. of the Negev
P.O.B. 653
IL-84105 Beer-Sheva, Israel
nat@menix.bgu.ac.il

Leonhard Kleiser*
Fluiddynamik, ML H 36
ETHZ
CH-8092 Zürich, Schweiz
kleiser@ifd.mavt.ethz.ch

Petros Koumoutsakos*
Fluiddynamik, ML
ETHZ
CH-8092 Zürich, Schweiz
petros@ifd.mavt.ethz.ch

Anti Kupiainen*
Mathematics Department
Helsinki University
PO Box 4
00014 Helsinki, Finland
ajkupiai@cc.helsinki.fi

R. Labbé†
Ecole Normale Supérieure de Lyon
CNRS URA 1325, 69364 Lyon
France

E. Lamballais†
L.E.G.I./Institut de Mécanique de
Grenoble
B.P. 53
38041 Grenoble Cedex 09, France

Vladimir V. Lebedev*†
Landau Inst. for Theor. Physics
Kosygina 2
117940 Moscow, Russia
lebede@landau.ac.ru

Antony Leonard*†
Graduate Aeronotical Laboratories
CALTECH
Pasadena, CA 91125, USA
tony@vorton.galcit.caltech.edu

M. Lesieur†
L.E.G.I./Institut de Mécanique de Grenoble,
B.P. 53
38041 Grenoble Cedex 09, France

E. Leveque†
CNRS, Laboratoire de Physique,
E.N.S. de Lyon
46 Allée d'Italie, 69364 Lyon, France

Erik Lindborg*
Dept. of Mechanics
KTH
S-100 44 Stockholm, Sweden
erikl@mech.kth.se

St. Lück†
FB 8 Physik
Universität Oldenburg
D-26111 Oldenburg, Deutschland

John L. Lumley*†
Sibley School of Mech. & Aerosp. Eng.
Ithaca, NY 14853-7501, USA
jll4@cornell.edu

Jakob Mann*
Wind Energy & Atm. Phys. Dept.
Risoe National Laboratory
DK-4000 Roskilde, Denmark
jakob.mann@risoe.dk

R. I. McLachlan†
Massey University, Palmerston North, New Zealand

James C. McWilliams*
Inst. Geophysics & Planetary Physics
Dept Atmospheric Sciences
University of California
Los Angeles, CA 90095, USA

Maurice Meneguzzi*
ASCI-CNRS, Bat 506
Université de Paris Sud
F-91405 Orsay, France
mngz@asci.fr

Charles Meneveau*†
Dept. Mech. Engineering
The John Hopkins University
Baltimore, MD 21218, USA
meneveau@titan.me.jhu.edu

H. Merava*
ETHZ
Lab. für Technische Chemie
CAB D41
CH-8092 Zürich

Olivier Metais*†
Charge de Recherches CNRS
LEGI/IMG
BP 53
F-38041 Grenoble Cedex 9, France
metais@frodon.hmg.inpg.fr

Igor Mezic*
Dynamical Systems & Control
Dept. Mech. & Envir. Engineering
University of
California Santa Barbara,
CA 93106-5070, USA
mezic@ana.ucsb.edu

F. Moisy†
Laboratoire de Physique
Statistique, ENS
24 rue Lhomond,
75231 Paris, France

Andreas Müller*
IHW
ETHZ
CH-8093 Zürich, Schweiz
mueller@ihw.baum.ethz.ch

Mark Nelkin*
3 Washington Square Village, 14-B
New York, NY 10012, USA
mn21@is4.nyu.edu

Evgeny A. Novikov*†
Inst. Nonlinear Science
University of California
San Diego LaJolla
CA 92093-0402, USA
novikov@gibbs.UCSD.edu

Martin Oberlack*†
Inst. f. Techn. Mechanik
Rheinisch-Westfälische TH
Templergraben 64
D-52056 Aachen, Deutschland
mo@itm.rwth-aachen.de

P. Odier†
Ecole Normale Supérieure de Lyon
CNRS URA 1325, 69364 Lyon
France

J. O'Neil†
Applied Physics Laboratory
The Johns Hopkins University,
Baltimore, Maryland 21218, USA

Paolo Orlandi*†
Univ. degli Studi di Roma "La Sapienza"
Via Eudossiana 18
I-00184 Roma, Italia
orlandi@orlandi.ing.uniroma1.it

Edward Ott*†
Dept. of Physics
University of Maryland
College Park, MD 20742, USA
kynam@Glue.umd.edu

J. M. Patton†
Center for Advanced Computing Research
California Institute of Technology

M. B. Parlange†
Dept. of Geography and
Environmental Engineering
The Johns Hopkins University,
Baltimore, Maryland 21218, USA

Joachim Peinke*†
FB 8 Physik
Universität Oldenburg
D-26111 Oldenburg, Deutschland
peinke@uni-oldenburg.de

Richard Pelz*†
Mech. & Aerospace Eng. Rutgers
The State Univ. of New Jersey
98 Brett Road, Room B225
Piscataway, NJ 08854-8058, USA
pelz@jove.rutgers.edu

P. Petitjeans*
Lab. de Phy. & Méc. des Mil. Hét.
URA CNRS 857
ES de Physique & de Chimie Indust.
10, rue Vauquelin
F-75005 Paris, France

Jean-François Pinton*†
ENS de Lyon
Lab. de Physique CNRS
46, Alleé d'Italie
F-69364 Lyon Cedex 07
pinton@physique.ens-lyon.fr

R. Piva†
Dip. di Meccanica e Aereonautica
Università "La Sapienza"
Via Eudossiana 18
I-00184, Roma, Italy

B. Podvin-Delarue†
ONERA, Chatillon, France

F. Porte-Agel†
Dept. of Geography and
Environmental Engineering
The Johns Hopkins University,
Baltimore, Maryland 21218, USA

A. Pumir*†
UMR 129 C.N.R.S. UNSA
Institut Non-Linéaire de Nice
1361, Route des Lucioles
F-06560 Vallbonne, France
pumir@inln.cnrs.fr

Birgit Reisner*†
FB 8 Physik
Universität Oldenburg
D-26111 Oldenburg, Deutschland

Dietmar Rempfer*†
Mech. & Aerospace Engineering
Cornell University, Upson Hall
Ithaca, NY 14853-7501, USA

Ch. Renner†
FB 8 Physik
Universität Oldenburg
D-26111 Oldenburg, Deutschland

I. Rogachevskii*†
Dept. Mech. Engineering
Ben-Gurion Univ. of the Negev
P.O.B. 653
IL-84105 Beer-Sheva, Israel
gary@menix.bgu.ac.il

Thomas Roesgen*
Fluiddynamik, ML H 33
ETHZ
CH-8092 Zürich, Schweiz
roesgen@ifd.mavt.ethz.ch

G. Ruiz-Chavarria†
Departamento de Física
Facultad de Ciencias, U.N.A.M.
04510 México D.F., Mexico

K. Schneider†
ICT, Karlsruhe Universität

Leonid Shtilman*
Faculty of Engineering
Tel-Aviv University
IL 69978 Tel-Aviv, Israel
shtilman@eng.tau.ac.il

M. A. Stremler†
Department of Theoretical &
Applied Mechanics
University of Illinois at
Urbana-Champaign

Heinrich Stüer*
IHW
ETHZ
CH-8093 Zürich, Schweiz
stueer@ihw.baum.ethz.ch

S. Succi†
Istituto Applicazioni Calcolo
"Mauro Picone",
V.le Policlinico 137, I-00161,
Roma, Italy

Patrick Tabeling*†
ENS Paris
Lab. de Physique Statistique
24, rue Lhomond
F-75231 Paris Cedex 05, France
Patrick.tabeling@peterpan.ens.fr

Federico Toschi*†
Università di Pisa
Piazza Torricelli 2
I-56126 Pisa, Italia
toschi@pcape2.pi.infn.it

Arkady Tsinober*
Faculty of Engineering
Tel-Aviv University
IL 69978 Tel-Aviv, Israel
tsinober@eng.tau.ac.il

G. Urbin†
CEA DRN/DTP/STR/LMTL
17, rue des Martyrs, 38050
Grenoble Cedex 09, France

D. L. Vainchtein†
Department of Theoretical &
Applied Mechanics
University of Illinois at
Urbana-Champaign

Willem van de Water*†
Physics Department
Eindhoven University of
Technology
P.O.Box 513, 5600 MB Eindhoven,
The Netherlands
willem@tnh.phys.tue.nl

Christos Vassilicos*†
DAMTP
University of Cambridge
Cambridge CB3 9EW, UK
J.C.Vassilicos@damtp.cam.ac.uk

Emmanuel Villermaux*†
LEGI-CNRS
BP 53 X
F-38041 Grenoble Cedex, France
villerma@img715.img.fr

Jose-Eduardo Wesfreid*
Lab. de Phy. & Méc. des Mil. Hét.
URA CNRS 857
ES de Physique & de Chimie
Indust.
10, rue Vauquelin
F-75005 Paris, France
wesfreid@manet.pmmh.espci.fr

H. Willaime†
Laboratoire de Physique
Statistique, École Normale
Supérieure
24 rue Lhomond, 75231 Paris,
France

P.K. Yeung*†
Scool of Aerospace Engineering
Georgia Inst. of Technology
Atlanta, GA 30332, USA
yeung@peach.ae.gatech.edu

Alistair Young*
Dept. Physics & Astronomy
University of Edinburgh
Edinburgh EH9 3JZ, Scotland
ajy@ph.ed.ac.uk

Vladimir Zeitlin*†
LMD, BP 99 Univ. P. et M. Curie
4, place Jussieu
F-75252 Paris Cedex 5, France
zeitlin@lmd.ens.fr

MIX
Papier aus verantwortungsvollen Quellen
Paper from responsible sources
FSC® C105338

If you have any concerns about our products,
you can contact us on
ProductSafety@springernature.com

In case Publisher is established outside the EU,
the EU authorized representative is:
Springer Nature Customer Service Center GmbH
Europaplatz 3, 69115 Heidelberg, Germany

Printed by Libri Plureos GmbH
in Hamburg, Germany